# The Mekong Delta System

Springer Environmental Science and Engineering

For further volumes:
http://www.springer.com/series/10177

Fabrice G. Renaud • Claudia Kuenzer
Editors

# The Mekong Delta System

## Interdisciplinary Analyses of a River Delta

Springer

*Editors*
Fabrice G. Renaud
Institute for Environment
and Human Security
United Nations University
Hermann-Ehlers-Str. 10
Bonn
Germany

Claudia Kuenzer
German Aerospace Center, DLR
Earth Observation Center, EOC
German Remote Sensing Data Center
Oberpfaffenhofen
82234 Wessling, Germany

ISBN 978-94-007-3961-1 ISBN 978-94-007-3962-8 (eBook)
DOI 10.1007/978-94-007-3962-8
Springer Dordrecht Heidelberg New York London

Library of Congress Control Number: 2012939334



Printed on acid-free paper

Springer is part of Springer Science+Business Media (www.springer.com)

# Foreword

The Mekong Delta in South Vietnam is home to nearly 18 million people. The region is characterized by a steady socio-economic transformation that has occurred over the last few decades. Agricultural practices in the delta are intensifying, aquaculture has rapidly expanded, and a few cities in the delta have experienced fast urban growth. The delta's natural and social systems face numerous challenges as a result of the socio-economic transformation, urbanization, related environmental degradation and climate change. Most of these challenges are related to water: too much or too little water, polluted water, and difficulties in accessing water are the problems that local inhabitants struggle with. Communities in the Mekong Delta are used to the annual fluctuations of floods and low flow periods, but the frequency of extreme events is increasing.

The delta lies at the vulnerable end of a river that runs 4,800 km and is heavily regulated upstream. The riparian countries China, Myanmar, Laos, Thailand and Cambodia have all started to erect dams on the Mekong or its tributaries. Vietnam is also involved in hydropower generation and is planning to build dams on Mekong tributaries. It also plans to import hydropower-generated electricity from northern neighbours. However, hydropower construction on the Mekong and its tributaries leads to changes in flood pulses and has severe impacts on the aquatic ecosystem biodiversity in the delta. Furthermore, agriculture, expanding industry, and urbanization both in the delta and upstream lead to a deteriorating water quality. On top of this, the delta is threatened by climate change. The sea level rise along the delta's coast aggravated the problem of salt water intrusion inland. This salt-water intrusion affects surface waters, aquifers, and soils. Close to the coast, rice harvests are diminishing and farmers have begun to experiment with more salt-tolerant rice varieties. As most mangrove belts and protective coastal forests have disappeared due to the expansion of aquaculture, the delta is slowly losing its protection against severe cyclones. According to projections, 20–50% of the Mekong Delta will be severely affected within the next 50 years by sea level rise if no counter-measures are taken.

In 2010, the German Federal Ministry of Education and Research (BMBF) initiated a funding programme entitled "Research for Sustainable Development – FONA", which includes the new funding priority "Sustainable Water Management – NaWaM".

These measures set the main framework for German research to contribute innovatively and effectively to the design and implementation of solutions for better water resource management. An important contribution is thus provided to integrated water resources management (IWRM), and knowledge and technology within this field are transferred to the developing and emerging partner countries. The development of planning instruments, water management concepts, and water technology – adapted to the differing natural and socio-economic conditions in the respective countries – actively contributes towards implementing the Millennium Development Goals.

The German-Vietnamese WISDOM Project, led by the German Aerospace Centre (DLR) on the German side, and the Southern Institute of Water Resources Research (SIWRR) on the Vietnamese side, started its activities in 2007 and will continue until the end of 2013. The goal of this bilateral project is to design and implement a water-related information system for the Mekong Delta. This system will not only contain all of the project research findings, but also a multitude of geo-data relevant to planning as well as further information from numerous scientific disciplines, such as environmental sciences, sociology and information technology. The project also undertakes large efforts in capacity building.

The project benefits PhD students and when it concludes in 2013, over 30 PhD students will have received a PhD degree from a German university – half of these students will be Vietnamese. Overall, in the context of WISDOM, intensive bilateral cooperation between science, industry, and administration has been achieved. The project is one of many contributions to closer scientific and technological cooperation between Germany and Vietnam.

On the road towards a successful IWRM strategy, sharing information and research results is a key component. Most of the chapters in this book present a selection of results from the first few years of the WISDOM project. For detailed information on the progress of the project and further results, please visit www.wisdom.caf.dlr.de. Further information and results from the BMBF-funded IWRM strategy can be found at www.bmbf.wasserresourcen-management.de.

I hope that I have awakened your interest in IWRM and in the WISDOM project, as well as in BMBF programmes aiming to preserve our water resources for the future.

Dr. Helmut Loewe, Deputy Head of Division Resources and Sustainability, German Federal Ministry of Education and Research

# Acknowledgements

Most of the results and findings presented in this book were generated within the framework of the BMBF-MOST funded bilateral research project WISDOM (Water related Information System for the Sustainable Development of the Mekong Delta). The work would not have been possible without the substantial funding of these two ministries. Therefore, we would like to express our sincere thanks to the German Ministry of Education and Research (BMBF), as well as to the Vietnamese Ministry of Science and Technology (MOST). Further thanks go to the 'Projektträger Jülich', PTJ, the project management office supervising the WISDOM project for their excellent support during the past 5 years, as well as to the WISDOM coordinators: the German Remote Sensing Data Centre (DFD), of the German Aerospace Centre (DLR), as well as the Southern Institute of Water Resources Research (SIWRR).

We are extremely grateful to the following experts who have volunteered their time and knowledge to peer-review the chapters in this book. Alphabetically, our sincerest thanks got to Hanne Bach (DCE-Danish Centre for Environment and Energy, Aarhus University, Denmark); Prof. Dr. Mathias Becker (Institut für Nutzpflanzenwissenschaften und Ressourcenschutz, Universität Bonn); Prof. Dr. Janos Bogardi (Global Water System Project, Center for Development Research – ZEF, Universität Bonn); Jake Brunner (IUCN – Vietnam); Prof. Dr. Bernd Diekkrüger (Geographisches Institut, Universität Bonn); Dr. Corrine Frey and Tim Funkenberg (University of Würzburg and German Aerospace Center, DLR); Matthias Garschagen (United Nations University, Institute for Environment and Human Security); Dr. Thomas Heege (EOMAP GmbH & Co. KG); Juliane Huth (University of Würzburg and German Aerospace Center, DLR); Kent Hortle (Fisheries and Environment Consultant); Pasi Lehmusluoto (Ministry of the Environment, Finland); Dr. Thomas Menkhoff (Lee Kong Chian School of Business, Singapore Management University); Dr. Fiona Miller (Department of Resource Management and Geography, The University of Melbourne); Dr. François Molle (Institut de Recherche pour le Dévelopement and International Water Management Institute); Koos Neefjes (Policy Advisor Climate Change, UNDP Viet Nam); Em Prof. Dr.-Ing E.h. Erich Plate (University of Karlsruhe); Prof. Dr. Torsten Schlurmann (Hydraulic, Waterways and Coastal Engineering, Leibniz Universität Hannover); Dr. Hannes Taubenböck (German Aerospace Center, DLR); Dr. Gabi Waibel (Center for Development Research – ZEF).

We are also grateful to Philipp Koch from the United Nations University – Institute for Environment and Human Security for his great editorial support during the preparation of this book.

# Contents

# Contributors

**Heiko Apel** Section 5.4 Hydrology, German Research Centre for Geoscience (GFZ), Potsdam, Germany

**Ute Arnold** Section Plant Nutrition, Institute of Crop Science and Resource Conservation (INRES), University of Bonn, Bonn, Germany

**Tatjana Bauer** Center for Development Research (ZEF), University of Bonn, Bonn, Germany

**Tran Thanh Be** Can Tho City Institute for Socio-Economic Development Studies (CIDS), Can Tho City, Vietnam

**Simon Benedikter** Center for Development Research (ZEF), University of Bonn, Bonn, Germany

**Nguyen Thanh Binh** United Nations University - Institute for Environment and Human Security (UNU-EHS), Bonn, Germany; Mekong Delta Development Research Institute (MDI), Can Tho University, Can Tho, Vietnam

**Jörn Birkmann** United Nations University - Institute for Environment and Human Security (UNU-EHS), Bonn, Germany

**Ian C. Campbell** School of Biological Sciences, Monash University and Principal River Scientist (GHD), Melbourne, Australia

**Bui The Cuong** Southern Institute of Sustainable Development (SISD), HCM City, Vietnam

**Jose Miguel Delgado** Section 5.4 Hydrology, German Research Centre for Geoscience (GFZ), Potsdam, Germany

**Javier Revilla Diez** Institut für Wirtschafts- und Kulturgeographie, Leibniz Universität Hannover, Hannover, Germany

**Judith Ehlert** Center for Development Research (ZEF), University of Bonn, Bonn, Germany

**Hans-Dieter Evers** Center for Development Research (ZEF), University of Bonn, Bonn, Germany

**Matthias Garschagen** United Nations University - Institute for Environment and Human Security (UNU-EHS), Bonn, Germany

**Steffen Gebhardt** Department of Remote Sensing, University of Würzburg, Wessling, Germany

**Sven Genschick** Center for Development Research (ZEF), University of Bonn, Bonn, Germany

**Solvay Gerke** Center for Development Research (ZEF), University of Bonn, Bonn, Germany

**Vo Thi Guong** Soil Science Department, Can Tho University (CTU), Can Tho City, Vietnam

**Nguyen My Hoa** Soil Science Department, Can Tho University (CTU), Can Tho City, Vietnam

**Nguyen Nghia Hung** Research Centre for Rural Technical Infrastructure Development, Southern Institute for Water Resources Research (SIWRR), Ho Chi Minh City, Vietnam

**Pham Cong Huu** Mekong Delta Development Research Institute (MDI), Can Tho University, Can Tho, Vietnam

**Verena Klinger** German Aerospace Center (DLR), Earth Observation Center (EOC), German Remote Sensing Data Center, Wessling, Germany

**Frauke Kraas** Geographisches Institut, Universität zu Köln, Köln, Germany

**Claudia Kuenzer** German Aerospace Center (DLR), Earth Observation Center (EOC), German Remote Sensing Data Center, Wessling, Germany

**Huong Thi Thu Le** United Nations University - Institute for Environment and Human Security (UNU-EHS), Bonn, Germany

**Patrick Leinenkugel** German Aerospace Center (DLR), Earth Observation Center (EOC), German Remote Sensing Data Center, Wessling, Germany

**Trinh Thi Long** Center for Environment Science Technolgy and Ecology, Southern Institute for Water Resources Research (SIWRR), Ho Chi Minh City, Vietnam

**Bruno Merz** Section 5.4 Hydrology, German Research Centre for Geoscience (GFZ), Potsdam, Germany

**Florian Moder** Bộ Khoa học và Công nghệ, Ministry of Science and Technology (MoST), 31 HanThuyen – Q. 1, TP. Ho Chi Minh, Vietnam

**Lam Dao Nguyen** GIS and Remote Sensing Research Center (GIRS), Ho Chi Minh City, Vietnam

**Loan Nguyen** Center for Development Research (ZEF), University of Bonn, Bonn, Germany

**Dang Kieu Nhan** Mekong Delta Development Research Institute (MDI), Can Tho University, Can Tho City, Vietnam

**Bui Van Quyen** Bộ Khoa học và Công nghệ, Ministry of Science and Technology (MoST), 31 HanThuyen – Q. 1, TP. Ho Chi Minh, Vietnam

**Nadine Reis** Center for Development Research (ZEF), University of Bonn, Bonn, Germany

**Fabrice G. Renaud** United Nations University - Institute for Environment and Human Security (UNU-EHS), Bonn, Germany

**Zita Sebesvari** United Nations University - Institute for Environment and Human Secuity (UNU-EHS), Bonn, Germany

**Pham Van Toan** United Nations University - Institute for Environment and Human Security (UNU-EHS), Bonn, Germany; College of Environment and Natural Resources, Can Tho University, Can Tho City, Vietnam

**Vuong Bui Tran** Technical Department, Division for Water Resources Planning and Investigation for the South (DWRPIS), Ho Chi Minh City, Vietnam

**Vo Khac Tri** Department of Training and International Cooperation, Southern Institute for Water Resources Research (SIWRR), Ho Chi Minh City, Vietnam

**Vo Van Tuan** United Nations University - Institute for Environment and Human Security (UNU-EHS), Bonn, Germany; Mekong Delta Development Research Institute (MDI), Can Tho University, Can Tho, Vietnam

**Frank Wagner** Sub-Department of Groundwater Resources – Quality and Dynamics, Federal Institute for Geosciences and Natural Resources (BGR), Hannover, Germany; Center for Water Resources Planning and Investigation (CWRPI), Long Bien District, Hanoi, Vietnam

**Gabi Waibel** Center for Development Research (ZEF), University of Bonn, Bonn, Germany

**Thilo Wehrmann** German Aerospace Center (DLR), Earth Observation Center (EOC), German Remote Sensing Data Center, Wessling, Germany

**Zengrang Xu** Institute of Geographic Sciences and Natural Resources Research, CAS, Beijing, China

# Part I
# Deltas Under Change

# Chapter 1
# Introduction

**Fabrice G. Renaud and Claudia Kuenzer**

The Mekong Delta in Vietnam (also known as the *Cuu Long* or "nine dragons") covers an area of 39,000 km$^2$ and is home to more than 17 million inhabitants. The region is familiarly known as the "rice bowl" of the country. Yet, although it is the principal rice-producing region in Vietnam, agricultural outputs go beyond rice production alone as the delta is also a main producer of fruits and vegetables as well as of aquaculture products. Economically, the delta is therefore very important for the country as a whole, however the region remains one of the poorest when compared to other regions in Vietnam. Despite the rapid economic growth of Vietnam in recent years and important improvements in agricultural systems in the region, many farmers in the delta have to deal with low profitability and high economic and environmental risks forcing them into insecure livelihoods. Key to the further development of the delta and to addressing part of the development barriers in the region is the management of the principal natural resource in the region: water. As for any delta, water plays a crucial role in shaping social-ecological systems in the Mekong Delta particularly for communities who depend on delta water resources directly for their livelihoods and daily subsistence. Water can also be directly and indirectly a threat to these livelihoods as, for example, large portions of the delta are flooded annually and although people have adapted to this flooding cycle, extreme floods (such as in 2000 and to a lesser extent in 2011) can be destructive. For decades now, national-level initiatives have shaped the delta to increase agricultural production with initiatives such as canal dredging for improved irrigation and drainage, dyke building to

F.G. Renaud (✉)
United Nations University - Institute for Environment and Human Security (UNU-EHS), UN Campus, Hermann-Ehlers-Str. 10, Bonn 53113, Germany
e-mail: renaud@ehs.unu.edu

C. Kuenzer
German Aerospace Center (DLR), Earth Observation Center (EOC), German Remote Sensing Data Center, Oberpfaffenhofen, 82234 Wessling, Germany
e-mail: claudia.kuenzer@dlr.de

F.G. Renaud and C. Kuenzer (eds.), *The Mekong Delta System: Interdisciplinary Analyses of a River Delta,* Springer Environmental Science and Engineering,
DOI 10.1007/978-94-007-3962-8_1, © Springer Science+Business Media Dordrecht 2012

protect specific areas from flooding and allowing triple rice cropping systems, and through the development of sluice gates to attempt limiting salinity intrusion in inland coastal areas. The last decades have therefore seen rapid transformations in the delta, both from a biophysical perspective and, through programmes of market liberalization (notable since *doi moi*), from social and economic perspectives.

The delta currently faces many water-related challenges and opportunities. External challenges consist in water regulation and land use changes in riparian countries which could alter the flow of the Mekong's branches in the delta. These changes, if managed collaboratively, could however turn into opportunities through e.g. reducing the occurrence of major flood events or increasing low-flows in the dry season. The delta is also seriously threatened by climate change induced sea-level rise which negatively impacts some of the existing production systems in coastal areas and further inland. Serious pollution problems, induced by agricultural, industrial, and urban activities, are also emerging in the delta, threatening the health of people and ecosystems. Groundwater is being increasingly utilized for domestic and irrigation purposes, but extraction is poorly managed and groundwater resources are threatened by overexploitation and pollution. Again some of these challenges could be transformed into developmental opportunities should natural, social, economic, and perhaps more importantly, institutional barriers be removed in order to allow for a more integrated and participative management of water resources in the region.

The objective of this book is to provide a broad overview of these water-related challenges and opportunities in the Mekong Delta through a multi-disciplinary lens. Most of the chapters are written by scientists who carried out research within the frame of the WISDOM[1] project (*Water-related information system for the sustainable development of the Mekong Delta in Vietnam*). The WISDOM project is a mutli- and cross-disciplinary research project with numerous research partner institutions in Germany and Vietnam, coordinated by the German Aerospace Center (DLR) in Germany and the Southern Institute for Water Resources Research (SIWRR) in Vietnam. In order to provide an even broader understanding of water-related themes in this volume, we invited a few experts to write chapters on topics not covered by the WISDOM project. These include Chap. 7 by Wagner et al. dealing with groundwater resources, Chap. 11 by Campbell on biodiversity, and Chap. 14 by Vo Thi Guong and Nguyen My Hoa on nutrient pollution.

The book is divided in four parts. The first part, *Deltas Under Change*, is composed of this introduction and Chap. 2 by Kuenzer and Renaud addressing the challenges and opportunities for delta systems worldwide to deal with environmental, climate, and social change.

The second part, *Water and Development in the Mekong Delta*, provides background information on the surface freshwater system (Chap. 3 by Vo Khac Tri) and the social and economic transformations of the delta (Chap. 4 by Garschagen et al.). This part also addresses the challenges in implementing Integrated Water Resources

[1]More information on the WISDOM project can be obtained at http://www.wisdom.caf.dlr.de/en

Management at the Mekong Basin scale (Chap. 5 by Moder et al.) and in the Mekong Delta per se (Chap. 6 by Waibel et al.).

The third part, *Water Resources and Related Risks*, deals with the threats to the freshwater resources system in the Mekong Delta and water-related risks to communities. Chapter 7 by Wagner et al. looks at groundwater resources which are being tapped at an increasing rate in the region, both promoting development but also bringing new risks. Apel et al. (Chap. 8) address sedimentation issues in the delta; Chap. 9 (Delgado et al.) looks at flood risks at the basin scale; and Chap. 10 (Birkmann et al.) looks at water-related vulnerabilities and risks with respect floods and salinity intrusion in urban and rural environments, framing the issue in the climate change discourse.

The fourth part, *Water and Environment*, addresses environmental threats in the Mekong Delta. Chapter 11 by Campbell provides an overview of the state of the biodiversity in the delta. Gebhardt et al. (Chap. 12) look at the status of mangrove ecosystems in the delta and discuss the usefulness of remote sensing tools to track mangrove related land use changes. This is followed by two chapters that deal with water pollution problems. Chapter 13 (Sebesvari et al.) provides a review of water pollution by agricultural activities in the delta for a wide range of contaminants and makes suggestions for mitigation measures while Chap. 14 (Vo Thi Guong and Nguyen My Hoa) addresses more specifically nutrient pollution in soil and water by agriculture and aquaculture.

The last part, *Water Knowledge and Information*, deals with the need to improve knowledge management and information-sharing on water-related matters in the Mekong Delta. Chapter 15 by Gerke et al. investigates the links between knowledge management and development in the delta. Chapter 16 by Klinger et al. presents the concepts behind the WISDOM information system and a description of its structure and functionality. Finally, Renaud and Kuenzer conclude this volume by synthesizing some of the key messages in the various chapters focusing in particular on the water-knowledge-development nexus (Chap. 17).

# Chapter 2
# Climate and Environmental Change in River Deltas Globally: Expected Impacts, Resilience, and Adaptation

**Claudia Kuenzer and Fabrice G. Renaud**

**Abstract** A large part of the world's population lives in coastal areas, and even though river deltas only contribute to 5% of the global land mass, over 500 million people live in these areas, where major rivers reach the ocean. Deltas have numerous advantages for societal development, such as a flat topography, available fresh and salt water resources, good transportation via waterways and the nearby coast, fertile soils for agriculture, and usually also a rich biodiversity and recreational value through, for example, wetland ecosystems, coastal forests, and beaches. However, at the same time, river deltas of the world belong to the most endangered ecosystems with respect to societal, environmental and climate change – the latter especially manifested through sea level rise (SLR). In this chapter we present the most common environmental impacts of society's economic development expressed via agricultural intensification, increasing urban sprawl, industrial activity, and infrastructure expansion in deltas, as well as the climate change and SLR related challenges that threaten delta development in many countries. We discuss means of mitigation and adaptation towards these negative impacts, which can be of educational, ecological, technological and political nature, and depict how delta populations with differing resilience address the challenges induced by environmental and climate change. We use the Mekong Delta in Vietnam as a case study by addressing the threats and possible adaptation options of this region.

---

C. Kuenzer (✉)
German Aerospace Center (DLR), Earth Observation Center (EOC), German Remote Sensing Data Center, Oberpfaffenhofen, 82234 Wessling, Germany
e-mail: claudia.kuenzer@dlr.de

F.G. Renaud
United Nations University - Institute for Environment and Human Security (UNU-EHS), UN Campus, Hermann-Ehlers-Str. 10, Bonn 53113, Germany
e-mail: renaud@ehs.unu.edu

F.G. Renaud and C. Kuenzer (eds.), *The Mekong Delta System: Interdisciplinary Analyses of a River Delta,* Springer Environmental Science and Engineering,
DOI 10.1007/978-94-007-3962-8_2, © Springer Science+Business Media Dordrecht 2012

## 2.1 River Deltas: Characteristics and Advantages for Socio-Economic Development

The modern river deltas of the world formed when global sea level stabilized within a few meters of the present level around 6,000 years ago. As the deltas developed, their main stem channel split into a series of distributary channels on flat terrain (Goudie 2007; Overeem and Syvitski 2009). Deltas are landforms naturally shaped by the forces of rivers, waves, and tides in areas where a river drains into a lake or ocean basin. They are characterized by low topography, usually high productivity, rich biodiversity, abundant natural channel systems, and general land-coast interaction. They act as filters, repositories, and reactors for a suite of continental materials, including carbon, on their way to the ocean. Through self-organization of channels and river mouths a dynamic response to controlling processes is given, which normally makes the systems resilient and able to adapt to hazards. However, through increased human interference, natural resources and populations in delta areas are nowadays highly vulnerable (Nicholls and Mimura 1998; Nicholls 2003; Overeem and Syvitski 2009; Kuenzer et al. 2011a). According to the World Wildlife Fund International Director James Leap: "Where the river meets the sea has always been one of the most important habitats for humanity, but we have done enormous damage to vibrant life in estuaries" (de Wit 2011: 12).

Many definitions for river deltas and river estuaries exist; however in this chapter, we adopt the common viewpoint that an estuary is a river mouth, where fresh and saltwater meet, creating a zone of brackish water. Deltas are formed where rivers and the coast interact, and where sediment is deposited, creating new land. Usually, a delta can be made up of a number of estuaries, and throughout the chapter we use the term 'deltas'.

Coastal zones are the preferred choice for settlement of humanity. Of the 20 largest mega cities worldwide, 13 are located directly on the coast: Tokyo (the largest city, with over 32 million inhabitants), Seoul, New York, Mumbai, Jakarta, Osaka-Kobe, Shanghai, Manila, Hong Kong-Shenzhen, Los Angeles, Kolkata, Buenos Aires and Karachi (the smallest with ca. 12 million inhabitants). Sao Paulo, Greater Cairo, and London, while not directly located at the coast, are considered to lie nearby and partially experience tidal influences. Many of these cities are located in delta areas. Aside from mega cities, other coastline areas and deltas are also fairly populated, including the deltas of the Mississippi (Törnqvist et al. 2010), Nile (Elsharkawy et al. 2009; Conway and Hulme 1996), Po, Rhine, Elbe (Holzwarth et al. 2010), Brahmaputra, Irrawaddy, Zambezi (Beilfuss and Dos Santos 2001), Paraná (Menéndez and Sarubbi 2007; Sarubbi 2007) or Mekong River, to name only a few. With some exceptions such as the Lena delta located north of the Polar Circle, deltas generally tend to attract larger parts of a country's population. Delta areas comprise only 5% of the land area globally, but over 500 million people live there. The Yangtze, Ganges and Nile deltas alone held a population of 230 million in 2000 (Overeem and Syvitski 2009). It is expected that – from the year 2000 on – there will be a 35% increase of population in the major world deltas by 2015 (Timmerman and White 1997; Ericson et al. 2006). Figure 2.1 presents an overview of the 40 largest and most well-known river deltas globally.

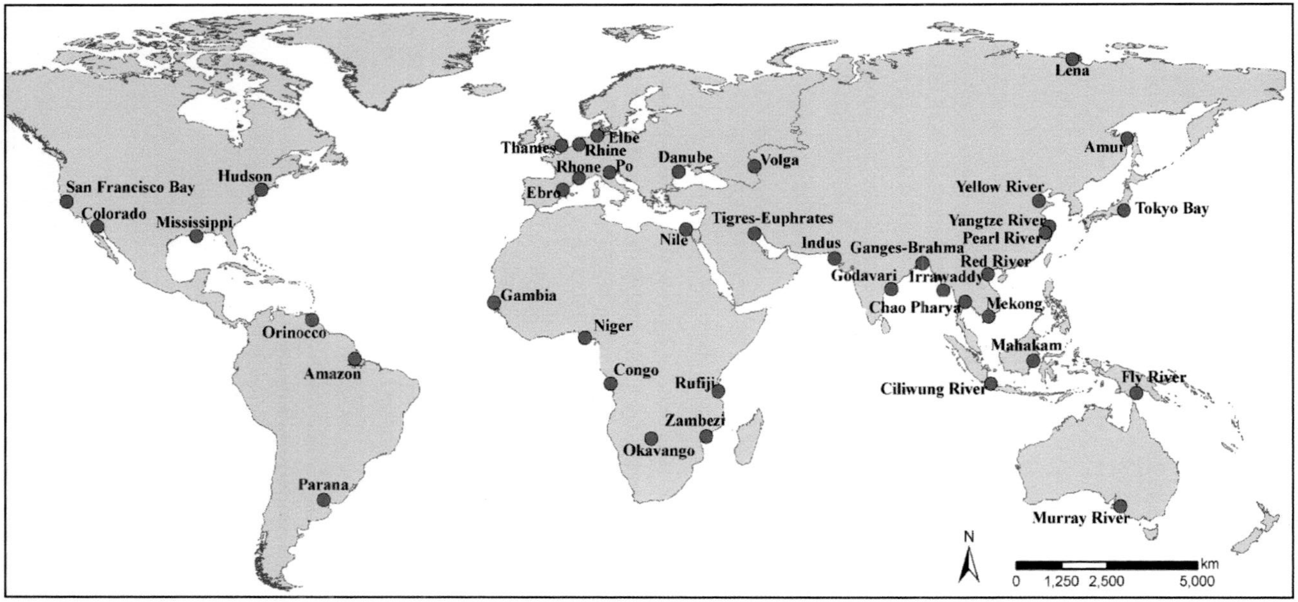


**Fig. 2.1** The 40 largest river deltas globally (This figure does not contain all the river deltas and estuaries that exist in the world; the 40 largest deltas (with respect to delta and river size) were selected)

Deltas often are the agricultural and/or industrial production base of a country, and are therefore zones of strategic importance for a nation. The Pearl River Delta and Yangtze River Delta in China contribute a large portion of the countries national Gross domestic production (GDP). For example, even though the Pearl River Delta only encompasses 0.5% of the Chinese territory and 4.5% of China's population, the delta accounts for 10% of the country's GDP. For the Yangtze delta this number is even larger – the delta contributes more than 20% to the national GDP. Less urbanized deltas are also of strategic importance. The Nile Delta is the source of food crops for most of Egypt (Abd El-Kawy et al. 2011; Elsharkawy et al. 2009) while the Po flats of the Italian Adria feed the whole country with vegetables (Breber 1993). Comparing the population densities in the city agglomerations of Hanoi or Ho Chi Minh City in Vietnam with that of the Mekong Delta, the latter is relatively sparsely populated, but at the same time the delta is the country's most important region in terms of agricultural production: the Mekong delta is often referred to as the 'rice bowl' of the country. From cultural and ethnologic perspectives, several deltas are the cradle of a country's civilization, and some still today host indigenous people – such as the Warao people in the Orinoco Delta (White et al. 2002).

Deltas are attractive territories due to features such as:

- Highly productive arable land, rich in fertile sediment and organic material for agricultural cultivation
- Marine and river resources for protein food security
- Frequent onshore and offshore oil and gas reserves near the delta
- Flat topography, which provides a favourable environment for agricultural, urban and industrial development as well as for transport
- Large area, enabling urban growth and multi-directional sprawl (in most cases)
- Function as a transport hub for incoming goods from abroad, or from inland
- Function as a transport hub for outgoing goods and manufacturing
- Usually ice free harbours due to river and ocean currents
- Fresh water availability from the river or aquifers
- Rich biodiversity of wetlands and protective ecosystems (e.g. mangroves or reed belts)
- Health advantages, as winds and tidal mixing dilute the solid, liquid and gaseous wastes
- Recreation and eco-tourism options

Some deltas are sparsely populated but might have outstanding ecosystem and biodiversity value, such as the Rufiji Delta, which has only about 150,000 inhabitants, but is home to the largest area of estuarine mangroves in East Africa (Doody and Hamerlynk 2003; Ochieng 2002), the Lena delta with its lush wetlands in summer, and extensive wildlife (Yang et al. 2002), or the Okavango inland delta, which is the main water source of regional wildlife in the dry season and also the largest breeding site for birds in Southern Africa (Burg 2007). Also the Danube delta – despite an increasing number of shipping canals and aquaculture – remains one of the best preserved temperate climate deltas (Overeem and Syvitski 2009; Panin 1999). However, especially in more densely settled deltas, socio-economic development goes along with urbanisation and industrialization or agricultural intensification impacting the delta environment severely. These environmental changes are addressed in Sect. 2.2.

The term 'vulnerability' in this paper is understood according to the IPCC (2007) as the degree to which a system is susceptible to, and unable to cope with, the adverse effects of climate change. Here, vulnerability is a function of the magnitude and rate of climate change. We extend this definition in so far as we include the adverse effects of societal and economic development in river deltas as well.

The term 'resilience' is defined differently by varying authors, depending on the focus of their research area (Gallopin 2006; Folke 2006; Walker et al. 2004; Garschagen 2011). The major three concepts are those of engineering resilience, ecological/ecosystem resilience, and social-ecological resilience. In this chapter we refer to the latter concept, where resilience is 'the capacity of a system to absorb disturbance and re-organise, while undergoing change so as to still retain the same function, structure, identity and feedbacks' (Walker et al. 2004).

'Adaptation' is understood as an adjustment to the effects induced by climate change (as well as the effects of societal and economic development), which minimizes harm and exploits beneficial opportunities. Adaptation can occur naturally (autonomous adaptation – e.g. ecological adaptation), or as the result of a planning process based on the awareness that conditions will or have changed and that action is required to return to, maintain, or achieve a desired state (IPCC 2007). In this chapter we usually refer to the latter, which can occur as proactive or responsive adaptation (before the conditions have changed in anticipation of this change, or as a response afterwards).

## 2.2 Environmental Change in Deltas as a Cause of Socio-Economic Development and Transformation

The increasing vulnerability of the socio-ecological system in delta areas arises from many factors such as socio-economic development and transformation (urban development, industrial development, agricultural intensification) and climate change-related impacts. Rapid urban growth and infrastructure development usually go along with:

- Urban and industrial sprawl, and associated waste water and solid substance pollution (Kuenzer 2007), as well as atmospheric emissions (in delta areas next to industry, oil and gas reserves also frequently exist, see Fig. 2.2d). Harbour areas in deltas are often threatened through contamination and infill (e.g. Hong Kong, Tokyo)
- Strong surface sealing (Leinenkugel et al. 2011), impairing natural infiltration and changes in the local soil water patterns; reducing the buffer capacity of soils for flood water retention. The replacement of natural wetland areas like mangroves (Kuenzer et al. 2011a), and the building of heavy structures such as skyscrapers on swampy, weak ground is causing further problems (e.g. land subsidence, reduced filtering capacities)

**Fig. 2.2** (**a**) Natural wetland in the Danube Delta, Romania; (**b**) aquaculture reaching the coastline and leading to the degradation of the last coastal forest fragments in Bac Lieu province, Vietnam; (**c**) intensive aquaculture and salt production in the Yellow River Delta, China; (**d**) Shengli Oilfield, China's second largest oilfield located at the coastline of the Yellow River Delta, China; (**e**) Rhine delta near Rotterdam – a densely populated and industrialized region; (**f**) Tokyo – one of the most densely populated delta areas globally (Source: Google)

- Changes in land use, often leading to erosion, deflation, and degradation (e.g. deforestation to assess new land for construction)
- Ground water pumping to cover the urban drinking water demand, often leading to land subsidence (and thus increased flood hazard)
- Drilling of oil and gas, also leading to subsidence (e.g. Po Delta subsidence was 40–60 mm/year, but decreased to 10–20 mm/year directly after extraction stopped in the late 1970s, Overeem and Syvitski 2009)
- Higher pressure on remaining recreational or protected areas; substance, noise and light pollution

In deltas with an equally spread rural population and an agricultural production focus (e.g. Mekong Delta), economic development usually goes along with the following changes, also leading to negative impacts on the environment and the local population:

- Intensification of agriculture, leading to increased application of fertilizer, pesticides, antibiotics, and thus increased water pollution
- Shift towards most profitable monoculture crops or increase in the number of annual harvests (rice, but also cotton, sugar cane, oil palms etc.) often leading to a loss in biodiversity
- Shift from grain crops or mixed crops towards intensive aquaculture, and conversion of natural protective ecosystems (mangroves, reed belts, shoreline forest) into aquaculture areas, usually going along with the heavy use of antibiotics and other chemicals (see Fig. 2.2b, c)
- Increased production of domestic animals, also associated with waste water- and solid substance pollution (livestock, but also aquaculture industry)
- Changes in water flows, water availability and sediment dynamics (e.g. water use for irrigation)
- Simultaneous growth of urban centres and expansion of peri-urban environments leading to the aforementioned transformations (see Fig. 2.2e, f)

In addition to the above mentioned impacts of urbanisation as well as agricultural intensification driven by economic development, many deltas are indirectly impacted by upstream-downstream conflicts. In many developing and emerging countries the sediment load of rivers increased during the last century due to deforestation, intensification of agriculture, urban sprawl, and related erosion and deflation in the upper and middle catchments. The sediment loads led to steady aggradations of new land in delta areas. However, in recent decades the construction of upstream dams in different catchments has led to a depletion of the supply of sediments (and sometimes also water) to many deltas, with increased coastal erosion becoming a widespread consequence (Lu 2004; Zhang et al. 2009; Yang et al. 2005; Beilfuss and Dos Santos 2001; Paw and Thia-Eng 1991). The Zambezi Delta for example shrunk in size by nearly half since the Kariba and Cahora Bassa dams were built. Wang et al. (2011) found that the Yellow and Yangtze Rivers showed a severe decline in sediment flux due to dams. In the Mekong River, sediment has declined in the midstream basin due to trapping in upper basin dams, while

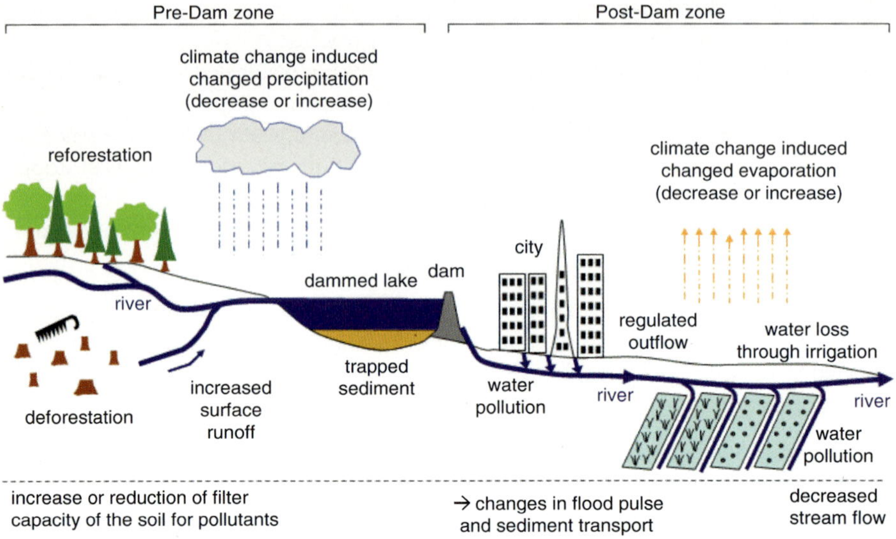


**Fig. 2.3** Sketch of upstream climate and environmental change, affecting the river delta downstream

sediment load in the lower Mekong is currently increasing due to human impacts such as mismanaged land reclamation (Wang et al. 2011). Negative consequences of upstream developments include:

- Lowered sediment rates further downstream and in the delta (impacting nutrient transfer and thus agricultural productivity)
- Increased coastal erosion (impacting wetlands, protective forest, local farmers, coastal defence in urban areas etc.) due to lowered sediment content
- Changed flood pulses and pulse variability over the year
- Downstream transfer of pollutants leading to water quality deterioration
- Changes in river ecology and biodiversity (impacting not only composition, but also fish catch quantity and local communities depending on the river) due to changes in water quantity and quality
- Changes in navigability due to changes in water quantity and barriers
- International conflict (in transboundary river basins)

Water transfer schemes from one catchment to another can also strongly impact a delta environment. The Godavari-Krishna water transfer scheme in India will lead to a water deficit in the delta compromising the potential of a second rice harvest (Bharati et al. 2009). The Yellow River Delta is also negatively impacted by local-scale water transfers (Overeem and Syvitski 2009). In addition climate change effects upstream further complicate the upstream impacts on the downstream development. Figure 2.3 illustrates climate change and human induced environmental impacts occurring within upper catchment areas, affecting delta regions further downstream.

This and all the above mentioned impacts impair the urban and agricultural delta ecosystems, as well as transitional ecosystem between land and ocean (Timmerman and White 1997). One region especially affected by all these challenges and threats mentioned above is Asia. Of the ca. 15% of the global population and ca. 20% of the urban population living along coastlines located less than 10 m above sea level, 75% of people in this zone are located in Asia. Population growth in these regions is rapid (Timmerman and White 1997; Milliman and Haq 1996; King and Adeel 2002; Nicholls 2003). In China alone, the low lying coastal zone forms 16.8% of China's total land mass, hosting 41.9% of the population, which produces 72.5% of China's Gross Domestic Product – GDP (Lau 2004; Le 2000). As depicted in Fig. 2.1, the density of large and often densely populated river deltas is especially high in Asia and Southeast Asia. Hence, this region plays a key role in the advancement of environmental and climate change focussed delta research and the development of novel approaches to improve resilience and adaptation.

## 2.3 Climate Change Related Impacts in River Deltas

Sea level rise (SLR) is mainly induced by the thermal expansion of ocean water, which has occurred frequently in geologic times (Liu and Milliman 2004) and can be aggravated by the melting of the Greenland and Antarctic Ice Caps. According to the IPCC (2007: 30), sea level rose at an average rate of 1.8 mm annually between 1961 and 2003 and at a faster average rate of 3.1 mm annually between 1993 and 2003. In the period from 1993 to 2003, rates in some regions were up to several times the global mean rise, while in other regions, sea levels fell. The main contributor to sea level rise in the last decade was thermal expansion (accounting for 57%) followed by the melting of glaciers and ice caps (accounting for 28%) and the loss of polar ice sheets (IPCC 2007: 30). Projections for the period 2090–2099 by the IPCC (2007: 45) put global average sea level rise at 0.18–0.59 m relative to the period 1980–1999, depending on the SRES (Special Report on Emission Scenario). Within the last 3 years, a rapid decline in ice mass balance from both Greenland and Antarctica has been observed, and it is predicted that sea-level rise by 2100 could easily reach 200 mm (Allison et al. 2009). In addition to such medium term sea level rise, short term sea level rise can occur due to events such as typhoons, storms, El Nino events, increased river runoff in delta areas, and tsunamis (WMO 2008; Ericson et al. 2006).

Modelling studies suggest that overall it is very likely that anthropogenic forcing – meaning the climate response to human activities – contributed to sea-level rise during the latter half of the twentieth century. According to Ericson et al. (2006) effective sea level rise (ESLR) in Asia was of 4.6 mm/year on average. For Northern America, South America, and Europe 4.5, 3.5 and 2.6 mm are expected respectively, while in Africa and Oceania 4.4 and 1.0 mm are expected annually, respectively over the next 50 years. The Tyndall Centre for Climate Change research in Britain underlined that Asia will be the region by far the most seriously affected by sea level rise. This is due to the region's vast coastline, a large number of mega deltas, big port cities, and a very dense population. Calculations assume that along with a 1 m rise in sea level would go the displacement of probably over 100 million people, and losses of>500 billion USD GDP annually. Southeast Asia, especially Vietnam, would be hard-hit with expected displacements of 1/10th of the population and a loss of 10% GDP (Richardson 2008). But also other deltas will have to face severe impacts. In the Nile delta a predicted sea level rise of 50 cm to 1 m until the end of the century would lead to the displacement of 6–10 Mio people, and large GDP losses (Elsharkawy et al. 2009).

The problem of sea-level rise is a challenge that most delta populations will have to face even if appropriate mitigation measures in terms of greenhouse gas emissions are taken immediately due to effects such as the "commitment to sea level rise". This effect describes the slow diffusion of heat into the deeper ocean layers: 'It may take thousands of years for the ocean temperature to reach equilibrium with a new stable climate' (Nicholls 2003: 10).

Based on the above, additional new challenges and threats with regard to climate change and associated SLR, complementing the challenges already listed in Sect. 2.2, exist for the deltas:

- Land inundation as most deltas are low-lying (IPCC 2007)
- Continuous salt water intrusion with the tides leads to an environment only certain species can accept. Valuable wetlands are and will be lost (Nicholls et al. 1999; Nicholls 2004). In case of sea-level-rise-induced intrusion of saltwater inland, agricultural practices might have to change, as the common crops might not tolerate the new salinity levels (Sebesvari et al. 2011). Furthermore, intruding salt water can destroy local underground fresh water aquifers and a higher water table can affect building foundations.
- Several authors predict a drastic increase in severe storm surge disaster events for many coasts worldwide (von Storch et al. 2008; Karim and Mimura 2008). Typhoon frequency in Asia has increased in past decades (Le 2000; Krishna 2009), and this is expected to aggravate the challenges already induced by SLR.
- Along with SLR will also come a loss of recreational natural and cultural resources, and a loss of tourism (beach erosion) and transportation functions (Kuenzer et al. 2011a). Milliman and Haq (1996), as well as Nicholls et al. (1999) estimate the loss of 40–80% of all delta wetlands.
- Due to the loss of property and appropriate agricultural land, population migration inland can be expected with rising sea level. This will lead to a new pressure on resources such as arable land, land for urban construction, and industrial settlement.

Countermeasures that need to be and will be taken against these threats (be they of technical or ecologic nature) will consume large amounts of an affected country's GDP. According to Lau (2004) the calculated losses through SLR by 2030 in the three main river deltas of China (Pearl River, Yangtze River, Yellow River) will total between 350 and 2,100 billion Chinese Yuan, for 30 cm and 1 m SLR scenarios, respectively. As another example, Jallow et al. (1996) underline that the shoreline retreat in the Gambia delta of western Africa will vary between 6.8 m on cliff coasts

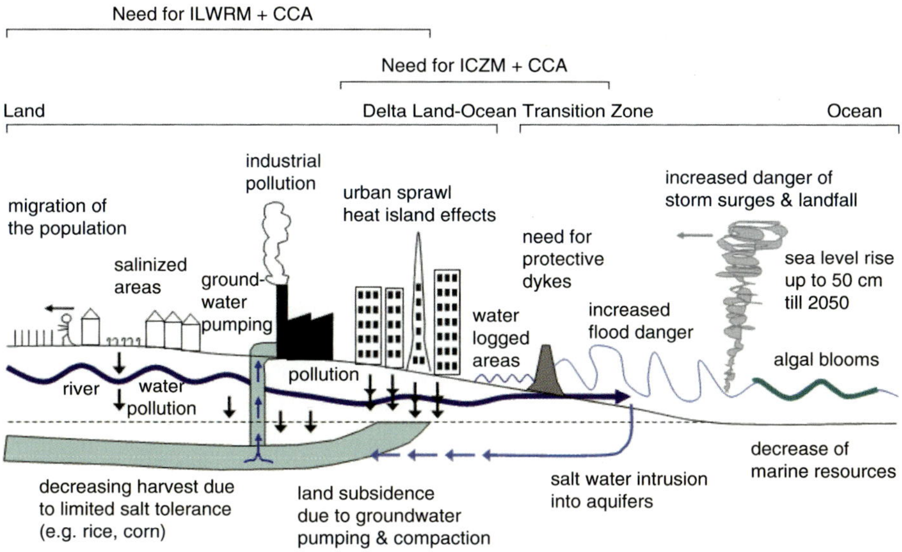


**Fig. 2.4** Sketch of most pressing challenges in delta areas globally. In urbanized deltas, all of the depicted challenges may occur while in less populated deltas only some of them might be applicable. *ILWRM* integrated land and water resources management, *CCA* climate change adaptation, *ICZM* integrated coastal zone management

and up to 880 m on flat, sandy coasts, leading to the loss of valuable mangrove ecosystems and other wildlife habitats, as well as complete inundation of the capital city Banjul. 42,000 people would have to be relocated, if severe protection measures are not undertaken. They investigated the costs of such losses for only a small fraction of the Delta including the capital area, and came to the conclusion that USD 217 million would be lost.

Figure 2.4 illustrates some of the threats to river delta environments affected by socio-economic development and transformation as well as sea level rise. Some of the changes affect the landside part of the delta only, while others affect the land-ocean transition zone, and others only affect the coastal zone. However, consequences of each can usually be felt in the entire region (e.g. algal blooms such as the 'red tide' kill coastal fish, affect fishermen's catch amounts, local market availability, price patterns, families incomes, and finally protein consumption, and household stability).

## 2.4 Climate Change Adaptation Options for Delta Populations

To a certain extent, the resilience of populations living in major deltas is dependent on the physical dynamics of the deltas themselves. Within a coupled social-ecological system, where the social system is highly dependent on the ecosystem for livelihoods, the rate at which environmental degradation, sea-level rise, and other hazards can affect natural resources such as land and water will determine the rate at which local populations will have to adapt to the new circumstances such as salinisation of surface- and groundwater systems as well as of soils and the potential exacerbated damages linked to coastal processes such as storm surges and coastal erosion (Smith and Barchiesi 2009). This is already underway in various larger deltas around the world including the Mekong Delta but also the Ganges-Brahmaputra-Meghna Delta in Bangladesh, where environmental changes combined with economic and policy drivers entice farmers to shift from rice production systems to aquaculture and in particular shrimp farming. This in turn poses another series of potential problems depending on the intensity of shrimp farming and the environmental stewardship put in place by the farmers. Resource dependency typically leads to low resilience levels and this is not a specific characteristic of delta regions. Adger et al (2001) illustrate this in the context of mangrove conversions in the Mekong Delta whereby households depending on the services provided by mangrove ecosystems see their livelihoods severely affected when mangroves are converted to other land uses which benefit wealthier households.

Bucx et al. (2010) compared the resilience and sustainability of ten deltas worldwide, including the Nile, Incomati, Ganges-Brahmaputra-Meghna, Yangtze, Ciliwung, Mekong, Rhine-Meuse, Danube, California-Bay, and Mississippi River. Resilience/sustainability was assessed using the DPSIR (driver-pressure-state-impact-response) approach with three criteria: "occupation layer", characterized by land and water use; "network layer", consisting of a description of the infrastructure; and "base layer", consisting of natural resources. Their overall assessment was that most deltas are under great pressure including an imbalance in demand for and supply of land and water, inadequate and/or ageing infrastructure, the prevalence of more extreme weather events and impacts on ecosystems (Bucx et al. 2010: 8). In their framework, the deltas ranking the lowest in terms of resilience/sustainability are the Ganges-Brahmaputra-Meghna, Ciliwung and Nile deltas. The Mekong Delta shows low resilience/sustainability in terms of the status of its natural resources and medium resilience/sustainability for the other indicators.

Adaptation to climate change in river deltas is realized through different means, depending on the geographic location of the delta, the prevailing challenges in the region (e.g. population density), the cultural background of the population, the level of development of the country, the prevailing paradigm concerning adaptation to climate change effects, the influence of the international community (e.g. introducing technology in developing countries), and many more. For example, a delta like the Yangtze River delta, which is densely populated and hosts the city of Shanghai and the business district of Pudong (both located below water level) must (and can afford to) employ technology-based solutions to safeguard the population, which – according to local and national considerations – has become too large to be resettled. In other river deltas, such as the comparatively less densely populated coastal Mekong, it is (still) possible to employ natural means of climate proofing, such as the reforestation of the shoreline with protective mangrove belts, experiments with salt-tolerant crop species, or the development of settlement and industry areas further inland. However, even within one country, river deltas can be managed in a completely different way.

"The Red River Delta and the Mekong River Delta are both big, flat areas, very fertile with predominantly rice planting and other agriculture activities, and a major city nearby. But the way they tackle flood is very different: The Red River Delta is protected by many dikes and polders whereas there are far fewer dikes in the Mekong Delta, and the land is more adapted to flooding" (de Wit 2011).

Nicholls (2003) underlined that, given the large number and growing concentration of the population along the coast and in the delta zones, autonomous climate change or SLR adaptation processes are unlikely to be sufficient. All levels of government and further institutions (science, NGOs etc.) need to be involved in three categories of adaptation, which commonly should be fostered in a hybrid mix:

- Planned retreat (natural system effects may occur and people retreat from the coast)
- Accommodation (natural system effects may occur and human use of the coastal zone is adapted to minimize impacts)
- Protection (natural system effects are controlled by hard and soft engineering, thus minimizing impacts on the population)

Adaptation is a continuous process, and only a well-chosen, cross-sectoral combination of measures can lead to an asymptotic approximation of the best installed adaptation potential at a given time. According to our understanding adaptation measures can be grouped into:

- Technological: coastal defence such as dykes, dams, sluice gates, levees, canals, and other protective constructions, securing harbours and densely populated water fronts, waste water treatment, water-saving technology, fresh water storage, installation of de-salinization plants, military equipment (disaster preparedness), early warning systems, sustainable 'green city' development (e.g. energy saving offices etc.), elevation of settlements, installation of evacuation shelters and routes, provision of adapted crop species varieties, stocks of water, food and medicine, etc.
- Ecological: Environmentally adapted utilization of resources, protection belts of coastal forests (mangroves, melaleuca), installation of protection zones, nature- and national parks, natural wetland renaturation, planting of salt-tolerant/drought-resistant/deep-rooting species, avoidance of practices leading to land subsidence, natural techniques minimizing evaporation, wind speed or current
- Educational: climate change related education (comprising geography, ecology, basic natural sciences, interactions of population and environment, first aid, medical preparedness, disease control, swimming lessons, strengthening awareness for the value of ecosystems, fostering information sharing, low-energy/low-carbon resource efficient behaviour etc.)
- Political: political measures (advisory panels, decrees, rules, laws, norms) must partially initiate, back-up, and increasingly foster the above three. Development of a re-insurance sector to promptly mitigate losses is still a necessity in many countries.

Table 2.1 presents climate change-related phenomena, their impacts in the natural environment as well as for the delta population, and lists possible counter measures

**Table 2.1** General climate change related phenomena, their impacts and possible counter measures in river deltas

| Climate change phenomena | | Impacts on the natural delta environment | Impacts on the delta's population | Possible adaptation measures (proactive and reactive) |
|---|---|---|---|---|
| **Temperature changes** | **Ocean temperature increase** | Increases of vibrio cholerae in water, increases of organic matter density, changes in coastal ecology and biodiversity | Increased chance of infection with bacteria, increase of illnesses, monetary losses in all sectors | Technological: Technology improving water quality, control of sewage release into the ocean, water quality monitoring networks, algae-killing bacteria, information & early warning systems |
| | | Faster algae growth, and lower oxygen supply, algal blooms, and harmful algal blooms killing fish, shellfish (e.g. "Red Tide"), and negative impacts on marine resources | Decrease in fish catch, economic losses in fishery sector, economic losses in transport sector (reduced navigability) and tourism sector, poverty threat for the already less-off | Ecological: restoration of coastal vegetation with buffer functionality (e.g. mangroves)<br>Educational: public education on water quality and health issues<br>Political: regulations/laws enforcing the long term sustainable implementation of the above |
| | | (Micro)climatic changes e.g. in coastal ocean-land and land-ocean wind systems, evaporation; exact manifestation difficult to foresee; in winter most likely higher temperature gradient and stronger winds while the reverse is expected in summer, chance of decreased air exchange, impact on air quality | Health impacts on the population due to air quality decrease/urbanizations in already hot regions get warmer/ economic losses; potential negative impacts on agricultural productivity | Technological: all technology minimizing air pollution (e.g. filters for factories, hybrid or electro-transport etc.), high-albedo roofs, early warning systems (air quality/temperature information systems)<br>Ecological: increase green belt areas (reforestation of coastal zones, urban greening, etc.)<br>Educational: promoting resource-efficient, low-energy and low-carbon behaviour; health education<br>Political: regulations/laws enforcing the long term sustainable implementation of the above |
| | | *Sea level rise through water expansion, and melting of ice caps and glaciers* | *Threats associated to SLR, see Sects. 2.2 and 2.5 and lower section of this table* | *Sections 2.2 and 2.5 and lower section of this table* |

| | | | |
|---|---|---|---|
| **Land surface and air temperature increase** | *Sea level rise through melting of ice caps and glaciers* | *Threats associated to SLR, see this paper and lower section of this table* | *See this paper and lower section of this table* |
| | Large scale heat waves and droughts, stress on crops and natural vegetation, stress on animals, increased fire danger, drying out of rivers and lakes<br>Acceleration of the urban heat island effect | Reduced or destroyed harvests, food shortages, limited navigability, increased anthropogenic water demand, increased energy demand (for cooling), heat related morbidity and mortality increases, large economic losses in all sectors, poverty threat for the already less-off | Technological: Heat tolerant species, precision-irrigation, minimizing evaporation and leakage, water-saving technology, water recycling, air conditioned or self-regulating housing, shaded building blocks, high albedo materials, inner city water surfaces, early warning systems, food supply stocks, water supply stocks, desalinization plants<br>Ecological: increase green belt areas (reforestation of coastal zones, urban greening with heat resistant species), avoiding urban irrigation, deflation control via vegetative cover<br>Educational: energy-saving, resource efficient behaviour, medical preparedness especially aiming at elderly and very young, new water efficiency standards<br>Political: regulations/laws enforcing the long term sustainable implementation of the above |
| | (Micro)climatic changes e.g. in coastal ocean-land and land-ocean wind systems, evaporation, soil water movement etc.; exact manifestation difficult to foresee | If air quality decreases/urbanizations in already hot regions get warmer/then impacts on delta population's health, economic losses, potential negative impacts on agricultural productivity | Same as above |

(continued)

**Table 2.1** (continued)

| Climate change phenomena | | Impacts on the natural delta environment | Impacts on the delta's population | Possible adaptation measures (proactive and reactive) |
|---|---|---|---|---|
| **Precipitation changes** | **Precipitation decrease** | Declined river flow, declined groundwater recharge, dry stress for crops and natural plants, vegetation and animals, potentially leading to droughts | Negative in areas, where precipitation is already low, increase in water prices, decline of irrigation water availability, drinking water shortages, impact on economy and infrastructure, decreased productivity, conflicts | Same as for heat waves and droughts |
| | | Increased soil erosion in fragile ecosystems with limited vegetative cover, loss of top soil, loss of soil fertility, decline of productivity, increase of sediment load | Decrease of arable land extent, decreased yields, crop diseases, possible damage to infrastructure | Technological: technical soil erosion counter measures, soil stabilizing technology, terracing, drainage, surface sealing control, stabilization of infrastructure (roads, settlement etc.) in erosion-prone areas<br>Ecological: planting of soil covering, deep-rooting species, renaturation of channelled rivers and retention basins<br>Educational: education on soil conserving practices and measures<br>Political: regulations/laws enforcing the long term sustainable implementation of the above |
| | | Increase of inundated 'standing water' surfaces, expansion of vector-borne diseases (e.g. malaria, dengue, schistosoma-spreading snails) | Increase in people infected with vector borne diseases, rising fatalities and health costs, decreased workforce productivity | Technological: disease prevention through host control (reduction of mosquito breeding places through coverage or movement of water surface), availability of sufficient medicine stocks<br>Ecological: strengthening natural enemies (food chain)<br>Educational: education of the affected population, early on in schools, health-, hygiene and prevention lessons<br>Political: enforcement of legislation |

| | | | |
|---|---|---|---|
| | Rise in flood frequency, increased surface run off and inundation | Threat to the local population through forces of water masses, increase of vector borne diseases, spread of pollutants, injuries and loss of lives, loss of livelihoods and homes, loss of agricultural ground, negative economic impacts, migration and potential conflicts | Technological: construction or improvement of dykes, dams, levees, sluice gates, and river canals, pumping stations, design of solid flood shelter centres and areas (solid construction of schools, stadiums, public buildings), flood disaster rescue hardware, flood early warning systems, elevation of settlements, infrastructure granting mobility also during flood times (water ways, boats, helicopters etc.)<br>Ecological: development of retention areas, renaturation of wetlands and rivers, natural flood water buffer zones, soil stabilizing, deep-rooting vegetation to minimize flood water's force and erosion, installation of natural reserves and nature parks, protection of breeding and spawning spaces<br>Educational: flood disaster training for the public and the emergency services, mandatory swimming lessons in school, disease control training<br>Political: regulations/laws enforcing the long term sustainable implementation of the above; and additionally especially proactive settlement and land use restrictions in high flood risk areas; possible relocation programmes, flood related insurances |

(continued)

**Table 2.1** (continued)

| Climate change phenomena | | Impacts on the natural delta environment | Impacts on the delta's population | Possible adaptation measures (proactive and reactive) |
|---|---|---|---|---|
| **Sea level rise** (actually an impact of temperature change – however, as it is the main climate change-related phenomenon in deltas, it is treated separately here) | **Inundation (floods, storm surges)** | Rise in flood probability, inundation | See above | See above |
| | | Storm surges, typhoons (cyclones, hurricanes) | Same as the above, but of special severity due to (usually) sudden onset nature of the events; destruction of coastal infrastructure and protection, inaccessibility to critical infrastructures, public infrastructure and drainage networks lose efficiency, loss of biodiversity/recreational/tourism functions, can destroy a complete country's economy for months or years to come, and lead to ten-thousands of fatalities (cyclone 'Nargis', 2008) | Same as the above. Improvement of typhoon forecasting and early warning |
| | **Salinisation** | Salinization of soil, surface- and groundwater aquifers | Loss of agricultural and aquaculture areas including surface water resources, decreased yield as plants and fish that are not salt tolerant, decreased availability of irrigation and drinking water, increase in water prices and further economic losses, food and supply shortages, unrest, migration | Technological: defence structures minimizing the intrusion of saline waters inland (dikes, sluice gates etc.) development and planting of salt resistant crops, plant species, and salt tolerant fish species; development of salt processing industry, de-salinization plants for drinking water supply, technologies fostering a downward soil-water stream; technology minimizing evaporation<br>Ecological: development of salt resistant species, recovery of natural coastal protection belts (e.g. reforestation of mangrove forests)<br>Educational: Education in the field of natural coastal protection, new agricultural methods, soil science and water movement, fostering resource-saving and water-saving behaviour<br>Political: enforcement of the above |

commonly undertaken. This table lists the most commonly implemented, reactive adaptation measures, grouped in the following categories: technological, ecological, educational and political. It is interesting that strategies from a range of countries and deltas frequently fail to address retreat scenarios of the local population further inland. However, in some cases this might be the only future option, calling for the development of urban areas at secure distances from coastal zones. The table does not imply that adaptation to climate change is finalized once an adaptation measure has been implemented.

With respect to political measures, the New York City Department of Environmental Protection, for example, established a Task Force in 2004 assessing the city's climate change vulnerability leading to the implementation of the New York City panel on climate change (NPCC) in 2008. Multiple layers of government and a wide range of public and private stakeholder and experts are involved in Task Force contribution and NPCC consultation (Major et al. 2010). Also many other cities are starting to consider how best to tackle the consequences of climate change, as discussed by Birkmann et al (2010), who compared climate change preparedness of the cities of Boston, Halifax, Ho Chi Minh, London, New York, Cape Town, Toronto, Rotterdam and Singapore. In New Zealand in the 1990s all coastal resources had already been brought under one piece of legislation to ensure integrated, multi-sectoral research and management (Timmerman and White 1997). The Shanghai Municipal Government addressed the delta threats as early as 1997, adopting counter measures against sea level rise in the years that followed, such as drainage improvement, the installation of pumping facilities, tidal barrier expansion and research on salt tolerant crops (Chen and Zong 1999). The Netherlands have extensive plans to withstand expected sea level rise in their formerly poldered areas, and government, science and private industry cooperate intensively to address the upcoming challenges (Wardekker et al. 2010).

However, many countries and delta cities (for example Jakarta), are lacking such initiatives. With 9 million inhabitants in 2007 and a predicted 20 million in 2025, and without any local laws for surface area management (e.g. creation of green zones, retention areas, shelter regions) the city will soon be too densely populated to have many possibilities for reaction. Not only in Jakarta, but in many developing country deltas, slums or so called 'shanty towns' have developed irrespective of planning guidelines and are usually home to the poorest people in the delta. Frequently, these agglomerations are even located in very low lying areas, obstructing water ways by solid waste accumulation (Paw and Thia-Eng 1991). Such settlements exist in the delta cities of Bangkok, Manila, Karatchi, or Lagos – to name only a few examples. These towns are often located in areas that would be affected by sea level rise – e.g. Recife in Brazil, where the illegal dwellings invaded the protective mangrove areas in the 1990s (a process still ongoing in the South of the city), which have nowadays mostly disappeared. Nicholls 1995 underlines the issue: "If law enforcement already did not work with illegal settling developed irrespective of planning and policies guidelines, how can then climate change related planning be more successful?" These are questions that need to be urgently addressed, as it is especially the political side of adaptation measures and the political backing through decrees and laws that ensure the sustainable long-term success of technological, ecological and educational measures. The gap between planning, legislation and execution of climate change proofing exists in many countries. The pure existence of a plan or even a law still does not guarantee that law enforcement will take place. For example, in China (Lau 2004), Vietnam (Chaps. 6 and 15) and many other emerging and developing countries there is a strong overlap of responsibilities of different ministries, who all want to keep key responsibilities for specific sectors. Until 2004, Coastal Zone Management in China, was still under the responsibility of the Ministries of Land and Resources, Agriculture, Science and Technology, Construction and Water Resources, as well as the State Environmental Protection Agency, the National Bureau of Forestry, the Chinese Academy of Sciences, the Maritime Safety Administration, the Development and Planning Commission, the Economic and Trade Commission, and the National Tourism Agency (Lau 2004). Doubled responsibilities and legislative gaps are common and can only be overcome by lengthy organizational and legal reform. Wolters et al. (2010: 73) emphasise that for the Nile: "…the issue of water scarcity has consequences … that can no longer be adequately addressed by any of the ministries alone. Many government departments and agencies must be involved and decisions will have to be made at the highest political level." Nurhidayah (2010: 97) states it even more frankly in the context of the Ciliwung Delta: "…it is not easy to implement integrated coastal zone management while there is still sectoral ego between institutions involved in the management of marine and coastal resources. At this stage ICZM in Indonesia still remains in infancy due to hard to resolve conflicting laws and legislation". Also in Latin America these issues are of major concern particularly in the context of delta cities. According to Zagare (2010: 133) the difficulties of "designing and implementing urban adaptation measures in a context with a lack of regional and metropolitan conscience, where politic-administrative fragmentation of the space leads to a polarization of resources, being an obstacle to the mitigation of impacts …". The challenge becomes even larger if several countries share one basin. Flessa et al. (2010: 87), concerned about the state of the Colorado Delta, state: "Finding water for nature is difficult because of over-allocation, existing treaties, interstate agreement and the limited geographic scope of the U.S. environmental laws. Nevertheless, new bi-national cooperative efforts among water agencies, universities, environmental agencies and environmental NGOs show promise for cross border effects".

In this context it is also interesting to analyse the proceedings of the 2010 conference 'Deltas in Time of Climate Change' held in Rotterdam, the Netherlands. Even though the conference was internationally visited and had over 1,000 attendees, in the section 'Governance and Economics of Climate Change Adaptation' 37 of the overall 40 papers stem from 'western' countries, while – in other thematic sections – there was a good international mix. This also stands exemplarily for a different focus on the adaptation-paradigm depending on a country's development stage and economic well-being. Some 15 years ago, Nicholls (1995) stated that "Abandonment of any part of the coastal zone in China is unthinkable" (p. 375). Here (technologically oriented) protection has been – and currently still is – the major response to climate change. Based on massive government based funding, large land reclamation projects e.g. in Hong Kong or Shanghai have been completed, to wrench more solid ground for construction from the ocean. In developing and emerging countries the major approach to adapt to climate change is a technology-focussed – not to say technocratic – one, while in developed countries especially social geographers are calling for 'soft approaches' for the improvement of resilience and adaptation and stronger involvement of non-structural (non technology oriented) methods to achieve this even though some believe that in practice, this approach fails to be seriously considered: Garschagen and Kraas (2010: 137) characterise this as a rather hollow discourse as they see a lack of clarity as to how conceptual notions can be translated into concrete adaptation strategies. Even though all 'soft' approaches, such as education and training, and the improvement of cross-sectoral political discourse are of importance and should always frame other climate change adaptation measures, they especially foster future pro-active response and adaptation options. These so-called soft approaches are important in terms of adaptation. However, one should also not lose sight that in many circumstances, physical infrastructure such as dikes, shelters, and related infrastructure save lives right now. Ideal solutions are perhaps not of the "either or" configuration but rather, depending on geographical and social circumstance, a mix of the two approaches. Birkmann et al. (2010) showed that for nine coastal cities – seven of those located in developed countries – adaptation is principally linked to infrastructure development. However, ecological, education and especially political non-structural approaches are still urgently needed when it comes to adaptation strategies. So, despite the often raised criticism of so-called technocratic approaches, even in the Netherlands, the USA, Canada or Germany, these are nowadays still the most important pillars of climate change adaptation. This does not mean that ecologic, educational and especially political non-structural approaches are not urgently needed to frame these actions. Despite the focus on structural approaches, it is striking that western countries consider retreat scenarios more often. Here, the Netherlands in particular set an example, where numerous national projects (Rozema et al. 2010; Stoop et al. 2010) focus rather on transition and a 'life with climate change' rather than a technological struggle against it. As Oome (2010) underlines: "…the philosophy is now different: work with nature rather than against it" (Oome 2010: 47).

In this process, cross-sectoral stakeholder contribution and involvement of further dialogue partners is a key element in the realization of adaptation and transformation processes. More and more scientists see this as the decisive aspect when addressing environmental and climate change related challenges in deltas. Cross-sectoral, cross-hierarchy dialogue mechanisms must be established, the availability and access to sufficient quantitative data and information on current and expected environmental and climate-change-related impacts in the river delta must be improved, and institutional flow and political processes should be as transparent as possible to ensure that science organisations, NGOs and consultants can supply their findings and results to the right stakeholders.

Major regional climate change related challenges and climate change adaptation measures currently under implementation are presented for selected major river delta areas in Asia in the following Table 2.2.

**Table 2.2** Major regional climate and environmental change related challenges and climate change adaptation measures currently under implementation within the largest Asian delta areas

| Delta area | Major regional environmental and climate change challenges | Currently implemented climate change adaptation and mitigation measures |
|---|---|---|
| *Asia* | | |
| Lena | SLR related delta retreat of about 4 m/year through coastal erosion | As population density in the delta is relatively small: based on geophysical and ecology/biology oriented research on changes in climate-, and GHG budget, microbial communities, coastal erosion: adapted natural resource/wildlife reserve management plans for local reserves and administrations. |
| | Methane and carbon dioxide release through thawing of permafrost soils, increased microbial activity | |
| | Increased number of natural forest fires in summer season due to global warming | |
| | Impacts on species composition in Lena Delta Wildlife Reserve through warming and thawing | |
| Yellow River | Annual temperature increase, stronger evaporation; decrease of flow; thus destruction of wetlands | Coastal defence and dike structures; wetland drainage and land reclamation |
| | On shoreline: increased coastal erosion through severe SLR (15 cm in last 10 years), expected increase in typhoon occurrence | Elevation of dike crest for already existing dikes |
| | Frequent algae blooms, especially toxic 'red tide', triggered also through diminished fresh water release into the coastal area and concentration of pollutants | Experiments with temperature increase- and salt tolerant crop species |
| | Additional environmental threats: reduced land accretion through water and soil conservation measures and water diversion upstream | Installation of two large wetland reserves undertaking the renaturation of formerly drained wetlands, artificially controlled flooding of former wetland areas, banning of new oil development in this coastal protection area |
| | Combination of human-induced and tectonic land subsidence of >5 cm/year | World's largest physical hydrologic/hydraulic Delta model; analyses of future river flow and sea level impacts under construction |
| | Wetland species decrease due to pollution and urban sprawl | Geo-environmental Information System for the support of local administration and decision makers in planning stage |
| Yangtze River | Severe SLR and temperature increase, increased coastal erosion, resulting salt intrusion into rivers, canals, soils, aquifers; resulting decrease of crop productivity; resulting economic losses | New coastal defence and dike structures; especially for Shanghai downtown (Pudong), mostly located below sea- and river level; artificial land (re)clamation |

| | | |
|---|---|---|
| | Due to water shortages in the North of China (near Beijing) China undertakes large engineering projects planning to divert water from the South to the North and channelling water from the lower Yangtze and its tributaries northwards. This will lead to lowered water amounts in the Delta, and increased coastal water intrusion and erosive power at the coast<br>With less water reaching the delta, concentration of pollutants increases – already existing algae bloom events will get more severe<br>Additional environmental threats: Severe land subsidence due to groundwater withdrawal and mega city induced ground compaction. (Chen and Zong 1999)<br>Heat islands in cities; regional cooling due to increased aerosol in sparser populated surrounding; increase of precipitation acidity impacting soil acidity | Elevation of dike crests for already existing dikes<br>Installation of pumping stations to safeguard the below sea level city areas in the case of storm surges<br>Experiments with temperature increase- and salt tolerant crop species<br>SLR rise reserve traps, trapping water before reaching more densely settled hinterland<br>Renaturation of wetlands and intensified wetland protection, such as in Chongming Dongtan Nature Reserve<br>Mainly engineering approaches to coastal defences and dike structures, artificial land reclamation from the sea, drainage systems; salt water purification plants; etc. "defensive infrastructure strategy" |
| Pearl River | Expected SLR of 10–30 cm by 2030, and 70–90 cm by 2100. At the same time sediment load arriving in the Delta decreased for 40% due to upstream hydropower dams. Also increased typhoon probability<br>Likelihood of erratic extreme events (precipitation and temperature) increasing<br>Delicate wetland ecosystems such as the World Heritage-listed Mai Po marshes will be invaded by seawater<br>Productivity is already strongly diminished (by 40% due to population pressure) agricultural area has decreased rapidly (rice, livestock productivity etc.)<br>Additional environmental threats: Severe land subsidence due to groundwater withdrawal and mega city induced ground compaction; water quality in PRD due to delta industry one of the most polluted in world; most vulnerable (poorest) parts of the population move to the most dangerous sections of the delta, because land for settlement reclamation is sparse (Han et al. 2000; Zhang et al. 2008) | Elevation of dike crests on already existing dikes<br>Slowly, policy makers and businesses in the delta start to incorporate climate change mitigation into all long-term policies, infrastructure projects, business plans and strategic investments; first larger workshop on climate change in the Pearl River Delta held in 2008 |

(continued)

**Table 2.2** (continued)

| Delta area | Major regional environmental and climate change challenges | Currently implemented climate change adaptation and mitigation measures |
|---|---|---|
| Red River | Increasing variability of flood and drought events and increased likelihood of extreme events; increased storm surge probability from tropical wind systems | Engineering solutions such as dike and canal systems; recently also water treatment plants |
| | Severely threatened food security: rice yield (dominant crop in delta) decreased 10% for each 1°C increase in growing season minimum temperature | Research on species with higher temperature (warming) and salt tolerance (short duration – high yield varieties) |
| | Loss of mangrove ecosystems (Quartel et al. 2007) | Forest restoration, mangrove reforestation |
| | Additional environmental threats: many chemical/paper-pulp plants along course of the lower reaches, large waste water amounts. | Improving cooperation among national, regional and local level administration (Ministry of Natural Resources and Environment – MONRE, DONRE, Peoples' Comities, etc.) |
| | | Ministerial water information database and monitoring network for the Red River Delta |
| Mekong | Sea level rise; resulting salt intrusion into rivers, canals, soils, aquifers; resulting decrease of crop productivity (yield decreases 10% for each 1°C increase in growing season minimum temperature); resulting economic losses and expected inland migration (1/5th to 1/3rd of Delta will be flooded by 2080 with no adaptation) | Maintaining the free flowing character of the lower Mekong; construction of delta-wide coastal dike structures along the southern ocean coast (several ministries involved, MARD, MPI, MOT) |
| | Increased probability for cyclones through regional wind pattern shifts; resulting stronger coastal erosion; resulting threat to mangrove ecosystems | Climate change and addressing the impacts of sea level rise are key areas in the National Target Plan on 'Climate Change'; and the 'Mekong Delta Plan'; identification of most vulnerable areas within the Delta |
| | Additional environmental threats: lowered sedimentation rates through upstream dams; water pollution through agriculture, households, industry; mangrove deforestation for aquaculture | Installation of a web-based Information System sharing relevant and up-to-date geodata for the delta |
| | | Mangrove reforestation programmes |
| | | Improving cooperation among national, regional and local level administration |
| Tokyo Bay | SLR related flooding threats to densely populated areas in Tokyo Bay, especially if SLR is accompanied by increasing occurrence of typhoons; especially in overpopulated Koto Ward, Chiba City, Funabashi, Yokohama | Coastal defence and dike structures |
| | | Artificial land (re)clamation |
| | Additional environmental threats: Urban heat islands in winter due to low sea breeze (alongside inversion layers and air pollution) | Development of new settlement areas further away from the coast |

| | | |
|---|---|---|
| | Increases of precipitation acidity; severe ground subsidence due to groundwater depletion | Counterbalancing local warming and air pollution: Tokyo Metropolitan Government released 10 year plan with goals to reduce $CO_2$ emissions of the city by 25% until 2020 (all via green technology, and a local carbon trading system, financed via Fund to Promote Measures against Climate Change with >4 billion USD funding every year) |
| Ciliwung River | Sea Level Rise in this heavily populated delta is endangering the floodplain, salinization of soil, canals and groundwater<br>Additional environmental threats:<br>Heavy illegal settlement spread of disadvantaged people in Ciliwung floodplain, and garbage accumulation leading to narrowing of river flow<br>Land subsidence of >5 cm/year | Extensive reforestation in the Ciliwung basin; mangrove reforestation on the coast<br>Information System set up under flood control 2015 innovation programme<br>Cleanup programme for the Ciliwung River and adjacent river with forced re-migrations of illegal settlers into the hinterland |
| Chao Phraya | Sea level rise of about 3–4 mm per year which is significantly faster than global average (aggravated through land subsidence of>1.5–4 cm/year); ground compaction through urban sprawl (water extraction from the clay horizons and compaction through city itself); extreme coastal erosion, especially on western side of Cha Phraya River mouth, of 5–10 m/year (Trisirisatayawong et al. 2011; Naejie et al. 2010)<br>Furthermore, decrease in sediment yield as a result of the construction of the Bhumipol and Sirikit dams, also endangering the delta<br>Increase wave variability and surges (e.g. typhoon Linda with wave heights of 4 m in 1997)<br>Salt intrusion into Bangkok's coastal aquifers<br>No costal protection forest of buffer zones left; complete estuary either densely populated, or intensified agriculture (aquaculture) to the forefront of the coastline<br>Average maximum temperature in Bangkok and Chao Pharya Delta has risen for>2.5 °C during the last 50 years | Testing of salt tolerant species<br>Extensive dike-, and embankment projects; several dykes protecting Bangkok metropolitan area from all sides, embankment and dikes around Suwanaphum International Airport<br>Installation of numerous pumping stations<br>Drainage canal network as well as flood water diversion projects (Vongvisessomjai 2007)<br>Dam at the river mouth being planned<br>Project on mangrove reforestation started – an intertidal mangrove belt of about 300–500 m is required to re-initiate sedimentation processes. Discussions on abandoning part of the aquaculture along the coast |

(continued)

**Table 2.2** (continued)

| Delta area | Major regional environmental and climate change challenges | Currently implemented climate change adaptation and mitigation measures |
|---|---|---|
| Irrawaddy | Densely populated with a rural population of 3,5 Mio rice farmers and fishermen<br>Irrawaddy delta currently growing at rate of 50 m/year, extending further into the Andaman Sea<br>Affected by cyclone Nargis in 2008: increased (natural) deforestation of mangroves and extensive land losses (6% of rice crop) due to Nargis and other severe cyclones (increased intensity in future)<br>Poor rice harvests due to sea level rise (salinization), cyclones<br>Future climate change related variation in water regimes (current glacier melting, reduction in the long term future), as well as anthropogenic impacts (7 dams planned upstream and on tributaries with 13,360 kW capacity) | Low adaptive capacity of the population (low education standards, limited infrastructure, limited funding of Myanmar's states, administrative divisions and communes)<br>Few storm shelters (recently, additional shelters are being build in the aftermath of Nargis)<br>No mangrove reforestation<br>Almost no financial aid<br>First embankments and rudimentary dikes<br>Against the former focus of the government favoring traditional low-yielding rice varieties now saline-resistant seed lines were requested from the International Rice Research Institute. First experiments with several of the resistant varieties ongoing |
| Ganges-Brahmaputra | Between 125 and 143 million people live on the delta (jute, tea, rice, fishing), population density exceed 200 people/km² creating pressure on resources, despite risks from floods caused by monsoons, heavy runoff from the melting snows of the Himalayas, and tropical cyclones<br>Inland flooding and saltwater intrusion especially during cyclones passing over, extreme bathymetry aggravates cyclone risk, natural land subsidence ongoing, leading to coastal erosion- severe land loss with even little SLR<br>Glacier melting and increase in precipitation (during monsoon) upstream changes the river flow; fresh water shortage during dry season | Currently undertaken and planned actions are not the result of CC adaptation strategies<br>Enhanced flow regime<br>Dredging to enhance freshwater supply<br>Decision Support Systems for the assessment of climate adaptation options (Zaman and Rahman 2010)<br>Mangrove reforestation suggested<br>Experiments with salt tolerant crops |
| Godavari | Very high risk of SLR (150 km² would be submerged with a SLR of 59 cm) (IPCC 2007). Higher spring tides and increased risk of storm surges (also further inland), increased salt water intrusion in soils an aquifers<br>Large land losses; 1,836 ha between 1975 and 2001 (Malini and Rao 2004), substantial loss of mangroves and shoreline retreat, mainly attributed to upstream dams leading to diminished sediment loads<br>Envisaged Godavari-Krishna water transfer schemes would aggravate these problems (Bharati et al. 2009) | |

Sources of information: literature of this article, public online media of the different countries, assessed in July 2011

There is no single best practice on how a river delta population's resilience could be increased via mitigation and adaptation measures. The resilience of a delta population largely depends on socio-economic and political circumstances such as average income, land ownerships, flexibility/mobility concerning living locality, population's opportunities to actively contribute to local political and decision making processes, cultural and religious background, social networks and insurance, demographic characteristics, to name only a few. At the same time their resilience is influenced by international, national and regional economic development status, global market prices, political decisions and framework, conflicts, or physio-geographical factors such as terrain, soil, water (is there any alternative place for (re) settlement, resource supply). Strictly speaking, one would have to differentiate resilience at different hierarchical levels and for different social groups.

To better understand what drives the changes, as well as the favoured adaptation processes to changes in the deltas, it is of utmost importance to analyse the following questions:

- How is the delta composed; where do people live; what are the different livelihood systems; how do people shape the delta and vice versa?
- What will be the consequences of environmental change (natural-, agricultural-, urban-, industrial- changes) in the river delta?
- What will be the consequences of climate change in the river delta?
- What will be the effects of upstream catchment environmental and climate change to the river delta?
- How do the manifold manifestations of the above three interact?
- Which components of the delta system (natural as well as population) are the most vulnerable with respect to different consequences and effects derived from the above?
- How can the resilience of the delta's natural resources as well as the population (especially of the most critical areas/groups) be improved, ensuring the maintenance of society's requirements?
- Which knowledge layers are available for the delta and how can they be combined?
- What is the institutional landscape in the delta; who are the stakeholders and decision makers and how are decision processes governed?
- Which other deltas experienced/experience similar challenges so that an exchange of knowledge and ideas can be strongly beneficial?
- Which adaptation measures from the different sectors (technological, ecological, educational, political) should be implemented, always considering protective and accommodative measures and not necessarily ruling out retreat measures from the beginning?
- How can such adaptation measures supporting the decrease of vulnerability and the increase of resilience be integrated into political frameworks, regulations and even laws, and how can law enforcement be ensured?

Answering the above questions should be framed by cross boundary and cross-sectoral dialogue. At an international level, South-South and North-South collaboration, understanding how other populations tackle climate-proofing, and getting new ideas and initiatives for the home country's delta area is often the first step in triggering political action followed by educational, eco-environmental, or technological measures – even if most deltas and cities face different challenges, and thus develop different adaptation pathways to climate change. For such an increased exchange among delta populations and delta cities, several global and regional initiatives fostering knowledge exchange and networking in the context of river deltas under climate change were initiated in recent years (Table 2.3).

In the sections above we have discussed environmental change and climate change impacts in deltas in general as well as indentified typical adaptation and mitigation measures considered in these deltas. Furthermore, the challenges on the pathway to the sound realization of adaptation measures – especially within the political domain – have been discussed. In the next section we focus our review on the Mekong Delta in Vietnam.

## 2.5 Climate and Environmental Change in the Mekong River Delta, Vietnam

The environment of the Mekong Delta has changed drastically since the mid-nineteenth century, when the canal system was extended, changing the landscape considerably. Depressions, which were not settled, were drained via these canal systems opening up land for rice cultivation, and resulting in a drastic reduction of the area extent of natural wetland ecosystems (Duong Van Ni et al. 2001). Another consequence of the drainage was soil acidification, which limited agricultural production, acidified surface waters, and has resulted in the loss of biodiversity and ecosystem services (Duong Van Ni et al. 2001). Other environmental concerns include salinity intrusion, which is constantly worsening, co-occurrence of extreme flooding in the rainy season, and extreme low flows during the dry season, as well as the degradation of fisheries, and pollution (Sneddon and Nguyen 2001; Chaps. 13 and 14) (Fig. 2.5). As for all regions of the country, the Mekong Delta (MD), also known as *Cuu Long*, has undergone major economic and political transformations initiated by the *Doi Moi* (or renovation) reforms which were officialised in 1986. Particularly relevant to the MD was the decollectivisation of agricultural land, which contributed to the redistribution of the land and which, in the process, was one of several factors triggering landlessness for many farmers (Garschagen 2010; see also Chap. 4). Through the renovation, agricultural activities have intensified, with some areas of the MD shifting from a single rice crop harvest to up to three annual rice crop harvests, and a rapid development of aquaculture, notably shrimp farming. In parallel, several cities in the MD have developed very rapidly (Leinenkugel et al. 2011), notably the city of Can Tho, which in the past few years has seen its infrastructure include a major bridge over the Hau River and the construction of an international airport whose traffic is steadily increasing. Agricultural and urban development has often been carried out with little consideration for the

**Table 2.3** Global delta initiatives, alliances and networks (June 2011)

| | World estuary alliance | Delta alliance | Connecting delta cities | USGS Dragon initiative | WWF Ecoregion Initiative |
|---|---|---|---|---|---|
| Website | www.estuary-alliance.org | www.delta-alliance.org | www.deltacities.com | http://deltas.usgs.gov/ | http://wwf.panda.org/about_our_earth/ecoregions/ecoregion_list/ |
| Goals | Awareness creation, developing and integrating knowledge, knowledge exchange, exchange of best practises, initiate joint projects, network building, connecting individuals and organizations across all sectors (individuals, science institutes, NGOs, policy makers); international events, conferences, and communication | | | ← See left, and analyzing the below river deltas in a comparable manner, especially utilizing earth observation data | Regional WWF projects with respective country to protect biodiversity and the environment |
| Deltas involved | Indus (Pakistan), Ganges-Brahmaputra (India, Bangladesh); Guadalquivir (Spain); Mekong (Vietnam); Rhine (Netherlands); Zambezi (Mozambique); | Ciliwung (Indonesia); Mahakam (Indonesia); Ganges-Brahma-Putra (India, Bangladesh); Mekong (Vietnam); Yangtze (China), Nile (Egypt); Rhine (Netherlands); Pantanal (Brazil); Mississippi (USA); San Francisco Bay (USA) | Ciliwung (Indonesia); Mekong (Vietnam); Pearl River (China); Tokyo Bay (Japan); Rotterdam/Rhine (Netherlands); Thames (UK); Hudson (USA); Mississippi (USA) | Irrawaddy (Myanmar), Chao Phraya (Thailand); Amazon (Brazil), Danube (Rumania), Ganges (India), Yellow River (China), Lena (Russia), Mekong (Vietnam), Mississippi (USA); Nile (Egypt), Okavango (Botswana); Rhine (Netherlands), Selenga (Russia), Volga (Russia), Yangtze (China) | Niger (Nigeria); Indus (Pakistan); Volga (Russia); Danube (Rumania); Lena (Russia), Lower large Rivers: Mekong (Vietnam), Congo (Congo); Lower Mississippi (USA); Amazon (Brazil), Orinoco (Venezuela) |

**Fig. 2.5** Impressions from the Mekong Delta in Vietnam. *Upper left*: urban settlement in Can Tho; people have to adapt to strongly fluctuating water levels; *Upper right*: tidal variations influence shipping; *Lower left*: many transecting river channels in deltas pose challenges for infrastructure construction; *Lower right*: valuable tide-influenced ecosystems need to be protected (Photographs: C. Kuenzer, January 2008)

environment, with increasing surface and groundwater pollution induced by agrochemicals and untreated urban and industrial effluents (Chap. 13), and with the destruction of previously well-established agroecosystems in peri-urban areas (Garschagen et al. 2011), or coastal ecosystems (Chap. 12, Kuenzer et al. 2011a). The Government of Vietnam readily recognises that the state of the environment in Vietnam in general and in the MD in particular is deteriorating and, in its current decadal planning, now links environmental protection to the sustainable development of the country (SRV 2003). In practice however, the problems of pollution remain a constant threat for the MD and elsewhere and articles to that effect appear regularly in the print press.

Superimposed on these environmental threats, is the fact that the MD is one of the most vulnerable regions of the world to the effects of climate change. Various studies have investigated the consequences of climate change in the MD and notably the amount of land inundated, proportion of population affected and potential economic impacts. These studies however provide different estimates both in absolute terms and in geographical extent. This is caused by the diversity of approaches and data used, resolution and quality of the data, and processes accounted for (Kuenzer et al. 2011b). Two broad conclusions from these studies can be drawn. The first is that Vietnam in general and the Mekong Delta in particular are likely to be heavily impacted by a 1 m sea level rise and other hydro-climatic hazards. The second is that the studies should only be considered as indicators of future problems as they are accompanied by various methodological and data uncertainties. Four such studies are detailed below.

The World Bank carried out a global survey of the impact of sea-level rise which indicated that for the East Asia region, Vietnam would be the country most impacted with the country's two deltas taking the brunt of the impact (Dasgupta et al. 2007). Disaggregated data is not available from the report, but a 1 m SLR would impact 10.8% of Vietnam's population (as it stood at the time of the survey) and it can be implied that a majority of this would be in the MD. Of the 84 developing countries covered by this study, Vietnam would rank second in terms of percentage land area impacted and percentage agricultural area impacted; and first in terms of percentage population impacted, percentage GDP impacted, percentage urban area impacted, and percentage wetlands impacted.

The Asian Development Bank estimated that an area of approximately 12,300 $km^2$, or 31%, of the total land area of the Mekong Delta will be inundated in case of a 1 m sea level rise by 2100 (ADB 2009). This would affect 9,800 $km^2$ of land used for agriculture and aquaculture as well as 4.8 million people and their livelihood.

Carew-Reid (2008) reported that a 1 m sea-level rise (which in their report is arbitrarily assumed for 2100) would seriously affect 12 of the MD provinces, as well as Ho Chi Minh City which lies just outside the MD, in terms of inundation extent, population affected and freshwater resources impacted. Of the 12 MD provinces, some such as Tra Vinh, Long An and Soc Trang will have more than 40% of their territory inundated, while for Ben Tre it will be more than 50%. The simulations showed that the MD will represent ca. 85% of the total inundated surface of Vietnam and over 25% of the MD population would be affected, representing more than 80% of the total affected population of the country. This will have serious implications for the agricultural and fishery sectors and therefore the majority of the population in the MD.

The last report discussed briefly here is from the Vietnamese Ministry of Natural Resources and Environment (MONRE). Over the period 1958–2007, the average annual temperature in Vietnam increased by between 0.5°C and 0.7°C and no general trends in terms of precipitation were noted (MONRE 2009). Of additional concern is the fact that more typhoons of high intensities are being recorded and that the tracks of these typhoons tend to move southwards (MONRE 2009), thus with a higher probability of affecting the MD. The MD has, in relative terms, so far been much less affected than the rest of the country, with the exception of Tropical Storm Linda, which had terrible consequences for the region (Kelly et al. 2001). However, when typhoons are combined with SLR (which has taken place at a rate of 3 mm/year over the period 1993–2008), severe consequences for communities living in coastal areas or along the main river channels can be expected as these communities are relatively unprepared (due to the low recurrence of this hazard) and highly exposed to this type of hazard. Scientifically though, it is not clear how climate change may affect the patterns of typhoons affecting Vietnam. The country has been affected by a number of such events in the past with serious consequences.

Projections to 2100 using low, medium and high emission scenarios were compared to the reference period 1980–1999 (MONRE 2009). Results indicate that for South Vietnam, mean annual temperature would increase by 1.4°C, 2.0°C, and 2.6°C, respectively; annual rainfall would increase by 1.0%, 1.5% and 1.9%, respectively but with large seasonal fluctuations, and in particular a decrease in rainfall during the dry season; and there would be an SLR of 0.65, 0.75 and 1 m, for the three scenarios respectively. Large portions of the MD are predicted to be flooded by a 1 m SLR. Although data on percentage extent inundation is not available from the MONRE report, visual inspection of the maps provided shows differences with the maps published by Carew-Reid (2008), highlighting the different approach and uncertainties underlying the simulations.

Although the various studies provide different specific outcomes as to the impacts of climate change in the MD, there is a convergence in that all studies predict severe impacts for the MD. The simulations described above do not present a worst case scenario, as none of them accounted for additional threats or complicating factors such as the effects of waves, tides, storm surges, and floods which, when combined with SLR, could affect many more people in the MD. They also do not account for demographic, socio-economic or political developments, which could have positive or negative effects in terms of exposure, vulnerability and/or adaptation of the social-ecological systems in the MD.

The Vietnamese Government has been very proactive in addressing climate change issues during this past decade and in 2008 published various official documents tackling the issue (Kuenzer et al. 2010). The first describes an action plan framework for adaptation to climate change in the agriculture and rural development sector for a period extending to 2020 with the Ministry for Agriculture and Rural Development (MARD) in the lead (SRV 2008a). This action plan framework is very broad in its scope and covers a number of thematic activities to be undertaken by 2020 including knowledge communication on climate change impacts and adaptation at various governance and stakeholder levels, capacity development activities, scientific investigations, policy development, and international collaboration for implementing identified key priorities. However, the document lacks the description of specific modalities to reach these various objectives, and information available from the website of the newly established Standing Office of the Steering Committee for Climate Change Adaptation and Mitigation (OCCA) remains very general and does not depict the progress made on the action plans.

The second document describes the National Target Programme to respond to climate change, with the Ministry of Natural Resources and Environment (MONRE) in the lead (SRV 2008b). These responses should follow the principles of sustainable development, focus on providing short and long-term impacts, and should originate and/or be developed by all scales of governance and various stakeholder groups. The three-phase plan is currently in its implementation phase (covering the period 2011–2015), which followed a so-called start-up phase (2009–2010), and which will be followed by a development phase after 2015. The principal objectives of the plan are to:

> assess climate change impacts on sectors and regions in specific periods and to develop feasible action plans to effectively respond to climate change in the short-term and long-term to ensure sustainable development of Viet Nam, to take opportunities to develop towards a low-carbon economy, and to join the international community's efforts in mitigating climate change and protecting the climatic system (SRV 2008a)

These objectives are supplemented by eight specific objectives, which address a broad range of topics such as generation of new knowledge on the effects of climate change and adaptation, as well as proposing concrete solution for adaptation including components of capacity development and knowledge sharing. As outlined in the policy document, the planning of activities is very centralised and sectoral, with MONRE taking the lead and the Ministry of Planning and Investment playing an important decision role, while other Ministries and Government Authorities, People's Committees and Civil Societies are principally in the role of "executing" agents. Several documents stating progress made so far in implementing the Government's climate change strategy are currently under preparation and should have been released in 2011 but were not available at the time of writing. However, it is anticipated that assessments would reveal limited progress in some areas due to lack of capacities at the local governance levels and lack of delta-wide, integrated planning (see Garschagen 2011).

The consequences of climate change can manifest themselves as new threats (such as sea level rise), by increasing the frequency of rare hydro-climatological events such as typhoons, and as a magnifier of existing threats such as floods and droughts. Disaster risk reduction and climate change adaptation therefore need to be considered conjointly. Since 2001, the Government of Vietnam has recognised that climate change is increasing the number and intensity of disasters in the country, a fact clearly spelt out in its strategy for disaster risk reduction (MARD 2001). According to this strategy (MARD 2001: 30), the MD is particularly exposed to floods, storms and storm surges, fires, saline intrusion and landslides. With respect to risk reduction, MARD (2001: 1–2) emphasises that contrary to the Red River Delta which is "protected against flooding by one of the world's major river dyke (levee) and sea dyke (coastal protection) systems", the situation in the MD is different as "very little has been done historically to protect against this flooding other than to learn to live with the floods".

However, the MD has since seen the development of many infrastructures to reduce flood and salinity intrusion risks, such as dykes and sluice gates, which has also allowed the intensification of agricultural production. Nevertheless, the review of the First (1994) National Strategy and Action Plan for Water Disaster Mitigation identified shortcomings such as the lack of anticipation of the increases in duration, intensity and complexity of water-related disasters (MARD 2001) as an outcome of both climate change and rapid urbanisation.

The combination of climate change, environmental degradation and environmental hazards is putting increased pressure on both rural and urban communities in the MD (see Garschagen 2011; Garschagen et al. 2011; Kuenzer et al. 2009, 2010, 2011b; ; Sebesvari et al. 2011). Policies have been put in place in the past decade but their effectiveness is either debatable (particularly environmental and disaster risk reduction policies), or only now being evaluated (climate change policies). These policies and actions need to be made more specific in their future iterations and further integrated with one another in order to protect the environment of the MD while at the same time encouraging adaptation and reducing disaster risks.

## 2.6 Conclusion

The advantages for societal development in river deltas comprise a flat topography, available fresh and salt water resources, good transportation via waterways and the nearby coast, fertile soils for agriculture, and usually also a rich biodiversity and recreational value through wetland ecosystems and beaches. However, at the same time, the river deltas of the world belong to the most endangered regions with respect to the negative consequences of environmental and climate change – the latter especially manifested through sea level rise. Environmental impairment through societal transformation related to urbanisation, agricultural intensification and industrialization includes: reduced sedimentation in the delta and changed flood pulses through dam construction further upstream; the accumulation of solid, liquid and gaseous wastes stemming from communities upstream as well as from the deltas themselves; the loss of natural delta (wet) land; land subsidence due to groundwater extraction and compaction through urbanization; increased surface sealing and runoff; a depletion of the buffer capacities of the natural ecosystems. All these effects are aggravated by heterogeneous planning, competing sectoral responsibilities, and limited law enforcement.

Climate change impacts in river deltas globally mainly manifest through: sea level rise, which leads to increased flood risk from the ocean (especially during storm surges); intrusion of salt water further inland into rivers and canals, as well as soils and groundwater aquifers; the resulting degradation of agricultural land and drinking water supplies; stronger coastal erosion (protective shelterbelts, wetlands, beaches etc.); a loss of ecosystem service functions and biodiversity; and last but not least a loss of large amounts of a country's GDP, which need to be invested in measures to mitigate the climate change effects. Such adaptive measures can be educational, eco-environmental, technological or political. In most cases, and especially in the heavily populated deltas of developing and emerging countries, a focus is put on technological counter measures such as elevated dykes, sluice gates, pumping stations, and drainage networks. This is then partially accompanied by ecological programmes such as mangrove shelter belt reforestation. With a few exceptions, the migration of the population inland is not usually a scenario that is being considered, even though in the long run this might be necessary in several deltas.

As the circumstances of each delta are unique, there are no generic solutions applicable to all world deltas in terms of environmentally-benign development pathways or adaptation to climate change. According to Nicholls (2004) to better understand the individual settings of the deltas, a joint evaluation of mitigation and adaptation based on probabilistic risk-based methods is urgently needed. Crucial actions for deltas are understanding consequences of environmental and climate change as well as the impact of upstream development. The most vulnerable and most exposed components of the delta (natural and societal) must be identified. Increasing resilience and the adaptive capacity of the populations should be prioritised and there must be an assessment of which proactive and responsive adaptation measures are most suitable for the given the circumstances. This should be implemented in a fully participatory manner with as many stakeholders as possible.

Information exchange and knowledge sharing must be facilitated locally, regionally and internationally. Efforts should be made to develop easy-to-use tools for the characterisation of vulnerability and resilience of complex delta social-ecological systems. Complex investigations of the many factors shaping vulnerability and resilience are needed to enable delta populations to not only react through implementation of technological structures to address environmental problems or climate change impacts, but also to develop comprehensive adaptation strategies that look at all aspects of people's livelihoods.

It is thus time to call for a transdisciplinary approach to assess the state of deltas with experts from disciplines such as geography, geology, hydrology, chemistry, physics, sociology, economy, ethnology, biology and many more, who must join forces to address this challenge. Such exercises should be undertaken jointly with national, regional and local authorities to ensure political backing leading to the development of strategies for adaptation. All expert knowledge and the translation into planning documents and adaptation concepts will not be beneficial and sustainable if a translation into political measures remains vacant. It is mainly via the integration into rules, decrees and laws, and especially the enforcement of the same, that transdisciplinary concepts aiming at the improvement of resilience and adaptation can lead to effective changes. Therefore, stakeholder support at highest hierarchical level is needed.

## References


ADB [Asian Development Bank] (2009) Technical assistance report. Socialist Republic of Viet Nam: climate change impact and adaptation study in the Mekong Delta. Manila, Philippines

Abd El-Kawy OR, Rød JK, Ismail HA, Suliman AS (2011) Land use and land cover change detection in the western Nile delta of Egypt using remote sensing data. Appl Geogr 31(2):483–494

Adger WN, Kelly PM, Nguyen Huu Ninh, Ngo Cam Thanh (2001) Property rights, institutions and resource management: coastal resources under *doimoi*. In: Adger WN, Kelly PM, Nguyen Huu Ninh (eds) Living with environmental change. Social vulnerability, adaptation and resilience in Vietnam. Routledge, London, pp 79–92

Allison I, Bindoff NL, Bindschadler RA, Cox PM, de Noblet N, England MH, Francis JE, Gruber N, Haywood AM, Karoly DJ, Kaser G, Le Quéré C, Lenton TM, Mann ME, McNeil BI, Pitman AJ, Rahmstorf S, Rignot E, Schellnhuber HJ, Schneider SH, Sherwood SC, Somerville RCJ, Steffen K, Steig EJ, Visbeck M, Weaver AJ (2009) The Copenhagen diagnosis, 2009: updating the world on the latest climate science. The University of New South Wales, Climate Change Research Centre (CCRC), Sydney, 60 p

Beilfuss R, Dos Santos D (2001) Patterns of hydrological change in the Zambezi delta, Mozambique. Working papers of the program for the sustainable management of Cahora Bassa Dam and the Lower Zambesi Valley

Bharati L, Anand BK, Smakhtin V (2009) Analysis of the Inter-basin water transfer scheme in India: a case study of the Godavari – Krishna Link. Strategic Analysis of the National River Linking Project (NRLP) on India 5:63–78

Birkmann J, Garschagen M, Kraas F, Quang N (2010) Adaptive urban governance: new challenges for the second generation of urban adaptation strategies to climate change. Sustain Sci 5(2):185–206. doi:10.1007/s11625-010-0111-3

Breber P (1993) The misguided policy of draining estuarine wetlands for agriculture: the case of Goro, Po Delta, Italy. Landsc Urban Plan 27(1):1–5

Bucx T, Marchand M, Makaske A, van de Guchte C (2010) Comparative assessment of the vulnerability and resilience of 10 deltas – synthesis report

Burg V (2007) Climate change affecting the Okavango Delta. Diploma thesis at the Swiss Federal Institute of Technology Zurich (ETH), 84 p

Carew-Reid J (2008) Rapid assessment of the extent and impact of sea level rise in Viet Nam. Climate change discussion paper 1, ICEM – International Centre for Environmental Management, Brisbane, Australia, 74 p

Chen X, Zong Y (1999) Major impacts of sea-level rise on agriculture in the Yangtze delta area around Shanghai. Appl Geogr 19(1):69–84

Conway D, Hulme M (1996) The impacts of climate variability and future climate change in the Nile Basin on water resources in Egypt. Water Resour Dev 1(3):277–296

Dasgupta S, Laplante B, Maisner C, Wheeler D, Yan J (2007) The impact of sea level rise on developing countries: a comparative analysis. World Bank Policy Research working paper 4136, Washington, DC, 51 p

De Wit R (2011) Vulnerable deltas, difficult choices. In: Schoenmaeckers B, van Rijnsoever M (eds) Deltas in times of climate change. A special issue of 'Change' magazine, SynergosCommunicatie, Netherlands

Doody K, Hamerlynk O (2003) Biodiversity of Rufiji District, Technical report no. 44, Rufiji Environmental Management project

Elsharkawy H, Rashed H, Rached I (2009) Climate change: the impacts of sea level rise on Egypt. In: The impacts of SLR on Egypt, 45th ISOCARP Congress 2009, Porto, Portugal

Ericson JP, Vörösmarty CJ, Dingman SL, Ward LG, Meybeck M (2006) Effective sea level rise and deltas: causes of change and human dimension implications. Glob Planet Chang 50:63–82

Flessa K, Zamora Aroya F, Aguilera L (2010) Two countries, one river, zero water: the Colorado River delta and estuary. In: Proceedings of 'Deltas in Times of Climate Change', 29 Sept–1 Oct 2010, Rotterdam, the Netherlands

Folke C (2006) Resilience: the emergence of a perspective for social-ecological systems analyses. Glob Environ Chang 16(3):253–267

Gallopin G (2006) Linkages between vulnerability, resilience, and adaptive capacity. Glob Environ Chang 16(3):293–303

Garschagen M (2010) Challenges and opportunities of climate change adaptation in high risk areas using the example of the Mekong Delta, Vietnam. In: Dölemeyer A, Zimmer J, Tetzlaff G (eds) Risk and planet earth: vulnerability, natural hazards, integrated adaptation strategies. Schweizerbart Science Publishers, Stuttgart, pp 56–65

Garschagen M (2011) Resilience and organisational institutionalism from a cross-cultural perspective: an exploration based on urban climate change adaptation in Vietnam. Nat Hazards (in press). doi:10.1007/s11069-011-9753-4

Garschagen M, Kraas F (2010) Urban climate change adaptation in the context of transformation: lessons from Vietnam. In: Otto-Zimmermann K (ed) Resilient cities: cities and adaptation to climate 131 change proceedings of the Global Forum 2010, Local sustainability 1. doi:10.1007/978-94-007-0785-6_13

Garschagen M, Renaud FG, Birkmann J (2011) Dynamic resilience of peri-urban agriculturalists in the Mekong Delta under pressures of socio-economic transformation and climate change. In: Stewart MA, Coclanis PA (eds) Environmental change and agricultural sustainability in the Mekong Delta, vol 45, Advances in global change research. Springer, Dordrecht, pp 141–163

Goudie A (2007) Phyische Geographie – Eine Einführung, 4th edn. Spektrum Akademischer Verlag, Heidelberg/München, 506 p
Han M-K, Wu L, Liu Y-F, Mimura N (2000) Impacts of sea-level rise and human activities on the evolution of the pearl River Delta, South China. Proc Mar Sci 2:237–246
Holzwarth I, Hessner F, Schulte-Rentropp A, Sehili A (2010) Impacts of rising sea level on German Estuaries. In: Proceedings of 'Deltas in Times of Climate Change'. 29 Sept–1 Oct 2010, Rotterdam, the Netherlands
IPCC (2007), Climate change 2007: synthesis report. Contribution of working groups I, II and III to the fourth assessment report of the Intergovernmental Panel on Climate Change. IPCC, Geneva, Switzerland, 104 p
Jallow B, Barrow M, Leatherman S (1996) Vulnerability of the coastal zone of The Gambia to sea level rise and development of response strategies and adaptation options. Clim Res 6:165–177
Karim MF, Mimura N (2008) Impacts of climate change and sea-level rise on cyclonic storm surge floods in Bangladesh. Glob Environ Chang 18(3):490–500
Kelly PM, Hoang Minh Hien, Tran Viet Lien (2001) Responding to El Niño and La Niña: averting tropical cyclone impacts. In: Adger WN, Kelly PM, Nguyen Huu Ninh (eds) Living with environmental change: social vulnerability, adaptation and resilience in Vietnam. Routledge, London, pp 154–181
King C, Adeel Z (2002) Strategies for sustainable coastal management in Asia and the Pacific – perspectives from a regional initiative. Glob Environ Chang 12:139–142
Krishna MK (2009) Intensifying tropical cyclones over the North Indian Ocean during summer monsoon—Global warming. Glob Planet Chang 65(1–2):12–16
Kuenzer C (2007) Water pollution, scarcity and distribution challenges. Bus Forum China,02/07, 63–65
Kuenzer C, Wehrmann T, Gebhardt S, Schmidt M, Mehl H (2009) A water related information system for the Mekong (Lancang), Proceedings of the 6th international symposium on digital earth (ISDE6), Sept 2009, Beijing, China, digital version
Kuenzer C, Vo Kakh T, Wehrmann T, Le Van T, Gebhardt S, Lam Dao N, Renaud F, Tran Thanh B, Waibel G, Huyng Ngoc T, Birkmann J, Dech S (2010) An information system for the sustainable development of the Mekong Delta. Deltas in times of climate change, 29 Sept–01 Oct 2010, Rotterdam, the Netherlands
Kuenzer C, Bluemel A, Gebhardt S, Vo Quoc T, Dech S (2011a) Remote sensing of mangrove ecosystems: a review. Remote Sens 3:878–928
Kuenzer C, Huth J, Gebhardt S, Leinenkugel P, Lam Dao N, Dech S (2011b) Environmental and climate change related trends in river deltas – examples from the Mekong. In: Proceedings of the 34th international symposium for remote sensing of the environment, 10–15 April 2011, Sydney, Australia
Lau M (2004) Küstenzonenmanagement in der Volksrepublik China und Anpassungsstrategien an den Meeresspiegelanstieg. In: Schernewski G, Dolch T (eds) Geographie der Meere und Küsten, Coastline Reports 1 (2004), ISSN 0928-2734, pp 213–224
Le K (2000) An analysis of the recent severs storm sureg disaster events in china. Nat Hazards 21:215–223
Leinenkugel P, Esch T, Kuenzer C (2011) Settlement detection and impervious surface estimation in the Mekong delta using optical and SAR data. Remote Sens Environ 115(12): 3007–3019
Liu JP, Milliman JD (2004) Reconsidering melt-water pulses 1A and 1B: global impacts of rapid sea level rise. J Ocean Univ China 3(2):183–190
Lu XX (2004) Vulnerability of water discharge of large Chinese rivers to environmental changes: and overview. Reg Environ Chang 4:182–191
Major D, Rosenzweig C, Solecki W, Horotn R, O'Grady M, Patrick L (2010) Climate change adaptation in New York City: risk management process & progress. In: Proceedings of 'Deltas in Times of Climate Change'. 29 Sept–1 Oct 2010, Rotterdam, the Netherlands
Malini BH, Rao KN (2004) Coastal erosion and habitat loss along the Godavari delta front – a fallout of dam construction (?). Curr Sci 87(9):1232–1236

MARD [Ministry of Agriculture and Rural Development] (2001) Second national strategy and action plan for disaster mitigation and management in Vietnam – 2001-2020. Ministry of Agriculture and Rural Development, Hanoi, 63 p
Menéndez AN, Sarubbi A (2007) A model to predict the Paraná delta front advancement. Workshop on Morphodynamic processes in large lowland rivers, Santa Fe, Argentina, Nov 2007
Milliman JD, Haq BU (1996) Sea-level rise and coastal subsidence – causes, consequences and strategies. Kluwer Academic Publishers, Dordrecht. ISBN 0-7923-3933-9
MONRE [Ministry of Natural Resources and Environment] (2009) Climate change, sea level rise scenarios for Vietnam. Ministry of Natural Resources and Environment, Hanoi, 34 p
Naejie M, Trisirisatayawong I, Simons W, Satirapord C, Snidvongs A (2010) Sea level change in the river delta areas of the Gulf of Thailand. In: Proceedings of 'Deltas in Times of Climate Change'. 29 Sept–1 Oct 2010, Rotterdam, the Netherlands
Duong Van Ni, Safford R, Maltby E (2001) Environmental change, ecosystem degradation and the value of wetland rehabilitation in the Mekong Delta. In: Adger WN, Kelly PM, Nguyen Huu Ninh (eds) Living with environmental change. Social vulnerability, adaptation and resilience in Vietnam. Routledge, London, pp 123–135
Nicholls RJ (1995) Coastal megacities and climate change. GeoJournal 37(3):369–379
Nicholls RJ (2003) Case study on sea level rise impacts. OECD Workshop report in the context of the workshop 'OECD workshop on the benefits of climate policy: improving information for policy makers', Dec 2002, Paris, 32 pp
Nicholls RJ (2004) Coastal flooding and wetland loss in the 21st century: changes under the SRES climate and socio-economic scenarios. Glob Environ Chang 14(1):69–86
Nicholls RJ, Mimura N (1998) Regional issues raised by sea level rise and their policy implications. Clim Res 11:5–18
Nicholls RJ, Hoozemans FMJ, Marchand M (1999) Increasing flood risk and wetland losses due to global sea-level rise: regional and global analyses. Glob Environ Chang 9(suppl 1):69–87
Nurhidayah L (2010) Coastal flooding in Jakarta: critical analysis on management of mangrove areas. In: Proceedings of 'Deltas in Times of Climate Change'. 29 Sept–1 Oct 2010, Rotterdam, the Netherlands
Ochieng CA (2002) Research master plan for the Rufiji Floodplain and Delta 2003 – 2013 Rufiji Environment Management Project – REMP. Technical report 28
Oome K (2010) The sand engine. In: Proceedings of 'Deltas in Times of Climate Change'. 29 Sept–1 Oct 2010, Rotterdam, the Netherlands
Overeem I, Syvitski JPM (2009) Dynamics and vulnerability of delta systems. LOICZ reports & studies no. 35. GKSS Research Center, Geesthacht, 54 p
Panin N (1999) Global changes sea level rising and the Danube Delta: risks and responses. In: Proceeding of international workshop on "Modern and Ancient Sedimentary Environments and Processes" in Moeciu, Romania, 8–15 Oct 1998, pp 19–29
Paw JN, Thia-Eng C (1991) Climate changes and sea level rise: implications on coastal area utilization and management in South-east Asia. Ocean Shorel Manag 15:205–232
Quartel S, Kroon A, Augustinus PGEF, Van Santen P, Tri NH (2007) Wave attenuation in coastal mangroves in the Red River Delta, Vietnam. J Asian Earth Sci 29(4):576–584
Richardson M (2008) Facing a rise in sea level. The Japan Times on 24 Sept 2008
Rozema J, van der Zee S, Bouksila F, Bahri A (2010) Perspective of saline agriculture for deltas in times of changing climate. In: Proceedings of 'Deltas in Times of Climate Change'. 29 Sept–1 Oct 2010, Rotterdam, the Netherlands
Sarubbi A (2007) Análisis del Avance del Frente del Delta del Río Paraná. Tesis de grado en Ingeniería Civil. Facultad de Ingeniería, Universidad de Buenos Aires, 2007
Sebesvari Z, Huong Le Thi Thu, Renaud FG (2011) Climate change adaptation and agrichemicals in the Mekong Delta, Vietnam. In: Stewart M, Coclanis PA (eds) Environmental change and agricultural sustainability in the Mekong Delta. Springer, Dordrecht, pp 219–239
Smith DM, Barchiesi S (2009) Environment as infrastructure – resilience to climate change impacts on water through investments in nature. IUCN report, Switzerland

Sneddon C, Nguyen Binh Thanh (2001) Politics, ecology and water: the Mekong Delta and development of the Lower Mekong Basin. In: Adger WN, Kelly PM, Nguyen HuuNinh (eds) Living with environmental change. Social vulnerability, adaptation and resilience in Vietnam. Routledge, London, pp 234–262

SRV [Socialist Republic of Vietnam] (2003) National Strategy for environmental protection until 2010 and vision toward 2020. Socialist Republic of Vietnam, Hanoi, 36 p

SRV [Socialist Republic of Vietnam] (2008a) Decision: issuance of the action plan framework for adaptation to climate change in the agriculture and rural development sector, period 2008–2020. Document no: 2730/QD-BNN-KHCN. Socialist Republic of Vietnam, Hanoi, 29 p. Available at http://www.presscenter.org.vn/en//images/QD-BNN-KHCN_EN.pdf

SRV [Socialist Republic of Vietnam] (2008b) Decision: on approval of the National Target Program to respond to climate change. Document no: 158/2008/QĐ-TTg. Socialist Republic of Vietnam, Hanoi, 23 p

Stoop H, Van Well E, Schengenga P, Akkermans R (2010) From salty foe to silty friend. In: Proceedings of 'Deltas in Times of Climate Change'. 29 Sept–1 Oct 2010, Rotterdam, the Netherlands

Timmerman P, White R (1997) Megahydropolis: coastal cities in the context of global environmental change. Glob Environ Chang 7(3):205–234

Törnqvist T, Yu S, González J, Hu P, Meffert D (2010) Sea-level rise and subsidence: a dual threat for the Mississippi Delta. In: Proceedings of 'Deltas in Times of Climate Change'. 29 Sept–1 Oct 2010, Rotterdam, the Netherlands

Trisirisatayawong I, Naeije M, Simons W, Fenoglio-Marc L (2011) Sea level change in the Gulf of Thailand from GPS-corrected tide gauge data and multi-satellite altimetry. Glob Planet Chang 76(3–4):137–151

von Storch H, Gönnert G, Meine M (2008) Storm surges – an option for Hamburg, Germany, to mitigate expected future aggravation of risk. Environ Sci Policy 11(8):735–742

Vongvisessomjai S (2007) Flood mitigation master plan for Chao Phraya Delta. In: 4th INWEPF steering meeting and symposium, Bangkok pp 1–15

Walker B, Holling CS, Carpenter SR, Kinzig A (2004) Resilience, adaptability and transformability in social–ecological systems. Ecol Soc 9(2):5

Wang H, Saito Y, Zhang Y, Bi N, Sun X, Yang Z (2011) Recent changes of sediment flux to the western Pacific Ocean from major rivers in East and Southeast Asia. Earth-Sci Rev 108(1–2):80–100

Wardekker JA, de Jong A, Knoop JM, van der Sluijs JP (2010) Operationalising a resilience approach to adapting an urban delta to uncertain climate change. Technol Forecast Soc Chang 77(6):987–998

White WA, Warne AG, Guevara EH, Aslan A, Tremblay TA, Raney JA (2002) Geo-environments of the Northwest Orinoco Delta, Venezuela. Inverciencia 27:521–528

WMO [World Meteorological Organization] (2008) WMO Bericht zum Zustand des Globalen Klimas 2007. Switzerland

Wolters W, El Guindy S, El Deen MS, Roest K, Froebrich J (2010) Integrated policies are a must to face future water scarcity in Egypt. In: Proceedings of 'Deltas in Times of Climate Change'. 29 Sept–1 Oct 2010, Rotterdam, the Netherlands

Yang D, Kane DL, Hinzman LD, Zhang X, Zhang T (2002) Siberian Lena River hydrologic regime and recent change. J Geophys Res 107:1–10

Yang S, Zhang J, Zhu J, Smith JP, Dai S, Gao A, Li P (2005) Impact of dams on Yangtze River sediment supply to the sea and delta intertidal wetland response. J Geophys Res 110. doi:10.1029/2004JF000271

Zagare V (2010) Climate change, urban dynamics and (lack of) governance: a Latin American case. In: Proceedings of 'Deltas in Times of Climate Change'. 29 Sept–1 Oct 2010, Rotterdam, the Netherlands

Zaman A, Rahman SM (2010) Testing of an IWRM DSS for assessing climate adaptation options in Bangladesh. In: Proceedings of 'Deltas in Times of Climate Change'. 29 Sept–1 Oct 2010, Rotterdam, the Netherlands

Zhang S, Lu XX, Higgitt DL, Chen C-TA, Han J, Sun H (2008) Recent changes of water discharge and sediment load in the Zhujiang (Pearl River) Basin, China. Glob Planet Chang 60(3–4):365–380

Zhang Q, Xu C, Chen Y, Jiang J (2009) Abrupt behaviours of streamflow of the Pearl River basin and implications for hydrological alterations across the Pearl River Delta China. J Hydrol. doi:10.1016/j.hydrol.2009.08.026

# Part II
# Water and Development in the Mekong Delta

# Chapter 3
# Hydrology and Hydraulic Infrastructure Systems in the Mekong Delta, Vietnam

**Vo Khac Tri**

**Abstract** The Mekong Delta with more than 700 km of coastline and countless estuaries connected to the East Sea (South China Sea) is known to be the breadbasket of Vietnam. This role has been promoted by the massive expansion of irrigated rice fields that rapidly increased the country's food production over the last decades. Floods during the rainy season are playing an important role in the agricultural system of the Delta but are also causing serious problems like yield losses, river bank erosion, destruction of infrastructure and the casualties. During the dry season, however, under the influence of tides, salinization and the lack of fresh water in coastal areas are the problems to be addressed. Moreover, tides are causing sedimentation in estuaries and canals which impedes the flow and increases the risk of flooding. Most recently and in the near future upstream water usage by neighboring countries, hydropower dams in the upper reaches and a higher frequency of extreme weather events due to the changing climate will have a major impact on the flow regime in the Mekong Delta. Recent studies by the Southern Institute of Water Resources Research are addressing these issues and attempt to provide solutions to promote the sustainable development of the Mekong Delta.

## 3.1 Overview of the Mekong Delta

The Mekong River flows from the Tibetan Plateau in China to the East Sea, passing through six countries on its way: China, Myanmar, Laos PDR, Thailand, Cambodia and Vietnam. At the end of this journey the river spreads into one of the largest deltas in the world, the Mekong Delta (MD) in southern Vietnam. It was formed by the

V.K. Tri (✉)
Department of Training and International Cooperation, Southern Institute for Water Resources Research (SIWRR), Ho Chi Minh City, Vietnam
e-mail: vokhac.tri7@gmail.com

F.G. Renaud and C. Kuenzer (eds.), *The Mekong Delta System: Interdisciplinary Analyses of a River Delta*, Springer Environmental Science and Engineering,
DOI 10.1007/978-94-007-3962-8_3, 

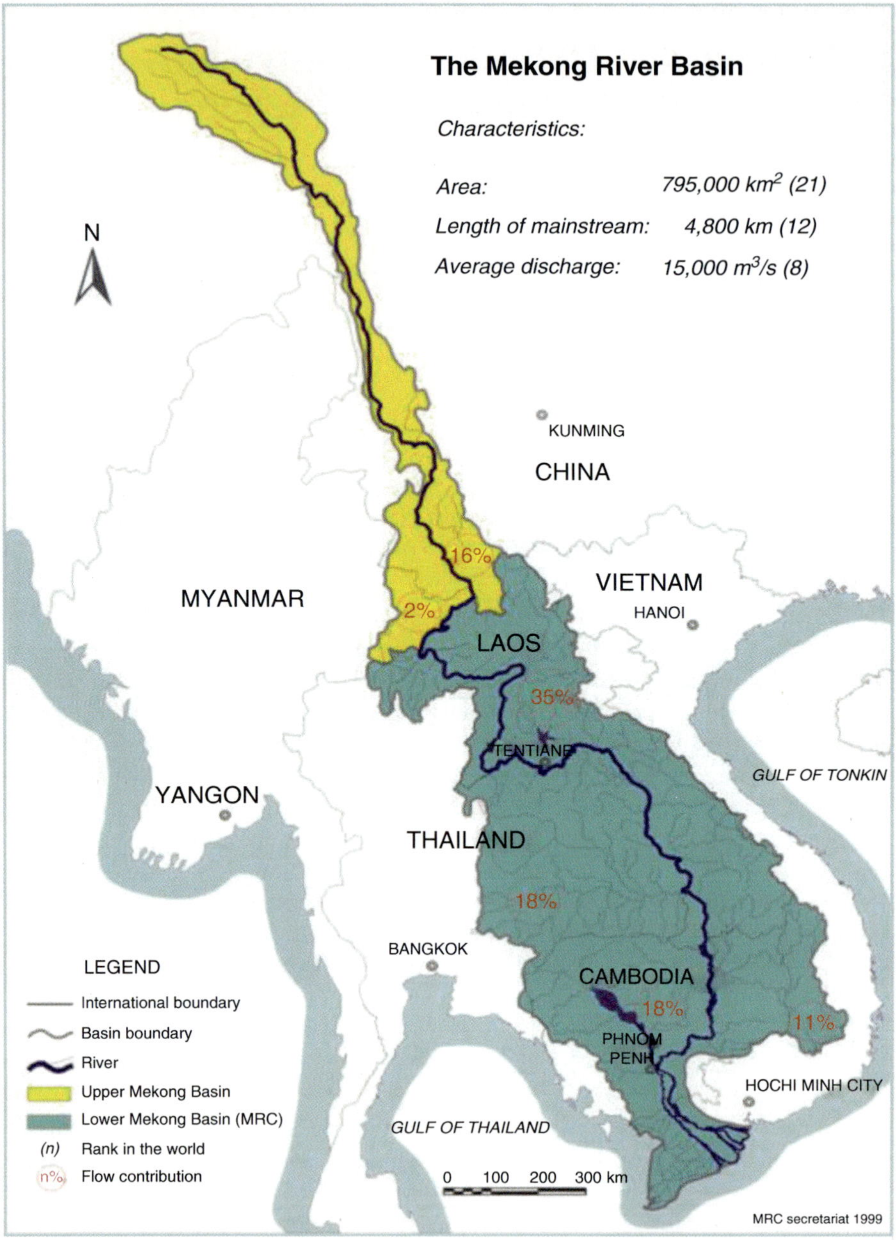


**Fig. 3.1** The Mekong River Basin (MRC 2005)

deposition of sediments from the Mekong River over thousands of years (Fig. 3.1; Chap. 8). The average elevation of the delta ranges from 0.5 to 1.2 m above sea level, except for some land areas along the Cambodian border, where the terrain reaches heights of up to 12 m; the elevation of the central part of the delta ranges from 1.0 to 1.5 m whereas coastal areas have an elevation of 0.3–0.7 m.

The MD covers an area of 39,700 $km^2$ or 12.7% of the total land area of Vietnam. In this region, the agricultural area covers 2.4 million hectares and the population is about 17 million. The MD plays a very important role in the agricultural and aquaculture production of Vietnam. The economy of the delta, which is the major agricultural production area of Vietnam, is oriented towards the primary sector. The delta contributes 27% of the total GDP of Vietnam, 40% of the agricultural production, and half of the rice production of the country. Rice production reaches 11 million tons, nearly 740 kg per capita. Although the population density is high, at nearly 400 people/$km^2$, the MD has become one of the biggest exporters of rice in the world. Rice and fishery products contribute significantly to the nation's export revenues in which rice represents approximately 85% of the total amount (Viet et al. 2005).

One of the main reasons for this success is the large-scale hydraulic infrastructure systems which have been developed very quickly over a few decades (Tri et al. 2005). Most of these systems aim to supply fresh water resources, and control flooding as well as salinity intrusion in urban, agricultural, and aquaculture areas.

The Mekong River flows into the MD through the Cambodian border in two major distributaries: the Mekong River (or Tien River) and the Bassac River (or Hau River). Both rivers split into nine branches that form the shape of the MD before running into the East Sea.[1]

The Mekong Delta river network is relatively dense and complex, and includes natural river systems as well as manmade canals (Fig. 3.2). The seasonal flow, the hydrological regime of the rivers and the canal network are affected strongly by the upstream flow, internal rainfall, and the tidal regime of the East Sea and West Sea.[2]

However, serious floods in the wet season and salinity intrusion in the dry season are natural disasters that have constrained the economic development of the MD. Other problems such as water pollution, erosion, and river sedimentation have also seriously affected the lives of MD residents. Recently, the hydrological regime of the MD has been changed critically by the development of hydropower dams and by changes in land use of the upstream countries. The main stream of the Mekong River supplies water and sediment to a natural reservoir, the Tonle Sap Lake in Cambodia, which has a storage capacity of 80 billion $m^3$ and can regulate the upstream flow in the flood and dry seasons. Since 2000, several reservoirs have been constructed in China, and four other reservoirs (Pakmun, Nam Ngum 1, Nam Ngun 2, Yali) with a capacity of 10 billion $m^3$ have been constructed on the tributaries. Further developments have been planned since (see Chap. 5) (Fig. 3.3) (MRC 2005).

[1] South China Sea.

[2] Gulf of Thailand.

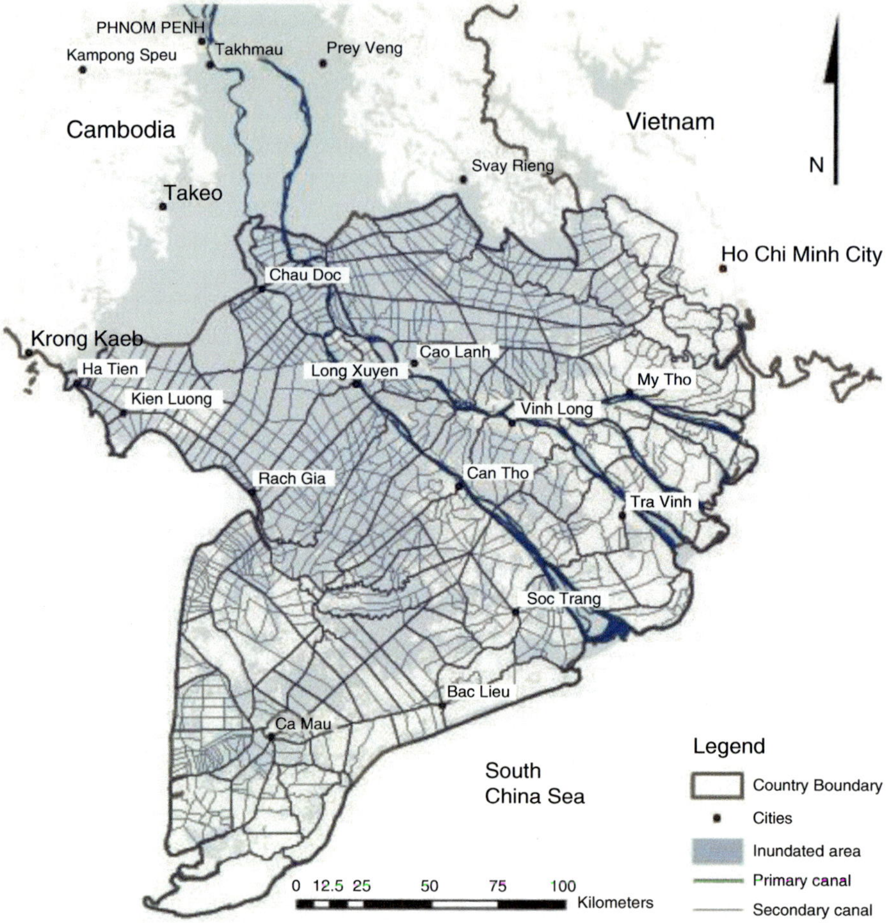


**Fig. 3.2** Map of river systems and canal networks in the MD (MRC 2005)

## 3.2 Hydrology of the Mekong Delta

The Mekong River flows into the Mekong Delta and drains into the sea through various channels in a complicated flow regime:

- Inflow: The major inflows are through the Mekong and Bassac River, as well as from overflow from the Long Xuyen Quadrangle and the Plain of Reeds (Fig. 3.4). Based on the result of simulations, the inflow rate from Cambodia into the Plain of Reeds is approximately 2,850 $m^3/s$.
- Outflow: Part of the Bassac River flow is diverted into the Long Xuyen Quadrangle, and this is supplemented by the inflow from Cambodia for this

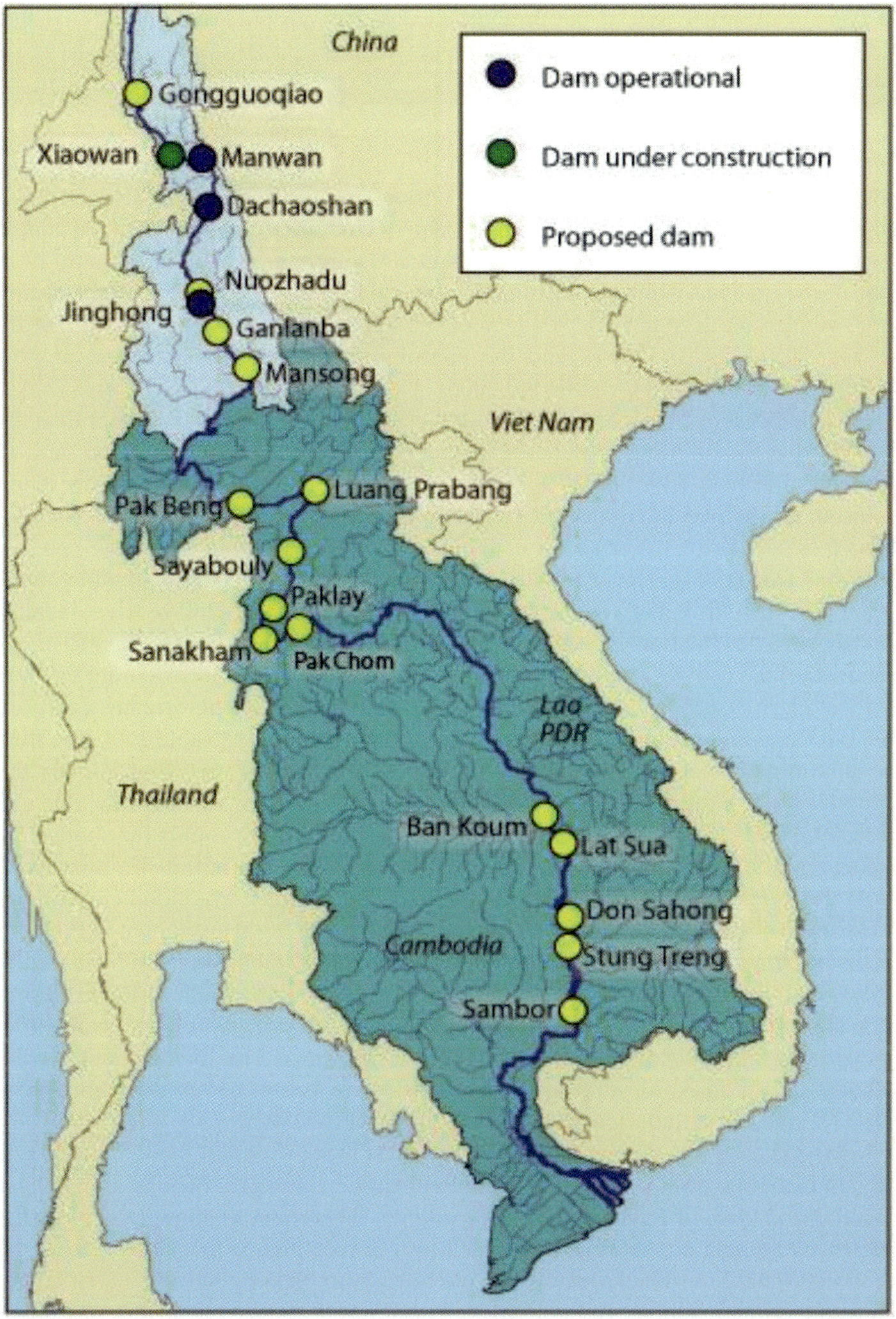


**Fig. 3.3** Map of locations of dams on the Mekong River (MRC 2005)

region. Another part of the flow drains directly into the West Sea through the canal network and the Cai Lon River, and the rest runs through the Bassac River into the East Sea (Lanh et al. 2005).

During the rainy season (May to November), high rainfall in the basin causes flooding in the main stream of the Mekong River and the Mekong Delta. The hydrology regime of the MD is affected rather strongly by the flow of the Mekong River, the tides of the East Sea, of the West Sea, and of the rainfall regime. The total annual water volume of the Mekong Delta is about 500 billion $m^3$ (ca. 422 billion $m^3$ from the upstream and 67 billion $m^3$ from the regional rainfall). The water volume during the wet season accounts for 80–83% of the annual rainfall, while that of the dry season accounts for 7–9%, and the contribution of the two transition months (April to May) is 10–11% (SIWRP 2005).

At the border of Vietnam, the flow of the Mekong River is much higher than the flow of the Bassac River, where the discharge at Tan Chau and Chau Doc stations account for about 83% and 17% of the annual flow, respectively. During the flood season, the distribution of flow at Tan Chau is about 80% and that of Chau Doc section is 20%. During the dry season, the flow at Tan Chau accounts for 85% and that of Chau Doc is 15%. However, due to the Vam Nao tributary (Fig. 3.4) which links the two rivers, the water volume of the Mekong River and the Bassac River is fairly balanced. Thus the flow rate at Can Tho (Hau River) and My Thuan (Tien River) is almost equal. The Vam Nao River plays an important role in re-distributing the flow of these two distributaries. The total average volume of the main streams observed at Tan Chau and Chau Doc is about 387 billion $m^3$ annually (Table 3.1). The flow is distributed unevenly throughout the year. The flow is very high during the flood season, particularly during September and October, where the flow rate can reach up to 25,500 $m^3/s$. The total flow in these 2 months accounts for about 17% of the annual flow (Table 3.1). In the year 2000, the flooding was historical and the average discharge in September was 29,624 $m^3/s$. The flood current and discharge distribution in the Mekong delta for the year 2000 are shown in Fig. 3.5. This illustrates that the flood water transfers from the Cambodian flood plains to the Mekong delta through the Long Xuyen Quadrangle and the Plain of Reeds during the flood season is rather high. Moreover, there was an unusually low flow in 1998, where the highest monthly flow was only 18,783 $m^3/s$. The lowest monthly discharge occurs in April, with the average discharge being 2,340 $m^3/s$, although in 1998 it was particularly low at 1,820 $m^3/s$ (SIWRP 2005).

The Mekong River is connected with the vast storage area of the Plain of Reeds by an elaborate river and canal system. When the river tide fluctuates, the water level of the Plain of Reeds changes accordingly. Therefore, the flow velocity in the downstream area of the Mekong River Basin is very high, particularly during the ebb tide. The flow velocity caused by the tide of the Mekong River is much higher than the critical velocity for soil erosion. As a result, the river banks in the downstream area of the Mekong River have been seriously eroded.

At the river mouth of the Bassac River, the flow velocity is also rather high at 1.5 m/s, even though the water exchange between the river and the delta is not so

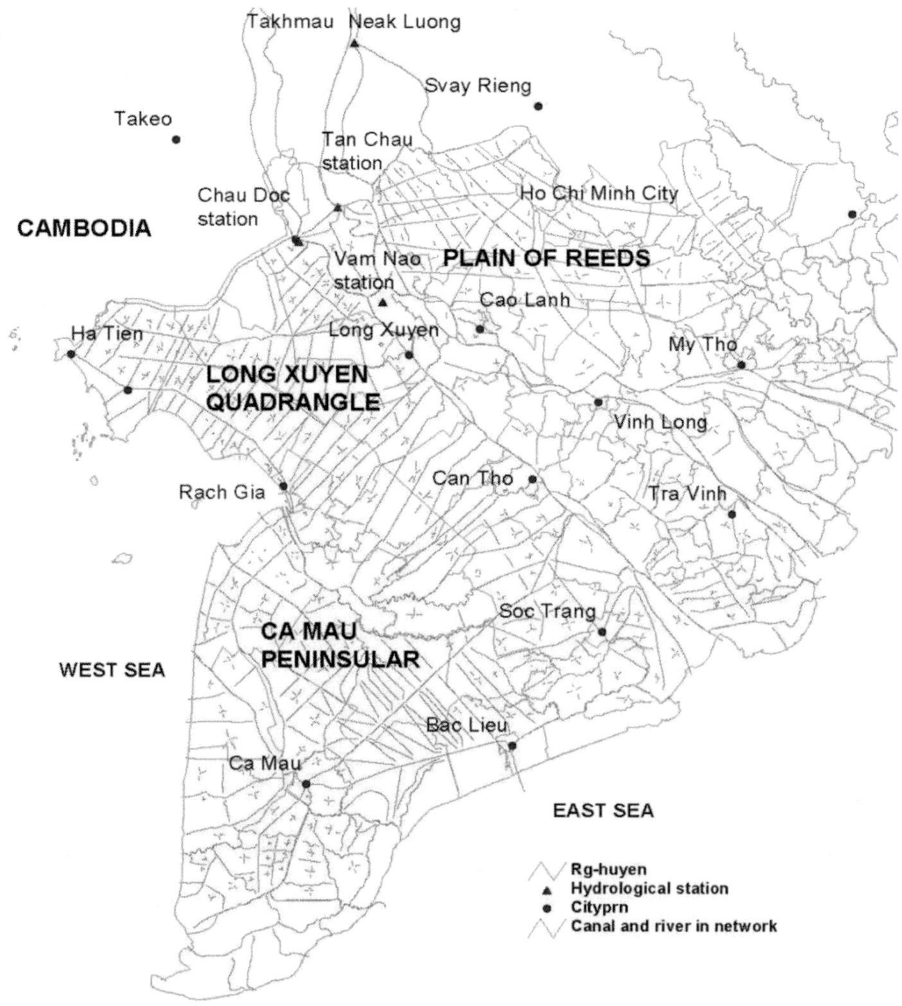


**Fig. 3.4** Hydrological stations in the main stream of Mekong River (Giam 2010)

high. Due to the high flow velocity which persists over a long period, the riverbank around the river mouth is eroded constantly to form individual deep gutters.

The flow during the dry season is rather low, with a mean discharge $Q < 6{,}000\ m^3/s$. The velocity of the inflow is relatively low, and the main stream axis runs along the riverside (Lanh et al. 2005).

The tidal regime of the East Sea includes two tidal cycles each day, each with rather high amplitude. Because of the condition of the wide and shallow riverbed, the tidal effects penetrate far inland. During the dry season, tidal fluctuations can be seen in Phnom Penh, about 300 km away from the coast. However, a reverse

**Table 3.1** Average monthly discharge at Tan Chau and Chau Doc (period 1977–1999)

| Month | Q Tan Chau ($m^3/s$) | Q Chau Doc ($m^3/s$) | Q total ($m^3/s$) |
|---|---|---|---|
| 1 | 6,220 | 1,360 | 7,580 |
| 2 | 3,720 | 700 | 4,420 |
| 3 | 2,600 | 420 | 3,020 |
| 4 | 2,010 | 330 | 2,340 |
| 5 | 2,640 | 460 | 3,100 |
| 6 | 7,180 | 1,450 | 8,630 |
| 7 | 11,270 | 2,390 | 13,660 |
| 8 | 16,390 | 3,970 | 20,360 |
| 9 | 20,140 | 5,290 | 25,430 |
| 10 | 20,340 | 5,480 | 25,820 |
| 11 | 15,260 | 4,700 | 19,960 |
| 12 | 10,180 | 2,710 | 12,890 |
| Average | 9,830 | 2,440 | 12,270 |
| Total discharge (billion $m^3$) | 310 | 77 | 387 |
| % total discharge | 80% | 20% | 100% |

*Source*: MRC, flood control planning for developing the Mekong Delta (MRC 2005)

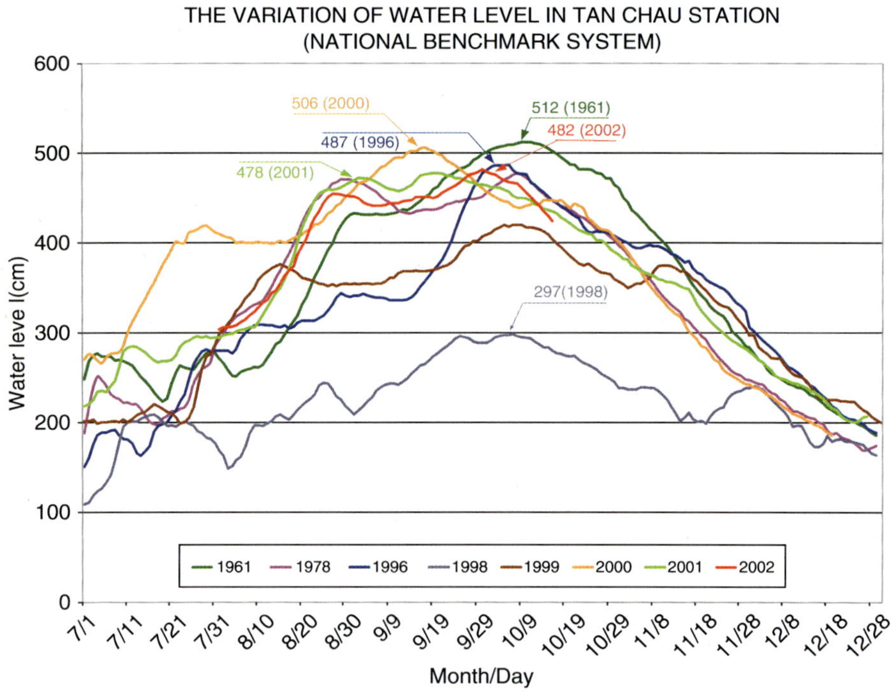


**Fig. 3.5** Yearly water level at Tan Chau station (Lanh et al. 2005)

flow may exist up to Tan Chau and Chau Doc for a very short time during the dry season. At the interface of tide (tides coming from the East Sea and from the West Sea), the flow velocity is very low in the canal network of the MD. This has caused the sedimentation of canals in these areas such as the Ca Mau Peninsula and the South Mang Thit area. This problem seriously affected the flow regime in the Mekong Delta.

Other river systems are contributing to the hydrological regime of the MD. The Vaico River system consists of two streams, East Vaico and the West Vaico, that originate from Cambodia and flow through the north-east of the MD. The Cai Lon and Cai Be River system consists of so-called tidal rivers that originate from the center of the Ca Mau peninsula and flow into the West Sea through the Cai Lon River mouth. This river mouth is very wide but shallow because of siltation. The My Thanh River and the Ganh Hao River are small river systems on the East Sea side. All river systems are connected to a network of interlacing excavated canals, which has completely changed the hydrology regime of the MD.

## 3.3 Floods in the Mekong Delta

### *3.3.1 Characteristics of Flood Flow*

During the wet season, the flood from the upstream Mekong River running into the MD can inundate more than 2 million ha of land. The floods of the MD have the following characteristics:

- The total volume of the annual flood is very high at approximately 400 billion $m^3$, while the storage of Tonle Sap lake is around 60–80 billion $m^3$.
- The duration of flooding is about 6 months from July to December.
- The variation of water levels in Tan Chau and Chau Doc ranges from 3 to 5 m (Fig. 3.5). Floods rise and withdraw slowly, the mean value ranges from 5 to 7 cm/day. High floods are about 10–12 cm/day, and the highest floods are 20–30 cm/day (Lanh et al. 2005).

### *3.3.2 Direction of Flood Flow in the Mekong Delta*

The transition of flooding from Phnom Penh to Tan Chau takes about 2–3 days. The velocity decreases gradually after it runs into the inland area of the MD, moving in two directions (Figs. 3.6 and 3.7):

- From the main stream it flows into the MD through the Mekong River and Bassac River

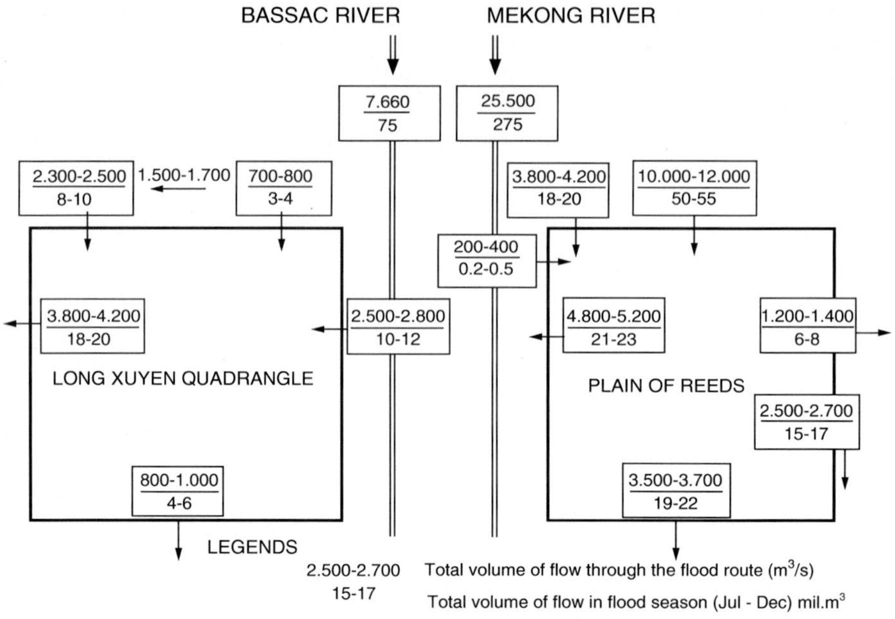


**Fig. 3.6** Estimated max discharge and total volume in the year 2000 (Lanh et al. 2005)

- From the left[3] side of the Mekong River it flows into the flood plain of Cambodia then it runs into the Plain of Reeds and from the right side of the Bassac River it runs into the shallow area of Takeo. After that it runs towards the plain of the Long Xuyen Quadrangle, Vietnam.

In the Plain of Reeds, the flooded area can be divided into two regions:

- Region 1 (from Duc Hoa District, Long An to Long Khot creek): This inundation area is caused by flooding events in Cambodia which lead to an increased water level in the Long Khot creek and the West Vaico River so that some areas between the two streams of Vaico River are inundated. Since flooding in this region has to pass through the horizontal canals of the Plain of reeds with a reduced velocity it occurs more slowly and 1.0–1.5 months later than in the Mekong River (at the end of August or mid-September).
- Region 2 (from the left side of Mekong River to Long Khot creek): After floods run over the flood plain of Cambodia, they flow into the Plain of Reeds from So Thuong – So Ha River, Cai Co – Long Khot creeks and the canal networks.

The tidal effect is limited during the flood season. However, it causes a substantial delay in terms of drainage to the sea, particularly on days of high tide. For peak

[3] Left and right refers to direction of flow.

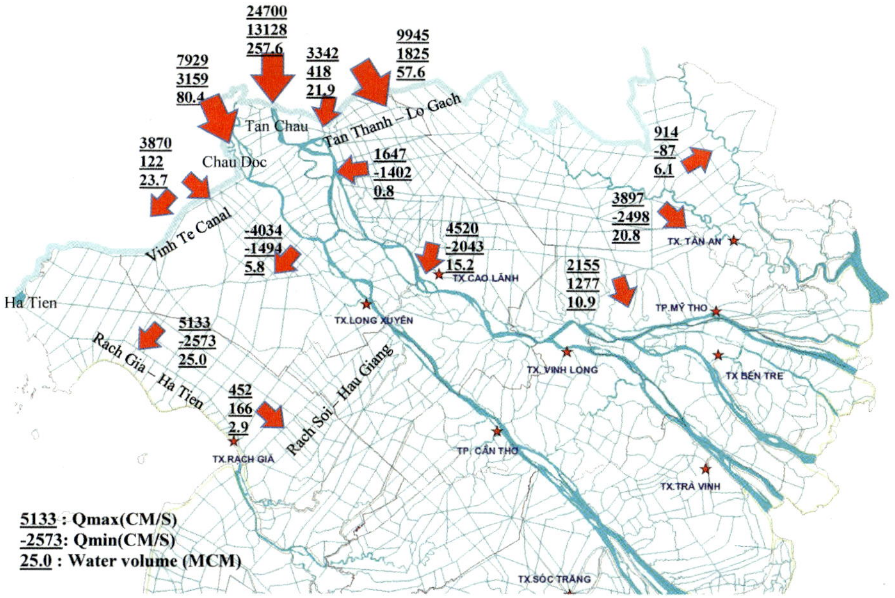


**Fig. 3.7** Flood water distribution in the Mekong Delta in the flood season of 2000 (Qmax/Qmin: discharge max./min. in $m^3/s$ and – Water volume: million $m^3$) (Lanh et al. 2005)

flood during the year, the flow was stagnant in the days of high tide, which increased the water level in the river and the inland area (up to Tan Chau and Chau Doc). This stagnation causes instant sedimentation in some river reaches and increases the flow velocity during ebb tides. Around the river mouths, the high tides may cause regression to decrease considerably the flow velocity. Therefore, most of the sediment is found around the river mouth.

The tides of the East Sea intrude deeply into the plain through the estuaries of the Mekong River, Bassac River and the Vaico River, which affects most of the Mekong delta areas, including the area left of the Mekong River, the area between the Mekong River and the Bassac River, almost all of the Long Xuyen Quadrangle and the Ca Mau Peninsula. In the coastal zone, the tidal amplitude is frequently high. However the amplitude in the interior field is much lower. In fact, the multi-directional transmission in the complicated canal network results in weak flow, low amplitude, and poor drainage capacity. The tidal intrusion of salty water has bad as well as good effects on the agricultural production and the livelihood of the people who live around the 1.7 million hectares of coastal zone and along the main river areas.

The floods of the MD depends on the upstream flood, the regulation by the Tonle Sap, the inundated area of Cambodia, the East Sea – West Sea tidal regime, the interior field rainfall regime, geomorphologic conditions of the inundated areas and anthropologic impacts across the whole basin. The main cause of flooding in the Mekong Delta is rainfall. The rainfall of the Mekong River basin is rather high compared to other large river basins in the world. Such a high rainfall is the result of the tropical monsoon climate, which is predominant in the Mekong River basin. Three main weather turbulences are responsible for high rainfall and floods in the basin: (1) turbulence caused by the south-west seasonal wind, (2) the tropical convergence band, and (3) tropical depressions. Based on the results of studies at Tan Chau station, it has been shown that the high water level usually occurs with the early floods and that the flood intensity is not usually higher than 10 cm/day for the main seasonal floods.

Flooding from the upstream flow in the Mekong River and the Bassac River discharge 92.6% of the total volume (about 30,000–32,000 $m^3/s$) into the East Sea partly to the West Sea of 4.6%, and 2.8% drained through the Vaico River. Of these 92.6%, about 82–86% is drained by the Mekong River and 14–18% by the Bassac River, which results in a water level of 4.4 m at Tan Chau and 3.88 m at Chau Doc. The remaining of 14–18% will overflow the boundary of floodplain in Cambodia into the Mekong Delta (about 8,000 $m^3/s$) and flows into the interior field at a discharge rate of 2,000 $m^3/s$ (Lanh et al. 2005).

### 3.3.3 Impacts of Floods

The floods inundating the Plain of Reeds is mainly caused by the floods that overflow across the Vietnam – Cambodia border (88%), and the remaining 12% is from the Mekong River to the Hong Ngu, Muong Lon, and An Binh channels. In addition, some of the flood water in the Plain of Reeds (ca. 3.5–4.0 billion $m^3$) originates from rainfall within the region itself. The flood water in the Plain of Reeds is drained out to the Mekong River and the Vaico Dong River. The inundation of the Long Xuyen Quadrangle is mainly caused by the floods in the Mekong River, which pass through the inundated areas of Cambodia, and from the Bassac River through channels connecting the Bassac River with the Rach Ha Mekong channel. In addition, the intra-field rainfall contributes partly to the floods.

During the last 40 years, the flood discharge from inundated areas of Cambodia into the Long Xuyen Quadrangle and the Plain of Reeds has increased. Comparing the average monthly water level of the period 1982–1985 and the period 1992–1995, the levels of the main stream and coastal area in September and October are not so different between the two periods. However, the water level in the field for the period 1992–1995 is higher than that for the period 1982–1985 by 0.2–0.4 m. This suggests that the flood in the main stream will be lower and the flood in the intra-field will be higher, if the overflow across the border into the Plain of Reeds and Long Xuyen Quadrangle is not controlled.

The livelihood of the people in inundated areas of the Mekong Delta depends mainly on agriculture, and to some extent, on aquaculture, forestry, industry, handicrafts, services and trade. In recent years, the economy of the Mekong Delta provinces has grown quickly, with a growth rate of 5–7%. The average GDP per capita is about 300 USD. However, the economy generally relies on traditional agriculture, in which the production depends heavily on natural conditions, and the livelihoods are vulnerable and uneven in the various regions. The quality of life in the remote areas remains poor, and the people face many difficulties.

Land transport infrastructures of the inundated areas are underdeveloped. Presently, there are 10 national highways with a total length of 901 km, and 71 inter-provincial roads with a total length of 1,579 km. However, the quality of the roads and highways is poor. Except for the important national highways, most of the roads are damaged and many sections are inundated during the flood season. The canals have been constructed with high density so navigation has also become very important in the inundated area of MD. An estimation puts at 70% the total goods transported via these canals.

The inundated area of the Mekong Delta includes 8 provinces, 60 districts and 1,069 communes and auxiliary areas. The population of these region is around 10 million people (26.8% urban and 73.2% rural). Most of the people live along the major canals and roads. The average population density of the region is 474 persons per $km^2$; but varies strongly between towns and townships (more than 2,000 persons per $km^2$), districts along the Mekong River and the Bassac River (about 500–1,000 persons per $km^2$) and remote areas (less than 100 persons per $km^2$).

The floods of the Mekong Delta are of relatively low intensity, but they persist for a long time, which causes erosion of river banks and canal banks, and serious damage to the infrastructure. For example, the 1994 flood caused 407 deaths and a value loss of 2,284 billion VND; the 1995 flood caused 199 deaths and a value loss of more than 700 billion VND; the 1996 flood caused 217 deaths and a value loss of 2,182 billion VND (Sam et al. 2004). The flooding causes difficulties for permanent settlements and the development of infrastructure, and leads to vulnerable and unstable livelihood. Mobility, school enrolment, and health care are greatly limited. The exploitation of natural resources, the development of rural and urban areas, as well as the industrialization and modernization process are all negatively affected by floods.

Nonetheless the floods of the MD also have positive aspects. First, a large volume of alluvium soils, mainly from the upstream basin (95%), are deposited in the MD. The ongoing deposition extends the Mekong Delta continuously towards the East Sea, improves the fertility of fields, provides a well-stocked fishery and creates very favorable conditions for aquaculture. The flood flow also improves the water quality, particularly by flushing acid sulfate soils and agricultural pests. In addition, this freshwater source is important for agriculture, aquaculture, and livelihoods, and provides a vast freshwater ecosystem for the plain.

## 3.4 Salinity Intrusion in the Mekong Delta

The MD is affected by the strongest salinity intrusion of the country over approximated 1.77 million ha. Salt water affects about 45% of the delta in the dry season. Salinity seriously affects social and economic development and the lives of residents. In April when the upstream flow is weaker and strong easterly winds (Gio Chuong) blow, salinity penetrates inland. Salinity intrusion itself is affected by the flow rate (discharge) from the upstream and the tide regime from the sea.

In the dry season from December to June, the average discharge of the Mekong River is about 6,000 $m^3/s$. From March to April it is at its lowest, at approximately 2,000 $m^3/s$; this leads to salinity intrusion far inland. The tide is the main factor affecting the ratio of flow distribution in the MD in the dry season. The tide regime of the East Sea is of an unsteady semi-diurnal type with two peaks and two lows as well as two flood tides and two low tides in the month. The highest tides are in December and January, the lowest in April. The amplitude of the tide is about 2.5–3.0 m in March and April. When the upstream flow decreases, the tide can have an effect up to 60–70 km from the estuaries of the Mekong River. The tide of the West Sea is an unsteady diurnal type with one peak and one low; the amplitude is about 0.8–1.0 m, which has a marginal impact on the MD (SIWRP 2005). It mainly affects the small canals and Ong Doc River and Cai Lon River system.

The effect of salinity in the MD can be recognized in four zones: (i) the estuary zone, (ii) the zone between the two Vaico Rivers, (iii) the Ca Mau peninsular zone, and (iv) the west coastal zone. The intrusion of salt water into the Mekong delta is shown in Fig. 3.8.

### *3.4.1 Estuary Zone*

Estuary areas are vastly affected by saline water intrusion from the East Sea with an amplitude of 3.5–4.0 m and two tidal cycles each day. The degree of salinity of the sea water in the East Sea varies from 32 to 33 g/l but decreases in the estuary zone (EZ) due to the dilution by fresh water from upstream (Nhan et al. 2008). The impact of the salinity intrusion on the estuary zone depends on the upstream water flow. The maximum salinity values at the monitored stations in the estuary zone from 1991 to 2004 are shown in Table 3.2.

### *3.4.2 Vai Co Rivers Zone*

The West Vai Co River is 325 km long with a catchment area of 6,000 $km^2$. The river has its origin in Cambodia, then flows past Long An province and drains to the East Sea through the Soai Rap mouth. The East Vai Co River runs from the Cambodian low mountains through the Tay Ninh and Long An province, and has a length of 283 km.

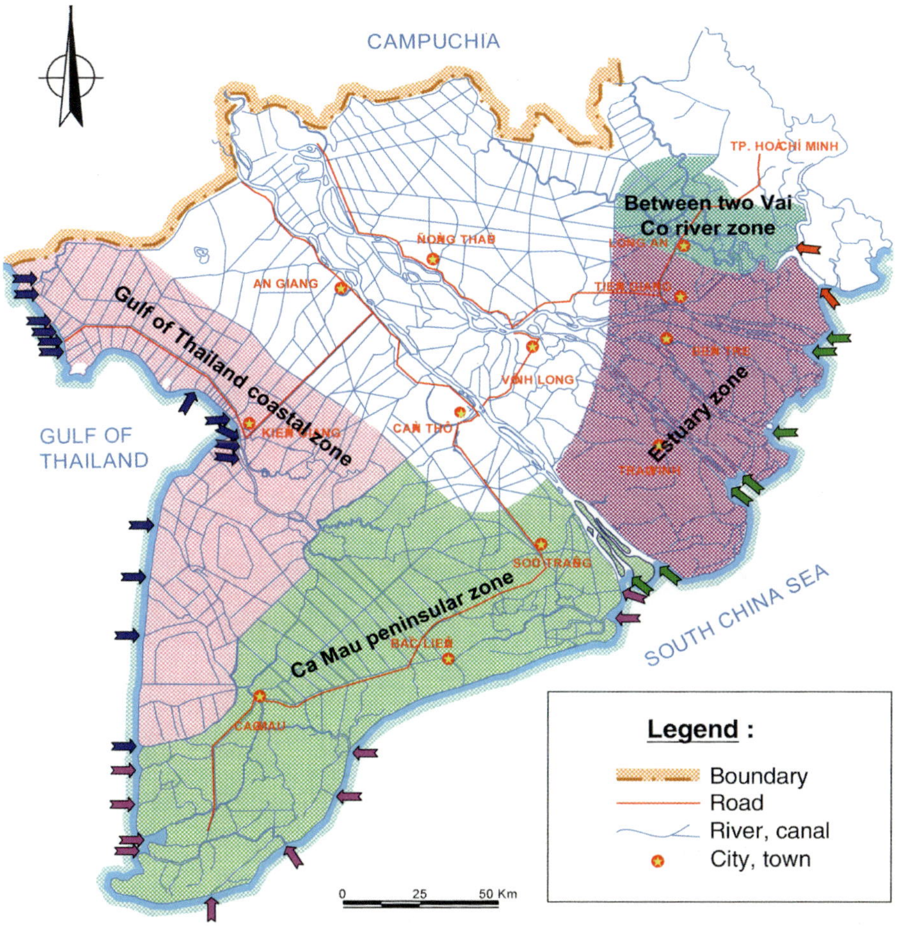


**Fig. 3.8** The main salinity intrusion pathways into the Mekong Delta (Sam et al. 2008). The *colored arrows* indicate the four different zones of salinity intrusion

The hydrological regime of the two rivers is affected by the flow from upstream and the tidal regime of the East Sea. In the dry season, the flow from upstream is often low, leading to strong salinity intrusion far inland and in the estuaries.

The maximum salinity values at monitoring stations in the two Vai Co estuary zones from 1991 to 2004 are shown in Table 3.3.

The water salinity values at the stations fluctuate monthly as well as yearly. The water storage in the fields is also affected by the salinity intrusion into this zone. Moreover, the increased water level in the river mouths due to the wind from the sea also increases the salinity intrusion capacity, and restricts the fresh water from flowing into the estuary areas. Measured results at the stations in this zone show that the annual salinity intrusion is highest in April, causing difficulties for agricultural development and domestic water supply in the estuary areas (Chaps. 7 and 13).

**Table 3.2** Maximum salinity values in the estuary zone (g/l) (SIWRR 2010)

| Station | Year | 1991 | 1992 | 1993 | 1994 | 1995 | 1996 | 1997 | 1998 | 1999 | 2000 | 2003 | 2004 |
|---|---|---|---|---|---|---|---|---|---|---|---|---|---|
| Vam | March | 20.8 | 15.0 | 32.4 | 31.0 | 27.4 | 27.0 | 27.3 | 23.5 | 23.3 | 19.6 | 27.1 | 26.4 |
| Kenh | April | 19.8 | 15.6 | 28.7 | 27.6 | 27.4 | 26.4 | 26.5 | 30.9 | 25.7 | 22.7 | 21.2 | 24.2 |
| Binh | March | 17.8 | 30.6 | 29.8 | 21.3 | 24.2 | 23.9 | 15.7 | 22.1 | 26.6 | 9.3 | 27.0 | 27.4 |
| Dai | April | 20.8 | 27.5 | 27.2 | 25.0 | 23.1 | 25.0 | 15.8 | 26.5 | 20.7 | 17.9 | 20.8 | 24.3 |
| Ben | March | 23.5 | 22.4 | 21.1 | 31.8 | 26.3 | 23.8 | 29.6 | 17.5 | 28.3 | 13.8 | 25.2 | 29.1 |
| Trai | April | 27.2 | 21.2 | 23.9 | 29.9 | 26.0 | 24.7 | 26.0 | 26.3 | 26.0 | 23.1 | 23.1 | 25.4 |

**Table 3.3** The maximum salinity values between the two Vai Co River zones (g/l) (SIWRP 2005)

| Station | Year | 1995 | 1996 | 1997 | 1998 | 1999 | 2000 | 2002 | 2003 | 2004 |
|---|---|---|---|---|---|---|---|---|---|---|
| Cau | March | 13.1 | 12.7 | 12.3 | 17.9 | 14.0 | 9.5 | 11.8 | 10.4 | 16.1 |
| Noi | April | 14.5 | 14.5 | 11.8 | 16.7 | 11.9 | 9.4 | 11.2 | 13.8 | 14.7 |
| Tan | March | 6.7 | 7.0 | 3.2 | 8.3 | 5.6 | 1.2 | 2.5 | 3.3 | 6.3 |
| An | April | 9.8 | 8.8 | 3.7 | 9.3 | 5.8 | 1.1 | 6.5 | 4.9 | 10.1 |
| Tuyen | March | 1.3 | 1.6 | 1.7 | 3.2 | 1.2 | 0.0 | – | – | – |
| Nhon | April | 5.8 | 3.3 | 3.4 | 6.5 | 1.9 | 0.0 | – | – | – |
| Doi ma | March | 7.0 | 9.1 | 4.5 | 9.5 | 6.0 | 1.5 | 4.7 | 5.0 | 9.6 |
| Sluice | April | 7.5 | 10.2 | 3.9 | 8.8 | 5.8 | 1.7 | 7.3 | 6.0 | 10.2 |
| Ben | March | 6.4 | 8.7 | 5.0 | 9.3 | 5.8 | 2.6 | 3.2 | 4.6 | 8.3 |
| Luc | April | 6.9 | 9.9 | 5.0 | 10.7 | 5.5 | 3.0 | 8.9 | 5.6 | 9.2 |
| An Ha | March | 2.2 | 4.5 | 1.0 | 0.8 | 0.0 | 0.0 | 0.6 | – | 1.3 |
| Bridge | April | 5.5 | 6.2 | 1.1 | 4.5 | 0.0 | 0.0 | 0.8 | – | 2.3 |

### 3.4.3 West Sea Coastal Zone

This zone covers the two provinces of Kien Giang and Ca Mau which are characterized by a low topography (0.2–0.6 m asl.). The maximum salinity values at stations in the West Sea coastal zone from 1991 to 2004 are shown in Table 3.4.

The water salinity values at the measurement stations fluctuate greatly, both monthly and yearly. This zone is affected by the tide of the West Sea, and is also supplied with fresh water from the Bassac River. The salinity intrusion of this zone is strongly affected by the tidal regime. Salinity intrusion in this zone does not spread too far inland compared to the two Vai Co river mouths and the estuaries of Mekong River. The salinity of the water gradually increases from March to April.

### 3.4.4 Ca Mau Peninsular Zone

The hydrological regime of this zone is vastly affected by the tide regimes of the East Sea and the West Sea, the Mekong River flow and local precipitation. The maximum salinity values at stations in the Ca Mau peninsula zone from 1991 to 2004 are shown in Table 3.5.

Saline water intrudes into the main canal and creek system of Ca Mau peninsula zone through My Thanh, Ganh Hao Rivers from the East Sea and through Ong Doc, Cai Lon, Cai Be Rivers and the canals from the West Sea.

### 3.4.5 Salinity Intrusion in the Estuaries of the Mekong River

Based on the salinity intrusion data recorded by the Southern Institute of Water Resources Research (SIWRR) and Tri et al. (2006) in the main estuaries of the Mekong River, graphs of maximum salinity intrusion from the sea into the river mouths were derived (Figs. 3.8, 3.9, 3.10, 3.11 and 3.12).

The extent of salinity intrusion at most river mouths reaches up to 60 km inland and has its strongest impact on the estuary zones in March and April. Figure 3.13 shows the average salinity intrusion in April 2000 in the MD.

The results of the studies on water balance in the MD show that the highest water requirements in December and January are approximately 650–700 $m^3/s$ but the average discharge at this time is greater than 6,500 $m^3/s$, enough to meet present and future needs of the population. The problem lies more in the development of water distribution networks. Fresh water supply in April will become increasingly difficult when the mean flow is down to about 2,000–2,300 $m^3/s$. Agricultural areas should be reduced to control the salinity intrusion inland.

**Table 3.4** The maximum values of salinity at typical stations in the West Sea coastal zone (g/l) (SIWRR 2010)

| Station | Year | 1994 | 1995 | 1996 | 1997 | 1998 | 1999 | 2000 | 2002 | 2003 | 2004 |
|---|---|---|---|---|---|---|---|---|---|---|---|
| Kien | March | 17.0 | 16.7 | 14.3 | 11.2 | 8.5 | 5.9 | 0.3 | 7.8 | 0.0 | 1.8 |
| Luong | April | 17.3 | 23.1 | 12.1 | 10.0 | 22.5 | 1.2 | 0.3 | 3.6 | 0.3 | 2.6 |
| Tam | March | 6.5 | 16.5 | 15.3 | 0.5 | 0.0 | 0.0 | 0.5 | 0.4 | 0.0 | 0.7 |
| Ngan | April | 6.0 | 20.8 | 13.3 | 0.4 | 10.5 | 1.0 | 0.5 | 3.0 | 0.0 | 0.9 |
| Rach | March | 10.1 | 11.8 | 19.1 | 11.7 | 17.4 | 12.2 | 16.7 | 14.8 | 13.4 | 23.2 |
| Gia | April | 12.7 | 20.2 | 17.4 | 11.6 | 21.0 | 9.9 | 15.8 | 23.0 | 21.0 | 23.5 |

**Table 3.5** Maximum salinity values at stations in the Ca Mau peninsular zone (g/l) (SIWRR 2010)

| Station | Year | 1994 | 1995 | 1996 | 1997 | 1998 | 1999 | 2000 | 2002 | 2003 | 2004 |
|---|---|---|---|---|---|---|---|---|---|---|---|
| Cai Xe | March | 4.6 | 5.2 | 2.5 | 1.3 | 4.5 | 6.4 | 4.0 | – | – | – |
| Sluice | April | 3.2 | 4.0 | 4.6 | 2.4 | 7.3 | 7.5 | 9.6 | – | – | – |
| Tiep | March | 6.5 | 10.3 | 10.7 | 7.0 | 7.6 | 5.6 | 5.5 | – | – | – |
| Nhat | April | 10.0 | 12.8 | 18.8 | 7.0 | 8.8 | 6.1 | 4.1 | – | – | – |
| Soc | March | 2.4 | 3.0 | 2.9 | 1.3 | 4.5 | 6.1 | 1.4 | – | 3.3 | 3.8 |
| Trang | April | 3.5 | 5.3 | 4.2 | 2.0 | 6.8 | 7.0 | 4.4 | – | 3.2 | 4.4 |
| Ngan | March | 21.0 | 37.6 | 23.2 | 1.6 | 8.8 | 1.3 | 0.6 | 1.3 | 0.0 | 0.5 |
| Dua | April | 20.5 | 34.7 | 28.3 | 1.6 | 8.9 | 1.2 | 0.6 | 1.7 | 0.5 | 6.0 |

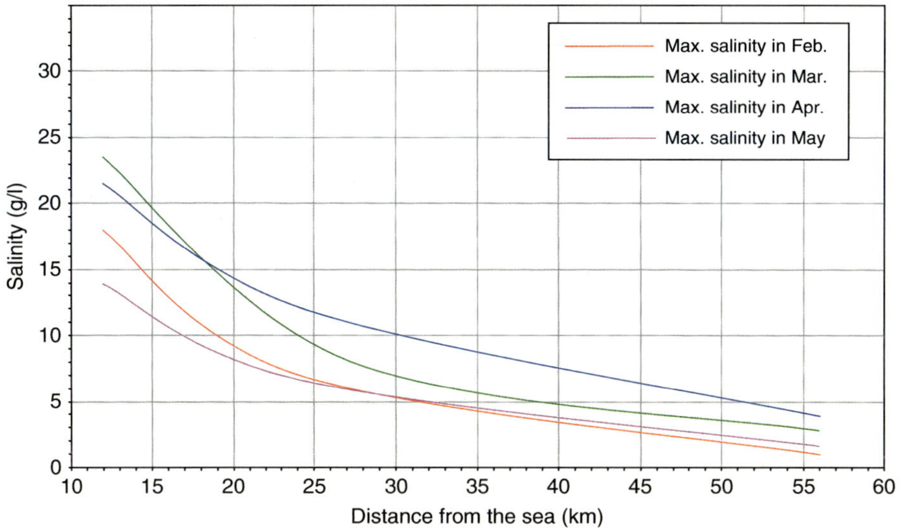


**Fig. 3.9** The distribution of salinity concentration in the Tieu River mouth (Tri et al. 2006)

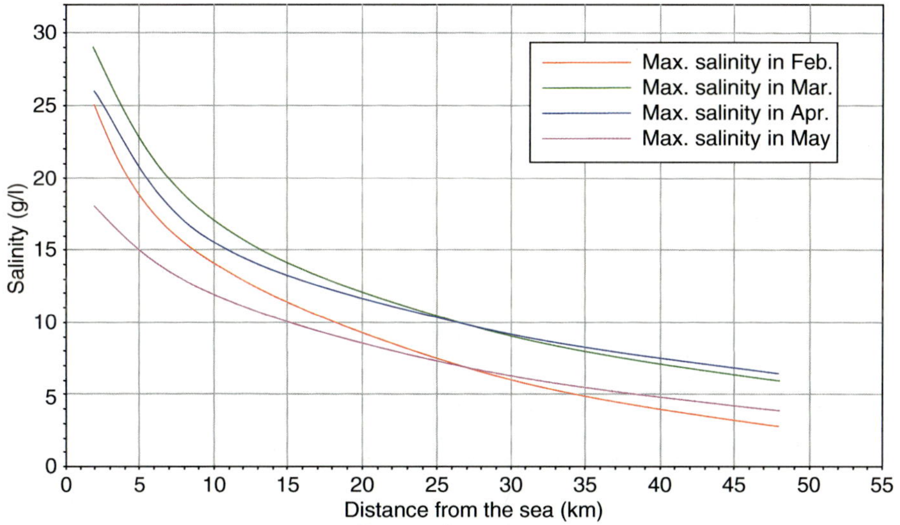


**Fig. 3.10** The distribution of salinity concentration in the Co Chien River mouth (Tri et al. 2006)

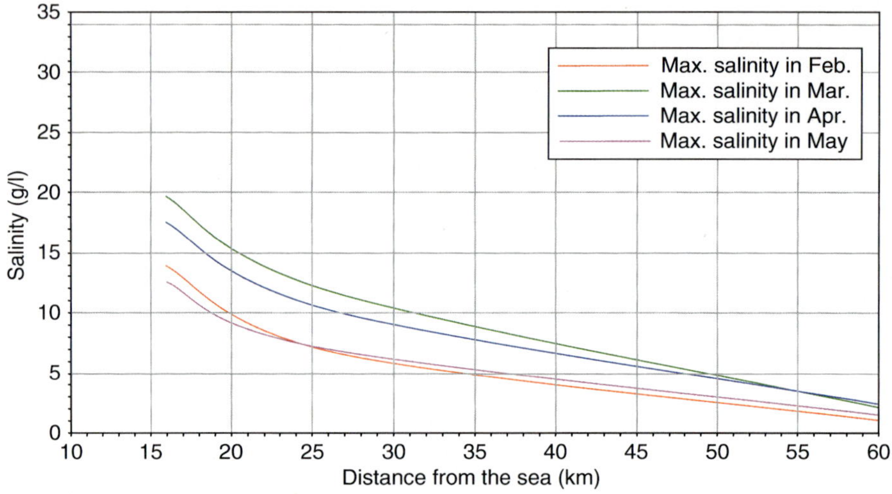


**Fig. 3.11** The distribution of salinity concentration in the Ham Luong River mouth (Tri et al. 2006)

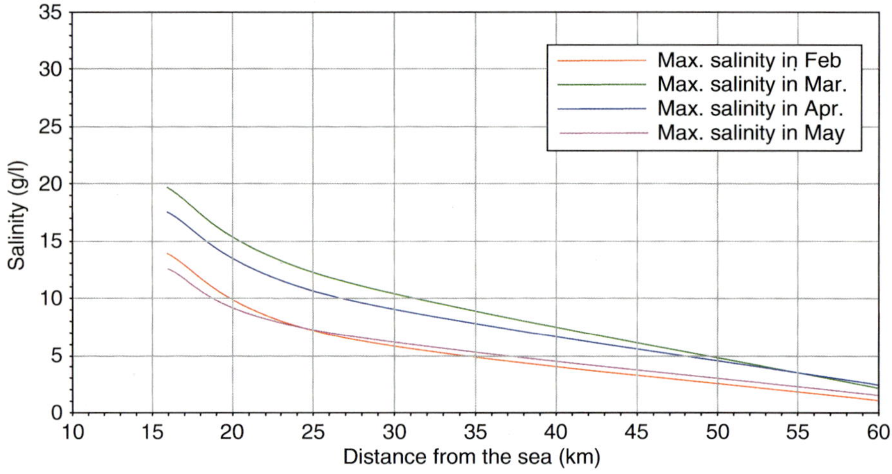


**Fig. 3.12** The distribution of salinity concentration in the Bassac River mouth (Tri et al. 2006)

## 3.5 Erosion and Sedimentation in the MD

In recent years, serious erosion has occurred along the Mekong River banks and estuaries. This has affected the socio-economic development and the security of the population in the region. The MD has alluvial soils and soft subsoil, so severe erosion has occurred on some river banks (Fig. 3.14).

The geological profile characteristics along the banks of the Mekong and Bassac rivers were described by Hung and Ngoc (2004). Based on geological profiles, soil

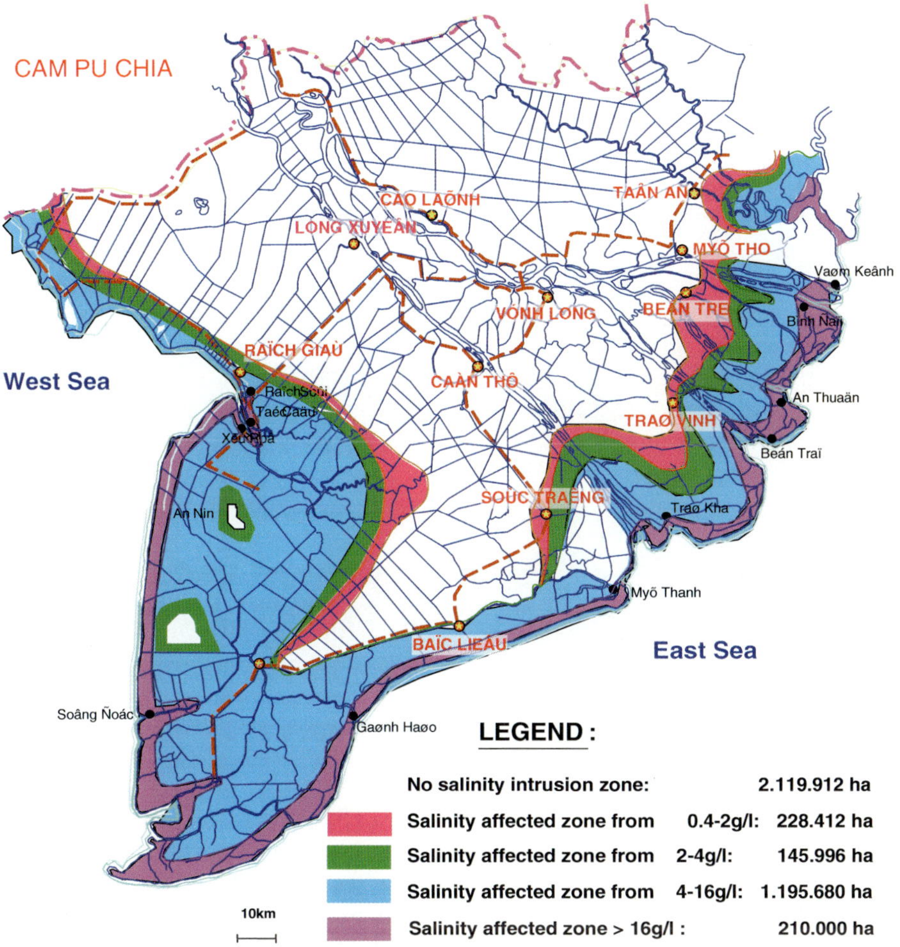


**Fig. 3.13** Areas affected by salinity in April in the MD (SIWRR 2010)

characteristics, the flow regime and the river bank geomorphology, researchers concluded that the erosion potential along the Mekong River banks is stronger than for the banks of the Bassac River (Hung and Phon 2004). The process of bank erosion mainly occurs during the flood season, in the high floods. Meanwhile the wet cross-sections of the Mekong River mouths have become wider, and at the same time, winds in the estuaries have become stronger, which has led to high waves that cause the collapse of the river banks. In addition, the exploitation of sand for construction materials in the Mekong and Bassac River has contributed to river bank erosion. The high density of ships in the Mekong River is also one of the causes of river bank erosion.

Bank erosion rates of some sites along the Mekong and Bassac River have accelerated (Table 3.6). The typical erosion and deposition patterns for the Mekong

**Fig. 3.14** River bank erosion in the Tan Chau District (Hung and Ngoc 2004)

River (in Thuong Phuoc) is shown in Fig. 3.15. The erosion rate at Thuong Phuoc – Thuong Thoi Tien on the left bank of the Mekong River is very severe, with an average movement of 34.7 m per year. The maximum width of river bank erosion from 1965 to 2000 has reached 1,250 m. There are seven cross-sections (MC1 to MC7) that have been established to measure the rate of river bank erosion from 1965 to 2000. These measurements have been made by the Southern Institute of Water Resources Research (SIWRR).

Based on the Mekong River banks erosion data recorded by SIWRR, there are several locations prone to erosion along the Mekong and Bassac Rivers. River bank erosion has seriously affected the socio-economic situation, agriculture, housing, infrastructure and human safety in the estuary areas.

## 3.6 Hydraulic Infrastructure Systems in the MD

In recent years, many large scale irrigation systems have been built across the Mekong Delta in Vietnam (Fig. 3.16, Table 3.7) aimed at rapid development of this fertile plain. The irrigation systems have contributed to the economic and social development of the MD. The Vietnamese Government has introduced many policies

**Table 3.6** The bank erosion rate of some sites along the Mekong and Bassac rivers (Data from 1966–2002) (Hung and Ngoc 2004)

| River | Erosion sites | Erosion length (km) | Max. erosion width to the bank (m) | Average erosion rate (m/year) | Movement rate of erosion center point (m/year) |
|---|---|---|---|---|---|
| Left bank of Mekong River | Thuong Phuoc – Thuong Thoi Tien | 6 | 1,250 | 34.7 | 33.3 |
| | Hong Ngu | 8 | 110 | 3.1 | 3.5 |
| | Lach market – Ben Tre | 4.5 | 400 | 11.1 | 3.2 |
| Right bank of Mekong River | My Luong – Long Dien | 4 | 120 | 3.3 | 6.8 |
| | Chau Thanh – Sa Dec – My Thuan | 10 | 1,200 | 33.3 | 38.1 |
| Vam Nao River | East My Hoi | 6.5 | 350 | 9.7 | 23.3 |
| Right bank of Bassac River | Khanh An – Khanh Binh | 3 | 300 | 8.3 | 2.8 |
| | An Chau – Long Xuyen | 2.6 | 100 | 2.8 | 3.1 |
| | Binh Thuy – Can Tho | 2.8 | 300 | 8.3 | 2.2 |

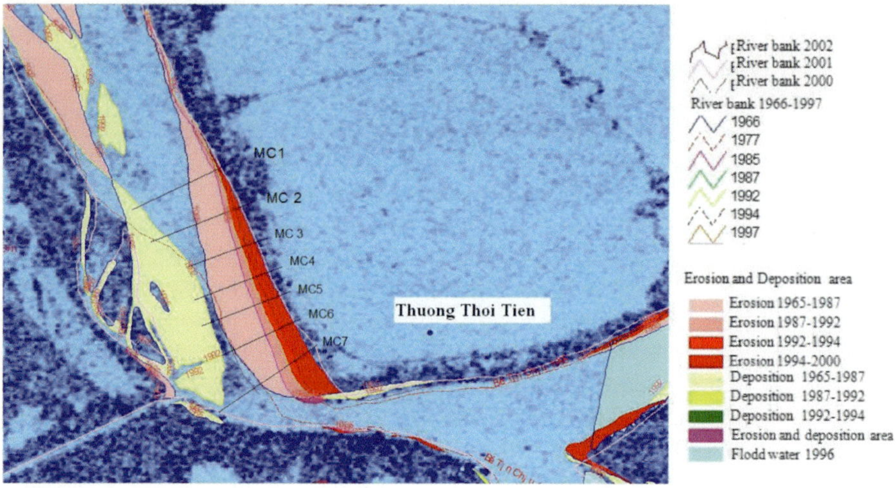


**Fig. 3.15** Typical erosion and deposition at Thuong Phuoc – Thuong Thoi Tien on the left bank of the Mekong River (Hung and Ngoc 2004)

to invest in and develop the MD, especially the agricultural sector (Table 3.7). These systems have performed very well for irrigation, and for flood and salinity control, and soil reclamation. The irrigation systems have become a driving force to promote the economic development of the MD, which is now one of the largest rice exporter in the world.

However, they have also generated various problems, whose negative effects have become obvious such as water pollution and ecological changes, which may become an obstacle to the economic development and social sustainability of the MD in the future. Therefore, these systems need to be adapted to the diversity of water services entailed by the development of the MD.

Unlike other irrigation systems in low lands, the canal systems of the MD are fully interconnected, with no separation between irrigation and drainage systems. The operation of the canal system is quite minimal but some canals with tidal meeting points require regular de-silting in order to maintain their depth and profiles. These tasks are usually organized on the basis of administrative boundaries (at the provincial or district level) rather than on the basis of hydrological or hydraulic units. Almost all of these canals are of a multi-purpose type, providing irrigation or water uses from the rivers during the dry season, and discharging water towards the rivers and the sea during the wet season. These canals are also intensively used for navigation and transportation year round. Except for some hydraulic structures in the intake/outfall canals in the coastal regions, the system is almost completely devoid of control structures. Water management is predominantly exercised at the tertiary level, where groups of farmers have established small water management units covering an area of 50–200 ha each. These units typically have their own intake/outfall structures and are surrounded by small dikes which provide some

**Fig. 3.16** Large scale surface irrigation systems developed in the Mekong Delta during the period of 1995–2010 (Tri et al. 2005)

degree for flood protection. The type of intake/outfall structure used depends on the opportunities for gravity irrigation and drainage, but typically is a combination of a sluice/flood-gate and pump station. The flow rates and flow directions in the secondary and main canals are largely determined by the tidal movement and by the gradient generated by the combined water intake and disposal activities exercised by the tertiary water management units. Based on the functions, hydraulic infrastructure systems can be divided in various categories of following structures.

**Table 3.7** Large scale irrigation systems developed in the Mekong Delta during the period of 1995–2010 (Tri et al. 2005)

| No. | Name of the irrigation system | Location | Service area (ha) | Functions |
|---|---|---|---|---|
| 1 | Go Cong | Tien Giang | 54,000 | Fresh water supply, salinity control |
| 2 | Tiep Nhat | Soc Trang | 53,910 | Fresh water supply, salinity control |
| 3 | South Mang Thit | Vinh Long, Tra Vinh | 225,682 | Fresh water supply, salinity control |
| 4 | Quan Lo – Phung Hiep | Soc Trang, Bac Lieu | 178,888 | Fresh water supply, salinity control |
| 5 | O Mon – Xa No | Kien Giang, Can Tho | 45,430 | Fresh water supply, salinity control |
| 6 | Nhat Tao Tan Tru | Long An | 13,320 | Fresh water supply, salinity control |
| 7 | Ba Lai | Ben Tre | 50,800 | Fresh water supply, salinity control |
| 8 | Bac Dong canal | Long An, Tien Giang | 40,400 | Soil reclamation, Flood control |
| 9 | Ba Rinh – Ta Liem | Soc Trang, Can Tho | 30,944 | Fresh water supply, salinity control |
| 10 | Hong Ngu canal | Dong Thap | 21,754 | Fresh water supply, Flood control |
| 11 | Huong My | Ben Tre | 17,000 | Fresh water supply, salinity control |
| 12 | Ba The – Tri Ton | An Giang, Kien Giang | 43,700 | Soil reclamation, Flood control |
| 13 | Can Tho – Long My | Can Tho | 50,000 | Fresh water supply, Flood control |
| 14 | Cai San – Thot Not | Can Tho, Kien Giang | 58,000 | Fresh water supply, Flood control |
| 15 | Ke Sach | Soc Trang, Can Tho | 32,000 | Fresh water supply, salinity control |

### 3.6.1 Structures for Irrigation and Water Supply

The canal network of the MD includes over 15,000 km of primary canals, nearly 27,000 km of secondary canals, 50,000 tertiary canals, 80 large culverts (e.g. the Lang The gate of 100 m, Ba Lai gate of 84 m); 800 medium size culverts, 50,000 small structures, over 1,000 large and medium electric pumping stations, and thousands of small pumps for active irrigation (Table 3.8 and Fig. 3.17) (SIWRP 2005).

### 3.6.2 Structures for Flood Control

In order to control flooding of the MD, systems of embankments and dykes have been constructed in the Plain of Reeds and Long Xuyen quadrangles with a total length of about 13,000 km, including 7,000 km of embankments for flood protection in the Summer-Autumn rice season. Also, over 200 km of dykes have been built to protect the national conservation parks and mangrove forests in Dong Thap and Kien Giang provinces against fire (Tran Nhu Hoi et al. 2005).

### 3.6.3 Structures for Controlling Salinity Intrusion

In the coastal regions of the MD, 450 km of sea dykes have been built as well as 1,290 km of river dykes. In addition, there are about 7,000 km of embankments along the primary canals to prevent salinity intrusion, high tide and wave surges. Almost all irrigation systems in the coastal zone have been closed with intakes/off takes to control saline intrusion for agriculture and aquaculture, except when the canal systems are opened and linked together in the Ca Mau Peninsula for both irrigation and navigation. This region systematically sees conflicts in water use between agriculture and aquaculture (MRC 2005).

### 3.6.4 Structures of Water Supply

The water supply for cities and urban areas is also an important function of the irrigation system in the MD. Almost all towns and large cities are supplied with processed clean water even though sometimes there may not be enough water because of saline intrusion. Meanwhile, only 40% of the rural population gets their clean water from treatment plants; the others have to use water directly from canals. The Vietnamese Government has planned to construct three large water supply plants for inter-provinces (Table 3.9) and 11 water supply plants for cities and peri-urban areas in the MD in the near future (Decision no. 2065 2005).

**Table 3.8** Statistics of the main hydraulic works in the Mekong Delta (SIWRP 2005)

| | | Whole MD | | Plain of Reeds | | Long Xuyen Quadrangle | | Ca Mau Peninsula | | Trans Bassac | |
|---|---|---|---|---|---|---|---|---|---|---|---|
| No | Structure | Project | L (km) | Project | L (km) | Project | L (km) | Project | L (km) | Project | L (km) |
| 1 | Main Canal | 133 | 3,190 | 45 | 1,068 | 64 | 1,056 | 36 | 633 | 32 | 1,039 |
| 2 | Canal-Level I | 1,015 | 10,961 | 343 | 3,116 | | | 428 | 5,294 | 200 | 1,945 |
| 3 | Canal-Level II | 7,656 | 26,894 | 2,187 | 6,742 | 2,313 | 7,374 | 3,297 | 13,689 | 1,072 | 3,363 |
| 4 | Canal-Level III | 36,853 | 50,019 | 3,400 | 7,200 | | | 7,467 | 16,692 | 24,773 | 21,853 |
| 5 | Large Sluice | 984 | – | 169 | – | 38 | – | 322 | – | 455 | – |
| 6 | Small Sluice | 20,517 | – | 2,491 | – | 1,915 | – | 6,000 | – | 10,111 | – |
| 7 | Flood control | – | 13,332 | – | 7,099 | – | 4,485 | – | – | – | 1,748 |
| 8 | River dyke | – | 281 | – | – | – | – | – | – | – | 281 |
| 9 | Sea dyke | – | 523 | – | 21 | – | 63 | – | 306 | – | 133 |
| 10 | Pumping station | 1,151 | – | 338 | – | 319 | – | – | – | 494 | – |

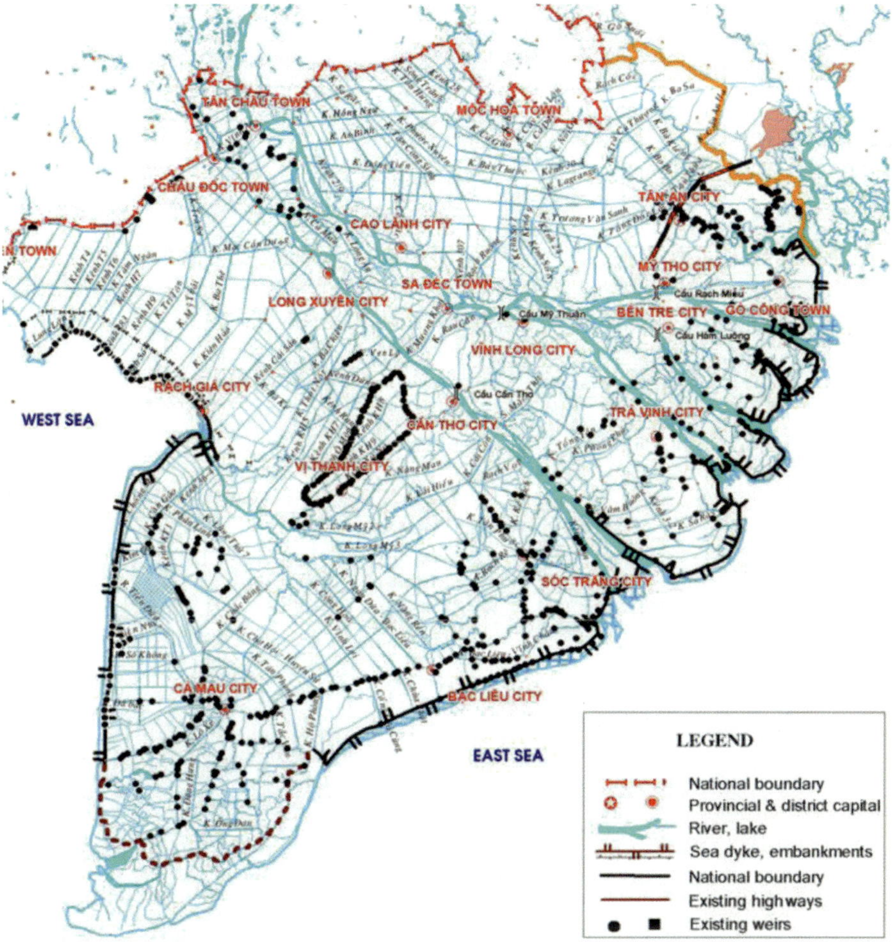


**Fig. 3.17** Hydraulic infrastructures in the Mekong Delta (SIWRP 2005)

## 3.7 Conclusions

Flood and saline water have caused severe damage in terms of production and infrastructure in an area of 1.2–1.8 million ha in the MD. Recently, the MD has had to face other challenges, namely climate change (Chap. 2) and the development of water uses in the upstream countries. These will have a strong effect on hydrological regimes of the river. Floods and drought have become more difficult to control in the MD, so it is necessary to find innovative solutions. One recent idea was to suggest closing river mouths to keep some fresh water during the flood season (Marchand et al. 2011). Careful consideration should be given to whether this would work or not. The hydraulic infrastructure is very important for the development of the MD now and in the future.

**Table 3.9** Water supply planning of the key economic zones in Mekong Delta in 2020 (Decision no. 2065, 2010)

| No. | Project | Location | Intended capacity ($m^3$/day) | Water sources | Service scope |
|---|---|---|---|---|---|
| 1 | Water supply plant Hau River I | Tan Thanh District, Can Tho City | Phase 1: 500,000<br>Phase 2: 1,000,000 | Hau River | Can Tho City, Soc Trang, corridor of Hau River and the north of Hau River (Ben Tre, Tra Vinh) |
| 2 | Water supply plant Hau River II | Chau Thanh District, An Giang province | Phase 1: 1,000,000<br>Phase 2: 2,000,000 | Hau River | Key economic zones (An Giang, Kien Giang, Ca Mau) and a part of Hau Giang, Bac Lieu provinces |
| 3 | Water supply plant Hau River III | Chau Doc District, An Giang province | Phase 1: 200,000<br>Phase 2: 300,000 | Hau River | An Giang, Kien Giang and urbans along the south-west boundary |

# References

Decision no. 2065/QD-TTg dated 12 Nov, 2010 (2010). Approving the water supply planning of the key economic zones in Mekong Delta in 2020, issued by the Prime Minister

Giam NM (2010) Suggestion of hydrological network for Mekong Delta. WISDOM project

Hung LM, Ngoc DTB (2004) Emperical formula to calculate the riverbank sliding velocity in the lower Mekong River at Thuong Phuoc, Dong Thap province. SIWRR J 2004:231–239

Hung LM, Phon HT (2004) Erosion and sedimentation in the lower Mekong River system and the protected measures. SIWRR J 2004:249–258

Lanh DT et al (2005) Study on integrated water resources utility and environmental flow aims to protect the ecology and the sustainable development of the Mekong Delta (in Vietnamese). WRSI Danida project

Marchand M, Dam R, Bucx T (2011) Towards a Mekong Delta plan. Synthesis of water sector assessment. Prepared by Deltares in commission of the Delta Coordinator of Directorate General Water in the framework of the Mekong Delta Programme. http://www.partnersvoorwater.nl/wp-content/uploads/2011/09/Annex5_mekongdelta-assesments_synthesis.pdf. Accessed 14 Nov 2011

MRC [Mekong River Commission] (2005) Overview of the hydrology of the Mekong Basin. Mekong River Commission, Vientiane, 73 pp

Nhan DK, Be NV, Trung NH (2008) Chapter 4: water use and competition in Mekong Delta Vietnam. In: Be TT, Sinh BT, Miller F (eds) Challenges to sustainable Development in the Mekong Delta: Regional and National Policy Issues and Research Needs. The Sustainable Mekong Research Network, Bangkok.

Sam L et al (2004) Study on the salinity intrusion to serve the economic – social development in the coastal zone of the Mekong Delta. The national program of the environmental protection and the natural disaster prevention, code KC.08–18

Sam L et al (2008) The salinity forecast background in the main rivers in the dry season 2008 of the coastal areas in the Mekong Delta. SIWRR J 2008:64–77

SIWRP (2005) An "Integrated planning of irrigation of the Mekong Delta", Southern Institute for Water Resources Planning

SIWRR (2010) The report on the monitoring of the Mekong Delta 2010. The program of the fundamental survey of 2010

Tran Nhu Hoi et al (2005) Study and suggestion on the scientific strategies of dyke structure for the sustainable development in the flooding area of Mekong Delta. SIWRR J 2005:319–327

Tri VK et al (2005) Study on the strategies of the management and operation of the surface water large scale irrigation systems to conserve and use effectively water resources in the Mekong Delta. SIWRR J 2005:202–210

Tri VK, Nghia PD, Kiem LV (2006) Forecasting the influences of upstream flow on the threshold of salinity intrusion in the Mekong Delta, Vietnam. In: Proceedings of the second international symposium on sustainable development in the Mekong River Basin. Phnom Penh, 16–18 Sept 2006

Viet NT, Tanaka H, Tri VK, Thang TD (2005) In-situ investigation of water quality in Hau River estuary. In: Proceedings of the second international symposium on sustainable development in the Mekong River Basin. Ho Chi Minh City, 6–7 Oct 2005

# Chapter 4
# Socio-Economic Development in the Mekong Delta: Between the Prospects for Progress and the Realms of Reality

**Matthias Garschagen, Javier Revilla Diez, Dang Kieu Nhan, and Frauke Kraas**

**Abstract** Socio-economic development in the Vietnamese Mekong Delta is shaped by a complex web of interacting and dynamic trends. Based on the analysis of statistical data, special reports, planning documents and scientific literature, the chapter examines the key dimensions of such trends, paying particular attention to agricultural transformation, industrialization, migration and urbanization. It is argued that changes in these fields have been producing ambiguous economic net-effects and socially stratified development outcomes over the last decades. On the one hand, the agricultural sector in the Mekong Delta has been experiencing profound production gains due to de-collectivization, expansion, intensification and diversification. This has contributed to overall poverty reduction in the Delta and to the economic progress of the entire country. On the other hand, the Mekong Delta lags behind the national average in terms of many development indicators in the socio-economic sphere (e.g. education levels or housing conditions). Under stress from multiple economic and environmental pressures and risks, small-scale farmers

M. Garschagen (✉)
United Nations University - Institute for Environment and Human Security (UNU-EHS), UN Campus, Hermann-Ehlers-Str. 10, Bonn 53113, Germany
e-mail: garschagen@ehs.unu.edu

J.R. Diez
Institut für Wirtschafts- und Kulturgeographie, Leibniz Universität Hannover, Schneiderberg 50, 30167 Hannover, Germany
e-mail: diez@wigeo.uni-hannover.de

D.K. Nhan
Mekong Delta Development Research Institute (MDI), Can Tho University, Can Tho City, Vietnam
e-mail: dknhan@ctu.edu.vn

F. Kraas
Geographisches Institut, Universität zu Köln, Albertus-Magnus-Platz, 50923 Köln, Germany
e-mail: f.kraas@uni-koeln.de

F.G. Renaud and C. Kuenzer (eds.), *The Mekong Delta System: Interdisciplinary Analyses of a River Delta,* Springer Environmental Science and Engineering, DOI 10.1007/978-94-007-3962-8_4, increasingly have difficulties securing a minimum level of profitability and a stable livelihood base. Rising inequalities, high incidences of landlessness, and labour migration, notably into urban areas, are among the most significant consequences. At the same time, industrial development falls short of earlier expectations. The Delta's secondary and tertiary sectors are presently unable to sufficiently absorb the former agricultural labour force. As a result, strong outmigration occurs, most importantly to Ho Chi Minh City and its neighbouring provinces. Guided by development theory we argue that next to the neoclassical expansion of conventional capital stocks for fostering endogenous growth potentials, development in the Mekong Delta heavily depends on institutional factors, enabling social and economic development. Aspects such as the need for improved access to land or for extended education and professional training, more integrated planning, and intensified promotion of economic innovations are discussed in detail.

## 4.1 Introduction

Discussing socio-economic development in the Mekong Delta is important not only in its own right but also due to the wider economic and demographic importance of the Delta for Vietnam as a whole. Nearly one quarter of the Vietnamese population lives in this region where 18% of the national GDP, 90% of the rice exports and 73% of the country's farmed aquatic products are generated (GSO 2011). But, economic and social development lags behind the high expectations fuelled at the beginning of the transformation process. Just after the introduction of *Doi Moi* the agricultural sector of the Delta performed very dynamically. Within a few years, the country had started to export rice, mainly from the Mekong Delta, which contributed heavily to the overall economic recovery and progress of the country. In that time, one quarter of the Vietnamese GDP was generated in the Mekong Delta. However, since then the importance of the Mekong Delta for the overall national GDP generation has diminished and the structural change proceeded less profoundly than envisaged. In terms of GDP contribution, industry and services succeeded to increase in importance, yet the majority of employees are still found in the agricultural sector. The last labour force survey shows that in 2009, 51% of the employed work force were involved in the agricultural sector (as compared to 47.6% for Vietnam) and only 12.2% in industry (14.5%) (GSO 2010b).

Similarly to the development trajectories of many other countries, particularly in South East Asia, growth and profitability gains in the Delta's agricultural sector are not keeping pace with the industry and service sectors. Albeit the fact that agriculture is more productive than in many other parts of Vietnam and will continue to play a central role for its economy, there is the need to improve agricultural productivity and to diversify economic activities away from agriculture in order to allow for sustainable livelihood foundations and economic gains at a larger scale. One might expect that in the long run agriculture will be intensively intertwined with higher quality food-related industries and services. Without such productivity gains and structural change processes in the Mekong Delta, the overall wellbeing of the large and on average young population is at risk of falling behind many other regions which succeed in climbing up the ladder of value-adding and economic progress.

One of the central contradictions of the socio-economic development in the Mekong Delta is that even though the Delta has achieved a relatively high GDP per capita, it is lagging behind Vietnam's other regions in important socio-economic aspects like education, qualification of the work force and poverty reduction (see next section). Against this background the Prime Minster has issued the establishment of a steering committee devoted to developing new ideas for strengthening the Delta's economic and social performance. The steering committee belongs to the party central committee and is supported by the Central Institute for Economic Management (CIEM). This institute was assigned by the Prime Minister to design a regional development plan. In a first step, the assessment of development potentials detected the weaknesses and strengths of the Mekong Delta. Bottlenecks are in particular seen with regards to transport infrastructure, education and strong regulation of the agricultural sector. Important links connecting the Mekong Delta with other regions outside the Delta but also within the Delta itself are found to be insufficient or in bad shape. Furthermore, improved hydraulic infrastructure is seen as being critical for the future progress of the region. The low educational level of the Delta's population poses an additional challenge. The share of adolescents attending a secondary school is fairly small compared to other Vietnamese regions. The figures are particularly low among ethnic minorities, notably among the 1.26 million Khmer living in the Delta. Furthermore the share of qualified and trained work force falls behind the actual needs of the region.

In addition, institutional barriers hamper the progress in agricultural productivity and profitability. The Vietnamese government still maintains a relatively tight control on agricultural production (notably rice), following the deeply rooted paradigm of food security.[1] Farmers often face external institutional pressures constraining their ability to diversify their agricultural production. Amongst the most important factors causing high levels of insecurity and economic risk are lack of land tenure, limited access to capital, bounded knowledge, poor market information, limited direct access to markets, price pressures resulting from increased international integration, limited storage capacities, slim financial buffers, and price dependencies in terms of input factors, middlemen and timing of sales. Despite the

[1] In this context, it is worthwile pointing to the different interpretations or dimension of food security. The Vietnamese policy thinking predominantly revolves around national self-sufficiency in terms of rice and other food staples. However, understanding food security more in terms of individual or socially stratified entitlements and access to sufficient levels of nutrition, it needs to be noted that Vietnam has been experiencing food poverty and nutrition challenges, notably amongst children and marginal population groups, even after having turned into one of the world's leading rice exporters. Hence, food security can be framed in many different ways and may in fact be understood as a highly contested concept in the context of Vietnam and the Mekong Delta in particular.

fact that agricultural land plots in the Mekong Delta belong to the largest in the country and although mechanisation has reached higher levels than in other regions of Vietnam, the Delta's productivity and profitability, therefore, lags behind other leading agricultural regions in Asia. Development potentials, on the other hand, are seen in the favourable natural conditions for agricultural production (mainly rice, fruits and aquaculture), the large number of young people entering the labour market, and the entrepreneurial drive of the local governments.

In the remaining part of this chapter, the aim is to reflect on these potentials and challenges indentified by the preliminary analysis of the newly established steering committee and to discuss the factors regulating socio-economic development in the Delta in more detail.

Theoretical discourses on the drivers of economic and social development can provide guidance for such analysis. During the last couple of years, two strands of argumentation have been highly debated in the literature.

On the one hand, the argumentation is based on further developments of the so called mainstream economics. Empirical studies of developing countries have shown that regional disparities persist as predicted in the simple Centre-Periphery Model developed by Krugman (1993). As Myrdal (1957) and Hirschman (1958) pointed out, regional disparities are furthered by vicious circles, i.e. circular and cumulative causation leading to low development. Based on Krugman's ideas, regional economic development tends to be unbalanced due to the strong economic advantages of concentration. The recent World Development Report "Reshaping Economic Geographies" concludes that "cities, migration, and trade have been the main catalysts of progress in the developed world over the past two centuries" (World Bank 2009). Under these circumstances the development perspectives of rural regions are heavily depending on endogenous growth potentials. In this context, the exploration of the spatial dimension of rural non-agricultural employment has been identified as a priority for future research (Haggbladc ct al. 2007).

On the other hand, institutional argumentations have gained considerable importance. Based on this alternative understanding of regional disparities the focus of policy shifted from expanding the capital stock in the neoclassical model towards an institutional approach. More and more economic theorists are incorporating the role of institutions within the economic theory of growth. According to Haggard (1999, p. 30) "institutions refer to the formal and informal rules and enforcement mechanism that influence the behaviour of organizations and individuals in society. They include constitutions, laws and regulations, and contracts, as well as trust, informal rules and social norms". The creation of opportunities for rural households is closely related to institutional issues like property rights, contract enforcement, private entitlement to land, etc. Being able to utilise land use rights as collateral, for instance, is crucial for inducing capital that can be utilised for stepping up productivity, particularly in the rural areas of Vietnam. Duncan and Pollard (2002) argue that if basic institutions for the creation of capital and full participation of the whole society in economic activity are not in place, neither investments in infrastructure, education, health, nor economic reforms or public sector reforms will be effective. Building roads and bridges or undertaking agricultural research will, for instance, not increase incomes as much as they could if people do not have secure property rights to farm lands. Education and health improvement projects will not promote income growth for the poor unless there is the generation of capital with which the healthier and better-educated labour force can work.

Against this background, the following sections explore the specific potentials and challenges for socio-economic development in the Mekong Delta. The analysis aims for an integrated multi-temporal perspective, by linking path-dependencies resulting from past developments with present conditions and potential future dynamics. Key trends of transformation in the Delta are examined and their institutional dimensions explored. Based on this analysis, the Delta's main drivers of development are identified and their interrelations as well as temporal and spatial dynamics investigated. The chapter concludes by embedding the discussion into the wider national and regional context. This allows assessing not only the development opportunities and challenges of the Delta – as being nested within a wider socio-economic and political fabric – but also its influence on Vietnam's transformation and development path as a whole.

## 4.2 Taking Stock – Key Trends and Current Situation in the Delta

The Mekong Delta plays a key role not only within the wider national and regional economic context (notably foodstuff exports[2]) but also within the demographic fabric of Vietnam. Hosting 17.2 million people, the Delta contributes more than 20% of the country's total population. This makes it an even larger contributor than the entire Greater South-Eastern Region which includes 14 million people in Ho Chi Minh City and its five neighbouring provinces including, for example, Dong Nai and Binh Duong. With its 40,500 km$^2$, the delta has a comparatively high population density of 425 inhabitants per square kilometre – compared to a national average of 260 (GSO 2011).

Yet, the Mekong Delta belongs to the regions with the lowest population growth in Vietnam. Since the year 2000, the annual growth rate has remained below 1%, dropping to 0.5% between 2008 and 2009 (GSO 2011). Only the Northern Midlands and Mountain Areas feature lower growth rates while the national average decreased from 1.4% to 1.1% over the last 10 years (GSO 2011). Within the Delta, Can Tho, Kien Giang, Bac Lieu and Long An have been experiencing the highest population increase, while the growth in other provinces almost stagnated or was even negative in the case of Ben Tre, mainly due to out-migration (see section on migration of this chapter and Fig. 4.11).

[2]In 2010, Vietnam for instance exported rice worth of 947 million USD to the Philippines, 346 million USD to Indonesia, 228 million USD to Singapore and 178 million USD Malaysia with the majority of which being produced in the Mekong Delta. On a similar note, fruits worth of 75 million USD have been exported to China in that year, cereal-related products worth 47 million USD to Cambodia, and seafood products worth 894 million USD to Japan, for example (GSO 2011).

Also called the rice bowl of Asia, the Mekong Delta is Vietnam's region with the highest percentage of agricultural land use, accounting for almost 65% of the deltas area. It hosts around 52% of Vietnam's entire paddy production area on which 53% of the national paddy is grown (GSO 2011). Between 1995 and 2009, the delta has increased its paddy production by almost 60% and – by doing so – was the main contributing region for supporting Vietnam's shift from a rice-importing nation to one of the largest rice exporters in the world – with the shift having taken place around 1989. These figures illustrate why the delta has been ascribed the role of the national guarantor for food-security – not only in the perception and rhetoric of the country's political leaders but also in actual policy making and regional development planning.

The primacy of the Mekong Delta in terms of Vietnam's food-production is even more obvious with respect to aquatic products, notably fish and shrimps. In 2009, it contributed around three quarters to the national production of both farmed fish and shrimps (GSO 2011). In combination with strongly increasing fruit and vegetable production, the delta plays a central role not only for the national food supply but also for the country's food-related export revenues and, hence, foreign trade balance.

The agricultural sector's importance for the Mekong Delta is also mirrored in the regional GDP profile and the sector-wise distribution of labor force in the Delta. On the national Vietnamese scale, the agricultural sector – including agriculture, forestry and fisheries – contributed only 21% to the overall GDP in 2009, with 40% contributed by industry and construction and the remaining 39% by the service sector (GSO 2011). Yet, in the Mekong Delta, agricultural production contributes 38% of GDP (GSO 2011). In parallel, a comparatively high share of the Delta's population is working in the agricultural sector. 51% of the work force had their main job within agriculture in 2009, compared to 17% in industry and construction and 32% in trade and services. The national average was at 48%, 22%, and 30%, respectively (GSO 2010b). Yet, the fact that around 75% of the workforce is involved in agricultural activities as their secondary occupation indicates the importance of additional and temporary work in agriculture to complement livelihoods (Fig. 4.1).

Judged on the basis of official income statistics, expenditure assessments, and poverty indices, the Mekong Delta could be considered reasonably successful in terms of socio-economic development. Official statistics suggests that in 2008 the monthly average income per capita was at 940,000 VND and the average monthly expenditure per capita at 624,000 VND (GSO 2011). This is below the national average (which is at 995,000 and 705,000 VND respectively), slightly lower than the levels in the Red River Delta and notably below the average levels of the Southeast (which are about twice as high) (ibid.). However, the Mekong Delta does considerably better than the remaining regions in Vietnam which feature lower levels, with up to roughly a one-third-difference. On a similar note, the Mekong Delta has experienced a substantial reduction in the general poverty rate[3] from 23% in 2002 to

[3]The general poverty rate is based on the assessment of average monthly per capita expenditure. The relevant threshold is set by the General Statistics Office (GSO) and the World Bank and is adjusted annually. The following thresholds were set for the last years: 160,000 VND for 2002; 173,000 VND for 2004; 213,000 VND for 2006; and 280,000 VND for 2008 (GSO 2009: 9.2).

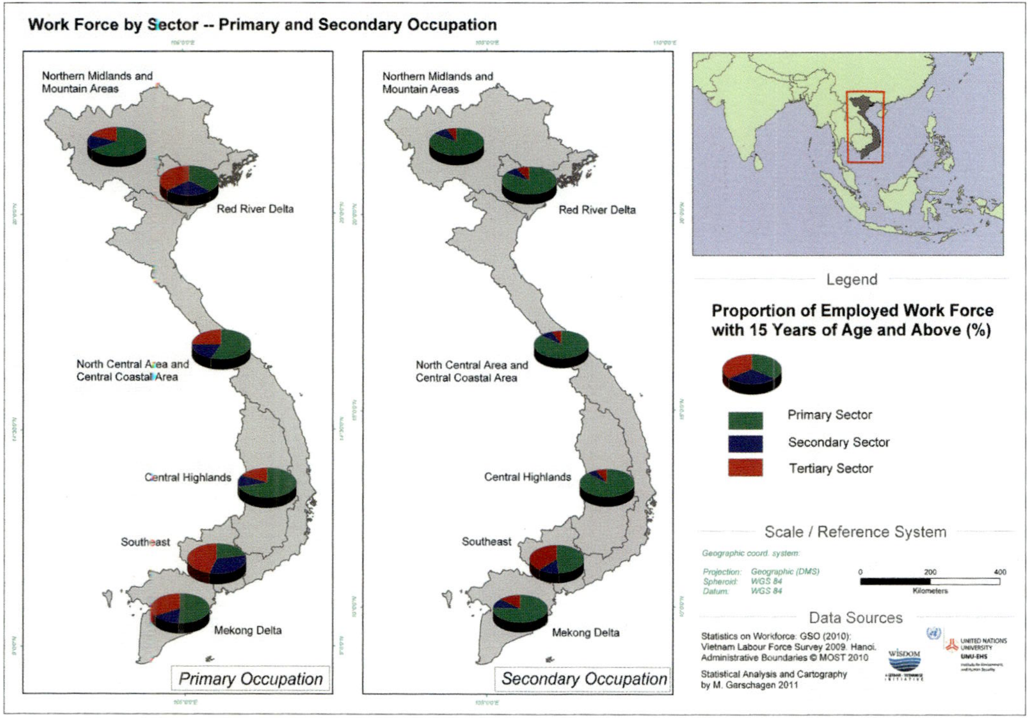


**Fig. 4.1** Workforce by sector in Vietnam and the Mekong Delta (Note: When comparing the maps throughout the chapter, caution needs to be given to the two different regional categorizations. While older statistical data in Vietnam was divided into eight regions (i.e. Red River Delta, North East, North West, North Central Coast, South Central Coast, Central Highlands, South East, and Mekong River Delta) younger data for the recent years is grouped according to the new categorization using six regions (i.e. Red River Delta, Northern Midland and Mountain Areas, North Central Area and Central Coastal Area, Central Highlands, South East, and Mekong River Delta). Combining the former administrative regions North East and North West into the Northern Midland and Mountain Areas results in good proxies. The same is true for combining formerly North Central Coast and South Central Coast into North Central Area and Central Coastal Areas. However, since the reforms of the region grouping overlays with changes at the level of provincial administrative boundaries, the figures do not match up completely)

12% in 2008 and it remains slightly below the national average (c.f. GSO 2009: 9.2; compare Fig. 4.2).

It needs to be noted though that there are substantial internal inequalities hidden behind these figures, most notably along the lines of ethnic minorities, a rural-urban gradient and land-title groups (see Luong 2003). The poverty rate amongst the 1.26 million Khmer people in the Delta – being by far the largest minority group – has decreased at a slower pace than amongst the Kinh majority and the ethnic Chinese and was around 20% above the national average around the turn of the century (Baulch et al. 2007). In the two strongholds Soc Trang and Tra Vinh province, where the Khmer population amounted to almost one third of the total population in 2002, the poverty rates where 55.1% (versus 30.7% among the rest of the population) and 45.6% (versus 27.4% among the rest of the population), respectively (UNDP and USAID 2004: 40).

In general, it should be mentioned that the validity and significance of such statistics on income, expenditure and poverty are controversially debated due to issues of data quality and particularly the high degree of informal or indirect generation of earnings and livelihood assets and their equally informal spending and trading. Both may be only partly captured by the calculations and over- or understated in the statistical survey. Such data is hard to collect and verify given the sensitive nature of the topic and multiple response effects. Furthermore, the general poverty thresholds may overestimate the actual purchasing power, particularly against the backdrop of the strong inflation and spatial disparities in price levels – despite the fact that the basket of goods and services considered for analysis is re-evaluated at the outset of most surveys. Poverty thresholds tend to create a dualistic impression and do not give information on the remaining vulnerabilities of the great number of people who have been pushed just above the poverty line, e.g. through mechanisms within the national programme 135[4] or the Comprehensive Poverty Reduction and Growth Strategy issued in 2002.

In addition, other data reveals that economic growth in the Mekong Delta has only partly translated into improved living conditions. In 2009, only 8% of the Delta's households were living in permanent houses, i.e. dwellings with the pier, the outer wall and the roof made of solid materials such as concrete, bricks or tiles. 22% of the households lived in houses with none of these elements made of solid materials, with the remaining 70% having houses being partly built of solid elements and partly of other materials (Fig. 4.3). These figures indicate that housing conditions in the Mekong Delta greatly lag behind the national average, according to which 47% of the Vietnamese households lived in fully solid houses in 2009 (GSO 2010a). This variation can to a certain extent be explained by the different climatic conditions in Vietnam's north and south as well as by the comparatively low typhoon occurrence in the Mekong Delta in the past, making solid houses less necessary than in the more cyclone-exposed areas in central and northern Vietnam. However, housing

[4] The national programme 135 was initiated in 1998 and focuses on poverty reduction amongst ethnic minorities and in remote rural areas.

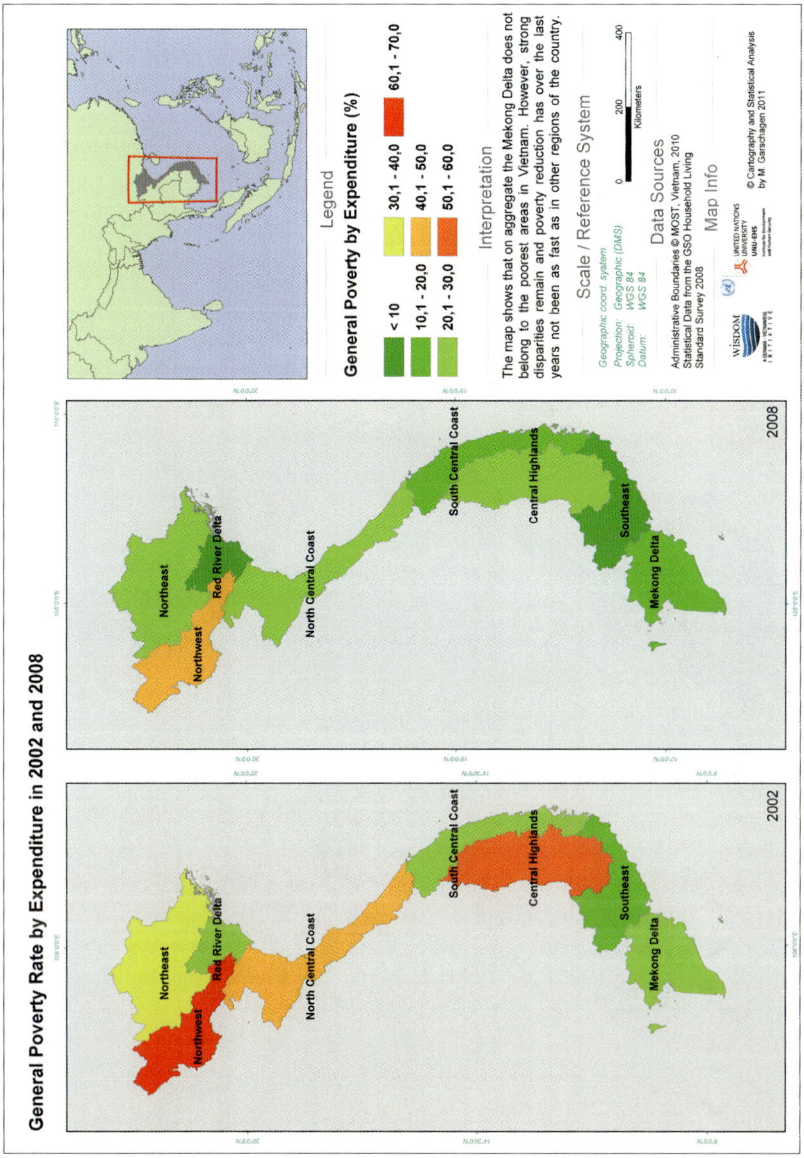


**Fig. 4.2** Poverty rate in Vietnam and the Mekong Delta (Source: own figure)

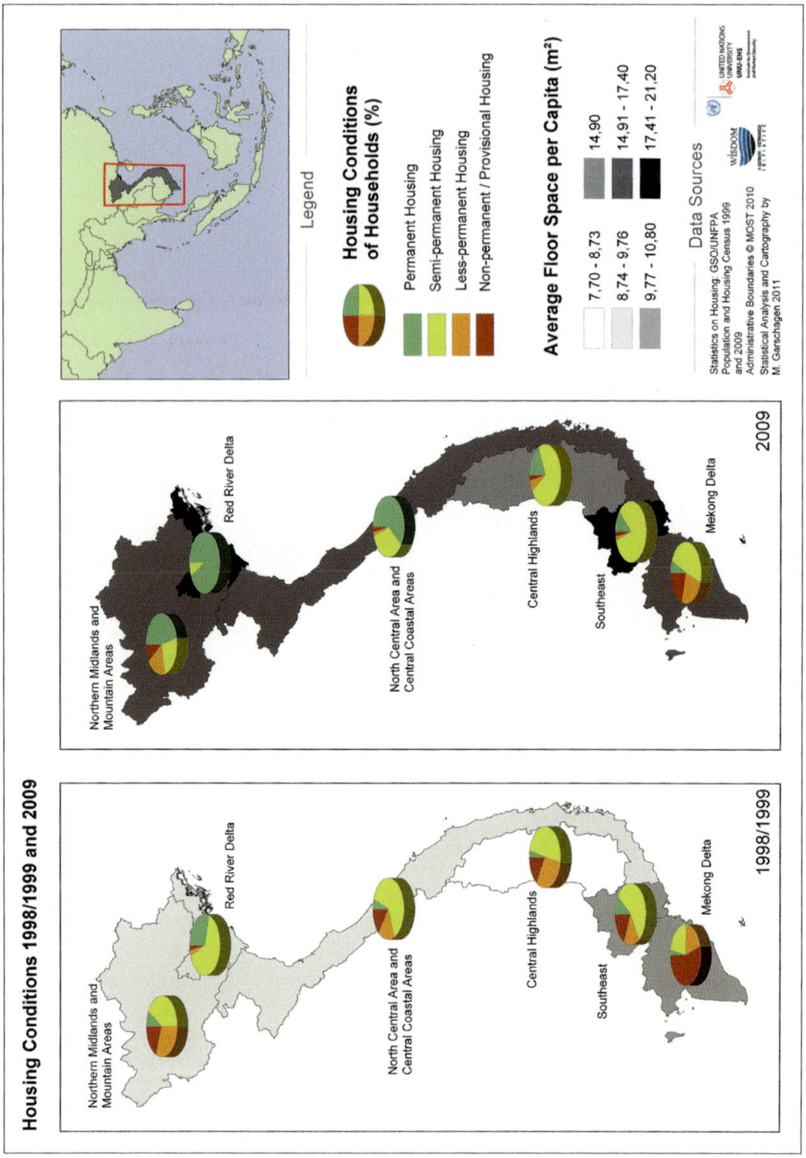


**Fig. 4.3** Housing conditions in Vietnam and the Mekong Delta 1999 and 2009 (Source: own figure)

**Table 4.1** Housing conditions in the Mekong Delta

| | Proportion of households by type of housing (%) | | | |
|---|---|---|---|---|
| | Solid house | Semi-solid house[a] | Less solid house[b] | Non-solid house |
| 2009 | | | | |
| Entire country | 46.7 | 38.2 | 7.8 | 7.4 |
| *Urban* | *41.4* | *52.7* | *3.3* | *2.6* |
| *Rural* | *49.0* | *31.7* | *9.8* | *9.5* |
| Mekong Delta | 8.3 | 48.5 | 21.4 | 21.8 |
| 1999 | | | | |
| Entire country | 12.8 | 50.4 | 14.1 | 22.7 |
| Mekong Delta | 3.3 | 22.3 | 17.9 | 56.5 |

Source: own draft based on GSO (2000a, 2010a: 162)

[a]Dwellings with two out of the three evaluated components (pier, outer walls, roof) being solid, i.e. being constructed of concrete, brick/stone, or steel/iron/durable wood for the pier and the outer walls, and concrete or tiles for the roof

[b]Dwellings with one out of the three evaluated components (pier, outer walls, roof) being solid

upgrades are foremost the manifestation of socio-economic progress. Hence, the fact that housing improvements in the Mekong Delta greatly fell behind the rest of the country between 1999 and 2009 (Table 4.1) hints towards the region's socio-economic development challenges.

On a similar note, the Mekong Delta lags behind other regions in terms of access to safe drinking water and sanitation infrastructure. With more than 22% of the Delta's households not having access to clean drinking water and 58% not having sanitised toilet facilities the Delta shows a significant difference to the national average (which in 2009 was at 13% and 46%, respectively) (GSO 2010a: 165). Detailed case study analysis suggests that the situation is even worse in reality than these figures may tell, particularly in the rural areas of the Mekong Delta (Reis 2012).

In addition, the figures representing medical infrastructure indicate significant health care deficits in the Delta. With 20.5 patient beds in medical facilities per 10,000 inhabitants, the Delta ranks last in comparison to the country's other regions (own calculations based on GSO 2011). Similarly, the number of doctors, nurses and midwives are also comparatively low (4.8, 5.0, and 2.6, respectively, per 10,000 inhabitants). The Mekong Delta lags far behind the national average in all these categories, and has the lowest regional values with respect to doctors and nurses (own calculations based on GSO 2011). Because of this, private providers of health care play an important role in the Mekong Delta. In 2008, 47% of the out-patient treatments were exercised in private health facilities, which is the highest value for the whole of Vietnam, the national average being at 38% (GSO 2009: 4.5).

Despite the fact that education in the Mekong Delta is often provided outside the formalised public education system (Taylor 2004), the Delta's poor educational performance is striking. According to a national survey in 2009, 6.6% of all inhabitants aged 5 or older had never attended school. This is a much higher share than in the regions of the Red River Delta (2.1%), the Southeast (3.0%), and the Central Coast

(1.3%) (own calculation based on GSO 2010a: 281). However, the Northern Midlands and Mountains as well as the Central Highlands experience even worse conditions (10.3% and 8.9%, respectively). Resulting from the low rates of school attendance, the Mekong Delta features a relatively high illiteracy rate amongst the population aged 15 and older, amounting to 8.4% according to a recent survey (GSO 2010a: 326). Yet, given that these figures represent average values, they hide the disparity between various social groups. In 1999, the primary school enrolment rate amongst Khmer was at 76.3% compared to 93.4% amongst the Kinh (Baulch et al. 2007: 1159). Secondary school enrolment of the Khmer was at 22.5% while the enrolment rate amongst the Kinh majority was about three times higher (Baulch et al. 2007: 1160). Additional inequalities exist in terms of gender. Out of the people identified as illiterates in 2009, 64% were female (GSO 2010a: 326). Along the same line, 62% of the above-mentioned people aged 5 or older without any school attendance were female (GSO 2010a: 281). Similar disparities can also be observed when comparing rural with urban areas. While 4.9% of the Delta's urban population aged 5 or older never attended school (GSO 2010a: 284), the figure for the rural areas amounts to 7.1% (GSO 2010a: 287). The rate of illiterates among the population aged 15 and older is 6% in the urban wards (GSO 2010a: 329) compared to 9% in rural communes of the Delta (GSO 2010a: 332).

The Delta is in an even more unfavourable situation when considering higher education and university attendance. Only 8.1% of its inhabitants born between 1987 and 1990 have ever attended university (own calculation based on GSO 2010a: 317). In the Red River Delta and the Southeast, the figures are 25%. Only the Northern Midlands and Mountains and the Central Highlands do worse than the Mekong Delta, with 5.7% and 7.0%, respectively (ibid.). Merely 9.7% of the Delta's economically active population aged 15 and older has ever completed a vocational or professional training within the formal educational system (with only 7.9% amongst women compared to 11.3% amongst men) (GSO 2010b: 11). Only 2.9% of the same cohort graduated from university (2.7% amongst women and 3.0% amongst the men) (ibid.). These values are the lowest in Vietnam with the national average being 17.6% in terms of professional training and 5.2% for university graduations (GSO 2010b: 11). Figure 4.4 depicts some major trends in education and training and relates them to unemployment and underemployment[5] levels in the Delta – as compared to other regions in Vietnam.

These observations for the Mekong Delta are backed up by the findings of other studies at national and sub-national scale. Epprecht et al. (2011) concluded that institutional factors are the main reasons hampering the successful participation of Vietnam's ethnic minorities in the mainstream economy – more important than, for example, the location of the communities. The authors, hence, argued that tackling these factors is more effective for reducing inequality and poverty (particularly in

[5]In the 2009 Labor Force Survey, underemployment refers to "persons aged 15 years and over who are working less than 35 h per week and are willing or available to engage in additional work." (GSO 2010b: II, 58).

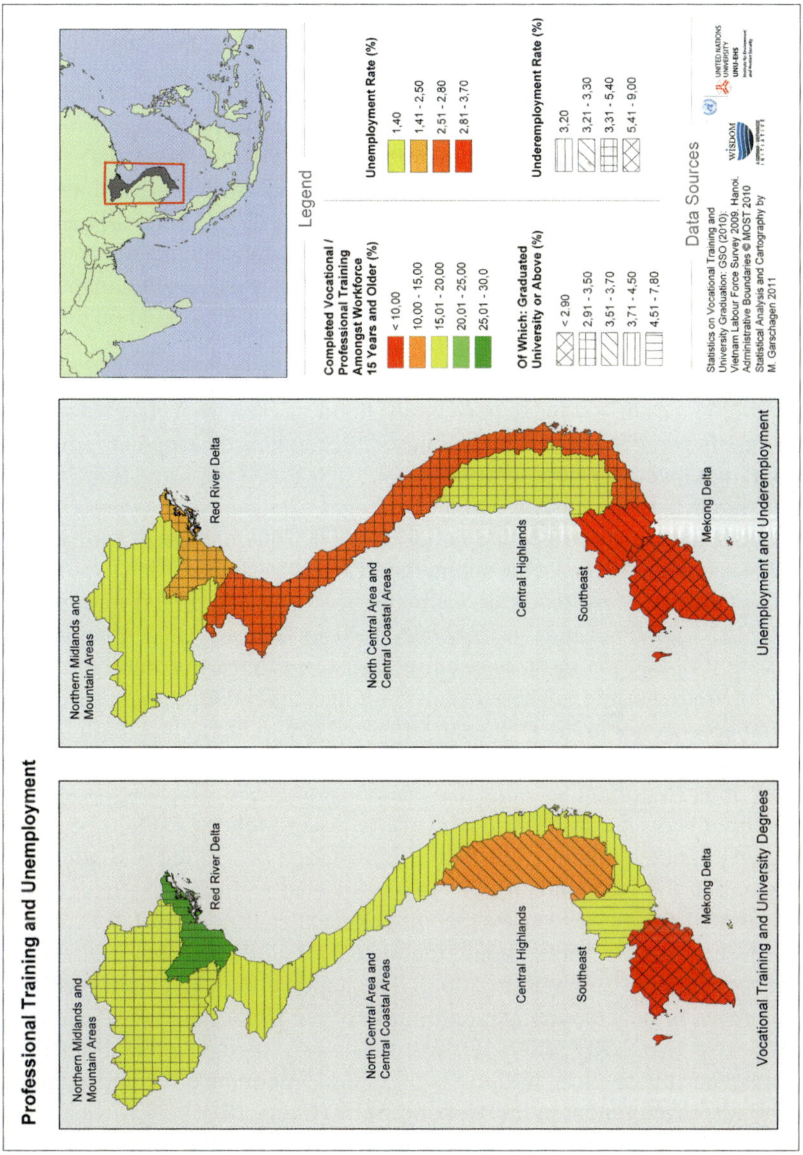


**Fig. 4.4** The link between professional training and underemployment in the formal sector (Source: own figure)

rural areas) than geographic targeting following the grid of administrative boundaries (ibid.). On a similar note, Taylor (2004) finds that certain conventional development interventions meant for reducing poverty and inequality in the Mekong Delta do in fact accentuate existing inequalities (notably between Kinh and Khmer). This is because these interventions are based on an institutional framework that plays into the hands of the economically more successful groups and does not account for localised patterns of economic, spatial, informational and cultural activity (ibid.). Extending roads into remote rural areas with high poverty rates, for example, has been observed to increase land prices along the road, in turn pushing former residents out of their land and increasing the profits of those actors who have sufficient resources and information to engage in land speculation. Such observations are in line with the findings by van de Walle and Gunewardena (2001) who have analysed the sources for ethnic inequality in northern Vietnam and concluded that poverty reduction programmes always need to be tailored towards the specific institutional problems in question and will not be effective if they are solely based on the mainstream economic model of the upper-end group.

On top of these challenges, the Mekong Delta is heavily exposed to natural hazards and climate change impacts (notably flooding, sea level rise, salinization and typhoons). In parallel, the deficiencies in socio-economic development contribute heavily to a high level of vulnerability to these hazards. At the same time, the risk of resulting disasters is amongst the prime challenges for sustainable socio-economic development. The severe consecutive floods of 2000, 2001 and 2002, for example, caused 1,044 casualties in the Delta, submerged 1.6 million houses, destroyed nearly 500,000 ha of rice and resulted in an estimated total damage of 300 million USD (SRV 2004: 15). Typhoon Linda in 1997 cost the lives of nearly 3,000 people in the Delta, making it one of the most devastating disasters related to natural hazards in the history of Vietnam (ibid.). These events have had a great impact on disaster risk management approaches in the Delta and triggered new policies including resettlement programmes, additional dyke systems and adaptation in agricultural production cycles. It is important to note that adapting to the difficult environmental conditions has always been a central part of life in the Mekong Delta. Local communities have, hence, gathered substantial knowledge on coping and adaptation mechanisms. Yet, the profound changes to be expected with climate change in the Delta may exceed current response capacities. Recent studies project, for instance, that a sea level rise of one meter would inundate 31% of the Mekong Delta, based on current topography (Carew-Reid 2008). The national climate change scenarios for Vietnam even project the inundation of 38% of the Delta's area for a one meter of sea level rise (MONRE 2009). Detailed Analysis of the vulnerabilities and response mechanisms to these hazards is of central importance for understanding socio-economic development in the Mekong Delta (Chap. 10).

On the back of this overview, the next section examines the four main factors regulating socio-economic development in the Mekong Delta. Particular attention is given to the theoretical perspectives introduced in the first section. The cumulating effects and internal growth potentials but also institutional framework conditions are assessed in detail.

## 4.3 Main Factors Regulating Socio-Economic Development in the Mekong Delta

The Mekong Delta is a highly complex system with a multiplicity of coupled social-ecological processes (e.g. Garschagen 2010a; Käkönen 2008) and with political, economic, cultural, and ecological dimensions interacting at various scales. Any attempt to ontologise these dimensions and their processes will therefore lead to incomplete or debatable results, providing only one selected frame out of the many possible. In accordance with our observation on the political and scientific discourses on development issues in the Mekong Delta as well as with the theoretical debate on economic development introduced above, we have decided to group the factors influencing socio-economic progress in the Delta into the following four fields: agricultural transformation, industrialisation, migration and urbanisation. They are in our eyes representing the most pressing and influential dimensions of socio-economic development.

### *4.3.1 Agricultural Transformation*

As already mentioned, agricultural production in the Delta has been contributing significantly to national food security as well as to national agricultural exports. The major land use patterns for agricultural production are primarily determined by soil and hydrological conditions. Over the last three decades, changes in agricultural land uses and agricultural development have been strongly influenced by a combination of drivers, including policy reforms, irrigation development, market-oriented development, technological advances, environmental changes, urbanisation and industrialisation. The following sections describe the impacts of these drivers on land use dynamics and the Delta's transformations with respect to rice and aquaculture production.

Land use changes and agricultural development can be divided into three major periods since 1975, that is, firstly, rice expansion (1975–1990), secondly rice intensification (1991–1999), and thirdly agricultural diversification (2000-present). The "*rice-first*" policy for subsistence coupled with substantial investments in irrigation development resulted in the expansion of rice cropping into natural floodplains, and acidic and saline areas during 1975–1990 (compare for more detailed reviews of the Delta's hydraulic infrastructure development Biggs et al. 2009; Evers and Benedikter 2009; Chap. 3). During this period, the irrigated area for rice increased on average by 85,000 ha annually (own calculation, based on GSO 2000c). Many farmers shifted from growing one crop of long-growth-duration traditional rice per year to two or three crops of high-yielding rice varieties. Consequently, the total cultivated area (i.e. rice land × crops) grew substantially during this period. Looking at the figures of specific crops, one can find that the annual growth rate of the dry-season crop was 13% for the cultivated area and 17% for production. The corresponding figures for the wet season crops are 9% and 11%, respectively, while those for traditional rice crop decreased by 3% and 1%, respectively (ibid.). These figures suggest rice intensification foremost as a result of irrigation development.

During 1991–1999, rice production experienced further intensification and increase in cultivated area and shifted from subsistence purposes to a clear cash-crop orientation. This was the result of a combination of factors: Firstly, there had been substantial policy reforms since the early 1980s, and notably since 1986, implying a shift from a central planning-orientated to a more market-oriented economy and from agricultural collectives to individual control over agricultural production. Secondly, the hydraulic infrastructure for drainage, irrigation and salinity control had been further extended through canals, dikes, pumping stations and sluice gates (Chap. 3). These interventions allowed farmers to continue the intensification of rice production through cultivating more rice crops per year (i.e. two or even three crops annually). Statistical data shows that rice cultivated area increased by 1.4 million ha while the surface area used for rice cultivation remained basically unchanged. Consequently, rice land use intensity (defined as the ratio of the rice cultivated area to land surface under rice cultivation) increased from 1.2 in 1990 to 1.9 in 1999. During this period, the annual growth rate in cultivated area and the corresponding production of rice was 7% and 8% respectively, for both dry season and wet season crops, while low-yielding traditional rice decreased by 5% annually in cultivated area and by 4% in terms of production (own calculation, based on GSO 2000c).

The expansion and intensification of high-yielding rice production contributed significantly to a big leap of rice production in Vietnam, turning the country into one of the leading rice exporters globally. However, during the following decade there have been great changes in agricultural land uses which were caused by reforms in policy on agricultural diversification since 2000. Recognizing the need for further improving rice farmers' incomes and the importance of aquaculture and fruit production for sustainable development of the agricultural sector, the Vietnamese government implemented the policy on agricultural diversification and sustainability (i.e. Decree no. 09/2000/NQ-CP dated 15th June 2000). Accordingly, many farmers have shifted their production from rice to aquaculture, fruits or vegetables or from rice mono-culture to a more diversified and integrated rice-based farming system with higher income generation possibilities (Nhan et al. 2007a, 2008a). In particular the focus on aquaculture was resulting from the experience that individual pioneers in this emerging sector had generated high profits, allowing them to increase their income compared to rice cultivation and to improve their livelihood base significantly. At the same time water and land uses for agriculture have evolved towards more adaptive resource management approaches (e.g. making use of flood and saline water) rather than resource control and exploitation (e.g. flood and salinity control) in the years before.

Despite these changes, some major restrictions remained with respect to the degree of diversification possible at farm level. The governmental agricultural land use plans for a 5 or 10 year terms envisage the main land use categories from commune, district to provincial level based upon pre-defined production targets of certain crops at each respective level. These plans set out the main land use categories (e.g. rice production, fruit tree orchards, aquaculture etc.) for a given spatial entity.

The farmers producing within this spatial unit are free to choose the detailed form of production under the dedicated category (i.e. they can for example decide which varieties of rice or which type of fruit trees they are planting). However, in theory they should not shift the production into an entirely different category divergent from the plans. Extension services hence have the mandate to support the farmers in the production of the dedicated type of produces. Yet, at the local level, land use decisions and the way to achieve production targets are much less standardized than the theoretical planning framework may suggest. They depend heavily on local consultation and negotiation processes at commune or village level. There is hence a substantial degree of diversity between different communes, which makes it difficult to make general statements about the level of freedom a farmer has with respect to the diversification of production. The assignments defined in the land use plans are usually based on considerations around food security, market availability, revenue prospects, and physical conditions (mainly soils, water quality, and hydraulic infrastructure systems). At the same time, researchers have been pointing out that farm-level agricultural diversification can be considered a risk-spreading strategy particularly for small farming households since dependencies on price fluctuations in single markets can be compensated for and other risks distributed. The latter include, for example crop-specific pest outbreaks or extraordinary hydro-meteorological events. It is argued that community-level agricultural diversification with the integration of off-farm with agricultural activities needs to be given more attention for modernization of agricultural activities, market engagement and the creation of rural job opportunities (Nhan et al. 2004, 2007a).

While the period 1975–1999 was considered the period of rice primacy, the 2000s are strongly associated with aquaculture production in parallel to rice cultivation. Nevertheless, rice production continued to intensify in the Delta, especially in irrigated and alluvial areas in its north-western parts. Between 2000 and 2008, the annual growth rate in production was 2.7%, while the cultivated area decreased by 0.3% annually (GSO 2011). This is a result of the expansion of triple rice cropping, relying on irrigation infrastructure, flood-control embankments, fast-growing rice varieties and other intensive farming practices. Over the same time period, however, the aquaculture sector grew much faster than that of rice and other crops. This growth amounted to 7% in terms of used area and 22% in terms of production per year, compared to 6% and 11%, respectively throughout the previous decade (GSO 2011). These figures suggest a high intensification level in aquaculture production, mainly for freshwater *Pangasius* catfish and brackish *Penaeus* shrimp cultures. The intensification mainly relies on deeper ponds, higher stocking rates, higher inputs of commercial feed, and higher water exchange (or aeration) rates, etc. (De Silva et al. 2010). Figures 4.5 and 4.6 illustrate major trends of rice and aquaculture production over the last years. A distinct geographic distribution within the Mekong Delta can be recognized, with the most intensive rice production in the upper Delta along the main branches of the Mekong and the most extended aquaculture areas in the coastal provinces utilizing brackish-water. Intensification can be observed throughout the Delta for both rice production and aquaculture (i.e. overall much stronger growth in production than in used or cultivated area).

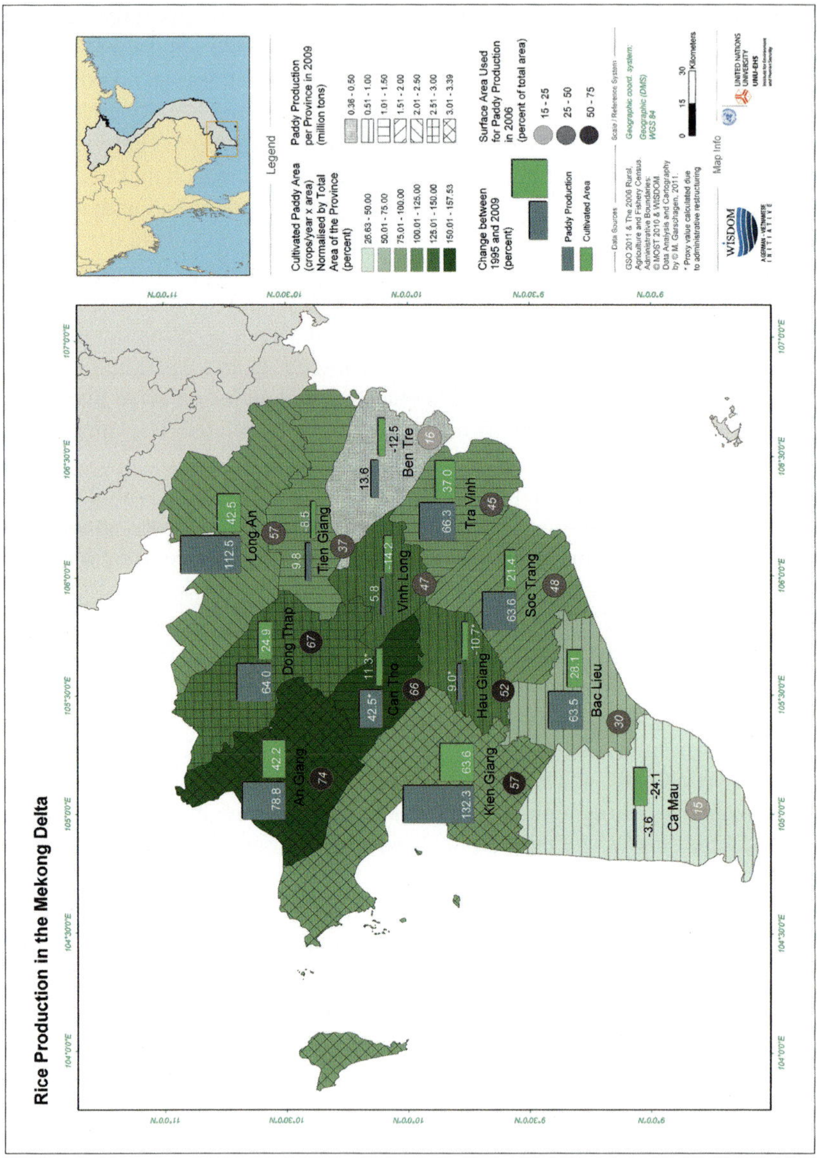


**Fig. 4.5** Rice production in the Mekong Delta (Source: own figure)

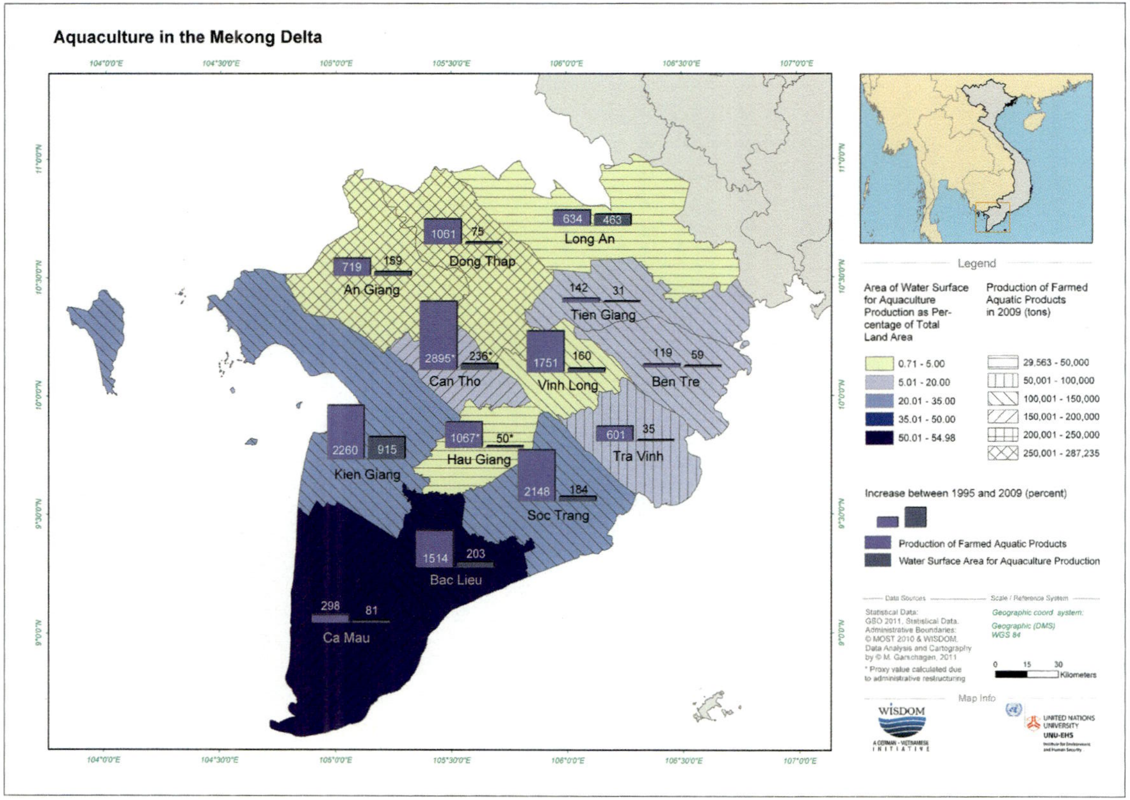


Fig. 4.6 Aquaculture in the Mekong Delta (Source: own figure)

However, the widespread early enthusiasm for increased profit prospects – particularly with regards to aquaculture – resulting from land use transition and intensification have over the recent years been challenged by the emerging recognition of problems related to such practices. Table 4.2 indicates that aquaculture production requires less labour input but more capital than most of the other rice and fruit based production patterns. In parallel, intensive aquaculture production allows for higher profit margins compared to rice production. In more detail, semi-intensive shrimp farming requires more capital input than semi-intensive fish farming due to the shorter cropping cycles (hence higher cropping frequency) and the higher costs for juveniles, specialized feed and necessary aeration of the ponds. Extensive shrimp production on the other hand requires less labour and capital input than semi-intensive shrimp farming. It therefore implies less economic risk but also a smaller gross margin. The higher labour input in fish farming compared to shrimp production results from the increased need for mud removal in fish ponds (due to river alluvial and run-off sediments from surrounding dikes particularly in flow-through systems) and a more labour intensive preparation of feed. Fruit production in mature cultivations has the highest benefit-cost ratio of all the production modes. However, fruit cultivation is linked to specific agro-ecological prerequisites since it is limited to production sites which have alluvial (fertile) soils, low flood occurrence and sufficient hydraulic infrastructure for irrigation and flood protection. In addition, it requires substantial amounts of up-front capital for the establishment and maturing of cultivations.

Even though, the numbers presented in Table 4.2 can only be taken as broad proxy values due to the significant complexity and variability for specific production modes, locations and times of assessment, they hint towards considerable economic and social challenges (Nhan 2009; De Silva et al. 2010; Joffre and Bosma 2009). The combination of land conversion with a reduction in labour demand generates a shortage of employment opportunities for the rural work force, notably for those farmers who lost their land (see below). The increased level of technology and the high costs for input variables force many farmers in new dependencies and often debts. Furthermore, they have to deal with price fluctuations in the target commodity markets and with an increasing pressure on prices given the growing integration into international markets. Pest outbreaks which can result from the wide-spread application of high yielding monocultures and can destroy entire crops pose an additional challenge. As a result, farmers who try to increase their profit gross margins through intensification at the same time face a reduction in investment efficiency (i.e. a lower benefit-cost ratio) due to increasing needs for capital input and diminishing marginal returns (e.g. with a third rice crop in the dry season). In combination with the increase of other economic risks coming along with intensification in monocultures (notably price dependency and pest outbreaks), farmers aiming for high intensification in fact have to negotiate more insecure profit margins, making sustainable income generation increasingly difficulties. Vietnamese and international scientists have since more than a decade warned against the negative consequences resulting from monocultures and sharp intensification. However, farmers and governmental agencies in Vietnam have only more recently started to accept their recommendation to

**Table 4.2** Capital and labour investments and economic returns of main farming patterns in the Mekong delta (means per one hectare farm per year)

| Farming patterns[a] | Total labour (man-days) | Total variable costs ($US)[b] | Gross margin ($US) | Own calculation from references |
|---|---|---|---|---|
| 2 rice crops[c] | 182 | 811 | 768 | Nhan et al. (2008a) |
| 3 rice crops[d] | 269 | 1,295 | 995 | Ha et al. (2008) |
| Fruit culture[e] | 145 | 721 | 3,980 | Thanh et al. (2008); own unpublished data of 2007 |
| Coastal improved extensive shrimp (one crop per year)[f] | 65 | 3,053 | 952 | Tam (2010) |
| Coastal semi-intensive shrimp (mono-culture with 2 crops in 8–9 months)[f] | 67 | 8,250 | 1,175 | Tam (2010) |
| Freshwater semi-intensive fish in pond (usually one crop per year)[g] | 87 | 777 | 1,817 | Own unpublished data in 2007 |

[a]Fruit culture: means of different fruit crops (longan, mango, orange, rambutan, etc.) excluding the non-productive phase of plantation establishment. Coastal improved extensive shrimp farming refers to low stocking rates (<10 juveniles/$m^2$), with supplemental home-made or pellet feed, without pond aeration and with yields ranging roughly between 500 and 600 kg/ha/crop of 4 months. Semi-intensive shrimp farming relates to farming at medium stocking rates (10–20 juveniles/$m^2$), relying on pellet feed and pond aeration and yielding from 0.8 to 2 tons/ha/crop of around 4 months

[b]Total variable costs include all material and labour (hired and family labour) inputs, renting costs and loan interests (if any), not considering taxes/fees and depreciation of infrastructure (which are considered to belong to the fixed costs). The following currency exchange rates have been used for calculation: 1 $US = 18,000 $VN (2007), 19,000 $VN (2008), 20,000 $VN (2010)

[c]Data period: 2007; study sites: Co Do and Thoi Lai districts (Can Tho)

[d]Data period: 2008; study site: Cho Moi district (An Giang)

[e]Data period: 2006–07; study sites: Long An, Vinh Long, Tien Giang and Can Tho

[f]Data period: 2008; study site: My Xuyen district (Soc Trang)

[g]Data period: 2007; study sites: O Mon (Can Tho), Tam Binh (Vinh Long), Cai Be (Tien Giang)

shift towards more integrated farming systems (e.g. rice-shrimp) in order to improve the long-term sustainability and security of agricultural practices.

In addition to the social and economic risks, the intensification of aquaculture poses significant degradation risks to the Delta's ecosystems (Chap. 13). Intensive *Pangasius* catfish yields range from 300 to 400 tons/ha/crop (Phan et al. 2009; Nhuan 2010). On average, water consumption per 1 kg of fish produced ranges from 6.4 to 7.2 $m^3$ (ibid.). De Silva et al. (2010) estimated that to produce one kilogram of fish about 46 g of nitrogen (N) and 14 g of phosphorus (P) from fish ponds are discharged into surrounding waters. Thus, to produce 375,500 tons of fish in 2005, 687,000 tons in 2007 and 1,095,000 tons in 2008 (Trong 2008; De Silva et al. 2010), *Pangasius* farms in the Delta discharged a quantity of nitrogen and phosphorus of 17,300 and 5,300 tons in 2005; 31,600 and 9,600 tons in 2007; and 50,400 and 15,300 tons in 2008, respectively, to the main canal and river systems. Impacts of effluent discharges from *Pangasius* ponds on other users have not yet been fully assessed. For *Penaeus* shrimp culture, average shrimp yields range between 77 and 240 kg/ha/year in extensive farming patterns and between 1,300–6,200 kg/ha/year in semi-intensive or intensive systems (Joffre and Bosma 2009). There are indications that the intensification of shrimp production pollutes surface water bodies and leads to soil salinisation, both degrading ecosystem services and constraining crop production, particularly rice (Binh et al. 2005; Thao et al. 2008; Hens et al. 2009; Chap. 13).

In addition to agriculture and aquaculture, inland fisheries targeting wild fish also play an important role for food and income generation of rural households in the Delta. The fisheries are of importance particularly to resource-poor households, whose fishing is considered as part of wider livelihood strategies (Ministry of Fisheries and The World Bank 2005), especially during the flood period when unemployment is more serious (Loc et al. 2007). However, changes in natural habitats of floodplains for wild fish (i.e. the conversion of natural wetlands into rice land, flood-control dike systems and water management) and intensive use of agrochemicals for rice production development have resulted in a substantial decrease in the catch of wild fish from floodplains (van Brakel et al. 2011). A particularly strong decline in inland fisheries has been observed in the upper Delta (ibid.).

Increasing accounts of landlessness amongst the rural population have grown into a serious problem for inclusive development in the Mekong Delta (Ravallion and van de Walle 2008a). This problem has emerged in interplay with insufficient profitability opportunities for small scale farmers, decollectivisation, privatization, individualization of land markets, transformations in the agricultural labour force, changes in land use patterns, and emerging aspirations for socio-economic progress. There are many feedbacks at work between these factors and it is often difficult or impossible to identify uni-directional causations. The rate of landless households in the Delta was estimated at 14% in 1993 and 23% in 2004, which is considerably higher than in the Red River Delta (2% and 6%, respectively) (Ravallion and van de Walle 2008b). Landlessness in the Mekong Delta does not only occur with particularly poor population but can – at a much lesser rate – also be observed amongst better-off households (ibid.).

Despite the complex causal relationships linked to landlessness, a couple of main contributing factors or drivers can be identified. First, a structural shift in the relationship between landlessness and living standards, mainly among the poor can be noted (Ravallion and van de Walle 2008a). This is indirectly the result of increased market liberalization and commercialization of agricultural production as well as land use changes (notably towards aquaculture which, as shown above, is more capital intensive and profitable but less work intensive than rice production; see Table 4.2 and Fig. 4.6). These factors are coupled with the establishment of de-facto land markets (i.e. the introduction of individual land use rights which can be exchanged and traded according to market-oriented prices) (Marsh and MacAulay 2006; Ravallion and van de Walle 2004; Dorward et al. 2009). A strong correlation can be observed between the poverty-level and land size amongst rural agricultural households in the Delta. Most of the poor[6] farming households have small land holdings between 0.1 and 0.3 ha. Given such small land holdings, these farmers have very limited possibilities to use their production inputs in an efficient manner and to increase mechanization. Simultaneously they face steep increases in the costs for fertilizers, agro-chemicals, seedlings, fingerlings etc. and often a reduction in price levels for their produces – in combination reducing their profitability (Taylor 2004). Their overall income is, hence, much lower than amongst wealthier farmers with more land (Nhan et al. 2007a, 2008b, 2010). This relation can even be observed within fruit crop cultivation (Thanh et al. 2008). As a very general rule of thumb, a poor farming household with a land holding as small as 0.3 ha could typically earn between 7 and 12 million VND per year with paddy or fruit production. Such income levels are not sufficient to move out of poverty, particularly if the respective livelihood relies solely on agriculture. Consequently, poorer farmers often sell their land to richer farmers and seek to find non-farm jobs (mostly unskilled labour), either within or outside their commune, hoping to achieve higher income (i.e. they step out of agriculture). In addition, many of the small-scale farming households have to accumulate debts in order to pay for the increased input factors of production coming along with intensification or to compensate for crop failures. Such debts can often not be re-paid by means other than selling parts of the land, decreasing the future prospect for productivity and profitability gains even further. As a result, many of these farming households eventually lose most or all of their land. It has been argued that farmers belonging to the Khmer minority are particularly susceptible to be caught in such a downward spiral, given their additional development challenges (c.f. Taylor 2004).

The second driver is the aspiration for improved living standards in a process of economic transition. This can entail a shift out of agriculture, which happens even amongst better-off farming households, who have relatively sufficient livelihood assets and tend to sell their land to take up new employment opportunities (Ravallion and van de Walle 2008a). This differs from a shift in the relationship between landlessness and poverty which has been explained above.

[6]Compare footnote 1 for the statistical definition of poverty.

There are debates over the impacts of the rising landlessness on overall poverty trends in the Mekong Delta: On the one hand, it can be argued that landlessness has increased rural poverty and income disparities. On the other hand, it can be seen as a result of the restructuring of rural labour markets and the economy, which allows for more rapid poverty reduction. Ravallion and van de Walle (2008a) have found that poverty reduction amongst landless population is in Vietnam on aggregate progressing slightly faster than amongst other groups. Yet the rate of poverty reduction for the Mekong Delta's landless is lower than for those with land (ibid.). The authors of this analysis have concluded that rising landlessness (allowing for land accumulation and more efficient production) has been a positive factor in poverty reduction in Vietnam as a whole, but that the Mekong Delta stands out as an exception. This can be explained by a combination of historical reasons. In the past, the South and the Mekong Delta in particular had a stronger market economy than the North and hence, stronger inequality (Ravallion and van de Walle 2008a, b). In addition, there are increased pressures for land consolidation in the Mekong Delta (Taylor 2004). Lastly, the low levels of education and training, hamper economic success in the non-farm sector. On the same token, there are signs of emerging rural class differentiation and increasing income disparities (Akram-Lodhi 2004). Investigating household income in the Delta's rural areas, Nhan et al. (2008b, 2010) and Thanh et al. (2008) found that a ratio of the average income of the rich group over the poor group ranged between 3 and 21; the ratio is higher with areas closer to urban centers. This highlights the importance of further improving human capacity and other livelihood-related capitals of the poor in the transition process of the rural economy in the Mekong delta.

In addition to the trends in land holdings, significant changes in the Delta's economic profile and agricultural labour force can be observed. These are heavily influenced by urbanisation, industrialization and rural-urban migration – all being simultaneously driver and result of labour transformations (see sections below). Although the gross output value of the agricultural sector was increasing continuously during 2000–2008, the share of this sector to the total gross output of the economy has gradually declined, from 47% in 2000 to 38% in 2008. Simultaneously, the share of the industrial and service sector has increased (Fig. 4.7). In the period of 2004–2008, labour force engaged with the agricultural sector decreased by 0.5 million people (2% per year) while the non-agricultural sectors experienced an increase of 1.3 million people (13% per year) (Fig. 4.8). Consequently, the share of agricultural labour in the total labour force of the economy decreased, from 68% in 2004 to 57% in 2008. In line with this trend, Thanh et al. (2008) found changes in the structure of rural household income in the Mekong Delta. In the period of 1993–2004, the share of income from agricultural activities in total household income decreased, while the share of income from wage activities increased significantly, particularly in areas closer to urban centres. These observations underline that agricultural activities provide only low incomes and poor livelihood bases due to the afore-mentioned problems with respect to low (and in many parts decreasing) profitability and economic uncertainty (see also Coclanis and Stewart 2011). Along the same line, strong correlations between income levels and land holdings can be identified for

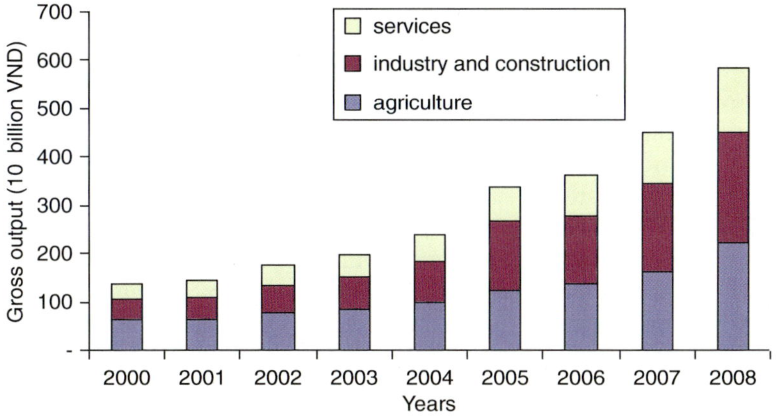


**Fig. 4.7** The contribution of agricultural, industrial, construction and service sectors to total gross output in the Mekong delta in the period of 2000–2008 (Source: own figure based on GSO 2010d)

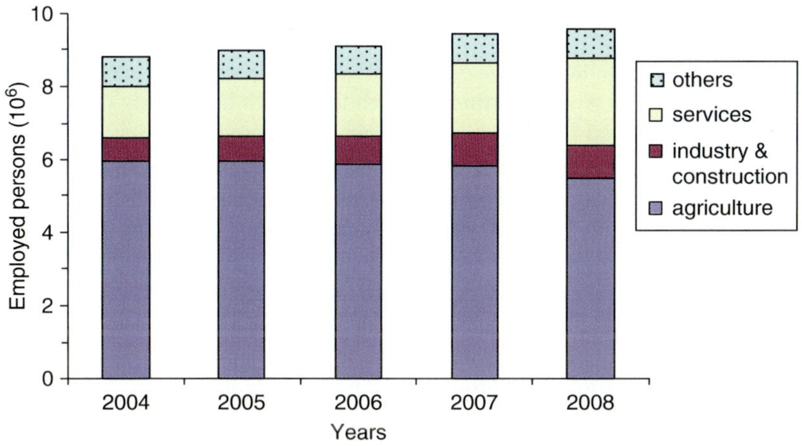


**Fig. 4.8** Employment distribution ($10^6$ persons) in economic sectors in the Mekong delta in the period of 2000–2008 (Source: own figure based on GSO 2010d)

those households that have not (yet) become landless. Analysis has shown that farming households holding 0.2–0.4 ha of land are in most cases not considered poor according to the statistical definition but in most cases achieve just enough production for daily subsistence (Thanh et al. 2008). In order to increase their income, households have to engage in on-farm wage labour for larger scale land holders or – more importantly – additional work outside the agricultural sector (ibid.). These findings reflect the increasing change of (former) agricultural labour into non-farm activities and the desire amongst many rural households to shift their livelihood out of agriculture (Taylor 2004).

The transition of rural labour implies opportunities as well as challenges for future development in the Delta. It can push the specialization and diversification amongst the rural labour force. Better-off farming households are likely to accumulate agricultural land and, thereby, modernize and commercialize agricultural activities for fostering market integration (i.e. stepping-up agriculture). At the same time, there is an increased need for expanding the rural non-agricultural infrastructure (i.e. stepping out of agriculture) (Dorward et al. 2009). In theory, this can create opportunities for non-agricultural employment, particularly amongst poor households, while also improving the situation of better-off farming households. However, matching the qualifications of former agriculturalists to the demands of other non-farm economic sectors remains a heavy challenge. This is because many of the former farmers do not have work experience and training outside the agricultural sector and lack sufficient education and vocational training. In addition, the development of non-agricultural industries and job opportunities lags behind demand. Substantial migration particularly of young labourers to urban and industrial centres – notably Ho Chi Minh City and the neighboring southeast provinces – could already be observed over the last years (Huy 2009, see also the section on migration below).

On a larger scale, the question whether or not the labour transition of rural labour force proves successful has influence on the sustainability of rural poverty reduction in the Mekong Delta as a whole. Farming households which lack young or skilled laborers, have been observed to often manage their agricultural activities at 'hanging-in' status (Dorward et al. 2009), looking at options for maintaining existing farming levels as temporary or secondary livelihood activities. In response to these challenges, the provinces in the Mekong Delta are currently implementing a governmental project on vocational training for rural workforce (Project 1956 – Decision 1956 of the Prime Minister), which aims at contributing to a more sustainable reduction of rural poverty through fostering diversification and value adding in the Delta's rural economy. Yet, the long-term effect of this project remains to be seen.

Besides the challenges with labour transition and landlessness, other emerging problems in the environmental sphere need to be acknowledged as well. The delta is still considered an important food production region for national food security towards 2030. Around 1.8 million hectares of agricultural land have been set aside for rice production to produce about 21 million tons of paddy annually for domestic consumption and exports. This implies that rice production will continue to be intensified. On the one hand, this is likely to cause severe competition over water between the upper and lower portions of the delta (Nhan et al. 2007b). During dry periods (Januray-May) the river flows are low while water volumes used for irrigation are high (ibid.). In April, when the river flows are lowest, water consumption for irrigation accounts for 29%, 34% and 39% of the total river flows in – statistically-spoken – wet, normal or dry years, respectively (own calculation based on Xo and Lam 2009). Irrigation water demands for rice production will become even higher with the further expansion of triple rice cropping areas and/or the earlier establishment of wet season rice crops, which is increasingly adopted in order to minimize harvest losses through the August floods in double rice cropping areas. Intensive water use for rice production and low river-flows are important determinants of saline water intrusion in downstream areas. On the other hand, further intensification of rice production would result in declined soil fertility, increased production costs, reduced crop water use efficiency and lower economic return (Nhan et al. 2007b; Nhan 2009). Reducing intensification of rice production through integrated farming systems (i.e. rice-aquaculture, rice-upland crop-livestock), applying innovative farming practices to reduce agricultural inputs while improving rice-grain quality, and, thereby, improving the net-value gains achieved by rice production are increasingly discussed as options for promoting sustainability of rice production in the Delta. Particularly the debate around the comparatively low quality of rice and, hence, the low revenue generated with rice exports – despite the great quantities – has gained some considerable momentum in the recent past and is being addressed by researchers and progressive politicians thereby challenging conventional quantity-oriented policies (e.g. Vietnam News 2005, 2009; Voice of Vietnam 2009; Vietnam Net 2008; Look At Vietnam 2009b; Saigon Times 2001).

Agricultural production and – to a lesser extent – aquaculture and fisheries are highly susceptible to abnormal changes in the Mekong's hydrology and the region's climatic conditions. The productivity of the Delta depends upon regular flooding cycles not only for crop irrigation but also for soil nutrient enrichment and soil acidity flushing in the upper and mid-Delta during the wet season as well as for the regulation of salinity intrusion in the lower Delta during the dry season. Pollution, urbanisation, hydropower development, and agricultural intensification upstream in the Mekong might result in increasing water abstraction and flow alterations (Hoanh et al. 2003; Kummu and Sarkkula 2008; ICEM 2010). Such changes might have negative impacts on agricultural production, fisheries and ecosystem services (e.g. the reduction of sediment and nutrition deposition or failures in salinity flushing). In addition, these stresses have critical effects on freshwater supply for other domestic uses. Climate change is likely to further enhance these stresses in the medium and long-term perspective (Chap. 2). In turn, these alterations will also have detrimental impacts on – not only but particularly – the livelihoods of those inhabitants, who directly depend on the natural resources in the Delta (Freitag 2003). In addition to such developments, numerous recent studies have suggested that climate change and especially the resulting weather variability and sea-level rise will have strong negative impacts on agricultural production. Next to broad scale assessments on climate change effect on agriculture in Southeast Asia and Vietnam (e.g. ADB 2009; Wassmann et al. 2009; IFAD 2009; MONRE 2009; Carew-Reid 2008; Chap. 2), Nhan et al. (2011) have assessed the impacts of temperature and rainfall anomalies on rice and shrimp production in the Mekong Delta. They have shown that vulnerability levels differ by crops, crop-development stages, cropping seasons, and regions. However, given the high remaining uncertainties, much is still to be done in terms of analyzing and projecting the effects of climate change on the Delta's agricultural production in detail.

Farmers in the Delta have a long experience in dealing with weather variability, floods and salinity intrusion. Yet, global climate change will enhance these risks and result in more frequent and severe impacts, which are likely to seriously challenge the adaptive capacities of farmers and governmental entities alike (Garschagen 2010a). The national Government has responded swiftly by promulgating the National Target Programme to Respond to Climate Change (Decision No.158/2008/QD-TTg issued on 2nd December 2008; see SRV 2008), requiring all sectors to formulate action plans for climate change adaptation and mitigation at different levels. In addition, the Ministry of Agriculture and Rural Development has issued a comprehensive adaptation master plan in 2008. However, the prospect and actual impact of these policy frameworks has to be interrogated critically (c.f. Garschagen 2010b).

### 4.3.2 Industrialisation

Overall, the industrial sector plays a major role in Vietnam's economic development process. *Doi Moi* has induced a structural change which was driven by a rapidly growing industrial sector. The most recent figures show that in Vietnam, industry accounts for 41% of the GDP, and employs around 20% of the work force (GSO 2011). Spatially, the industrial production is unevenly distributed showing a clear dominance of the southern part of the country. Nearly 45% of the industrial production is concentrated in the South East Region with Ho Chi Minh City as the most important national economic center, followed by the Red River Delta with the country's capital Hanoi with 27%. The third most important industrial region in Vietnam is the Mekong Delta contributing roughly 10% of the industrial production (GSO 2011). Yet, since the mid-1990s the Mekong Delta was not able to increase its relative contribution to the overall industrial production of the country. Instead, compared to 1995 the contribution of the Mekong Delta diminished slightly (from 11.8% in 1995 to 9.9% in 2008). Consequently, the contribution of the industrial sector to the regional GDP in the Mekong Delta is still far below the respective contribution at national level, meaning that it can only to a limited extend unfold the potential to absorb workforce released in the agricultural sector following intensification and transformation. Currently, around 30% of the Delta's GDP is generated by industry (GSO 2011). The most important industrial sector is the food-related industry, especially the processing of food (from aquaculture, fisheries, rice), the production of agricultural and aquaculture inputs (e.g. fertilizers, pesticides, seeds, fish feed or fingerlings) and related industries in equipment and machinery. Besides that, textile industries and building material production and other rather low-technology manufacturing also play a major role.

Within the Mekong Delta, the industrial production is highly concentrated. The three most important provinces Can Tho (18%), Long An (16%), and Kien Giang (11%) account for nearly half of the industrial production. Interestingly, the Mekong Delta is not able to profit from the proximity to Ho Chi Minh City (HCMC) in the same way as the provinces located north east of the megacity (Figs. 4.9 and 4.10).

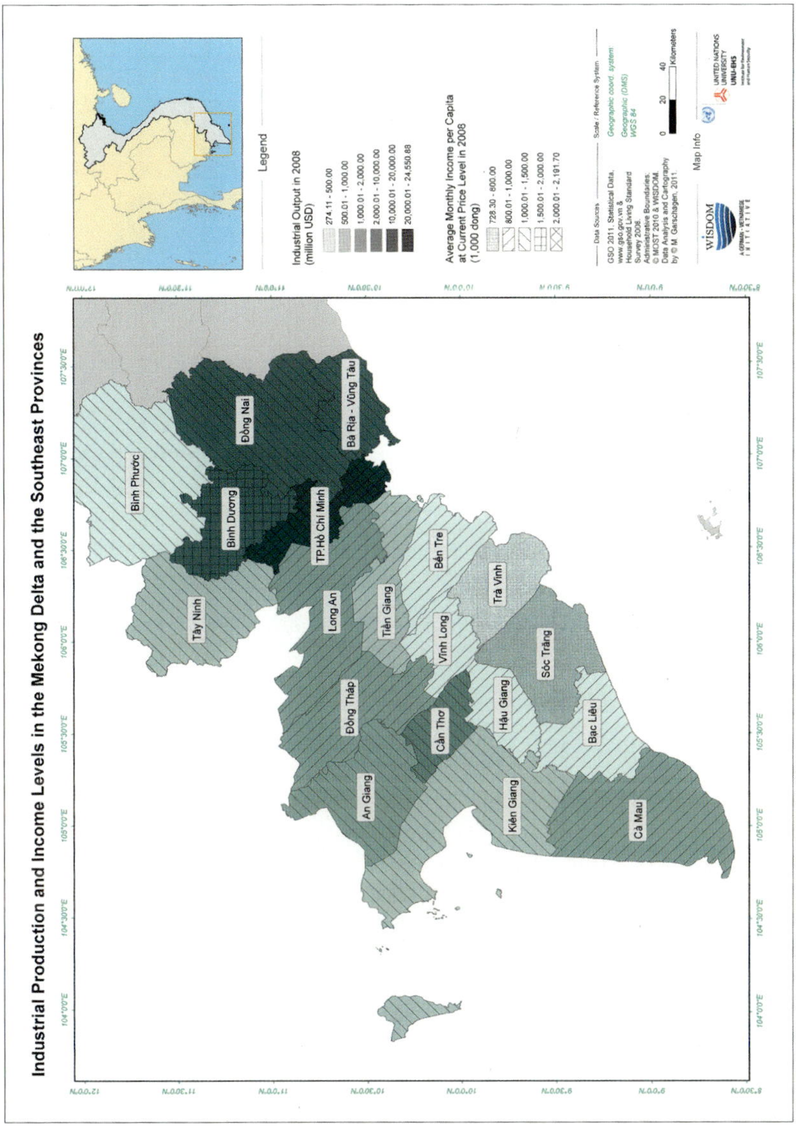


**Fig. 4.9** Industrial production and income levels in the Mekong Delta and the Southeast Provinces (Source: own figure)

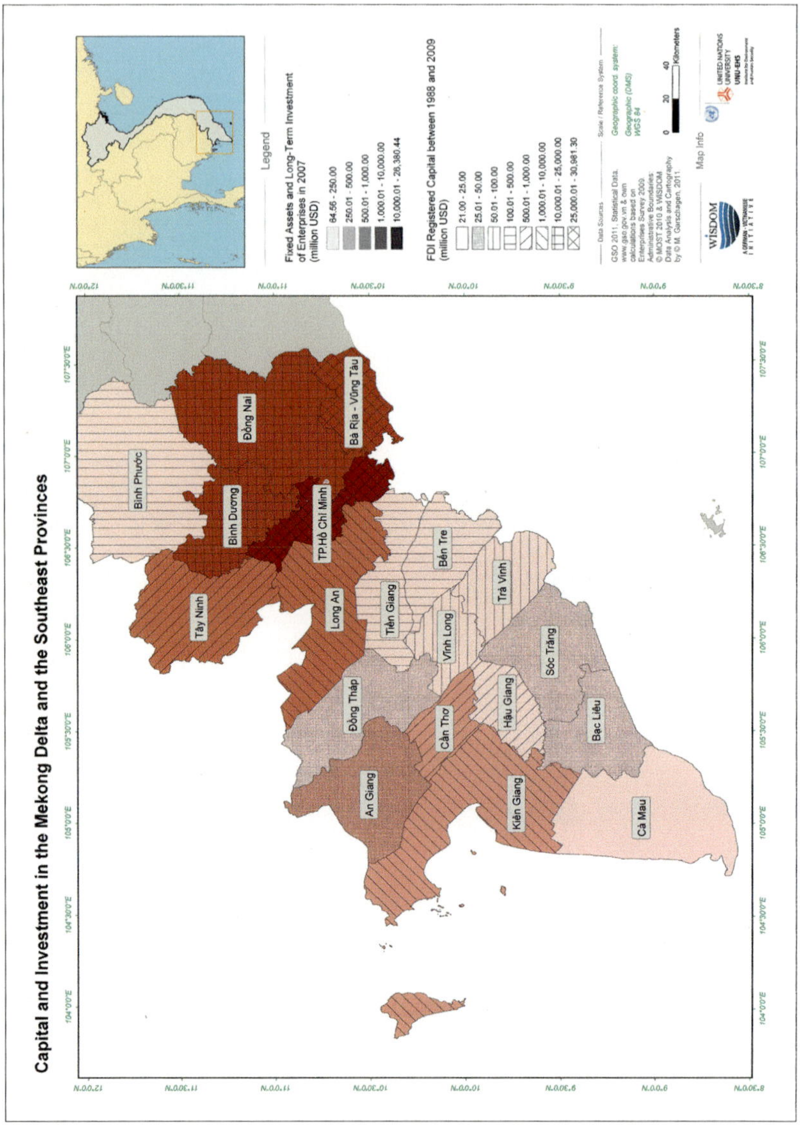


**Fig. 4.10** Capital and investment in the Mekong Delta and the Southeast Provinces (Source: own figure)

**Table 4.3** Industrial output by ownership in 1995 and 2008 in percent

| | 1995 | | | 2008 | | |
|---|---|---|---|---|---|---|
| | State | Non-state | FDI | State | Non-state | FDI |
| Whole country | 50.3 | 24.6 | 25.1 | 24.9 | 34.9 | 40.3 |
| Red River Delta | 60.9 | 24.5 | 14.6 | 22.9 | 34.7 | 42.4 |
| Northern midlands and mountain areas | 72.7 | 15.1 | 9.2 | 44.8 | 41.9 | 13.3 |
| Northern Central area and central coastal area | 58.6 | 35.3 | 6.1 | 27.8 | 51.8 | 20.4 |
| Central Highlands | 34.3 | 58.6 | 7.1 | 16.4 | 71.7 | 11.9 |
| South East | 38.8 | 19.7 | 41.5 | 18.3 | 27 | 54.7 |
| Mekong River Delta | 45.7 | 46.6 | 7.7 | 20.3 | 64.4 | 15.3 |

Source: own calculation based on GSO (2011)

For example, the relocation of industrial sites due to increasing agglomeration costs in HCMC (e.g. shortage of serviced land, overcrowded infrastructure, environmental problems) largely seems to neglect the Mekong Delta. The main reasons for this are disputed and may vary depending on the industrial sectors and provinces in focus. However, based on the analysis of other development factors (laid out above and below), it appears that the main restrictions are to be found in the lack of positive agglomeration or cluster advantages (such as innovation networks and physical proximity for intensive interaction), insufficient transportation infrastructure, insecurity with respect to the availability of qualified work force, comparatively less progressive local governments, and rather low levels of investment experiences in the Delta, when compared to the HCMC metropolitan area.

As in the rest of Vietnam, the ownership structure of industrial enterprises underwent a drastic change due to *Doi Moi* reforms. In 1995 nearly half of the Delta's industrial output was still contributed by stated-owned enterprises. Yet, by 2008, this share had diminished to roughly 20%. Instead, the domestic private sector gained in importance and represented 64% of the industrial production (see Table 4.3). Although the Mekong Delta received only 3.1% of the registered non-oil foreign direct investments during 1988 until 2006, foreign companies' share in the industrial production has reached 15% in 2008 (see Table 4.3).

The trends presented above demonstrate that the industrial development falls behind earlier expectations formulated in development master plans for the Delta after *Doi Moi*. The advantageous conditions of the Mekong Delta at the beginning of the transformation process were not able to initiate a substantial structural change towards industrialization and modernization of the production system. In order to prosper in a sustainable and competitive way compensating for the afore-mentioned challenges in the agricultural sector, the Mekong Delta has to deal with a number of tremendous challenges in the field of industrialization. These challenges are closely interrelated.

Firstly, the potential of the private sector to act as engine for industrial development has not yet been tapped sufficiently. Besides the positive impact of the private companies in terms of industrial output and job creation, most of these companies remained operating at a low level of technology and without sufficiently exploiting potential economies of scale (Khoi et al. 2008). Compared to the rest of the country, the majority of private companies are very small in terms of employment and registered capital. In 2008 around 33% of the private companies had less than 5 employees (35% in 2002) and 40% of the companies had less than 1 billion VND registered capital (80% in 2002) (GSO 2010c; Khoi et al. 2008). Several studies confirm that private companies face problems getting financial support from the government and from credit institutions. Private companies often have to face a lot of red-tape processes when trying to obtain land use rights – which they can then also use as collateral for bank credits. According to Khoi et al. (2008: 20) "land issues remain one of the biggest challenges for private companies".

Secondly, the further development of human capital is crucial. Not only would the qualification of the work force need to be improved. Even the education of the business owners would deserve increased attention, especially in the field of management skills. As indicated above, the general educational level in the Mekong Delta is falling behind the national level. Only 9.7% of the population aged 15 and over have undergone and completed a vocational or professional training, representing by far the lowest value in the entire country (with the national average being at 17.6%) (GSO 2010b: 11). Taylor (2004) notes that there are nevertheless multiple rather informal mechanisms for transferring relevant education, e.g. through community schools within ethnic minority groups or through day-to-day training within family businesses. Yet, it can be argued that the knowledge and training communicated there primarily supports and reinforces the low-profile industrial production systems, i.e. small enterprises, often lacking sufficient innovative potential and economies of scale. Attracting substantial domestic and foreign investment would, however, require an increase of available work force which has undergone more formalized and standardized education and training which links to broader standards and expectations.

Thirdly, the level of technology and innovation would need to be upgraded. As assessed by the first Vietnam Competiveness Report in 2010 the agro-processing cluster in the Mekong Delta is characterized by agricultural exports largely comprising unprocessed or semi-processed goods which have low value added (Ketels et al. 2010). Besides the weak human capital base, financial capital constraints hinder the companies to buy new technology. Consequently, the local companies remain rather passive when it comes to innovation. So far, learning from multinationals is also limited due to the small number of foreign investors in the Mekong Delta. Existing research and development (R&D) institutions would need to further enhance their capabilities in order to produce good and applicable research output (Chap. 15). In general, it can be observed that cooperation between – on the one hand – different companies for supply, processing, or trading with – on the other hand – research institutions, banks, and administrative organisations, is still in its infancy. Yet, such clusters can be considered the precondition for developing regional innovation systems. Incubators would, hence, need to be developed and exchange and synchronization between different actors fostered.

Forth, investments in physical infrastructure are needed. Today, the Mekong Delta has a large network of rivers and canals which for decades have been by far the most important backbone for transportation. Extended most progressively – but not exclusively – by the early settlers, the French and later the re-unified national Vietnamese government as well as international donors, this network has over the centuries catered for multiple purposes including in particular military control, irrigation, drainage, domestic fresh water supply and last but not least transportation. Until today, the network has grown to include almost 15,000 km of main and primary canals, around 27,000 km of secondary and 50,000 km of tertiary canals (SIWRP 2010: 15). Together with the natural rivers, the main and primary canals provide over 25,000 km of transportation lines for heavy weight cargo vessels. There are an estimated 500,000 mid-sized and large vessels (i.e. with a dead weight tonnage of at least 1 ton) operating in the Delta every day (Look At Vietnam 2009a).

However, in the recent decades, notably since *Doi Moi*, water transportation and the entire orientation of societal and economic activity has been experiencing a progressive shift away from rivers and canals and towards roads (c.f. Taylor 2006). While bulk goods such as food staples or construction materials are still largely transported on water ways in the Mekong Delta – in fact with an increase of 177% between 2000 and 2008 – there is drastically increased demand for road transportation from the more diversified emerging industrial sectors which need fast, direct and decentralized transportation of input factors and produce. The amount of goods transported on the Delta's roads, hence, increased from 11 million tons in 2000 to over 23 million tons in 2008 (GSO 2011). Additionally, motorized road passenger transportation has been increasing rapidly over the last years since it has become affordable to a growing number of people and is in general much faster and more convenient than travelling by boat.

As a result, the road network has increasingly become strained beyond its capacities since infrastructure development and upgrading has not been able to keep pace with the rocketing demand and has been suffering from poor management and planning as well as from insufficient financial means (World Bank 2006). Substantial infrastructure development programmes have therefore been planned and partially implemented over the last years. In 2010 a new bridge crossing the Hau River at Can Tho City has been opened, now allowing crossing the Delta from Ca Mau to Ho Chi Minh City without time-consuming ferry crossing. A second East-West corridor connecting Rach Gia and Ho Chi Minh City with a highway has been proposed by the Asian Development Bank, yet if and when such a project will be approved and implemented is far from certain at the time of writing. Also the opening of Can Tho Airport in 2010 and Ca Mau airport in 2004 increase the Delta's connectivity particularly with respect to business and conference travelling.

However, it remains arguable if these infrastructure upgrades provide sufficient and reliable interregional connections to the dynamic growth centres, notably Ho Chi Minh City, but also Phnom Penh in the north. Yet such connections are crucial for accessing, among others, markets, human capital, and research institutions and for introducing more diversified production modes targeting higher value adding (such as just-in-time-production with dense networks of contributory producers and consumers). This holds particularly true when considering the need for better connections between growth centers within the Mekong Delta in order to foster endogenous development potential. If the realization of the agro-processing cluster is envisaged, a network between relevant stakeholders such as farmers, traders, manufacturers, retailers, exporters and other service providers will need also improved physical connection between them allowing for face to face communication and trust building.

In addition to these transportation considerations, the amount and particularly the quality of industrial zones would need to be enhanced in the Mekong Delta. So far, 61 out of the 550 Vietnamese industrial zones were established in the Mekong Delta provinces. In general, industrial zones aim at fulfilling two basic functions. Within a well designed industrial policy scheme, industrial parks could help to coordinate industrialization processes in the Mekong Delta. However, as the Delta also represents a vulnerable ecosystem, industrialization cannot take place everywhere. Aspects like emissions, wastewater, availability of suitable land and freshwater need to be considered for a sustainable industrialization process alongside other economic factors such as the availability of capable work force and transport infrastructure. Another function of industrial zones is to provide necessary preconditions for production. As the overall quantity and quality of the infrastructure in the Mekong Delta is still low, industrial parks offer basic infrastructure for companies like energy and water supply, road connectivity, etc. But, as stated in the Vietnam Competitiveness Report "[to] date, industrial parks just play the role of an industrial estate solution rather than a platform for forming clusters" (Ketels et al. 2010: 91). Industrial zones will only be successful in the long run if they fulfill the needs of the companies, either local or foreign ones. Especially the local private actors with their weak financial background need special support for moving into industrial parks or for establishing entirely new enterprises under incubator conditions. A stronger enforcement of regionally synchronized industrial clusters in the Mekong Delta appears essential in this context in order to achieve the most efficient use of resources and the strongest agglomeration of attractive potential.

### *4.3.3 Migration*

The role of migration for socio-economic development is heatedly debated – not only in the Mekong Delta – since migration can imply opportunities as well as barriers to the development of specific locations and larger regions as a whole. Migration can be the cause or result of transformation processes – or both simultaneously. It can take various forms ranging from temporary to permanent, from within a commune or district to cross-regional or even international migration, and from single persons to entire households or even communities. Migration usually results from the interaction between push and pull factors as well as between externally 'forced' and internally aspired drivers (c.f. Zhang et al. 2006).

Migration has always played an important role for the Mekong Delta, facilitating its colonisation and cultivation, influencing its ethnic composition and political affiliation, interfacing with its numerous wars, cumulating in large-scale resettlement programmes, and affecting social and economic development. After centuries of changing demographic influence and political rule of particularly the Funan kingdom and later the Khmer empire, Kinh (or Vietnamese) settlements began to expand southwards into the Mekong Delta since the seventeenth century under the Nguyen dynasty. These migrants thereby boosted the previously sparse settlement of the Delta and expedited its agricultural opening, notably through the trenching of drainage canals (Biggs 2004; Miller 2003). Subsequently to France's Cochinchina campaign in the 1860s, the Mekong Delta became a centrepiece of France's first colony in Vietnam. This resulted in high migration activity in the Delta, taking various forms, but being mainly the result of the changed political and economic profile, in turn affecting the type and geographical distribution of work force (Zhang et al. 2006). During the Second Indochina War,[7] armed conflicts in large parts of the Delta led to substantial population movements, particularly from the countryside into urban areas or so-called 'strategic hamlets' of the South-Vietnamese forces (Dang et al. 1997; Thrift and Forbes 1986). The first decade following the war was characterised by the extension of socialist resettlement programmes to the South, pushing for rustication or de-urbanisation (often coupled with 're-education' measures) and population resettlement into new economic zones (Zhang et al. 2006). Such policies were backed by the household registration system (*ho khau*) which had been developed in Northern Vietnam from the mid-1950s following the Chinese model of *hu kou*. The implementation of the resettlement programmes, however, remained below the envisaged results mainly due to restricted livelihood prospects and the risk of depreciating social networks (Desbarats 1987; Zhang et al. 2006: 1073) as well as due to the fact that the *ho khau*-system was not realised stringently in the South of Vietnam (Hardy 2001). Following the initiation of *Doi Moi*-reforms, migration in Vietnam and the Mekong Delta has become more dynamic and multifaceted. The most important changes facilitating increased migration were the decollectivisation of land, progressive individualisation in land tenure, the easing of the household registration system, increased economic growth, increased regional disparities, and soaring rural underemployment due to agricultural intensification and land use transformation.

Over the last years, the Mekong Delta has experienced a drastically increasing trend towards cross-regional out-migration, particularly into the South-Eastern Region including Ho Chi Minh City (Fig. 4.11). According to a national survey in 2009, 46 out of 1,000 inhabitants of the Delta had out-migrated within a five year period prior to the survey, which amounts to a total of 734,000 people. With only 70,000 people migrating into the Delta over the same period of time, the Delta experienced a net-migration rate of –42 per 1,000 inhabitants, equalling to roughly 664,000 people. Ten years earlier, at the census of 1999 the figure had been only 144,000, translating into a net-migration rate of –10‰ (GSO 2010a: 80). This means that the Mekong Delta is not only Vietnam's region with the highest out-migration rate and the highest deficit in terms of net-migration. It is

[7]The Second Indochina War is in the West better known as the Vietnam War and in Vietnam as the American War.

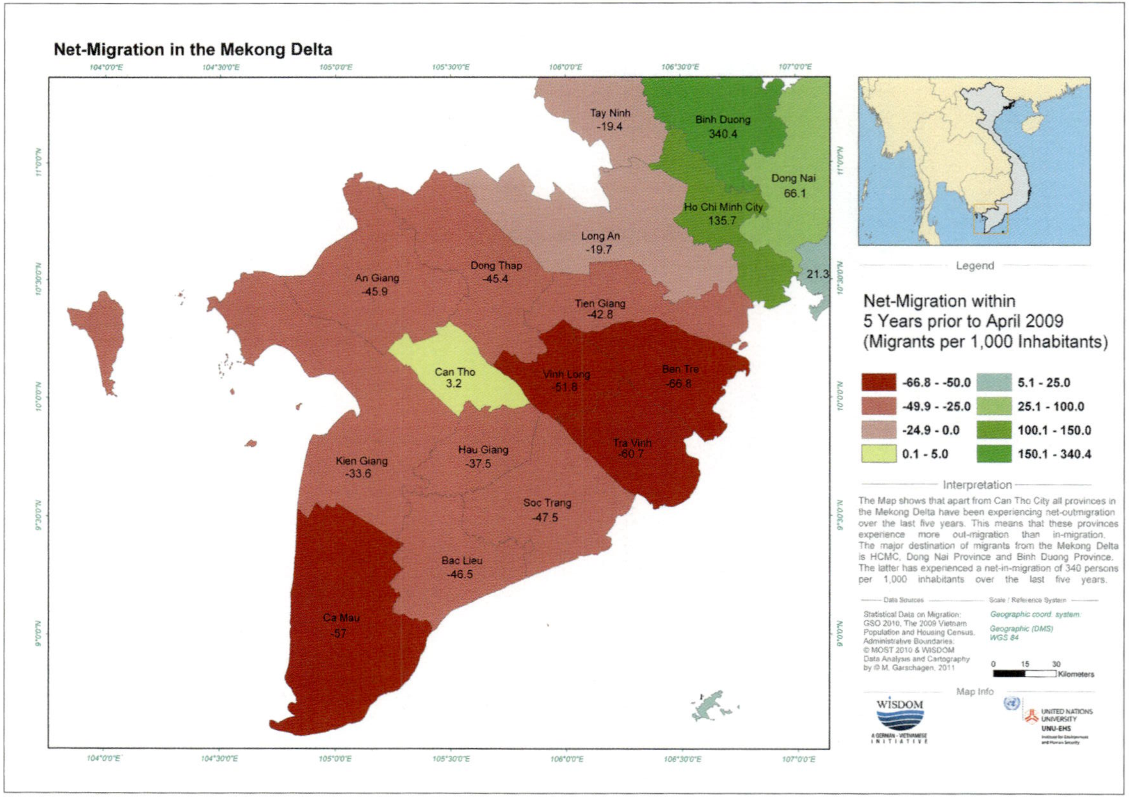


**Fig. 4.11** Migration rates in the Mekong Delta (Source: own figure)

also the region with by far the largest increase in net-migration deficits when compared to the survey of 1999 (increasing more than four-fold). In this way, the outmigration of the Mekong Delta contributed heavily to the overall national trend of increasingly dynamic migration with the strongest growth observable in inter-regional migration (GSO 2010a: 76). The vast majority – roughly 97% – of the inter-regional out-migrants from the Mekong Delta migrate to the Southeast, notably Ho Chi Minh City in search for employment while remaining relatively close to their home provinces (GSO 2010a: 81). Within the Mekong Delta, Can Tho City is the only province experiencing net in-migration, while many of the Delta's other more rural provinces experience some of the highest provincial net out-migration rates nationwide, notably Ben Tre (−67‰), Tra Vinh (−61‰), Ca Mau (−57‰) and Vinh Long (−52‰) (GSO 2010a: 82).

The role of the strong out-migration for the development of the Mekong Delta is highly disputed. On the one hand, it results to a large degree from the profound economic transformation processes in the rural and agricultural provinces. As laid out in the sections above, these processes have forced many farmers into landlessness and have, additionally, released high numbers of on-farm work force due to intensification or shifts towards less work-intensive production (e.g. from rice to aquaculture). For such households, migration – of either some household members or the entire household and either seasonal or permanently – can be understood as one of the major strategies to cope with such unfortunate conditions. Seen from this perspective, migration is often considered a manifestation of undesirable developments. On the other hand, Zhang et al. (2006) remind us that migration decisions are not necessarily perceived negatively as much of the conventional view suggests (p. 1078). Rather, migration is often perceived as a positive opportunity to improve the living conditions or to support the livelihood of household members, for example, through the transfer of remittances. When farmers in the Mekong Delta are being asked for their aspirations or dreams for the future,[8] one of the most prominent answers is the wish that their children should have a life with better living and income-earning conditions, often explicitly mentioning migration to economically more promising regions as envisaged means. Already today, remittances are a significant factor for fuelling livelihoods in the Mekong Delta. Even though statistics on inter-regional remittances within Vietnam are sparse, figures on international remittance flows indicate the strong potential they have in terms of socio-economic development. In 2004, the Mekong Delta received 22% of the total inbound remittances from overseas Vietnamese (Pfau and Long 2006), amounting to some 704 million USD (own calculation based on Thao et al. 2008: 15). With the national figure having risen from 3.2 to 7.2 billion USD between 2004 and 2010 (World Bank 2011) and with the Delta accounting for a growing share compared to the other regions within Vietnam, the importance of remittances is very likely to have even increased since then.

[8]This has, for example, been done in an organized manner within the framework of the WISDOM project, by conducting semi-structured as well as standardized household interviews with roughly 1,000 farming households in Can Tho City, Dong Thap and Tra Vinh (Birkmann et al. 2010b).

### 4.3.4 Urbanisation

Urban centres and increasing urbanisation in the Mekong Delta can be understood as simultaneously being a key driver and a prime manifestation of socio-economic transformation and development. The Delta's major urban centres such as Can Tho, Long Xuyen, My Tho, Vinh Long or Rach Gia are nodal points for processing and trading the Delta's agricultural produces but also for political administration as well as social and cultural functions. Furthermore, industrial zones can be mostly found in the outskirts of such cities along the main transportation veins.

The prospects for non-farm income generation and higher living standards in urban areas in combination with adverse conditions for specific population groups in the countryside have provided push and pull factors for driving-up urbanisation throughout several phases of the Delta's more recent history. As indicated above, the de-urbanisation targets after the end of the Second Indochina War[8] and the re-unification could only partially be met in the South. This is due in particular to the fact that many farmers who had been forced evacuees during the war had gotten used to life in the cities and did not want to resettle back to the countryside due to high uncertainties and the harsher living conditions there (Turley 1977). In addition, the later *Doi Moi* reforms provided institutional changes for propelling rural-urban migration in the Delta. This is due, firstly, to the gradual abolishment of the *ho khau* registration system (see above) – the introduction of which had been greatly driven by the intent to limit rural-urban migration. Secondly and in parallel, the increasing demand for formal as well as informal labour force in the growing urban economies attracted labour migrants. Thirdly, in combination with the previous points, the agricultural intensification and restructuring provided increasing numbers of farmers who had lost their land and/or livelihood base and were ready to migrate into urban areas (compare section above).

Overall, it has been argued that the urban areas in Vietnam have benefited more strongly from the economic growth than their rural counterparts, resulting in spatial disparities in living conditions which are widening particularly since *Doi Moi* (Luong 2003; Revilla Diez 1999). While these observations are especially true for comparing HCMC and the lesser developed rural provinces of the South, similar rural-urban disparities can also be observed within the Mekong Delta. In 1997/98, the annual mean real income per capita[9] in the Delta was 8.2 million VND in urban areas compared to 2.7 million in rural parts, meaning that the mean urban income was three times higher than the rural income. Similar differences can also be found

[9]The income was in the Viet Nam Living Standard Survey 1997–1998 calculated as the combination of agricultural income; non-agricultural income; formal wage and salary; income from pension, subsidies and scholarships; rents, loan interests and other income sources (GSO 2000b:297). Income has been adjusted with rural and urban price levels to derive real income levels. The authors of the survey argue that urban incomes may even be an underestimate due to the avoidant responses with respect to other sensitive income sources prominent in urban areas (GSO 2000b: 299).

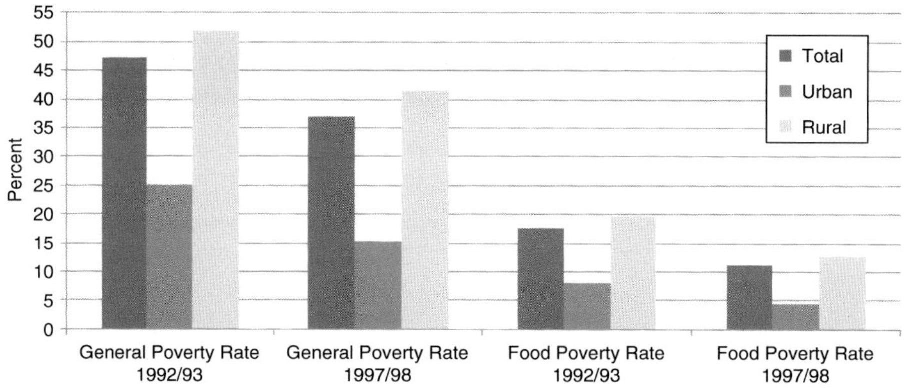


**Fig. 4.12** Urban and rural poverty rates in the Mekong Delta (Source: own draft based on GSO 2000b: 289)

with regards to poverty rates and the reduction thereof. While the Vietnam living standard survey of 1992/93 found 25% of the Delta's urban population in general poverty[10] (compared to 52% in the Delta's rural areas) and 8% in food poverty (20% in rural areas), the figures had fallen to 15% and 4.5%, respectively, in the year 1997/98 (with the rural figures at 42% and 13% in that year) (compare Fig. 4.12; GSO 2000b: 289).

As a result, the Mekong Delta's substantial urbanisation since *Doi Moi* and particularly since the 1990s has been mainly fuelled by rural-urban migration (Sang 2007). While in 1995 around 2.4 million inhabitants of the Delta were living in areas classified as urban,[11] the figure has risen to 3.9 million in 2009 (GSO 2011). This equals an increase in the share of the Delta's urban population from 16% to 23% over the same period of time, compared to a national average of 21% and 30%, respectively. Although these figures mirror the accelerating drive towards urbanisation, they also represent the remaining agricultural (and rural) dominance in the Mekong Delta.

This strong urban growth has in most of the Delta's cities exceeded the capacity for adequate infrastructure planning and development, particularly after the war and during the early post *Doi Moi* years (Coulthart et al. 2006; Lang 2006). In addition, the political cadres and administrative staff who had after the re-unification been appointed to manage and plan the Delta's urban centres were not sufficiently

[10] The general poverty lines for those years were 1,160,410 VND for 1992/93 and 1,789,871 VND for 1997/98. The food poverty line was 749,722 VND for 1992/93 and 1,286,833 VND for 1997/98 (GSO 2000b: 260).

[11] The demarcation of rural vs. urban areas is in Vietnam regulated by legislation on classifying administrative units (Decree 15/2007/ND-CP, issued in January 2007). According to this legislative framework, provinces and districts are graded into different levels, depending on their population size, land use profile, and particularities such as special service functions or population density.

experienced and equipped to integrate the increasingly unleashed market forces with the socialism-oriented administration norms. As a consequence, insufficiently integrated planning, a neglect of socially inclusive urban development and an insufficient legal framework for regulating land speculation can until today be considered the main institutional challenges for sustainable urban development in the Delta (see for parallels to the national level also van Etten 2007; Kim 2011; Yeung 2007; Laquian 1996; McGee 2008: Smith and Scarpaci 2000). If not responded to adequately, these challenges could jeopardise the Delta's urban development progress of the last years.

The current master plan on the socio-economic development of Can Tho City until 2020, for example, envisages three major industrial zones for the city, despite the fact that the management board currently faces substantial difficulties in even finding sufficient capital investment for the existing zones. And even more industrial zones are planned in the neighbouring provinces, for example, in Hau Giang just next to Can Tho City. Integrated and demand-oriented planning seems to have been neglected. The single provinces – or even districts within single provinces – are rather competing against each other, thereby, running the risk to waste resources and remain – as a region – behind the attractive potential of other regions.

In addition, the case of Can Tho illustrates how the virtually uncontrolled real estate speculation has become one of the major challenges for sustainable urban development. The master plan of Cai Rang district – one of the most dynamically developing peri-urban districts of Can Tho City – sets out to establish new housing units for 120,000–150,000 new inhabitants until 2025 (with the current population of Can Tho's urban core amounting to roughly 350,000 people). Yet, due to insufficient planning as well as thriving – poorly regulated – real estate development, the currently planned and permitted projects of private developers amount to providing residential units for 278,000 people (Willets et al. 2010: 11). Most of the land earmarked for this development has already been cleared and the original population resettled – leading to major livelihood challenges for the affected population and to the conversion of otherwise productive agricultural land (Garschagen et al. 2011). However, whether the envisaged projects will prove economically sustainable or will even be implemented currently remains questionable due to a divergence between rocketing price levels and actual demand profiles. The vast majority of projects are designed to target the upper middle class segment – implying corresponding price levels. Yet, rural-urban migrants – who are most likely to contribute the bulk of Can Tho's future population growth – are rather in need of affordable housing in the lower price segments (for parallels to the national situation, see Yip and Tran 2008). In addition, the current master plans only cater for a very limited number of state-subsidised social housing schemes, mainly for university students. Hence, the vision of developing Can Tho into an economically prosperous urban centre with a rapid increase of mainly upper and middle class population does not seem to match some of the major dynamics in terms of demography and migration. Climate change induced environmental change in the Delta's rural areas may in future even amplify this divergence since it is likely to contribute to driving up the number of migrants – misleadingly often referred to as environmental refugees (see for a detailed discussion on contested terminology Dun and Gemenne 2008; Renaud et al. 2007, 2011 and for flood triggered migration in the Mekong Delta Dun 2011). The majority of these migrants will probably not be able to enter into the currently developed housing market and will likely have difficulties to find infrastructures meeting their needs. In addition, climate change will increasingly uncover the shortcomings resulting from the insufficiently integrated and static system of urban planning and management with comparatively high barriers to innovation (Garschagen 2011; Garschagen and Kraas 2011; Birkmann et al. 2010a).

## 4.4 Conclusions and Outlook: Navigating Towards the Realisation of Opportunities

The above analysis has attempted to sketch out the main dynamics of and drivers for socio-economic development in the Mekong Delta, particularly since *Doi Moi*. Transformations in the agricultural sector have played a central role for overall poverty reduction and economic growth. Yet, they also propelled socio-economic disparities, expressed particularly in income inequalities and high incidences of landlessness. The most important factors for this development have been the shift towards market oriented and de-collectivised production systems, substantial intensification and changes in land use and production modes, notably towards aquaculture. Growth in the industrial sector has predominantly been related to food processing industries. However, the gains in value adding as well as the development of other industrial sectors have remained below the expectations. The comparatively low profitability and high livelihood insecurity[12] in the agricultural sector coupled with insufficient industrial growth have led to significant outmigration, most importantly directed towards the thriving economic development in and around Ho Chi Minh City. Urbanisation in the Mekong Delta has been important not only for providing central places with social and cultural functions (such as health care and higher education) but also for facilitating economic growth and the transition towards the secondary and tertiary sectors. Yet, this transition has remained below expectations. Social inequalities and a lack of integrated cross-sectoral and inter-provincial urban planning pose challenges for sustainable urban development in the Delta and hamper the full exploitation of the urban potential.

---

[12]The Government of Vietnam has recently launched a pilot project to reduce livelihood insecurity amongst agricultural producers. This policy project sets out to provide, test and adjust agricultural insurance against losses due to natural hazards, epidemics and pest outbreaks in seven pilot provinces in the country, including An Giang and Dong Thap in the Mekong Delta. (Decision 315/2011/QD-TTg, see SRV 2011). While such initiatives are in general heading into the right direction, the new policy has already been criticized for linking the access to insurance to too rigid production guidelines for cultivation and breeding, which necessitate very capital intensive production, thus discouraging many farmers from subscribing (Vietnam News 2011; Vietnam Business News 2011). The long term success of such policy and insurance models has therefore to be seen.

Altogether, the analysis has pointed towards the opportunities as well as challenges and barriers for a more progressive and sustainable development of the Delta – and through its economic and demographic importance to the development of Vietnam as a whole. On the one hand, the Delta is still seen to play a major role for securing the food security not only of the country but also more regionally. In addition, the (mostly agricultural) export revenues generated in the Mekong Delta make an important contribution to the national trade balance. Moreover, providing prospects for socio-economic development to the population in the Mekong Delta can reduce out-migration, which is hoped to work towards achieving the political aim of the central government to limit the growth of Vietnam's urban centres beyond control, notably of the emerging mega-city Ho Chi Minh City (see the national strategy for the development of urban centres to 2020, Decision 10/1998/QD-TTG). In accordance with these considerations, the Delta is likely to receive continued political attention and commitment, particularly from the central level.

On the other hand, the Mekong Delta has been underperforming in terms of profitability and economic value added – particularly in the industrial and agricultural sectors – and it lags behind the expected progress in terms of social development. The potential for socio-economic progress is only insufficiently capitalised on, which is to great extent due to flaws in the institutional framework conditions with respect to, for example, economic stimuli, educational systems, socio-economic planning mechanisms, and land or housing markets. In addition, the Delta increasingly has to face intensifying or emerging challenges in the fields of ecosystem degradation, international competition, and climate change. Due to the strong intensification of agricultural practices, environmental degradation poses a serious threat to the long-term economic sustainability of agricultural production in the Delta (Ni et al. 2001). Such degradations include, for example, the disruption of natural nutrient replenishment, the contamination of water bodies with excess agrochemicals, or soil salinization due to poor hydraulic management (Chaps. 13 and 14). On top of this, the Mekong Delta is highly exposed to the projected effects of climate change, most notably sea level rise, changing precipitation patterns and increasing typhoon activity, resulting in increased risk of flooding and extended salinization (e.g. MONRE 2009; Carew-Reid 2008; Dasgupta et al. 2007; Delgado et al. 2010; Chaudhry and Ruysschaert 2007; Wassmann et al. 2004; Garschagen 2010a). These effects could soon become a serious threat not only for agricultural production but also for cities and industrial premises which are – due to infrastructure reasons – often located along the Delta's costs and riverbanks or in exposed flood plains (Birkmann et al. 2010a; Garschagen 2009). This calls for more strategic and regionally integrated spatial planning in future, defining priority areas for protective infrastructure and land use transitions. Yet, recognising and debating also the financial, physical and cultural limits of climate change adaptation in the Mekong Delta will be necessary.

Furthermore, the Delta faces challenges with respect to demographic and social transformation processes. Notably the expected further urbanisation needs to be managed in a socially inclusive manner accommodating in particular the needs of resource-poor labour migrants. Attention also needs to be paid to the increasing social and ethnic disparities in the Delta in order to foster inclusive development and avoid the solidification of less progressive groups whose potential for contributing to the wider socio-economic development is not tapped. The provisioning of adequate social infrastructure is of key importance in this context, notably with respect to education, housing, health and transportation infrastructure.

In order to negotiate these challenges and to realise the development potential of the Delta increased attention needs to be given to integrated and anticipatory planning and management that synchronises not only horizontal and vertical administrative units but also mediates between the interests of different governmental and non-governmental stakeholders, notably state agencies, the population and private investors. In order to use public investments more efficiently and to establish stronger attractors for domestic and international investment, medium- and long-term economic planning needs to strengthened on a regional (i.e. Delta-wide) scale. The current competition for resources between different governmental sectors as well as between single provinces and districts may lead to a sharpening of specific local profiles, however, it also leads to a rather fragmented economic landscape and – from a broader perspective – to insufficient use of resources. From a regional economic policy perspective, it seems that the concentration of economic activities in the larger cities of the Delta needs to be fostered in order to create positive agglomeration effects for the local businesses and households. In this context, the Delta provinces should also strive for benefiting more intensively from the proximity to the Ho Chi Minh City. It is striking that so far the Mekong Delta has been benefiting much less from this proximity than the provinces north and east of the metropolis. Increasing negative agglomeration effects are likely to trigger a considerable relocation of economic activities out of Ho Chi Minh City which can potentially foster the economic development of the Mekong Delta – if the institutional conditions are provided. The recently approved Construction Master Plan for the Mekong Delta to 2020 and Vision to 2050 (Government Decision 158/2009/QD-TTg) foresees a multi-polar structure with a number of growth hubs throughout the Delta. Yet, there is reasonable concern that this will remain a visionary plan which shows impressive intentions but has limited prospects for implementation.[13] This is because the institutional mechanisms for more integrated regional (i.e. cross-provincial) planning are not provided or insufficiently enforced – hence, not tackling the aforementioned fragmentation of planning at provincial and even district level.

In addition, socio-economic planning for the Delta needs to be less driven by idealised visions of the future – as can be observed in current socio-economic master plans and sector specific master plans – but deserves thorough integrated scenario planning across different scales. Such planning needs to be based on actual needs assessments and projected trends in demographic and economic patterns in a liberalised economy and society. For instance, the above indicated link of to-be-expected labour transition, rural-urban migration and urbanisation needs to be acknowledged and translated

[13] Such plans are locally often referred to as *quy hoach treo*.

into adequate policies and action with respect to infrastructure development. The necessary implementation of public social housing schemes is one of the most pressing issues in this respect. Another instructive example is the link between the poor performance in terms of education or vocational training and slow progress in industrialisation and value adding. From a systemic point of view, the provisioning of relevant education and training would over the next decades be needed in order to absorb such work force which is increasingly losing its livelihood base in agricultural production. In addition the educational system needs to be designed in a way that it supplies the required highly qualified work force, allowing the Delta to climb up the ladder of industrial value added. Special attention needs in this respect to be given to the remaining inequalities, particularly with regards to the comparatively poorer educational performance of the Khmer ethnic minority and of women. Institutional barriers in the field of education need to be overcome through, for example, multi-language approaches and family support policies targeting young women.

Furthermore, the economic security and profitability of small and mid-sized farming households deserves to be enhanced through improving the institutional framework conditions. This could, for example, include improved access to capital from state-owned banks, a decrease in the role of land title as collateral requirement, tighter regulation on land speculation to limit the rapid growth of land prices, fostering public support and assistance for the diversification of production, or installing stronger mechanisms to buffer price fluctuations at a larger scale. Also the paradigm of producing large quantities of high yield agricultural goods – driven by the deeply rooted paradigm for food security – would deserve some critical rethinking. Given that the Delta and Vietnam have for long been a large net-exporter of foodstuffs, export revenues from the agricultural sector could today be drastically increased by focusing on more diversified production aiming at high quality products and, hence, higher price segments and revenues. Besides the envisaged diversification in agricultural production, more emphasis has to be paid to higher value added activities in food processing and other industrial sectors. Business promotion needs to be much more oriented towards the strengthening of local private companies in order to make full use of the endogenous development potential. So far, private companies remain rather small and operate on low levels of technology.

Such cases illustrate the need for fostering integrated and future-oriented planning and management, in order to shape institutional framework conditions that allow for converting the socio-economic potentials of the Delta into comprehensive economic and socially inclusive progress. Particularly the fields of agricultural transformation, industrialisation, migration, and urbanisation and the links between them deserve increased attention in this respect. The main challenge in this context is to improve the fit between – on the one hand – the thriving and continuing liberalisation, privatisation and global integration of economic activity in the Delta with – on the other hand – the established governmental planning and administration system which is not only overstrained in terms of resources but is also still based heavily on top-down management paradigms coupled with poor horizontal and vertical synchronisation and limited experience with respect to steering the flourishing market forces.

## References

ADB [Asian Development Bank] (2009) The economics of climate change in Southeast Asia: a regional review. ADB, Manila

Akram-Lodhi AH (2004) Are 'landlords taking back the land'? An essay on the agrarian transition in Vietnam. Eur J Dev Res 16:757–789

Baulch B, Chuyen TTK, Haughton D, Haughton J (2007) Ethnic minority development in Vietnam. J Dev Stud 43(7):1151–1176

Biggs D (2004) Between rivers and tides: A hydraulic history of the Mekong Delta. Ph.D. Dissertation, University of Washington, pp 1820–1975

Biggs D, Miller F, Hoanh CT, Molle F (2009) The delta machine: water management in the Vietnamese Mekong Delta in historical and contemporary perspectives. In: Molle F, Foran T, Käkönen M (eds) Contested waterscapes in the Mekong Region. Hydropower, livelihoods and governance. Earthscan, London, pp 203–225

Binh TNKD, Vromant N, Hung NT, Hens L, Boon EK (2005) Land cover changes between 1968 and 2003 in Cai Nuoc, Ca Mau Peninsula, Vietnam. Environ Dev Sustain 7:519–536

Birkmann J, Garschagen M, Kraas F, Quang N (2010a) Adaptive urban governance: new challenges for the second generation of urban adaptation strategies to climate change. Sustain Sci 5(2):185–206

Birkmann J, Garschagen M, Tuan VV, Binh NT (2010b) Vulnerability profiles for present and future water-related hazards in the Vietnamese Mekong Delta – providing the information-base for successful coping and adaptation within the framework of integrated water resources management. In: WISDOM Phase I – Final project report. Presented to the Federal Ministry of Education and Research, Germany

Carew-Reid J (2008) Rapid assessment of the extent and impact of sea level rise in Viet Nam, climate change discussion paper 1, ICEM –International Centre for Environmental Management, Brisbane, Australia

Chaudhry P, Ruysschaert G (2007) Climate change and human development in Viet Nam. Occasional paper to the human development report 2007/2008. UNDP, Hanoi

Coclanis P, Stewart M (2011) Precarious paddies: the uncertain, unstable, and insecure lives of rice farmers in the Mekong Delta. In: Coclanis P, Stewart M (eds) Environmental change and agricultural sustainability in the Mekong Delta. Springer, New York, pp 103–114

Coulthart A, Quang N, Sharpe H (2006) Urban development strategy. Meeting the challenges of rapid urbanization and the transition to a market oriented economy. World Bank, Hanoi

Dang A, Goldstein S, Nally J (1997) Internal migration and development in Vietnam. International Migration Review 31(2):312–337

Dasgupta S, Laplante B, Meisner C, Wheeler D, Jianping Y (2007) The impact of sea level rise on developing countries: a comparative analysis. World Bank, Washington, DC

De Silva SS, Ingram GJ, Nguyen TP, Bui MT, Gooley GJ, Turchini GM (2010) Estimation of nitrogen and phosphorus in effluent from the striped catfish farming sector in the Mekong Delta, Vietnam. Ambio 39:504–514

Delgado JM, Apel H, Merz B (2010) Flood trends and variability in the Mekong river. Hydrol Earth Syst Sci 14(3):407–418. doi:10.5194/hess-14-407-2010

Desbarats J (1987) Population redistribution in the socialist republic of Vietnam. Popul Dev Rev 13(1):43–76

Dorward A, Anderson S, Bernal YN, Vera ES, Rushton J, Pattison J, Paz R (2009) Hanging in, stepping up and stepping out: livelihood aspirations and strategies of the poor. Dev Pract 19:240–247

Dun O (2011) Migration and displacement triggered by floods in the Mekong Delta. Int Migr 49(S1):200–223

Dun O, Gemenne F (2008) Defining environmental migration. Forced Migr Rev 31:10–11

Duncan R, Pollard S (2002) A framework for establishing priorities in a country poverty reduction strategy. ERD working paper series no. 15. Asian Development Bank, Manila

Epprecht M, Müller D, Minot N (2011) How remote are Vietnam's ethnic minorities? An analysis of spatial patterns of poverty and inequality. Ann Reg Sci 46(2):349–368

Evers H-D, Benedikter S (2009) Hydraulic bureaucracy in a modern hydraulic society – strategic group formation in the Mekong Delta, Vietnam. Water Altern 2(3):416–439

Freitag U (2003) Erschließung und Entwicklung der Greater Mekong Subregion. Geograph Rundsch 55(1):20–25

Garschagen M (2009) Urban climate change adaptation in Vietnam – Institutional challenges and research agenda. Paper presented at the International Workshop on the Impacts of Climate Change on Urban Flooding, June 2009, Ho Chi Minh City, pp 24–25

Garschagen M (2010a) Potential humanitarian crises and climate change adaptation in the coupled social-ecological systems of the Mekong Delta, Vietnam. In: Shen X, Downing TE, Hamza M (eds) Tipping points in humanitarian crises: from hot spots to hot systems. United Nations University, Institute for Environment and Human Security, Bonn, pp 45–55, Source, 13:45–55

Garschagen M (2010b) Opportunities and challenges of climate change adaptation in high risk areas using the example of the Mekong Delta, Vietnam. In: Dölemeyer A, Zimmer J, Tetzlaff G (eds) Risk and planet earth. Vulnerability, natural hazards, integrated adaptation strategies. Schweizerbart, Stuttgart, pp 56–65

Garschagen M (2011) Resilience and organisational institutionalism from a cross-cultural perspective – an exploration based on urban climate change adaptation in Vietnam. Nat Hazards. doi:10.1007/s11069-011-9753-4

Garschagen M, Kraas F (2011) Urban climate change adaptation in the context of transformation – lessons learned from Vietnam. In: Otto-Zimmermann K (ed) Resilient cities: cities and adaptation to climate change. Springer, Dordrecht, pp 131–139

Garschagen M, Renaud F, Birkmann J (2011) Dynamic resilience of peri-urban agriculturalists in the Mekong Delta under pressures of climate change and socio-economic transformation. In: Steward M, Coclanis P (eds) Environmental change and agricultural sustainability in the Mekong Delta. Springer, Dordrecht

GSO [General Statistics Office] (2000a) The 1999 Vietnam population and housing census: number of households having houses by province and number of houses. Statistical Publishing Office, Hanoi

GSO [General Statistics Office] (2000b) Vietnam living standard survey 1997–1998. Statistical Publishing Office, Hanoi

GSO [General Statistical Office] (2000) Statistical year book of Vietnam. Statistical Publishing House, Hanoi

GSO [General Statistics Office] (2009) Results of the survey on household living standards 2008. Statistical Publishing House, Hanoi.

GSO [General Statistics Office] (2010a) The 2009 Vietnam population and housing census. Statistical Publishing House, Hanoi

GSO [General Statistics Office] (2010b) Report on the labour force survey of Viet Nam, 01.09.2009. Tabulated tables. Nha Xuat Ban Thong Ke, Hanoi

GSO [General Statistics Office] (2010c) The real situation of enterprises through the results of surveys conducted in 2007, 2008, 2009. Statistical Publishing House, Hanoi

GSO [General Statistical Office] (2010d) Statistical year book of Vietnam. Statistical Publishing House, Hanoi

GSO [General Statistics Office] (2011) Statistical data on Vietnam. Database of the GSO website. Accessed 13 Mar 2011

Ha VV, Nhan DK, Be TT (2008) The potential role of integrated farming for poverty alleviation: assessment of farmer based networks. A technical report (project SEI project no. 6591)

Haggard S (1999) Governance and growth: lessons from the Asian crisis. Asian Pac Econ Lit 13(2):30–42

Haggblade S, Hazell PBR, Reardon T (2007) Research perspectives and prospectives on the rural nonfarm economy. In: Haggblade S, Hazell PBR, Reardon T (eds) Transforming the rural nonfarm economy. Opportunities and threats in the developing world. Johns Hopkins University Press, Baltimore, pp 381–395

Hardy A (2001) Rules and resources: negotiating the household registration system in Vietnam under reform. Sojourn 16(2):187–212

Hens L, Nico V, Tho N, Hung NT (2009) Salination of surface water, groundwater, and soils in the shrimp farming areas of the coastal Cai Nuoc district, South Vietnam. Int J Environ Stud 66:69–81

Hirschman AO (1958) The strategy of economic development. Yale University Press, New Haven

Hoanh CT, Guttman H, Droogers P, Aerts J (2003) Water, climate, food, and environment in the Mekong Basin in Southeast Asia. Final report to the ADAPT project, adaptation strategies to changing environments. International Water Management Institute, Mekong River Commission Secretariat, Institute of Environmental Studies

Huy HT (2009) Rural to urban migration as a household decision: experimental evidences from the Mekong Delta, Vietnam. DEPOCEN working paper series no. 2009/20. http://www.depocenwp.org

ICEM [International Centre for Environmental Management] (2010) MRC Strategic Environmental Assessment (SEA) of hydropower on the Mekong mainstream: summary of the final report, Hanoi, Viet Nam

IFAD [International Fund for Agricultural Development] (2009) Climate change impacts – South East Asia. Paper for the Asia and Pacific annual performance review workshop 2009

Joffre OM, Bosma RH (2009) Typology of shrimp farming in Bac Lieu Province, Mekong Delta, using multivariate statistics. Agric Ecosyst Environ 132:153–159

Käkönen M (2008) Mekong Delta at the crossroads: more control or adaptation. AVISO 37(3):205–212

Ketels C, Cung ND, Anh NTT, Hanh DH (2010) Vietnam competitiveness report 2010. Central Institute for Economic Management & Lee Kuan Yew School of Public Policy, National University of Singapore, Singapore

Khoi PD, Loc TD, Danh VT (2008) An overview of the development of private enterprise economy in the Mekong Delta. In: Van Nam M, Lensink R (eds) Economic development of the Mekong Delta in Vietnam. CDS research reports CDS-27. University of Groningen, Centre for Development Studies (CDS), Groningen, pp 7–30

Kim A (2011) Talking back: the role of narrative in Vietnam's recent land compensation changes. Urban Stud 48(3):493–508

Krugman PR (1993) Geography and trade. MIT Press, Cambridge/London

Kummu M, Sarkkula J (2008) Impact of the Mekong flow alteration on the Tonle Sap flood pulse. Ambio 37:185–192

Lang N (2006) Viet Nam. In: ADB (ed) Urbanization and sustainability in Asia: case studies of good practice. http://www.adb.org/Documents/Books/Urbanization-Sustainability/default.asp. Accessed 28 July 2011

Laquian A (1996) China and Vietnam: urban strategies in societies in transition. TWPR 16(1):iii–xii

Loc VTT, Sinh LX, Bush SR (2007) Transboundary challenges for fisheries policy in the Mekong Delta, Vietnam: implications for economic growth and food security. In: Be TT, Sinh BT, Miller F (eds) Challenges to sustainable development in the Mekong Delta: regional and national policy issues and research needs. Sustainable Mekong Research Network (Sumernet), Bangkok, pp 99–142

Look At Vietnam (2009a) Mekong Delta waterway projects need investment, 19 Jan 2009

Look At Vietnam (2009b) Vietnamese market awash with low-quality rice, 29 May 2009

Luong HV (2003) Wealth, power, and inequality: global market, the state, and local sociocultural dynamics. In: Luong HV (ed) Postwar Vietnam: dynamics of a transforming society. ISEAS, Singapore, pp 81–106

Marsh S, MacAulay T (2006) Land reform and the development of commercial agricultural in Vietnam: policy and issues. http://www.agrifood.info/review/2002/Marsh.pdf. Accessed 25 Mar 2009

McGee TG (2008) The urban future of Vietnam reconsidered. JJ Ritsumeikan Geogr Soc 12:1–18

Miller F (2003) Society-water relations in the Mekong Delta: a political ecology of risk. PhD Dissertation, University of Sydney, Department of Geography

Ministry of Fisheries & World Bank (2005) Vietnam fisheries and aquaculture sector study. The final report

MONRE [Ministry of Natural Resources and Environment of Vietnam] (2009) Climate change and sea level rise scenarios for Vietnam, Hanoi

Myrdal G (1957) Economic theory and underdeveloped regions. Harper & Row, London

Nhan DK (2009) Determinants of rice yields and economic returns in the Mekong delta in the period 1995–2006. Sci J Can Tho Univ 4:119–128 (in Vietnamese with English abstract)

Nhan DK, Duong, LT, Thanh Sanh NV, Verdegem MJC (2004) Development of "VAC" integrated farming systems in the Mekong Delta: a view of a systems and a participatory approach. In: Yamada R (ed) The development of agriculture and sustainable farming systems in the Mekong Delta. Tre Publishing House, Ho Chi Minh City, pp 101–125

Nhan DK, Phong LT, Verdegem MJC, Duong LT, Bosma RH, Little DC (2007a) Integrated freshwater aquaculture, crop and animal production in the Mekong Delta, Vietnam: determinants and the role of the pond. Agr Syst 94:445–458

Nhan DK, Be NV, Trung NH (2007b) Water use and competition in the Mekong Delta. In: Be TT, Sinh BT, Miller F (eds) Challenges to sustainable development in the Mekong delta: regional and national policy issues and research needs. Sustainable Mekong Research Network (Sumernet), Bangkok, pp 143–188

Nhan DK, Duong LT, Thanh DN, Bosma RH, Verdegem MJC (2008a) On-farm integration of freshwater agriculture and aquaculture in the Mekong delta of Vietnam: the role of the pond and its effect on livelihoods of resource-poor farmers. In: Bank TW (ed) Sustainable land management source book. World Bank Publications, Washington, DC, pp 71–75

Nhan DK, Son NN, Be TT (2008b) Raising basin level agricultural water productivity for poverty reduction: Mekong Delta case study. Mekong Basin focal project report

Nhan DK, Son NN, Be TT (2010) Characterisation and improvement of farming systems in the buffer zone of the U Minh Thuong National Park, Kien Giang, Vietnam: multivariate analysis and multi-criteria decision making approaches. A technical report (project B2007-16-74)

Nhan DK, Trung NH, Sanh NV (2011) The impact of weather variability on rice and aquaculture production in the Mekong Delta. In: Stewart MA, Coclanis PA (eds) Environmental change and agricultural sustainability in the Mekong Delta. Advances in global change research. Springer, Dordrecht 45:437–451

Ni DV, Safford R, Maltby E (2001) Environmental change, ecosystem degradation and the value of wetland rehabilitation in the Mekong Delta. In: Adger N, Kelly M, Ninh NH (eds) Living with environmental change: social vulnerability, adaptation and resilience in Vietnam. Routledge, London, pp 122–135

Pfau DW, Long GT (2006) The growing role of international remittances in the Vietnamese economy: evidence from the Vietnam (Household) living standard surveys. Paper presented at the conference on global movements in the Asia Pacific Ritsumeikan Asia-Pacific University (APU), Oita, Japan, 17–18 Nov 2006

Phan LT, Bui MT, Nguyen TTT, Gooley GJ, Ingram BA, Nguyen VH, Nguyen TP, De Silva SS (2009) Current status of farming practices of striped catfish, Pangasianodon hypophthalmus in the Mekong Delta, Vietnam. Aquaculture 296:227–236

Ravallion M, van de Walle D (2004) Breaking up the collective farms: welfare outcomes of Vietnam's massive land privatization. Econ Transit 12:201–236

Ravallion M, van de Walle D (2008a) Does rising landlessness signal success or failure for Vietnam's agrarian transition. J Dev Econ 87:191–209

Ravallion M, van de Walle D (2008b) Land in transition: reform and poverty in Rural Vietnam. The World Bank, Washington, DC

Reis N (2012) Tracing and making the state. Policy practices and domestic water supply in the Mekong Delta, Vietnam. Lit, Berlin (in press)

Renaud F, Bogardi J, Dun O, Warner K (2007) Control, adapt or flee. How to face environmental migration? vol 5, Intersections. United Nations University, Bonn

Renaud FG, Dun O, Warner K, Bogardi J (2011) A decision framework for environmentally induced migration. Int Migr 49(S1):5–29

Revilla Diez J (1999) Vietnam: addressing profound regional disparities. Southeast Asian Aff 358–374

Saigon Times (2001) Vietnam well short of export-quality rice, 13 Dec 2001

Sang LT (2007) Urban migration in pre- and post-reform Viet Nam: macro patterns and determinants of urbanward migration, the 1984–1989 and 1994–1999 periods. Paper presented at the 8th APMRN international conference on migration, development and poverty reduction, 26–29 May 2007

SIWRP [Southern Institute for Water Resources Planning] (2010) Draft report on water resources infrastructure assessment study in the Mekong Delta. Vietnam-Netherlands Mekong Delta master plan project.

Smith D, Scarpaci J (2000) Urbanization in transitional societies: an overview of Vietnam and Hanoi. Urban Geogr 21(8):745–757

SRV [Socialist Republic of Vietnam] (2004) National report on disaster reduction in Vietnam for the world conference on disaster reduction in Kobe-Hyogo, Hanoi, 18–22 Jan 2005

SRV [Socialist Republic of Vietnam] (2008) Decision on the approval of the national target program to respond to climate change, decision no. 158/2008/QD-TTg

SRV [Socialist Republic of Vietnam] (2011) Decision on the pilot provision of agricultural insurance during 2011–2013. Decision no. 315/2011/QD-TTg

Tam NT (2010) Changes in coastal shrimp farming systems in Soc Trang province. MSc. thesis, Mekong Delta Development Institute

Taylor P (2004) Redressing disadvantages or re-arranging inequalities? Development interventions and local response in the Mekong Delta. In: Taylor P (ed) Social inequality in Vietnam and challenges to reform. ISEAS, Singapore, pp 236–269

Taylor P (2006) Rivers into roads: the terrestrialisation of a south-east Asian River Delta. In: Leybourne M (ed) Water: histories, cultures, ecologies. University of Western Australia Press, Crawley, pp 38–52

Thanh HX, Huong DTT, Phuong NT, Tacoli C (2008) Urbanization and rural development in Vietnam's Mekong Delta Livelihood transformations in three fruit-growing settlements. Working paper 14. Working paper series on rural-urban interactions and livelihood strategies. International Institute for Environment and Development, London

Thao N, Vromant N, Hung NT, Hen L (2008) Soil salinity and sodicity in a shrimp farming coastal area of the Mekong Delta, Vietnam. Environ Geol 54:1739–1746

Thrift N, Forbes D (1986) The price of war. Urbanization in Vietnam 1954-85. Allen and Unwin, London

Trong NV (2008) Vietnam – Working toward the production of safe and high-quality aquaculture foods. Food & Fertilizer Technology Center. http://www.agnet.org/library/bc/55005/. Accessed on 29 July 2011

Turley W (1977) Urban transformation in South Vietnam. Pac Aff 49(4):607–624

UNDP & AUSAID [United Nations Development Programme & Australian Agency for International Development] (2004) The Regional Poverty Assessment Mekong River Region. UNDP & AUSAID, Hanoi

Van Brakel M, Hambrey J, Bunting S (2011) Mekong – inland fisheries and aquaculture. Regional case study: R6.UK Government's foresight food and farming project

Van de Walle D, Gunewardena D (2001) Sources of ethnic inequality in Viet Nam. J Dev Econ 65(1):177–207

Van Etten J (2007) Urban transition in Vietnam: implications for urban management. Aladin 8(4):3–4

Vietnam Business News (2011) Agricultural insurance needs a firm foundation to succeed, 26 May 2011

Vietnam Net (2008) Rice exports in 2007: good business but poor quality, 28 Jan 2008

Vietnam News (2005) Quality new name of the game for rice, 22 Jan 2005

Vietnam News (2009) Mekong Delta farmers focus on quality, 14 Nov 2009

Vietnam News (2011) Farm insurance hard to understand, 30 Aug 2011
Voice of Vietnam (2009) Vietnam must promote rice brand names abroad, 5 Aug 2009
Wassmann R, Hien NX, Hoanh CT, Tuong TP (2004) Sea level rise affecting the Vietnamese Mekong delta: water elevation in the flood season and implications for rice production. Clim Chang 66:89–107
Wassmann R, Jagadish SVK, Sumfleth K, Pathak H, Howell G, Ismail A, Serraj R, Redoña E, Singh RK, Heuer S (2009) Regional vulnerability of climate change impacts on Asian rice production and scope for adaptation. Adv Agron 102:91–133
Willets J, Carrard N, Retamal M, Nam NDG, Paddon M, Thuy DX, Trung NH, Mitchell C (2010) Selecting sanitation options. A case study of South Can Tho. Technical report. Institute for Sustainable Futures, University of Technology, Sydney
World Bank (2006) Transport strategy: transition, reform and sustainable management. World Bank, Washington, DC
World Bank (2009) World development report 2009 – reshaping economic geography. World Bank, Washington, DC
World Bank (2011) The migration and remittance factbook 2011. World Bank, Washington, DC
Xo LQ, Lam DT (2009) Updating and inventorying water resources in sub-area 10. A draft report of the Basin Development Plan Programme. Vietnamese Mekong River Committee, Ha Noi (in Vietnamese)
Yeung YM (2007) Vietnam: two decades of urban development. Eurasian Geogr Econ 48(3):269–288
Yip N, Tran H (2008) Urban housing reform and state capacity in Vietnam. Pac Rev 21(2):189–210
Zhang HX, Kelly PM, Locke C, Winkels A, Adger WN (2006) Migration in a transitional economy: beyond the planned and spontaneous dichotomy in Vietnam. Geoforum 37(6):1066–1081

# Chapter 5
# IWRM for the Mekong Basin

**Florian Moder, Claudia Kuenzer, Zengrang Xu, Patrick Leinenkugel, and Bui Van Quyen**

**Abstract** The Mekong Basin is facing its most pronounced human induced changes in history. Development on the river's mainstream, climate change, and economic growth of the riparian countries are taking a toll on the highly sensitive aquatic ecosystems and affect the livelihoods of millions of people in the region. Integrated Water Resources Management (IWRM) can provide a holistic approach to address the emerging challenges in the Mekong Basin. However, transboundary discords between the riparian countries, scattered responsibilities at national levels, power asymmetries, and the absence of legal frameworks and commitment to an IWRM approach at a basin scale make it difficult to implement a strategy that ensures a stable economic growth, sustainable management of natural resources as well as social well-being. On the other hand almost all riparian countries of the Mekong have the potential for a positive economic growth and progressing technical development in the next years. If this development is framed by an appropriate ecological awareness the Mekong Basin could develop in an ecologically and socially sustainable manner.

F. Moder (✉) • B. Van Quyen
Bộ Khoa học và Công nghệ, Ministry of Science and Technology (MoST),
31 HanThuyen – Q. 1, TP. Ho Chi Minh, Vietnam
e-mail: florian.moder@cimonline.de

C. Kuenzer • P. Leinenkugel
German Aerospace Center (DLR), Earth Observation Center (EOC),
German Remote Sensing Data Center, Oberpfaffenhofen,
82234 Wessling, Germany
e-mail: claudia.kuenzer@dlr.de; Patrick.Leinenkugel@dlr.de

Z. Xu
Institute of Geographic Sciences and Natural Resources Research, CAS,
11A, Datun Road, Chaoyang District, Beijing 100101, China

F.G. Renaud and C. Kuenzer (eds.), *The Mekong Delta System: Interdisciplinary Analyses of a River Delta,* Springer Environmental Science and Engineering,
DOI 10.1007/978-94-007-3962-8_5, 

## 5.1 Basics of Integrated Water Resources Management

### 5.1.1 *Brief History*

The first "Integrated Water Resources Management" (IWRM) approach was developed and implemented by the Tennessee Valley Authority in 1933. It combined navigation, flood control, and power production with erosion control, recreation, public health, and welfare. It already comprised many components of the current understanding of IWRM, such as the combination of natural resource utilisation with the economic, social, and environmental needs. The IWRM approach was further shaped during the United Nations Conference on Water, which was held in Mar del Plata, Argentina in 1977, some 40 years later. The conference clearly addressed the issue for a better coordinated approach within the water sector by assessing the status of water resources, the need to increase water sufficiency and to develop concepts to avoid a global water crisis before the end of the twentieth century. The coordination of the water sector was seen during that time as a clear task of national governments, assisted in financial terms and on technical issues by bilateral or multilateral donors.

Another important conference was the 1992 Dublin conference in Ireland. This conference set the basis for the current understanding of IWRM, acknowledging that the management of water resources is an integral part of a nation's social and economic development and that maintaining healthy water ecosystems is essential for sustainable water management. The conference finally issued four "Dublin guiding principles" which are still applied today (ICWE 1992:1). The "Dublin guiding principles" acknowledges that water resources are finite, vulnerable, and are essential to sustain life, development and the environment. Water development and management should be based on a participatory approach, involving users, planners and policymakers at all levels. The principles identify the outstanding role of women in the provision, management and safeguarding of water resources. The last statement of the declaration recommends that water has an economic value in all its competing uses and should be recognised as an economic good. The fourth principle is subject to intensive debates, which have lasted for years and are far from being closed, since basic human right suppose considering water as a commodity. However, the four Dublin principles built the foundations for the Agenda 21 resolution in Rio de Janeiro in1992 during the United Nations Conference on Environment and Development (UNCED) (Rahaman and Varis 2005). In Rio de Janeiro a fifth principle was added, stating that integrated water resources management is based on the equitable and efficient management and sustainable use of water. All five principles are referred hereafter by the "Dublin-Rio Principles". Several other conferences followed the Dublin conference, such as the Second World Water Forum and Ministerial Conference in the Hague in 2000, the International Conference on Freshwater in Bonn in 2001, the World Summit on Sustainable Development in Johannesburg in 2002 and the Third World Water Forum in Kyoto in 2003 and the 5th one in Istanbul in 2009. All these conferences had in common the acknowledgement that the role of stakeholders had to be strengthened and emphasized the need to raise awareness and the importance of sustainable water management

Looking at the history of the last 100 years, water management experienced a transition from a partly implicit integrative and rather sectoral water management approach, considering the resource "water" as isolated, towards an explicit water management approach as given by the definition of IWRM as a systematic process, influencing environmental, social and economic issues. All components of this process are linked to each other and have complex dependencies. Realizing the complexity of water management by approaching it holistically, the effects of water management on economy, society and ecology were acknowledged. Therefore, water resources management has the potential to induce economic, social and ecological changes. All different social, environmental, and economic constraints have to be considered equally in the context of IWRM to achieve sustainability and to gain a proper balance between the different needs and user groups of limited resources.

### 5.1.2 *Definition of the Term IWRM*

Proper management of water resources sets the basis for sustainable economic growth as well as environmental and social well-being. The management of IWRM has to be understood in this context in a wider sense. It accentuates the need for the management for a sustainable and long term development of water resources by a more comprehensive approach. This means that IWRM has to integrate land cover, aquifers, surface water, as well as coastal areas and other natural resources. All of these components are dependent on each other through the water cycle, so that changes of one component will have effects on others. Therefore, a proper risk management is necessary in order to respond and mitigate future changes and disasters on an even unknown and unpredictable scale. Management of water resources can therefore never be described as a static approach, since water resources and its dependencies are too complex. New aspects in the context of IWRM emerge and disappear over time, such as new technologies, new socio- economic structures, as well as changing environmental conditions. The management of water resources must consider these uncertainties, handle them properly and in the best-case approach, provide forward-looking solutions Therefore, IWRM is described as an open end process, developing in a spiral manner over time (UNESCO 2009). Risk mitigation and adaptation play a crucial role.

Several attempts to define the term "IWRM" have been undertaken, the most acknowledged definition had been given by the Global Water Partnership (GWP). GWP defines IWRM as "a process which promotes the coordinated development and management of water, land and related resources, in order to maximize the resultant economic and social welfare in an equitable manner without compromising the sustainability of vital ecosystems" (GWP 2000). The definition clearly mirrors the commitment to the Dublin-Rio Principles.

### 5.1.3 IWRM in Transboundary Basins

Implementing IWRM for rivers makes it necessary to choose appropriate geographical and hydrological units in order to meet the holistic approach of IWRM. The basin level as the coarsest unit in terms of geographical and hydrological characteristics of a river is one of the most evident unit. This was confirmed in the United Nations Conference on Environment and Development (UNCED) Agenda 21. It links the management of water, land, and other resources to comprise the needs of IWRM at the basin level, which is also referred to as "Integrated River Basin Management" (IRBM). IRBM allows the integration of processes, which are up- and down-stream at the widest scale as well as their aquifers.

Many rivers, lakes, and aquifers are shared by two or more nations. In 2002 UNESCO listed 263 transboundary basins, covering nearly half of the earth surface. About 40% of the world's population relies on these shared water sources, which provide about 60% of the total freshwater resources available to humankind (Wolf et al. 1999). A total of 145 nations have their territory within international basins. Nineteen basins involve five or more different countries, among which is the Mekong River Basin with six countries (China, Myanmar, Lao PDR, Thailand, Cambodia and Vietnam). The Mekong River Basin covers an area of about 795,000 km$^2$, which includes some of the poorest regions in the world, in which more than 40% of the population lives in poverty (MRC 2010a). The Mekong River is used for agriculture, aquaculture, hydropower as well as transportation and provides income and food security for 60 million people. Since the catchment area doesn't follow political borders the Mekong River experiences increasing, and often conflicting demands along the 4,800 km long flow from the source in China until the Delta in Vietnam. Emerging conflict potential for resources in the transboundary Mekong Basin is induced by changing environmental conditions and altering demographic patterns, which goes hand in hand with the change of local and global economic developments of the riparian countries. The riparian countries of the Mekong do not follow the same development paths, due to different availability of natural resources, different economic development strategies, and financial capabilities of the countries as well as demographic patterns (Keskinen 2008). A further reason for power asymmetries between the different riparian countries is the location of the country along the river itself. Countries, which are located downstream, depend on water management decisions of upstream countries. However, all riparian countries acknowledged that the economic growth through the utilisation of natural resources offers the opportunity to address the inequitable sharing of resources and benefits, which could be approached by implementing IWRM on a Basin Scale.

### 5.1.4 Climate Change and Impacts on IWRM in the Mekong Basin

The international scientific community agrees that there is an anthropogenic influence on global climate caused by the emission of Green House Gases (GHG) into the earth atmosphere, causing rising temperatures on a global scale. Increasing temperature influences the physical environment, such as rainfall patterns, wind, sea level, ecosystem processes, as well an increased occurrence of severe storm events. Water is the primary medium through which climate change influences the Earth's ecosystems and therefore people's livelihoods and well-being(UN 2009:1). IPCC declared in 2007 that freshwater systems are among the most vulnerable to climate change (IPCC 2007). Impacts of climate change for water are predicted to be on water availability, temporal occurrence, quality, and demand.

Several studies to predict climate change impacts in the Mekong Basin have been undertaken with slightly different results, due to different scenarios, models, as well as different applied geographical scales. All studies concluded that the temperature rises at least by a minimum of 0.01°C/year (Ruosteenoja et al. 2003) up to a maximum of 0.036°C/year (Hoanh et al. 2010). Seasonal precipitation patterns are predicted to change as well. However, the intensity as well as the geographical coverage depends strongly on what atmospheric $CO_2$ level will be reached and which prediction method is applied. All studies point out that the Mekong region will experience a longer dry season, which goes most likely along with a change in the flow of the river and its tributaries, as well as with a higher possibility for extreme weather events such as typhoons (MRC 2010b). Sea level rise, caused by thermal expansion of the water body, predicted by IPCC could reach up to 1 m and have a high potential to affect especially the Mekong Delta by flooding and salt-water intrusion into the ground as well surface water (Chap. 2). All studies agree that there will be a change in climate with severe impacts on the water balance of the Mekong Basin. However, the intensity of climate change effects itself on the Mekong Basin remain uncertain, since all attempts to predict future climate and related effects are based to some extent on assumptions, which have to be taken due to unknowns in the equations for predicting future climate and the complexity of climatic interactions themselves. Since the predicted effects of climate change are most likely to vary in the different areas of the Mekong Basin a detailed discussion of effects due to climate change in different parts of the basin are given in the section "Challenges of IWRM for the Mekong" below.

The question which arises at the basin scale is how those predicted changes with different magnitudes affect IWRM in the Mekong Basin and how to deal with the uncertainty of climate predictions and its effects on the water regime. IWRM provides tools to identify as well as to implement adaptation and mitigation strategies, addressing climate change related risks and uncertainties, which can be applied for the Mekong Basin as well. IWRM tools deal with topics such as enabling environment, analysing institutional roles and change instruments (GWP 2011). Adaptation measures are the response to climate change that seeks to reduce the vulnerability of natural and human systems against actual or expected climate change effects, whereas mitigation measures targets on reducing the emission of GHG to reduce the risk at the hazard level. Mitigation as well as adaptation measures are discussed in detail in Chap. 2.

### *5.1.5 Legal Framework in the Mekong Basin*

The Mekong Committee and a management framework for the Mekong Basin were created in 1957 through the signature of a treaty by the four lower basin countries

(Cambodia, Laos Thailand, and South Vietnam). The treaty intended to establish a committee for investigation, planning, and development projects in the Mekong Basin. Cambodia left the Mekong Committee in 1977 and re-joined in 1995, which was the beginning of the Mekong River Commission (MRC) as it is known today. The agreement was signed between the governments of Cambodia, Lao PDR, Thailand, and Vietnam, with the purpose of jointly managing shared water resources and developing the economic potential of the river applying IWRM principles. The Mekong Basin Agreement does not allocate water, but requires that the countries "utilize the waters of the Mekong River system in a reasonable and equitable manner" (MRC 1995). China and Myanmar are "Dialogue Partners", meaning that they are only working within a limited cooperation framework with the MRC, which ultimately jeopardizes basic IWRM principles at the basin scale. In accordance with the IWRM principles, the MRC defines the Basin Development Plan (BDP) as an underlying development strategy for the Mekong Basin, which provides a comprehensive plan for water resources and transboundary governance. The BDP defines long term strategies, current objectives, and guidance for the implementation of IWRM at basin, national, and sub-basin levels in a sustainable and mutual beneficial way with all stakeholders. Recent challenges which have been identified by MRC for the time period 2011–2014 are the need for energy and raw material. This trend stands in contrast to low public investment in water management infrastructure and has the high potential for severe impacts on the Mekong Basin environment, ecosystems, and fishery with cascading effects on the people's livelihood. Sustainable implementation of hydropower dams by preserving vital ecosystems has been stated as one of the major challenges for IWRM related issues of the MRC in the on-going years (MRC 2010b). Activities of the MRC have not been consistent so far and have been shaped by stop-go policies between development aid focusing on environmental as well as social issues and development of large infrastructure projects. The alignment of MRC depended strongly on the preferences set by the Chief Executive Officer (CEO) of the MRC. At the beginning of the MRC, the institution was confronted with criticism due to improper communication amongst governments and the public, a lack of public engagement and missing mechanisms to enforce implementation of activities, recommendations and decisions (Ha 2011). However, recent activities of the MRC indicate a recognition of the problem and the institution responded by releasing numerous policy studies and meeting notes, and organizing stakeholder forums to improve information sharing and communication. MRC has widened its asset as a knowledge institution holding technical and scientific data. The "evolution" of the MRC in the last 15 years has therefore showed the ability of the institution to reform itself, being confronted with numerous challenges in the past. Topics such as a stronger focus on coordination, better articulation at national level, stronger results-based focus, improved ownership, better harmonization and a reduced dependency on the needs of individual donors were identified and are included in the Strategic Plan of the MRC 2011–2015. Although the Strategic Plan of MRC gives a clear guidance on future development, the shift towards "riparianization" which implies the transition of leading positions led and directed by citizens of riparian countries intended to strengthen ownership might have adverse effects. The objectivity of the MRC might be alleviated by individual interests of single member countries trying to occupy leading positions in order to push and influence decisions in favour of their country.

Apart from the MRC there is also an initiative called the "Greater Mekong Subregion" (GMS), which was set up in 1992 by all six riparian countries of the Mekong, with assistance from the Asian Development Bank (ADB) and other donors. The GMS program promotes regional development in the riparian Mekong states by fostering transport, energy, telecommunications, environment, agriculture, tourism, trade facilitation, investment, and human resource development on sub regional scale. The concept of GMS focuses more on regional economic cooperation and integration rather than water resources management.

## 5.2 Challenges of IWRM for the Mekong

According to Gupta 2009 the Mekong Basin can be divided in seven larger landform units, of which the first one is the 'Mountainous Panhandle' in China (Upper reaches), followed by the 'Mountains of northern Lao PDR and Thailand', which borders the Laotian-Vietnamese Annamite Range in the Southeast, the Mekong Lowland in the Centre, and the Korat Upland in eastern Thailand in the Southwest of the 'Mountains of northern Lao PDR and Thailand' area (see Fig. 5.1), all of which we consider the 'middle reaches of the Mekong' (except for the Mekong Lowland, here only the northern panhandle of this landscape unit is counted to the middle reaches). The lower reaches of the Mekong Delta is located downstream of the Korat Plateau represented mainly by the Great Cambodian Plain and the Mekong Delta, which splits into two main branches in Phnom Penh: the Mekong and the Bassac Rivers, entering the Mekong Delta in South Vietnam.

### *5.2.1 Upper Reaches (China)*

The upper reaches of the Mekong River to the boundary of China is called "Lancang River". The Lancang River originates from Zaduo county (5,200 m above sea level), located in the Qinghai province. The river flows southward and leaves China flowing through the tropic rainforest of the Nanla Bayout located in Yunnan Province. The Lancang River is over 2,160 km long, with a basin area of more than 167,000 km$^2$. The average annual runoff of Lancang River reaches 74 billion m$^3$ per year. In general, there are two reasons causing changes in terms of ecological and hydrological aspects in the upper catchment of the Mekong River: one is human activity, especially hydropower development; the other is climate change.

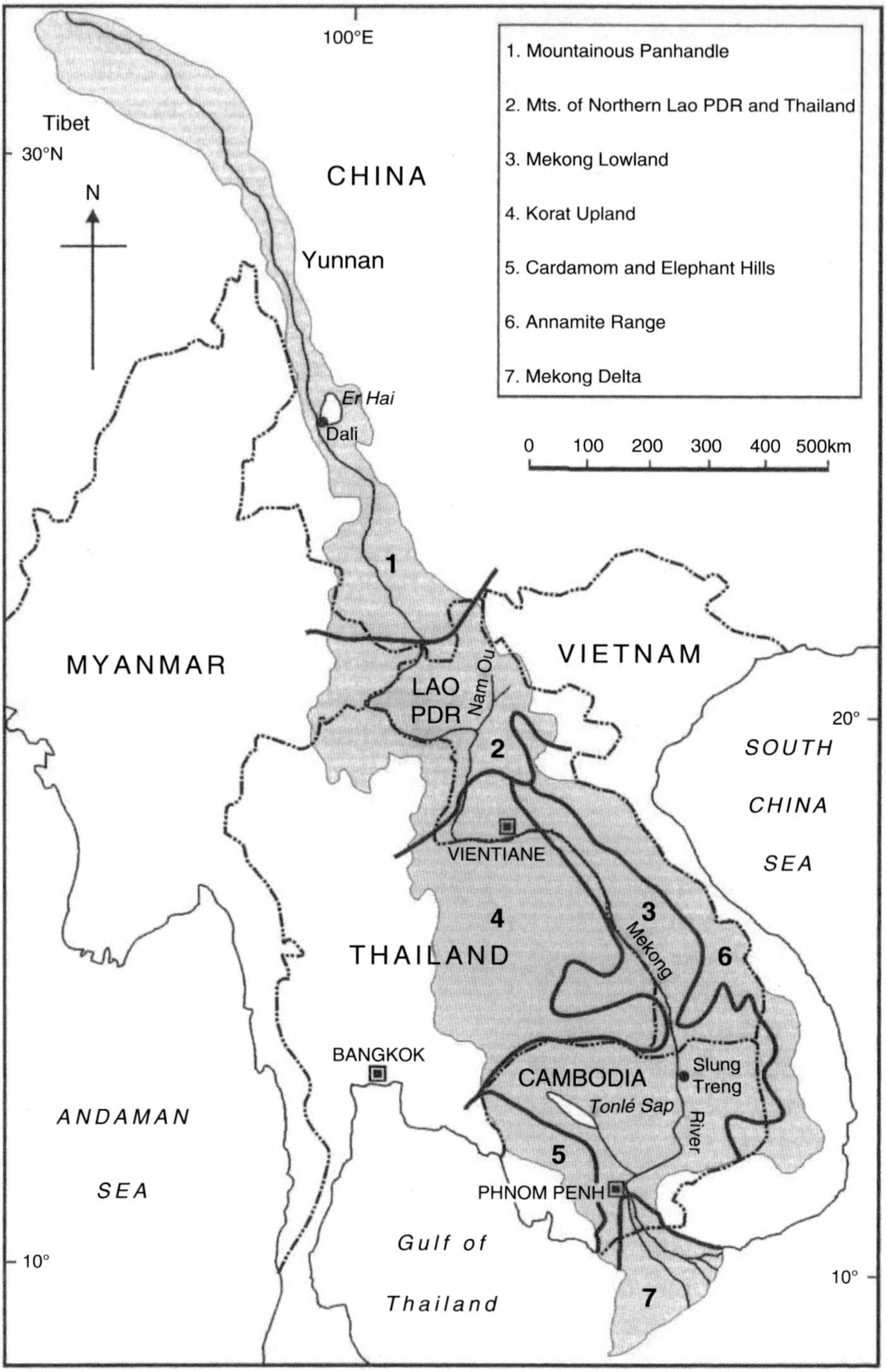


**Fig. 5.1** The Mekong Basin: physical units (Gupta 2009)

#### 5.2.1.1 Challenges Caused by Climatic and Environmental Changes

Droughts and floods are tending to occur more frequently due to climate change. The Lancang River Basin became warmer and drier in the past 40 years (He and Zhang 2005). Current studies suggest an increase in the mean temperature of approximately 0.8°C by 2030 with a range of potential impacts on the Lancang River such as melting glaciers in the Himalaya Mountains which will increasingly contribute water to the river system in the upper reaches of the Mekong Basin. Furthermore, an increase of annual precipitation with more than 13.5% (equivalent to 0.2 m) is expected, leading in certain areas to an increased flood risk. This goes along with a decrease of precipitation in the dry season in some areas, making parts of the basin more drought prone (such as the drought in Yunnan and other four provinces in China in spring 2010 (Liu 2010)).

Precipitation in the upper reaches of the Lancang River has decreased since 1960–2005, however extreme rainfalls events have increased. This resulted in more frequent drought and flood events, with crucial effects on the provision of surface water in areas of the middle- and lower-reaches of Lancang River. The water quality in the Lancang River remains in general in a good state (MRC 2010a).

#### 5.2.1.2 The Challenge of Balancing Economic and Environmental Needs

Water shortage, water pollution, and environmental degradation are caused by human intervention such as irrigation, industrial development, resident water consumption, while watercourse fragmentation and aquatic ecosystem fragmentation as well as changes in sedimentation patterns have to be assigned to damming.

In the late 1980s, China began to construct eight hydroelectric dams, on the Lancang's lower reaches, aiming to supply 15.6 GWh of electricity annually (see Fig. 5.2). Four have been completed so far, including the tallest with a height of 292 m, named Xiaowan. The magnitude of hydrologic and ecologic impacts such as low water levels of these hydropower dams have been subject of intensive discussion, such as the low water level of the Mekong River during the year 2003–2004. However, it was proved that the low water levels were caused by reduced rainfall rather than by retention through the dams in China (MRC 2004). Manwan and Dachaoshan dams have only a slight impact on the water environment and aquatic ecosystem. However, after completion of the Xiaowan dam, impacts are predicted to be more severe, mainly due to the increased runoff in the dry season and reduced runoff in wet season (He and Zhang 2005). Changes of transboundary runoff of the Lancang River are mainly caused by changed rainfall intensity in Yunnan. Dams on the lower Lancang reduce runoff only during the rainy season, when reservoirs are being filled. Dry season water releases is expected to increase the river volume by 35%. "There are a lot of accusations that the dams in China are exacerbating the current low water levels, but the Chinese have informed downstream nations that they will not fill any reservoir during the dry season," says Roger Mollot, a fisheries expert with the World Wide Fund for Nature in Vientiane, Laos (Stone 2010). Things may

**Fig. 5.2** Hydropower projects in the lancing river basin (Daming 2010)

get worse due to climate change. After examining weather and tree ring data, scientists found out, that in the past 40 years Yunnan became warmer and drier—a trend that started long before the dams were built (Stone 2010).

China's part of the territory of the Mekong and the impact of reduced runoff made only a minor fraction of the overall Mekong flow volume in the lower reaches of the Mekong Basin. "Before the construction of the dams of the Lancang Cascade, about 18% of water, and half of the sediment used to arrive from the panhandle in China. The rest of Mekong's annual water discharge (82%) mainly comes from 4 sources: (a) mountains of northern Lao PDR through a number of tributaries, (b) the Southern Mountains via the San, Kong and Srepok (c) the Mun Chi System, draining a large part of the Korat Upland, and (d) the drainage outflow from the Tonle Sap" (Gupta 2009:47). In addition, another researcher noted that only about 16% of the total discharge of the Lower Mekong River comes from China and 2% from Myanmar (Adamson et al. 2009:57).

Even if China would completely block all flows originating from within their boundaries, the major loss downstream would be a maximum 16% of the water. This means, the dams in China can temporary withhold a maximum of 3–5% of the water that usually reaches in the South. This still represents a large volume of water and should not be underestimated; however, it is still a relatively small fraction of the overall stream flow volume of the Mekong. This however does not consider the impact in the middle reaches of the Mekong Basin. China gives priority to its national interests, rather than considering impacts of their water management on downstream countries. This is contrary to the idea of a basin-wide integrated water management. Hence, during the dry season in 2004, public media in countries of the middle and lower Mekong reaches blamed quickly China as being responsible for the drought, which caused widespread public anger at China. However, low flows in Thailand, Laos and Cambodia on that occasion were not caused by the operation of dams in China (Campbell and Manusthiparom 2004) but were the consequence of water extraction for intensive irrigation in these countries. Adamson et al. (2009:67) estimated that in the lower plains at Phnom Penh in the month February, March, and April, 60%, 45% and 40% (respectively) of the flows is abstracted for irrigation, leading to unprecedented low levels of water. The discussion on impacts of upstream hydropower dams was quite emotional rather than based on facts, caused by improper information sharing policy of China and the lack of field data, creating conflict potential with the downstream countries.

#### 5.2.1.3 Challenges for Institutional Transboundary Cooperation for IWRM at Basin Level

Transboundary hydrological effects of hydropower dams, which have been built and are under construction in upper catchments of the Mekong River are claimed to have the following effects:

- water level change
- water re-allocation
- water temperature change
- watercourse change
- sediment change

As discussed in the section above, reasons for transboundary river management and ecological disputes are caused by improper information sharing between upstream and downstream countries in Mekong River Basin. Reasons for that might be that the information obtained from different groups, institutions and individuals are targeting different objectives. There is no data quality standard defined and data sources are tending to be inconsistent and are not standardized. This makes it difficult to share and to especially compare data between upstream and downstream stakeholders and others. Furthermore, there is lack of enough quantitative data in the whole Mekong Basin.

Transboundary problems arise as well from the fact, that there is no integrated basin-wide Environmental Impact Assessment (EIA) approach. There are quite a few models, which could be used to assess cumulated environmental impacts and the loss of ecological functions of aquatic ecosystems, caused for example by hydropower dams. However, no integrated model could analyse the multiple interactions between dam building, climate change, and land use change at a basin-wide level so far. Recent years showed that there are no suitable multi-level international cooperation mechanisms at the basin wide level in place. The absence of cooperation mechanisms hindered transboundary community participation, joint studies and information exchange between the upper and lower reaches of the Mekong River.

#### 5.2.1.4 Perspectives for IWRM at the Upper Reaches

China claims that all mainstream dams in the upper Mekong region are committed to be operated based on the "eco-hydrological process" in order to maintain the transboundary eco-security. China defined eco-security by limiting the minimum outflow of 728 $m^3/s$ during the storing period in Xiaowan reservoir and the outflow process has to be consistent and oriented on the natural flow process as much as possible. Furthermore, a reduction of peak discharge in the outbound river section is intended. Before regulation measures via eco-hydrological operation of dams have taken place, the water level variation changed about 1 m in 1 h. The implementation of hydropower dams caused a decrease of the variation to 0.88 m in 1 h, which is helpful for navigation and flood control purposes.

An integrated basin wide plan for sustainable development is not yet in place. Great differences of resources, environment, social, and economic conditions in all riparian Mekong States make a basin wide cooperation difficult. However, an integrated basin wide plan will be the key to develop the Mekong Basin sustainably. If the riparian Mekong countries could coordinate and use the integrated plan, the mismatch and conflict over water resources and environment could be reduced by harmonization at the regional, national and transboundary level.

In a river system, usually, the resources and good access to the river, flat arable bank topography, sediments and navigation in the lower reaches are much better than those in the upper area and the development history in the downstream region tends to be longer than that in the upper region. This is not reflected by international water law, which requires a revision and improvement. International laws' principles tend to protect downstream country's benefits and ignore the need for a proper development of upstream areas. In fact, international rivers' development is a "Bidirectional Road". This means upstream water utilization causes impacts downstream and vice versa (runoff change, sediment change, water temperature change etc.). This position was reflected during the general assembly of the United Nations in 1997, which was held to adapt the "Convention on the Law of the Non-navigational Uses of International Watercourses" (ILC) introduced by the UN in 1995 (United Nations 1995). China claimed that the ILC fails to reflect the principle of the territorial sovereignty of a watercourse, highlighting that a State has indisputable sovereignty of a watercourse flowing through its territory. Further, ILC shows clearly an imbalance between the rights and obligations of upstream and downstream States. Hence, ILC does not follow the basic principles set by the United Nations Charter according to Chinas point of view (United Nations 2007). As a result, China rather preferred to negotiate issues bilaterally with neighboring countries and voted against the adaption of ILC in 1997, which has been also not yet ratified by the UN in 2008 since only 16 of necessary 35 countries voted in 2008 for the implementation into international law. One reason for this might be the fact that river basin's upstream area are usually remote areas at high altitude, which are sparsely settled and thus no voices are raised to represent upstream needs. Apart from the fact that upstream countries have a much better geopolitical position, in the upper reaches of the Mekong (Lancang) River increased urban development is ongoing. The cities of Xishuangbanna or Menghanzhen are in need of water for electricity production, urban and agricultural water supply and navigation. Regional development equity is the key for international rivers' sustainability and cooperation. The basic principles of major international water laws should be revised and improved to include regional (mainly between downstream and upstream countries) fair articles, from which not only the Mekong will benefit in terms of sustainability but also international rivers around the world (Daming 2010). Scientific institutes in China could play a key role in this process and facilitate the dialogue towards a sustainable development of the Lancang-Mekong River. For instance, the Asia International River Center of Yunnan University, the Center for International Trans-boundary Water and Eco-Security of Tsinghua University, the Yunnan Key Lab of International Rivers and Transboundary Eco-Security, to name only a few, could serve as a platform in technology and science innovation, data sharing, and experts consult related Lancang-Mekong river issues. For this purpose there is a need to establish and promote multiple platforms for research, consulting and information sharing, which would help to:

- Study market-oriented mechanisms as an incentive for water resource and aquatic ecosystem conservation
- Work with communities to maximize overall conservation efforts
- Develop close and reciprocal ties with government officials, from local to national levels
- Cooperate and collaborate directly with large-scale industries, from energy to agriculture to tourism, that affect natural resource use
- Develop methods, tools and guidelines to assess ecosystem vulnerabilities, valuate ecosystem services for adaptation and its payment mechanisms, and to identify technologies and good practices to restore ecosystem adaptation functions and trans-boundary ecosystem management (Chen 2010).

### 5.2.2 *Middle Reaches*

The last bigger city that the Lancang/Mekong passes in China is Menghanzhen, in Yunnan province, before – in a steep meander – it leaves the country to the South forming the border between Myanmar in the West and Yunnan Province in the East (China) for about 30 km, and then forming the border between Myanmar in the West and Laos in the East for about 250 km dissecting forested mountainous terrain before it also leaves Myanmar at the 'Golden Triangle', where the borders of Laos, Myanmar and Thailand meet at Sop Ruak. Within this stretch no major cities are located along the river's banks, and also additional water input through larger rivers is small. From the 'Golden Triangle' the river flows further South for over 150 km, before entering the sharp eastward turn towards Louang Phrabang. On this course through the mountains of Laos and Thailand in the North also forms the border of the two countries. In this section of alternate mountain ridges and steep valley up to the stretches behind Louang Phrabang the only larger tributary is the Nam Ou river; smaller tributaries include the Nam Mae Kham, Nam Mae Kok, Nam Me Lao Nam Mae Ing (from the Thailand side), Nam Tha and Nam Beng, Nam Ou, Nam Soung, Nam Pa and Nam Khan (from the Laos side).

#### 5.2.2.1 Challenges Caused by Climatic and Environmental Changes

As already stated above, climate change will result in a basin wide temperature increase of approximately 0.8°C until 2030 and annual precipitation increases of 0.2 m. The increase in precipitation, however, will be largely attributed to an increase in wet season precipitation. For most areas in the middle reaches of the basin (Tha Ngon, Nakhon Phanom, Mukdahan, Ban Keng Done, Yasothon, Ubon Ratchathani, and Pakse) dry season precipitation is projected to decrease (MRC 2009a). Consequently, the impact of climate change on the hydrological regime in the middle reaches is characterised by an increased difference between dry season and wet season discharge. Daily average discharges at Kratie are expected to increase by 5.14% in the rainy season and decrease by 2.18% in the dry season during the period of 2010–2049 when compared to the baseline decade of 1995–2004 (TKK & SEA START RC 2008). Although increased precipitation will result in greater basin wide annual runoff, which will improve overall water availability in the basin, some areas in north-eastern Thailand and Lao PDR will continue to struggle with high levels of water stress. For those areas with projected precipitation decreases in the dry season, climate change will inevitably result in changes in rice productivity. Particularly within the sub-basins of Northern and Central Lao (Moung Nouy, Louang Phrabang, Tha Ngon) food scarcity is projected to increase (MRC 2009a). Basin-wide, the requirements for irrigated agriculture will increase resulting in an expected decrease of crop yields by 2% when irrigation applications were maintained at the status quo (Eastham et al. 2008). Furthermore, increased storm intensity, duration and frequency in Lao PDR and Thailand will result in losses of agricultural yields and infrastructure. Against this background and the increasing food demand of the growing population in the Basin, substantial investment in the agricultural sector will be necessary (construction of irrigation systems, purchase of water pumps etc.).

#### 5.2.2.2 The Challenge of Balancing Economic and Environmental Needs

Two dams are planned in the region between the mountains of northern Lao PDR and Thailand: The Pak Beng and the Louang Phrabang Dam. However, to the South many further dams are planned within the Korat Upland section, the Annamite Range section as well as the very northern 'Mekong Lowland' section. These are the Sayabouly,

Paklay, Sanakham, Pak Chom, Ban Koum, Lat Sua, Don Sahong, Stung Treng and Sambor Dams – so overall a minimum of 11 dams directly situated on the main stream of the Mekong – not even considering the numerous other planned dams on tributaries flowing into the Mekong from the Korat Upland or the Annamite Range. It is however, especially in this section that the annual water yield for the Mekong is coming from: its middle reaches (Adamson et al. 2009) (see Fig. 5.3). Just like in China, the planning of these dams is motivated by the need for societal development, which leads to demands for energy, irrigation, flood and drought control, technological development, transportation and navigation. The building of these Thai and Laotian, as well as partially central Vietnamese dams will have the biggest influence on Mekong downstream riparian countries at the lower reaches (Fig. 5.3).

Independent of future dam related effects of water redistribution and changes in flow and sedimentation patterns in the Mekong, numerous environmental changes occur in the area. Shifting cultivation practices and government-sponsored forest clearing for cash crops in Thailand and Laos have caused permanent forest losses. Alone on the Korat Plateau in Thailand contributing the Mun and Chi tributary systems, the forested parts of the landscape were diminished from 42% in 1961 to 13% in 1993 (Adamson et al. 2009). Logging pressure in the Myanmar, Thailand and Laotian sections of the Mekong are high. Such landcover and landuse changes in the area lead to a decreased water holding capacity of soils, which according to the MRC (2005) will lead to decreased dry season flows. Furthermore the comparatively high urbanised provinces of north-east Thailand with the highest population concentrations in Udon Thani and Nakon Rachasimaas as well as the Laotian capital Vientiane, pose with their growing development of polluting industries an increasing threat to natural resources in the region. In 1992, for example, a poisonous spill from a sugar factory in Thailand drifted down the Phong, Chi, and Mun Rivers leaving behind 500 tons of dead fish (Valbo-Jorgensen et al. 2009). For these areas the challenge of promoting further economic growth while preserving natural resources is most challenging.

#### 5.2.2.3 Challenges for Institutional Transboundary Cooperation for IWRM at Basin Level

One of the main challenges for transboundary cooperation in the Mekong Basin is to reconcile the many different demands, interests and strategies of the various institutions and stakeholders of the member states. The government driven needs for greater economic growth and their individual strategies for environmental conservation and water resource management differ considerably and, with a focus on the individual hydropower development plans, even conflict each other. The role of Thailand and Laos in this respect is rather ambivalent. On the one hand the countries of the middle reaches are dependent on the inflow of the Mekong upper reaches and its tributaries and thus are directly affected by the changing flow patterns attributable to the development of the Lancang cascade in China. On the other hand Laos and Thailand plan to develop hydropower dams on the mainstream and its tributaries,

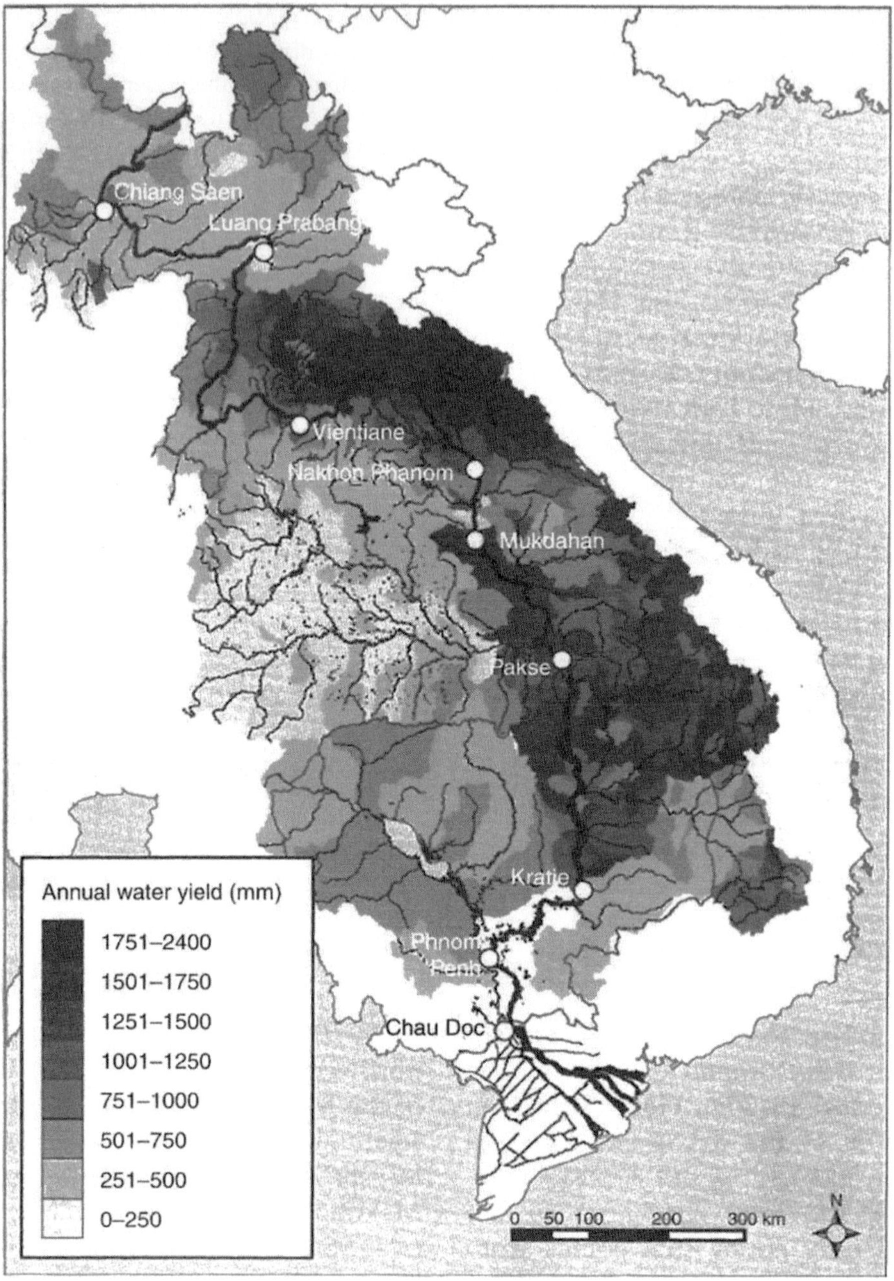


**Fig. 5.3** Annual water yield of the Lower Mekong Basin (Adamson et al. 2009)

which is likely to be accompanied by severe impacts on the river flow and flood schemes in Cambodia and Vietnam. The countries of the middle reaches thus have to enforce their interests at two contradicting fronts with their northern and southern neighbours. The unequal socio-economic development stages and the partially contrasting biophysical environmental conditions of the riparian countries of the Mekong states further exacerbate the challenges of harmonising national development plans and strategies.

The promotion and coordination of sustainable management and development of water and related resources for the countries' mutual benefit and people's wellbeing is the vision of MRC ( 2010b). However, the MRC failed to fulfil this vision so far. A reason for this might be the lack of common standards for basin-wide EIA approaches related to the development of the hydropower sector. Furthermore, insufficient communication and information sharing between the member states does not contribute to a closer cooperation in the field of IWRM (Kuenzer 2010). In addition, the MRC's ineffectiveness in jointly managing the shared water resources is hampered by the fact that the 1995 Agreement lacks real enforcement mechanisms (Ha 2011). The MRC is unable to influence the national policies of its member countries and is rather powerless against bilateral agreements on water resource strategies between neighbouring countries. However, the MRC has been, and will also constitute in future, the only organisational body in the region with the potential of promoting transboundary and basin wide strategies in the field of IWRM and currently attempts to redefine its tasks from a implementation-oriented to a coordination-oriented Commission (Schmeier 2011).

#### 5.2.2.4 Perspectives for IWRM in the Middle Reaches

Although the Governments of Lao PDR and Thailand have implemented a number of activities within the framework of IWRM as integral part of national adaptation and mitigation measures, still a number of shortcomings have to be addressed in the future. Particularly at the administrative and planning levels, data and information on local as well as regional climate change impacts is limited, and is furthermore rarely exchanged. In this respect, facilitating access to harmonized and standardized information sharing would be helpful to fulfil the criteria of IWRM in the Mekong Basin (Kuenzer 2010). With respect to the above-mentioned challenges of transboundary cooperation in the field of IWRM, the middle reaches countries may in future play also a more prominent role for facilitating collaboration amongst the Mekong Basin countries and serve as a mediator. In this respect, Laos as geographical center of the Mekong Basin and host of one of the MRC offices in Vientiane (OSV) may take over a more active role in basin-wide IWRM coordination by providing a more efficient link with the various institutions and stakeholders of the Mekong Basin.

### 5.2.3 *Lower Reaches*

The lower reaches of the Mekong River start downstream of the Korat Plateau then pass the Laotian city Pakse, flow through the Khone Falls and enter the Great Cambodian Plain with its numerous back swamps. The Mekong splits into two main branches in Phnom Penh: the Mekong and the Bassac Rivers, entering the Mekong Delta in South Vietnam before being discharged into the South China Sea. The Mekong Delta, as part of the lower reaches, is treated separately in Sect. 2.4 because of its economic importance for Vietnam and its geophysical distinct characteristics.

A unique feature of the Great Cambodian Plain is the Tonle Sap Lake, which is the biggest freshwater lake in Southeast Asia. The Tonle Sap Lake is connected to the Mekong by the Tonle Sap River. During the peak of its annual flow in the annual flood season, the water flow reverses its direction towards the Tonle Sap Lake. During the dry season the lake drains and plays a significant role in limiting the salinity intrusion of surface water in the Mekong Delta. The inundation area in an average rainy season (May–November) is over 1.2 million ha, almost five times the dry season's coverage and with a water level of almost 10 m (ADB 2006). Hence, fishery plays a crucial role in Cambodia and is the primary source of protein. Agriculture is also the second biggest source of income for the Cambodian economy, with a share of 35% of the national Gross Domestic Product (GDP). Cambodia's main income comes from the service sector with 42% and industry with 23% of GDP. The agriculture sector increased at a growth rate of 5% in 2009 (World Bank 2009). The population growth rate is predicted to be moderate at 1.6% for 2011 (the average population growth rate in South East Asia is 2.2%). Even though Cambodia contributes low GHGs emissions, the effects of global warming have the potential to affect the country seriously. Since fishery plays an important role in Cambodia and the country has a limited capacity to adapt to the potential impacts of climate change, Cambodian economy and society are both highly vulnerable to the effects of climate change.

#### 5.2.3.1 Challenges Caused by Climatic and Environmental Changes

Cambodia, which has a strong focus on agriculture production, is especially vulnerable to weather-related disasters. Climate change is predicted to lead to increase of the average annual temperature of 0.7–2.7°C by 2060, and 1.4–4.3°C by 2090 (McSweeney et al. 2008). The frequency of hot days and hot nights has increased by 13% and 17% since the 1960s. Rising air and water temperatures will affect aquatic ecosystems. Apart from potential eutrophication, fish reproduction, growth and migration patterns are likely to be altered due to a changed distribution and abundance of aquatic competitors and predators (MRC 2010c). Agriculture as the second most important economic sector is also very likely to be affected, especially by prolonged dry seasons.

In the lower reaches of the Mekong, a change in rainfall patterns and intensity caused by climate change is very likely in the next 100 years. Generally, the precipitation will increase up to 14% in the rainy season. The increased precipitation will go along with more flood events, which are stated to rise from 5% to 76% probability (MRC 2010c) and affect peak flows, flood duration, and results in larger flooded areas through increased runoff. Recent flood events such as in 1998 and 2002 caused rice production losses of 70% and 350 human causalities. The economic loss due to damages on infrastructure and crops was estimated to be around 150 million USD. The reduced rainfall in the dry seasons tends to decrease runoff and results in a higher probability of water stress.

Increased variation in precipitation between the dry and wet season will increase the water level of the Tonle Sap Lake by about 2.3 m each year during the wet season, resulting in an increase of the lake size of about 20%. The wet season will start earlier and last longer and the dry season rainfall and runoff will decrease although minimum flows during the dry season are predicted to increase (MRC 2010c).

Further environmental changes such as decreasing soil and water quality as well as the decline in the availability of natural resources have been observed by Nuorteva et al (2010). Deforestation especially increases the sediment load in the rivers due to increased erosion. FAO (2005) estimates an annual loss of forest by 2% in recent years. A further 2.7 million ha are predicted to be converted by 2020, by extraction of high value timber and conversion of forest areas into cash crop plantations. Increasing global demand for agriculture products such as coffee, rubber, tea etc. accelerate this process.

#### 5.2.3.2 The Challenge of Balancing Economic and Environmental Needs

In Cambodia, climate change will most likely affect fishery the most important protein source in the country. The aquatic ecosystem will undergo significant changes affecting habitat, ecosystem productivity, as well the abundance of aquatic competitors and predators. A reduction of fish stocks as a result of increasing water temperatures in the Tonle Sap Lake is another scenario drawn by scientists. Higher temperatures reduce oxygen solubility, which raises the demand of oxygen and food intake of the fish and results in an increase of uptake of aquatic pollutants (MRC 2009b). Increased water temperatures can also favour the survival of parasites and bacteria. These factors could lead to a reduction of fish stocks, especially affecting species with relatively narrow thermal tolerances, which will be replaced by more temperature tolerant species (Ficke et al. 2007). The impact of climate change might however not only be negative for fish stocks. Higher nutrition content in the water, caused by higher runoff rates leading to increased sediment and organic load, and a raised water level of the Tonle Sap Lake will also increase the feeding ground and provide space for more species. This could also lead to increased fish stocks in the lower Mekong reaches (MRC 2010c), assuming that current conditions are projected into the future without consideration of further human interventions such as building more hydropower dams upstream of the Mekong and in its tributaries. Vietnam is the front-runner with respect to develop tributary hydropower potential in the Central Highlands tapping the Se San and Sre Pok Mekong tributaries. The extent of ecological impacts of hydropower dams on the aquatic ecosystems is yet unknown in the Mekong Basin. However, extreme water level fluctuations during the dry season are very likely. Apart from water level fluctuations, sediment trapping behind the dams is a serious concern, since the upper reaches of the Mekong and the Se San and Sre Pok Mekong tributaries are the main sediment contributors to the Lower Mekong Catchment. The sediment balance in the Mekong River is not only influenced by hydropower dams, but also by land conversion practices taking place in the riparian countries of the Mekong. On the one hand, higher runoff, as an impact of climate change, and changed land use practices, increase the sediment load in the lower reaches of the Mekong River. On the other hand, hydropower dams trap sediments in the middle and upper reaches. Therefore, effects on sediment load for the middle reaches of the Mekong Basin remain uncertain, since dependencies on human intervention are complex and the full "equation" has not yet been understood completely.

#### 5.2.3.3 Challenges for Institutional Transboundary Cooperation for IWRM at Basin Level

Upstream development activities on the Mekong River have principally adverse effects on Cambodia. Hence, Cambodia positions itself is quite critical in regard to hydropower dams upstream and in the tributaries of the Mekong River. Recent incidents such as the unannounced release of water into the Se San River by Vietnam in 2002, which claimed human lives and caused economic losses in Cambodia's Ratanakiri Province, does not contribute to mutual trust between neighbours. However, Cambodia plans to establish hydropower dams as well, even though there are only a few suitable locations for this.

MRC has been called upon to facilitate dialogue among governments and other stakeholders to ensure that transboundary impacts of hydropower dams are considered. By signing the 1995 Agreement that established the Mekong River Commission, all member countries agreed that before any hydro-project can be built on the mainstream of the Mekong, all four members must reach an agreement. Although MRC members are only required to notify each other of their intention to develop a tributary, the 1995 Agreement forces members to ensure that no harmful effects will occur downstream in neighbouring countries. Informing each other and avoiding harm to downstream countries, however, leaves room for interpretation. Agreements between neighbouring countries on transboundary water issues are negotiated directly, without the direct involvement of MRC.

#### 5.2.3.4 Perspectives for IWRM at the Lower Reaches

Transboundary issues, especially in terms of hydropower dams, seem to be rather unsolved and the benefits to Cambodia itself are not visible. Cambodia faces challenging times in the next few years, caused by the changes to climate and river flow.

As previously stated, Cambodia is highly vulnerable as it does not have any adaptation strategies in place. The Government of Cambodia has addressed this by introducing a national framework response. The National Strategic Development Plan for example includes climate change aspects and IWRM perspectives and considers IWRM as a tool for climate change adaptation in the country (Cambo et al. 2007).

### 5.2.4 *Mekong Delta*

The Mekong Delta as a part of the lower reaches of the river basin starts near Phnom Penh, where the largest distributary river channel, the Bassac River, splits from the mainstream. South of Phnom Penh the delta expands rapidly and covers an area of 70,000 $km^2$ and has two major distributaries channels, the Mekong and the Bassac Rivers. Both rivers split in numerous smaller channels. The two mainstream rivers and the small channels play an important role for the livelihood of the local population, transportation, aqua- and agriculture in the Mekong Delta. It is home to 22% of Vietnam's population, produces half the nation's rice output of 49 million tons a year, 60% of seafood, and 80% of fruit crops in Vietnam (Brown 2009). Seasonal floods during the wet season bring fertile sediments, necessary for agriculture. However, the probability of extreme flood and drought has increased in recent years (Delgado et al. 2010; Chap. 9). Damage caused by high level floods is aggravated due to increasing population density, intensified farming/aquaculture practices as well emerging industrial sites along the river and its branches. The Mekong Delta encounters further challenges and problems, such as deteriorating water quality due to strong pesticide use in rice farming areas, increased hormone and antibiotics concentration due to unfiltered discharge of urban and industrial wastewater as well as intensification of aquaculture practices (Chap. 13). Changing climatic conditions and on-going rises in sea level result in salinization and acidification of soils, aquifers and surface water, and species/habitats decrease. These changes are also shaped by regulatory measures (hydropower) happening at the upper reaches of the Mekong.

#### 5.2.4.1 Challenges Caused by Climatic and Environmental Changes

As one of the most certain consequences of global warming, sea level rise (SLR) will have severe consequences for Vietnam with a hotspot being the low-lying Mekong Delta. The most imminent impact of SLR on the Mekong Delta will be the increased risk of inundation. A review of climate change impacts in the Delta is provided in Chap. 2 but salinity intrusion is discussed briefly below.

The intensity of salinity intrusion into the surface water of the Mekong Delta is influenced by tidal effects in the South China Sea/Gulf of Thailand and the water storage of the Tonle Sap, the river/canal systems and natural water retention areas such as wet lands in the Mekong Delta. In terms of the flow regime in the lower reaches of the Mekong it can be observed that the higher the discharge of the Mekong, the narrower the affected area of salinity intrusion. Keeping that in mind, recent developments indicate a disturbance of the balance between the fresh-water outflow from the Mekong River and salt-water inflow from the sea. Especially during the dry season, salinity intrusion, which is caused by tidal influence, reaches 40–60 km upstream (Wolanski et al. 1998). Reasons for the changed flow regime and resulting salinity intrusion are attributed to SLR, over-extraction of ground water, severe droughts and reduced runoff during the dry season caused by upstream regulatory measures of the Mekong. About 2.1 million ha were affected by saline water of 1 g/l and about 1.7 million ha of 4 g/l (Sam 2006). A change of salt-water concentration in the Mekong River directly affects agriculture and local aquatic ecosystems and indirectly affects local livelihood and economy by causing a shortage of freshwater, which is increasingly in demand due to a rapidly growing population, intense agriculture and industry in the Mekong Delta.

Acidification of soils is one of the increasing problems in the acid sulphate soil zone of the Mekong Delta (Chap. 14). Acidification is caused by oxidisation of sediments containing sulphides, which have been deposited during the last 10,000 years. Under anaerobic conditions, oxidation processes cannot take place, however, if those soils are exposed to air by droughts or after drainage for agriculture, a conversion of sulphide minerals to sulphuric acid takes place. Those sediments are then described as sulphate soils. Sulphate soil water solutions tend to dissolve aluminium, iron, manganese, and other heavy metals (Willett et al. 1993). Heavy metals can accumulate in aquatic ecosystems and agricultural products such as rice. The Mekong Delta has 2.3 million ha of recent sulphuric sediments (Minh 1995), representing an enormous 15% of the coastal sulphide sediments throughout the world. Most acid sulphate soils have been converted into agriculture areas in the 1970s in the Mekong Delta and are located in low-lying back swamp areas far from main distributaries in the Mekong Delta, such as the Long Xuyen Quadrangle, the Plain of Reeds and the Ca Mau Peninsula. pH values of the surface water lower than 5 during the month July in the Plain of Reeds were observed (Husson et al. 2000) and a study on the quality of canal and shallow ground water in the Long Xuyen Quadrangle during May and July found that the pH values of surface water were below 4 in acid soils (Dang et al. 2007). This proved, that the surface water in acid sulphate zones are heavily polluted by soil leaching, especially in the rainy season in which the highest amounts of released aluminium in the canals and low river discharges were observed (Minh et al. 1997). Land use as well as irrigation practices determine the intensity and extent of water acidification and pollution. Raised beds of upland crop for example, causes higher concentrations of acidic substances into the rice field due to oxidation processes, whereas low raised-bed field practice increase the pH value, resulting in aluminium precipitation (Minh et al. 1997) (Chap. 14).

Temperature rise due to global warming also affects the Mekong Delta. Globally, an average temperature increase of 0.1–0.2°C per decade is expected (IPCC 2007). According to the IPCC, South East Asia will follow this global trend. In Vietnam, the annual average temperature increased by 0.1°C per decade from 1900 to 2000, and 0.7°C during 1951–2000, suggesting that the temperature rose faster in the latter half of the century. Summers have become hotter in recent years, with average monthly temperatures increasing by 0.1–0.3°C per decade with longer and more severe dry seasons. Temperature rise has manifold effects. In particular, aquatic ecosystems are highly sensitive to increasing temperatures and are predicted to react by eutrophication. This would mean that complete ecosystem processes undergo modification due to increased oxygen consumption (MRC 2010d). Land ecosystems are very likely to react as well. Species distribution is very likely to change since important ecosystem processes, such as primary production, circulation of nutrients and degradation of organic matter, are directly affected by changing climate conditions (Kiem et al. 2008).

Apart from natural ecosystems, agriculture such as rice production will experience losses due to increasing temperatures. Experts estimate that an increase of 1°C will decrease rice production by 10% (Nguyen Binh Thin 2009). Precipitation patterns and intensity are also very likely to change by longer and dryer dry seasons where only 20–30% of yearly water will be available. Wet seasons will tend to be wetter and much shorter. The annual average rainfall will however remain about the same (Chinavanno 2009). Temporal distribution of the runoff comes along with changed rainfall intensity and patterns. About 80% of 505 billion $m^3$ annual runoff are discharged to the South China Sea in the months of May and June. In the event that no regulatory measures are taken, it is predicted that this discharge will increase by 21% in the rainy season and decrease downstream in the dry season (MRC 2010b). As a result, saline surface water will reach further distances from the river mouth, if no adaptation and mitigation measures are taken.

Environmental changes have to be considered in the context of rapid population growth in the Mekong Delta. The population density in 2005 was estimated to be 465 people per $km^2$, which is significantly higher than the national average of 424 people per $km^2$ in Vietnam (Brown 2009). Increasingly limited water availability causes more severe droughts, salinization, and acidification of soils, causing loses in agriculture and aquaculture. Lower water availability has the potential to sharpen the conflict between different water users. Changed flow and rain patterns increase the risk of severe flood events in the Mekong Delta, gaining in importance because of the combination of a predicted sea level rise and an increased risk of the occurrence of severe storms. This results in an as yet unrealised threat for vast areas of the Mekong Delta to be heavily impacted by floods with severe risks for the livelihood of local population, the local and national economies and highly sensitive aquatic ecosystems. In terms of the impact of climate change, the Mekong Delta can be described as one of the most vulnerable areas in the world (Chap. 2).

#### 5.2.4.2 The Challenge of Balancing Economic and Environmental Needs

The Mekong Delta benefits from the rapid economic growth of Vietnam with a GDP growth rate of about 7.5%, holding the potential to double every 10 years. This rapid economic growth sets enormous pressure on natural resources and on water resources in particular, by requiring large quantities of reliable water as well as energy resources. This results in increased wastewater generation, which is currently, in 80–90% of all cases, discharged untreated in Vietnam (ADB 2009a). This has severe impacts for the river ecology and livelihood of the local population. Besides the trend towards industrialisation and a shift of the population from rural to urban areas in the Mekong Delta, agriculture remains the most important economic factor with a share of 46% (Brown 2009) resulting in (over) application of herbicides, pesticides and fertiliser with negative consequences for the water quality (Chap. 13). Furthermore, agriculture is responsible for more than 80% of water consumption, which is higher than the worldwide trend of 70% (ADB 2009a). A shortage of fresh water, as it is predicted especially by a longer dry season in the Mekong Delta, makes it vitally necessary to consider options to increase water use efficiency in agriculture, proper wastewater treatment, and promoting alternative water use concepts for the industry, such as internalisation of the water cycles.

#### 5.2.4.3 Challenges for Institutional Transboundary Cooperation for IWRM at Basin Level

For the Mekong Basin and in particular for the Mekong Delta, the most challenging issue in terms of IWRM is the transboundary management of the Mekong River. Vietnam depends heavily on the management of natural resources in the upstream countries of the Mekong. Effects of their management and resulting ecological as well social impacts all accumulate and affect Vietnam. This includes the management of water retention areas, plans and activities to alternate the mainstream flow by setting up hydropower dams, land use change in the upstream countries, changing discharge patterns, and water quality. Dams built upstream will have admittedly severe impacts on the flow regime of the Mekong in the Delta. Vietnam might have difficulties arguing against those projects, since Vietnam itself had the fasted development in recent years and has an existing installed capacity of 1,204 Megawatt (MW), 1,015 MW under construction and further 299 MW in plan, tapping the Se San and Sre Pok Mekong tributary rivers (MRC 2010d) for hydropower production. In the Se San and Sre Pok tributaries shared by Vietnam, Lao PDR and Cambodia, 17 hydropower projects are already operational or at various stages of construction and eight further projects are at a detailed design stage. The national Power Development Plan identified 16 further projects and scheduled construction before 2025 (MRC 2010d). A reduced sediment supply, caused by several upstream hydropower dams will seriously affect the Mekong Delta causing an increased erosion of the coastal areas.

Transboundary water management requires fundamental data from upstream areas to support decision-making processes that have to be taken downstream. However, apart from the efforts of the MRC, which has established a data exchange platform, there seems to be hardly any willingness to exchange data due to territorial, political and interoperability issues between the different Mekong riparian countries.

#### 5.2.4.4 Perspectives for IWRM in the Mekong Delta

The Mekong Delta has a high potential for further positive economic development, since Vietnam itself will soon reach the status of a middle-income country (MIC). The World Bank defines MIC countries by a per capita income between $400 and $4,000 per year. The GDP has almost tripled in the last 10 years in Vietnam (Brown 2009) and is expected to grow further. A change from a low- to a middle-income country is caused by a strong growth rate in the industry and service sectors, resulting in significant demands on natural resources.

Sustainable economic growth in the Mekong Delta necessitates an implementation of IWRM principles, such as a holistic water management approach under participation of stakeholders and other involved water user groups. Finalizing and most importantly implementing the legal framework on water management is a prerequisite. The legal framework defines standards and norms for water resources management and is mirrored in the 'National Water Resources Strategy to 2020', the 'National Target Plans for Water Resources Management' and 'Climate Adaptation' released by the Ministry of Natural Resources and Environment (MONRE). However, currently, responsibilities for water management are fragmented at both national and subnational levels over several institutions and there is almost no coordinated approach towards an IWRM (Waibel 2010; Chap. 6). Decisions, which are taken on regional and local levels, should be adjusted according to the implemented national framework. As a core requirement of IWRM legal and institutional arrangements, settings for River Basin Institutions should be clarified and the role of River Basin Institutions itself must be strengthened at national and transboundary levels. A clear legal framework should both address and coordinate the numerous efforts of foreign supported projects, funded international governmental agencies, and NGOs to support Vietnam to reduce its vulnerability to the effects of climate change. Better coordination, methodological and thematic linkage between the different projects would significantly improve current efforts to increase coping capacities in the Delta.

Prediction and evaluation of environmental impacts arising from climate change and infrastructure development in the Mekong Delta are often hindered by the lack of adequate data, for example highly detailed topographic maps vitally necessary for hydrological modelling. Where data for baseline studies does exist, it is somehow scattered amongst different agencies. However there is no standard on data quality, or data sharing policy available. Addressing this issue in the context of a legal framework would be helpful to implementing IWRM in the Mekong Delta.

Flood events and effects of artificial alteration of the mainstream flow by hydropower dams for example, can only be addressed on a basin scale. In this context it is also necessary to focus on transboundary integrated river basin management which strengthens the institutional framework and through revision and application of international water law. All measures should aim to minimize the impacts of water related hazards and also ensure appropriate economic growth.

Climate change and its effects are predicted to affect the Mekong Delta severely, through sea water level rise, salt-water intrusion, changed seasonal rain patterns, extreme droughts, and storm events. The effects of climate change such as more severe drought and flood events will steadily increase vulnerability of the local population in the Mekong Delta. This vulnerability can only be reduced through a holistic water management approach. This means that for example coastal management, land use planning, urban development and socio-economic structures, as well as related changes, have to be considered and addressed as part of IWRM and strengthened by the national framework. This would mean in that Vietnam, different ministries and their sub institutes such as Vietnam Mekong Commission (VNMC) of MONRE and Ministry of Agriculture and Rural Development (MARD) on national as well as on provincial level have to cooperate and work together on structural and non-structural measures, within the frame of a legal framework. Vietnam and especially the Mekong Delta must make huge efforts to improve and adapt structural measures for flood control and mitigation such as dams, canals, reservoirs, embankment structures for prioritized areas, promote elevated residential areas, introduce salinity control structure such as sluices and river-sea dyke system for example. Non-structural measures such as a definition of a Delta Master Plan for response, adaptation and mitigation, inhabitant resettlement, implementation of a forecasting and early warning system, improvement of regional and bilateral cooperation and institutional strengthening have to be addressed.

The challenges which lie ahead for the Mekong Delta can be only counteracted by an integrated and coordinated approach as suggested by the basic IWRM principles. Structural and non-structural measures should be addressed equally and implemented step-by-step to improve resilience towards climate change. Resilience in this context defines the capacity of linked social-ecological systems to absorb recurrent disturbances such as hurricanes or floods so as to retain essential structures, processes, and feedbacks (Walker et al. 2004). Positive economic development and increasing prosperity in Vietnam might offer the financial scope to implement structural and non-structural measures on a large scale and help to reduce vulnerability, keeping in mind that reduced vulnerability paves the way for future social and economic development.

## 5.3 Identified Needs for Action in the Mekong Basin

### *5.3.1 The Technical Context*

Mitigation measures to reduce SLR and flooding caused by severe runoff and severe droughts due to climate change have to be addressed globally by reducing the emission of GHGs. Southeast Asia produced 12% of the world's GHGs in 2000. An increase in emissions is most likely, due to the expanding population and economies (ADB 2009b). However, smaller countries, such as most of the riparian countries of the Mekong, can contribute as well. Measures can be taken by increasing energy efficiency and by tapping renewable energy sources, such as biomass, solar, wind, geothermal energy and includes also the controversial discussed hydroelectric energy. Increased energy efficiency would lead to a decreased demand for energy, also from hydro power plants. Vietnam for example recognized the need to act and set up an energy efficiency programme. It aims to reduce the national energy consumption by 5–8% between 2011 and 2015, through technical measures. However, the total amount of energy required will rise. Therefore, a proper balance of economic needs and ecological as well as social impacts of $CO_2$ neutral technologies such as hydropower dams should be considered on a basin scale.

Integrated Flood Management (IFM) is increasingly important, since the occurrence of severe floods is very likely to increase due to climate change. IFM is closely related to IWRM, since flood disasters negatively affect the sustainability of planned developments formulated in the context of IWRM. IFM proposes managing the water cycle as a whole while considering all floods, including extreme ones, and integrating land and water management, as both have impacts on flood magnitudes and flood risks (WMO 2009). The Mekong Basin is in need of such an approach. In 2005 the MRC attempted to move in this direction, establishing a Regional Flood Centre in Phnom Penh called "Integrated Flood and Mitigation Programme". It addresses land-use planning measures, structural measures, flood preparedness measures and flood emergency measures in compliance with the guiding principles of IFM. However, the problem here seems to be the limited links of the MRC with China. As indicated above, China, as the most upstream country of the Mekong contributes about 16% to the downstream flow of the river. Hydropower dams serve the primary goal to produce reliable electricity according to the public demand. Hence, the management of hydropower dams does not meet the needs of flood control and an implementation of IFM for the complete Mekong is difficult, since the key data from hydropower dams in China are not communicated to the downstream countries. Here an incorporation of China and formulation and practice of a real basin-wide IFM should be of the highest priority for the benefit of all riparian countries.

Direct and indirect effects of global warming such as SLR, and floods coming from upstream as well as the occurrence of severe droughts all combine to aggravate the situation in the Mekong Delta in Vietnam. Hence, adaptation measures gain in importance and have been addressed by the Vietnamese Government, by intensifying efforts to control the hydraulic regime (Garschagen 2009). The Vietnamese Prime Minister approved about 1.1 billion USD in 2008 to set up sea dikes from central Quang Ngai province to Kien Giang province. This confirms the intention to focus on structural measures to counteract the impacts of climate change. However, in the sense of IWRM as a holistic approach it is necessary to consider non-structural measures as well. Structural measures are very cost-intensive and reduce vulnerability only to the extent to which change is expected. Unexpected changes in terms of intensity and timing cannot be addressed in this way. Therefore, there is a need to shift from flood control towards flood management.

Agriculture and aquaculture increasingly become a major cause of qualitative degradation of surface and groundwater resources through erosion and chemical runoff (FAO 1996). There is a strong need to systematically assess water pollution. Studies undertaken in the Mekong Delta suggest that the concentration of endocrine disruptors, various herbicides and pesticides have the potential to harm humans' health and the environment severely (Sebesvari et al. 2009). In terms of water quality, farming on sulphate soils and expansion of agriculture areas on it should be reconsidered in any case. There are some options available for neutralizing acids in soils. However, heavy metals released through chemical processes threaten aquatic ecosystems and can cause significant damage to human health.

Addressing water quality is one approach to help reduce the bottleneck of water availability, especially during the dry season. Improving water usage efficiency especially in the agricultural sector would also contribute significantly to improving the situation, especially during periods of water shortages. Agriculture as one of the most water consuming and water-dependent sectors is also one of the most vulnerable. Water shortage as predicted will severely affect agriculture and the livelihood of the population, since more than 70% of the Mekong Basin's population depend on water resources. The agriculture sector can drastically save water. Therefore, common practices in terms of irrigation, fertilization, crop cycle etc. should be reconsidered and financial incentives as well technical assistance to reduce water consumption and reduce pollution should be given. Saved water quantities in agriculture can help especially during the dry seasons, without having adverse effects on food security.

Protection and rehabilitation of valuable ecosystems are rewarded twofold. On the one hand, they help to conserve bio diverse ecosystems and related flora and fauna and on the other hand, they help to reduce predicted impacts of climate change through providing ecosystem services. Wetlands are natural water retention areas, which buffer extreme water flows. This helps to ally discharge peaks and reduces damaging effects of flood events. There is a need to identify valuable water retention areas, in terms of their water holding capacity and biodiversity. Those areas should be subject to restricted use and be protected consequently by national law.

### *5.3.2 The Political Context*

IWRM in the Mekong Basin has been implemented to a limited extent. MRC, which has subscribed to the principles of IWRM, covers the lower reaches of the Mekong. Upper reaches, which are essential for a complete picture, are missing. Furthermore, short-sighted acting of riparian countries is hindering the implementation of sustainable development principles, since these can contradict the economic needs and intentions of the country. This became especially evident in the recent discussion about hydropower dams and their impact on the flow regime of the Mekong. There seems to be no coherent strategy in place for the use of hydropower on the Mekong Basin. There is no open discussion on the benefits, risks and possible solutions which should result in a basin-wide hydropower management plan. This doesn't apply only to the hydropower dams in the upper reaches of the Mekong, it also applies to the lower reaches (for example hydropower dams in the central highlands of Vietnam). The economic potential to develop towards a more industrialized and service oriented society makes it even more necessary to address water management on a basin scale. Ignoring this fact and acting as if it's "business as usual" has the potential to negatively affect the country's economy, food security, and livelihood. Hence, unsolved transboundary issues result in an increased conflict potential.

The implementation of IWRM at national level has already been addressed and has been partly implemented in some Mekong riparian countries such as Vietnam and Cambodia. This demonstrates an initial commitment towards the IWRM principles. However, it seems that implementation is not yet properly in place (see Chap. 6). In other words, water management is, to a certain extent, in place in the upper reaches and separated in the lower reaches. Some countries have transferred IWRM principles into national strategies. Nevertheless, there is no overall coordinated IWRM approach at the basin scale. Admittedly, this is a big challenge but also the only chance for finding the balance between economic needs and sustainable development as well as utilization of water resources at the basin scale.

## 5.4 Overall Perspectives

The Mekong is the lifeline for six countries, which are all in very different initial positions and facing different social, economic, and environmental challenges. The question is how to balance these economic needs and at the same time ensure sustainability in terms of ecological and social development, targeting poverty reduction. Most of those challenges are of transboundary nature but are emerging also at national level in the different riparian countries. A balance between economic growth, which goes along with sustainable social and ecological development, requires effective river management on a basin scale including all riparian countries of the Mekong. Riparian states must work together, supported by technical and financial support from other countries. There should be a clear commitment by all riparian Mekong States towards transparency and a willingness to share and contribute, which is necessary for an open dialog between the different countries and stakeholders in order to achieve most objective decisions. This will be only possible if there is a legal framework of overriding importance in place, acknowledged by all Mekong Riparian States. Ideally spoken, participation of upstream countries should be addressed by proper incentives, which can be of political and economic nature and by assisting in increasing capacities and knowledge transfer.

Recent economic developments in South East Asia are showing an upward trend, which provides a chance for the least developed countries to achieve the Millennium Development Goals (MDG) by expanding industry/construction, agriculture as well as the service sectors. This however goes hand in hand with an increased demand for power and natural resources. Sustainable use of those resources must be ensured to gain long term stability in the countries, and to minimize ecological and social costs. IWRM provides an appropriate and holistic approach for this development.

Impacts of climate change are affecting all riparian states of the Mekong. The extent, intensity and type of effects vary according to the region and depend on the global effort to reduce GHG emissions in the future. Predicted changes, which are very likely to impact livelihoods on different scales, must be addressed prior to these changes becoming effective to reduce vulnerability and avoid causalities. This has to happen through an adequate and coordinated approach. Water as the main component of the climate cycle therefore plays a crucial role, which can be only addressed by IWRM.

Water resources development in such a river basin like the Mekong contains risks and faces several problems due to conflicting interests, ecological degradation and transboundary disputes. However, the current economic situation of the parties involved and progressing technical development framed by increasing ecological awareness provides an opportunity to develop the Mekong Basin, in compliance with the basic IWRM principles, in an ecologically and socially sustainable manner.

## References

Adamson PT, Rutherfurd ID, Murray CP, Iwona AC (2009) The hydrology of the Mekong River. In: Campbell IC (ed) The Mekong. Biophysical environment of an international river basin. Elsevier, New York, pp 53–76

ADB [Asian Development Bank] (2006) Future solutions now, the Tonle Sap initiative – reconciling multiple demands with basin management organizations. ADB, Manila. http://www.adb.org/Projects/Tonle_Sap/Tonle-Sap-200605.pdf. Accessed 1 July 2011

ADB [Asian Development Bank] (2009a) Technical assistance report. Socialist Republic of Viet Nam: climate change impact and adaptation study in the Mekong Delta. ADB, Manila

ADB [Asian Development Bank] (2009b) The economics of climate change in Southeast Asia: a regional review. ADB, Mandaluyong City

Brown K (2009) Water sector review project final report, National Water Resources Council on behalf of the Government of Viet Nam and a number of international development partners led by the Asian Development Bank, Australia 2009

Cambo WP, Cambodia Water Partnership & CNMC [Cambodia National Mekong Committee] (2007) IWRM in Cambodia – where are we, and where do we want to go? http://www.cambowp.org/download.php?file=CamboWP-0709-IWRM-review-0c.pdf. Accessed 1 July 2011

Campbell IC, Manusthiparom C (2004) Technical report on rainfall and discharge in the lower Mekong Basin in 2003–2004. Mekong River Commission Secretariat, Vientiane

Chen J (2010) Integrative conservation and development in GMS. International workshop on Resources and Environment Towards Sustainable Development in the Great Mekong Sub-regions, Kunming, China, 28–31 Oct 2010

Chinavanno S (2009) Information for sustainable development in light of climate change in the Mekong River. Southeast Asia START Regional Center, Bangkok. http://203.159.5.16/digital_gms/Proceedings/A77_SUPPAKORN_CHINAVANNO.pdf. Accessed 16 Mar 2012

Daming HE (2010) Trans-boundary water utilization and eco-security towards sustainability in the Great Mekong Subregion. International workshop on Resources and Environment Towards Sustainable Development in the Great Mekong Sub-regions, Kunming, China, 28–31 Oct 2010

Dang KN, Nguyen Van Be, Nguyen Hieu Trung (2007) Water use and competition in the Mekong delta, Vietnam. In: Tran Thanh Be, Bach Tan Sinh, Miller F (eds) Literature analysis: challenges to sustainable development in the Mekong delta: regional and national policy issues and research needs. The Sustainable Mekong Research Network Publisher, Bangkok

Delgado JM, Apel H, Merz B (2010) Flood trends and variability in the Mekong River. Hydrol Earth Syst Sci 14:407–418

Eastham J, Mpelasoka F, Mainuddin M, Thicehurst C, Dyce C, Hodgson G, Ali R, Kirby M (2008) Mekong River Basin water resources assessment: impacts of climate change. Water for a Healthy Country National Research Flagship, CSIRO

FAO [Food and Agriculture Organization] (1996) Control of water pollution from agriculture. FAO irrigation and drainage paper, vol 55. FAO, Rome

FAO [Food and Agriculture Organization] (2005) Global forest resources assessment: progress towards sustainable forest management. FAO, Rome

Ficke AD, Myrick CA, Hansen LJ (2007) Potential impacts of global climate change on freshwater fisheries. Rev Fish Biol Fish 17:581–613

Garschagen M (2009) Crises prevention and climate change adaptation in the coupled social-ecological systems of the Mekong Delta, Vietnam: the need for Rethinking concepts and policies paper for presentation at the fourth Munich Re Foundation Summer Academy on Social Vulnerability "Tipping Points in Humanitarian Crises", Hohekammer, Germany, 26 July–1 August 2009

Gupta A (2009) Geology and landforms of the Mekong Basin. In: Campbell IC (ed) The Mekong. Biophysical environment of an international river basin. Elsevier, New York, pp 29–52

GWP [Global Water Partnership] (2000) Integrated water resources management. Tac back ground papers no 4. Stockholm, Sweden

GWP [Global Water Partnership] (2011) Global water partnership toolbox. http://www.gwptoolbox.org/. Accessed 1 July 2011

Ha ML (2011) The role of regional institutions in sustainable development: a review of the Mekong River commission's first 15 years. J Sustain Dev 5(1):125–140

He Y, Zhang Y (2005) Climate change from 1960 to 2000 in the Lancang River Valley, China. Mount Res Dev 25(4):341–348

Hoanh CT, Jirayoot K, Locambe G, Srienetr V (2010) Impacts of climate change and development on Mekong flow regime. First assessment–2009. MRC technical paper. Mekong River Commission, Vientiane, Lao PDR

Husson O, Hanhart K, Phung MT, Bouma J (2000) Water management for rice cultivation on acid sulphate soils in the Plain of Reeds, Vietnam. Agric Water Manag 46:91–109

ICWE [International Conference on Water Environment] (1992) The Dublin Statement on Water and Sustainable Development. http://www.un-documents.net/h2o-dub.htm. Accessed 30 Mar 2011

IPCC (2007) Summary for policymakers. In: Parry ML, Canziani OF, Palutikof JP, van der Linden PJ, Hanson CE (eds) Climate change 2007: impacts, adaptation and vulnerability. Contribution of working group II to the fourth assessment report of the intergovernmental panel on climate change. Cambridge University Press, Cambridge, UK, pp 7–22. http://www.ipcc.ch/pdf/assessment-report/ar4/wg2/ar4-wg2-spm.pdf. Accessed 30 Mar 2011

Keskinen M (2008) Population, natural resources & development in the Mekong: does high population density hinder development. In: Kummu M, Keskinen M, Varis O (eds) Modern myths of the Mekong. Helsinki University of Technology, Helsinki, pp 107–121

Kiem AS, Ishidaira H, Hapuarachchi HP, Zhou MC, Hirabayashi Y, Takeuchi K (2008) Future hydroclimatology of the Mekong River Basin simulated using the high-resolution Japan Meteorological Agency (JMA) AGCM. Hydrol Process 22:1382–1394

Kuenzer C (2010) WISDOM: water related information system for the sustainable development of the Mekong Delta – thoughts on sustainability and long term implementation. In: Proceedings of the 3rd Indo-German conference on research for sustainability: water and waste management, New Delhi, India, 3–4 Feb 2010

Liu J (2010) The role of ecosystem management in climate change adaptation in MRB. International workshop on Resources and Environment Towards Sustainable Development in the Great Mekong Sub-regions, Kunming, China, 28–31 Oct 2010

McSweeney C, New M, Lizcano G (2008) Cambodia climate change profile. Oxford University, Oxford

Minh VQ (1995) Use of soil and agrohydrological characteristics in developing technology extrapolation methodology: a case study of the Mekong Delta, Vietnam. MSc thesis, Graduate School, University of the Philippines, Los Baños

Minh LQ, Tuong TP, van Mensvoort MEF, Bouma J (1997) Contamination of surface water as affected by land use in acid sulphate soils in the Mekong Delta, Vietnam. Agric Ecosyst Environ 61:19–27

MRC [Mekong River Commission] (1995) Mekong River Commission. Agreement on the cooperation for the sustainable development of the Mekong River Basin, 5 Apr 1995. http://www.mrcmekong.org/agreement_95/agreement_95.htm#intro. Accessed 30 Mar 2011

MRC [Mekong River Commission] (2004) Mekong's low flows linked to drought, says MRC study. In: Campbell IC (ed) MRC No. 03/04. Phnom Penh. http://www.mrcmekong.org/MRC_news/press04/26-mar-04.htm. Accessed 15 June 2011

MRC [Mekong River Commission] (2005) Overview of the hydrology of the Mekong Basin. Mekong River Commission, Vientiane

MRC [Mekong River Commission] (2009a) Mekong River Commission. Adaptation to climate change in the countries of the Lower Mekong Basin: regional synthesis report. MRC technical paper no 24, Vientiane. http://www.mrcmekong.org/assets/Publications/technical/tech-No24-adaptation-to-climate-change.pdf. Accessed 16 Mar 2012

MRC [Mekong River Commission] (2009b) Catch and culture. Fish Res Dev Mekong Reg. 15(1). http://ns1.mrcmekong.org/download/programmes/fisheries/Catch_Culture_vol15.1.pdf. Accessed 1 July 2011

MRC [Mekong River Commission] (2010a) Mekong River Commission. Strategic plan 2011–2015. MRC Secretariat Phnom Penh, Cambodia

MRC [Mekong River Commission] (2010b) Mekong River Commission. Impacts of climate change and development on Mekong flow regimes. First assessment – 2009 meeting the needs, keeping the balance, MRC technical paper. MRC Secretariat Phnom Penh, Cambodia

MRC [Mekong River Commission] (2010c) MRC SEA for hydropower on the Mekong Mainstream, Climate change baseline assessment working paper. Mekong River Commission, Vientiane, Lao PDR

MRC [Mekong River Commission] (2010d) State of the basin report 2010. Mekong River Commission, Vientiane, Lao PDR

Nguyen Binh Thin (2009) Report on implementation of Action program to the adaptation of climate change of the Ministry of Agriculture and Rural Development, Hanoi

Nuorteva P, Keskinen M, Varis O (2010) Water, livelihoods and climate change adaptation in the Tonle Sap Lake area, Cambodia: learning from the past to understand the future. J Water Clim Change 1(1):87–101. http://www.iwaponline.com/jwc/001/0087/0010087.pdf. Accessed 1 July 2010

Rahaman MM, Varis O (2005) Integrated water resources management. Evolution, prospects and future challenges. Water Resources Laboratory, Helsinki University of Technology, Espoo

Ruosteenoja K, Carter TR, Jylhä K, Tuomenvirta H (2003) Future climate in world regions: an intercomparison of model-based projections for the new IPCC emissions scenarios. Helsinki Finnish Environment Institute

Sam L (2006) Results of study on salinity intrusion for socio-economic development in the Mekong Delta. In: Sam L, Hoi TN, Nien NA, Tho NV, Huan HV, Hung LM, Thang TD (eds) Results of science and technology 2005, Agriculture Press. Ho Chi Minh City, Vietnam, pp 104–112

Schmeier S (2011) Resilience to climate change – induced challenges in the Mekong River Basin – the role of the MRC. World Bank water paper. http://www-wds.worldbank.org/external/default/WDSContentServer/WDSP/IB/2011/05/13/000386194_20110513013039/Rendered/PDF/618100WP0wp0200Box358352B00PUBLIC00.pdf. Accessed 15 July 2011

Sebesvari Z, Le TTH, Renaud F (2009) Impacts of climate change on pesticide fate in the Mekong Delta, Vietnam. Pesticide behaviour in soils, water and air. York, UK, 14–16 Sept 2009

Stone R (2010) Ecology. Severe drought puts spotlight on Chinese dams. Science 327(5971):1311

TKK & SEA START RC (2008) Water and climate change in the lower Mekong Basin: diagnosis & recommendations for adaptation, interim report. http://www.water.tkk.fi/English/wr/research/global/material/water&cc_2009_final-report.pdf. Accessed 1 July 2011

UNESCO (2009) IWRM guidelines at river basin level. Part 1 principles. UNESCO, Paris

United Nations (1995) Convention on the law of the Non-navigational uses of international watercourses. http://untreaty.un.org/ilc/texts/instruments/english/conventions/8_3_1997.pdf. Accessed 1 July 2011

United Nations (2007) General assembly adopts convention on the law of the Non-navigational uses of international watercourses. Press Release GA/9248. http://www.un.org/News/Press/docs/1997/19970521.ga9248.html. Accessed 14 Mar 2011

United Nations (2009). UN water policy brief for COP15. http://www.unwater.org/downloads/UNWclimatechange_EN.pdf. Accessed 01 Jun 2011

Valbo-Jorgensen J, Coates D, Hortle K (2009) Fish diversity in the Mekong River Basin. In: Campbell IC (ed) The Mekong. Biophysical environment of an international river basin. Elsevier, New York, pp 161–196

Waibel G (2010) Water resources management organisations and administration in Vietnam, ZEF working paper series. Center for Development Research (ZEF), University of Bonn, Bonn

Walker B, Holling CS, Carpenter SR, Kinzig A (2004) Resilience, adaptability and transformability in social–ecological systems. Ecol Soc 9(2):5. http://www.ecologyandsociety.org/vol9/iss2/art5/. Accessed 1 July 2011

Willett IR, Melville MD, White I (1993) Acid drainwaters from potential acid sulphate soils and their impact on estuarine ecosystems. In: Dent DL, van Mensvoort MEF (eds) Selected papers from the Ho Chi Minh City symposium on acid sulphate soils, Mar 1992. ILRI Pub. No. 53. International Institute for Land Reclamation and Improvement/ILRI, Wageningen, pp 419–425

WMO [World Meteorological Organization] (2009) Integrated flood management – concept paper. WMO, Geneva

Wolanski E, Nguyen HN, Spagnol S (1998) Sediment dynamics during low flow conditions in the Mekong River estuary, Vietnam. J Coast Res 14:472–482

Wolf AT, Natharius JA, Danielson JJ, Ward BS, Pender JK (1999) International river basins of the world. Int J Water Resour Dev 15(4):387–427

World Bank (2009) Cambodia at a glance. http://devdata.worldbank.org/AAG/khm_aag.pdf. Accessed 30 Mar 2011

# Chapter 6
# Water Governance Under Renovation? Concepts and Practices of IWRM in the Mekong Delta, Vietnam

**Gabi Waibel, Simon Benedikter, Nadine Reis, Sven Genschick, Loan Nguyen, Pham Cong Huu, and Tran Thanh Be**

**Abstract** The objective of this chapter is to explore Integrated Water Resources Management (IWRM) practices in Vietnam. Based on intensive empirical research in the Mekong Delta, the daily reality of water resources management is investigated in several sub-sectors. It will be shown that water management practices deviate to a large extent from the existing legal frameworks, policies and strategies commonly based on IWRM principles. It will be argued that the gap between official policy and actual practices is not the outcome of lacking capacity or resource scarcity, as often assumed in donor and government reports. Rather, it is a result of the peculiar structural features of the contemporary state in Vietnam.

## 6.1 Towards IWRM: Water Sector Reforms in Retrospect

Integrated Water Resources Management (IWRM) is defined as "a process which promotes the coordinated development and management of water, land and related resources, in order to maximise the resultant economic and social welfare in an

G. Waibel (✉) • S. Benedikter • N. Reis • S. Genschick • L. Nguyen
Center for Development Research (ZEF), University of Bonn,
Walter-Flex-Str. 3, 53113 Bonn, Germany
e-mail: gwaibel@uni-bonn.de; Simon.Benedikter@uni-bonn.de; Nadine.Reis@uni-bonn.de; sven.genschick@uni-bonn.de; loan.nguyen.vn@gmx.com

P.C. Huu
Mekong Delta Development Research Institute, Can Tho University,
3/2 Street, Can Tho City, Vietnam
e-mail: pham@uni-bonn.de

T.T. Be
Can Tho City Institute for Socio-Economic Development Studies (CIDS),
02 Hoa Binh Avenue, Can Tho City, Vietnam
e-mail: ttbe@cids.org.vn

F.G. Renaud and C. Kuenzer (eds.), *The Mekong Delta System: Interdisciplinary Analyses of a River Delta*, Springer Environmental Science and Engineering,
DOI 10.1007/978-94-007-3962-8_6, 

equitable manner without compromising the sustainability of vital ecosystems" (Global Water Partnership 2000:22).

In Vietnam, the first national workshop on IWRM was organised in 2001, but the concept has been promoted in the country for the past 15 years. It was in the context of the first comprehensive "Water Resources Sector Review" (World Bank et al. 1996) of the post *Doi Moi* period that the process was formally initiated. The review was, to a large extent, inspired by the global water discourse and supported by ODA (Overseas Development Aid) funding. In the following, a number of water sector reforms and the development of a legal framework for the water sector were undertaken. The promulgation of the first Law on Water Resources in 1998 constituted an important step, as it opts for the adoption of IWRM principles and provides for related policy development. In the following years, a series of institutional reforms attempted to strengthen the framework for IWRM implementation (World Bank et al. 2003). The issuance of the "National Water Resources Strategy towards the year 2020"[1] by the Ministry of Natural Resources and Environment (MONRE) in 2006 represents a milestone in this regard, as the policy paper was the first to set out "guiding principles, objectives, missions and implementation measures regarding the protection, exploitation, use and development of water resources, as well as the prevention and mitigation of adverse impact caused by water" (MONRE 2006:6). It therefore aims at providing an overarching framework for all water-related policies and implementation plans co-existing in Vietnam today. The new strategy paper confirms the (IWRM) approach with the river basin as the planning unit and suggests that river basin plans should be established for all regions (MONRE 2006:33, 44).

In addition to the challenge of integrating various water sub-sectors and regional concerns across administrative boundaries, the IWRM approach is underpinned with a series of guiding principles: effective water governance should be open and transparent, inclusive and communicative, coherent and integrative, equitable and ethical, and in its performance and operation it should be accountable, efficient, responsible and sustainable (Rogers and Hall 2003:27 ff.). These principles need to be implemented by concepts and measures such as participatory irrigation management, cost recovery, water rights, 'free basis water principles' and the set-up of institutions operating at the river basin level (Perret 2006:xxii; Conca 2006:2). IWRM can therefore be understood as a complex and ambitious normative concept, and its implementation a process of adaptation, innovation, contestation and learning. The local conditions for IWRM implementation also vary, since they are historically embedded in specific water governance systems.

In the case of Vietnam, water resources management has traditionally focused on flood control and the provision of freshwater for (mainly agricultural) production. In fact, the Sino-Vietnamese term '*thủy lợi*' (which can still be found in many water sector-related tasks and documents) literally means "water resources development for exploitation". The term illustrates that technical water engineering was the dominant approach in the sector. Water resources protection, in contrast, was long ignored, although new demands for water allocation development continued unabated and water pollution increasingly occurred. With the arrival of

[1] Approved by the Prime Minister's Decision No. 81/2006/QD-TTg on 14.04.2006.

IWRM concepts in Vietnam, a new term, '*Quản lý tài nguyên nước*', emerged in the country's water discourse, which can be translated as 'water resources management' (Molle and Hoanh 2008:31). Interestingly, both terms are in use today, each of which refers to a different set of concepts and policy approaches.[2]

The case of the Mekong Delta, located in Southwest Vietnam, provides a good example of this historical understanding of water resources management. The region's importance for national food security and agricultural export production is outstanding, yet it's recent and most successful economic growth depended to a large extent on the continuous development of technological innovations in water control (Käkönen 2008; To Van Truong and Nguyen Ngoc Anh 2002). In fact, although large-scale water engineering in the delta dates back to the eighteenth century (Biggs 2004:43 f.), efforts to effectively tap the region's water resources were strongly reinforced only after the country's reunification (1976). As part of these efforts, the new socialist state mobilised the local population to serve at the 'irrigation front' (*mặt trận thủy lợi*), which means that the irrigation, salinity and flood control infrastructure was literally built by manual labour. The *hydraulic mission*[3] aimed at ensuring the availability of freshwater for rural production, and therefore provided the ground for the intensification of farming systems (green revolution). Under the guidance of the by then existing Ministry of Water (*Bộ Thủy lợi*) and its sub-departments, the socialist government began to systematically plan water resources development for different ecological areas of the delta.

The first delta-wide master planning started in Vietnam in the late 1960s (Biggs et al. 2009:207). In the south of the country, various water resource studies and proposals were designed by Sogreah (France) and the Resource Development Company of the United States and the Netherlands (1974). During the early 1990s, the UNDP and the Netherlands prepared the NEDECO Master Plan for the Vietnamese part of the Mekong Delta, which was considered the first multi-purpose master plan of a national river basin in the country (CCFSC and MARD 2001:44). The NEDECO main document consists of a long-term development plan, based on the results of more than 50 scientific consultancy reports, covering a whole range of aspects of water use and water management. Strongly committed to national development targets, various scenarios were calculated and assessed (NEDECO 1993), and the finally presented results still constitute the basis of water resources management planning in the Mekong Delta today.

In order to operationalise the different plans, state management agencies, planning and research institutes, state-owned enterprises (SOE) and irrigation and drainage management companies (IDMC) were established in Ho Chi Minh City and

---

[2] While the Ministry of Agriculture and Rural Development (MARD) adheres to the Sino-Vietnamese term 'thủy lợi', the more recently established Ministry of Natural Resources and Environment (MONRE) purposely avoids the traditional terminology and uses 'tài nguyên nước' to describe its tasks and responsibilities. As a result, the deep friction between the two ministries, whose tasks have been shifted and redefined, finds a wording expression (Molle and Hoanh 2008:31).

[3] Borrowed from Molle et al. (2009:332).

throughout the Mekong Delta. These specialised agencies had the mandate to develop water resources on behalf of the people and in the common good. As qualified staff were basically missing, hundreds of hydraulic engineers were transferred from the northern part of the country to occupy leading positions in the new water bureaucracies of the south. With this, concepts of water resources engineering, developed in the Red River Delta, were transferred to the south and thereafter served as a blueprint for water resources management in the Mekong Delta (Evers and Benedikter 2009a). The fact that both deltas featured quite distinctive hydraulic conditions was consequently often ignored.

In 1986, the Vietnamese government launched its Renovation Policy (*Doi Moi*),[4] which aimed at shifting from central planning towards a market-based economy. The resulting reforms included, among other things, economic liberalisation, decentralisation and "socialisation",[5] as well as comprehensive changes to state management institutions. The resulting modifications to the state apparatus and government practices also had a strong impact on the policies and provisions for water resources management (Waibel 2010:34–39).

In the context of economic liberalisation, for instance, water control and supply services were partly privatised from the mid-1980s (Fontenelle 2001:5f.), and the dissolution of state-managed agricultural cooperatives led to the evolvement of new-style cooperatives (registered as private businesses) offering pumping and drainage services. The funding and management of many state-owned enterprises and planning and research institutes changed, and the new 'National Water Resources Strategy towards the Year 2020' (MONRE 2006) explicitly encouraged the mobilisation and development of private sector investments.

In the social sphere, the conditions of operation for non-state actors became more moderate. Nowadays, the one-party state tolerates and encourages certain forms of civil organisation, as long as their activities are strictly confined to social fields and development work (Wischermann 1999; Gray 2003). In line with this policy, and based on the promulgation of the Grassroots Democracy Decree (1998), concepts of community-based development and the idea of public participation in planning processes have received more recognition in policy formulation, in particular with regard to communal affairs (Bach Tan Sinh 2002:122; Zingerli 2004:56). Considering water resources management, these concepts have been translated into approaches such as 'Participatory Irrigation Management', yet not in all provinces (Nguyen Xuan Tiep 2008); in addition, new community-based organisations have emerged including voluntary water user associations, micro-credit groups engaging in water supply and sanitation and production groups taking care of small-scale irrigation works maintenance (Reis and Mollinga 2009; Benedikter and Waibel forthcoming 2012).

---

[4] By the mid 1980s, Vietnam faced rapid economic meltdown due to the failure of the centralised and subsidised system of a command economy the socialist government had been adhering to since 1954. In response to this severe crisis, the Vietnamese government reconsidered the country's economic system and, eventually, promulgated the Renovation Policy (*Chính sách Đổi mới*) in 1986 (the Sixth Party Congress) as a paradigm change towards economic liberalisation and integration.

[5] "Socialisation" (*xã hội hóa*) implies the outsourcing of former state functions to non-state entities, mainly to the private sector, households and civic organisations.

In administrative terms, the national leadership recognised the necessity to draw a clearer division between state functions and the role of the party. In order to create a 'modern' governmental system that would follow principles of the 'rule of law', several reform measures were initiated. Whereas under 'socialist legality' laws were used as an instrument of rule (Benedikter and Waibel forthcoming 2012), under the new slogan "socialist state ruled by law", governance and state management should be based on laws (Pham Duy Nghia 2005:85). In line with this, attempts were undertaken to divide and balance power between executive and legislative state bodies by empowering the National Assembly and the People's Councils at local levels (UNDP 2001). With regard to decentralisation, administrative and fiscal responsibilities, as well as decision-making power and functions, were increasingly delegated to the lower levels of administration (Fritzen 2006). Moreover, the comprehensive Public Administrative Reform Programme 2001–2010, which aimed at improving the efficiency and transparency of state agencies and combat corruption (UNDP 2001:29 ff.; Buhmann 2007:241; World Bank 2009), had considerable effects on the organisation and prescriptions of state management functions, including those of the water sector. (Post-) *Doi Moi* reforms therefore provided the constitutional preconditions and rationale for the development of the water policy and legal framework.

Following the enactment of the Law on Water Resources (No. 08/1998/QH10) in 1998, more than 300 water-related regulations on the guidance and implementation of the Law on Water Resources were issued and often amended to meet the changing requirements of the country's development. To date, Vietnam's legislation on the water sector consists of a complex system of legal documents ordered in many levels and adopted by different state authorities. It is thus complicated and, in a number of cases, lacks harmonisation with other laws and secondary regulations (Nguyen Thi Phuong Loan 2010a). Nevertheless, the legal framework defines the institutional set up of water resources management, which is summarised in Fig. 6.1.

Since 1995, the Ministry of Agriculture and Rural Development (MARD), with its subordinate state management organisation and planning institutes, has been in charge of hydraulic engineering, water service delivery and flood and storm control. The ministry's remit includes responsibility for the planning, construction and maintenance of all kinds of hydraulic works such as dykes, irrigation schemes, sluices, reservoirs, pumping stations and hydropower plants (Nguyen Thi Phuong Loan 2010a). In an attempt to separate water resources management functions from the responsibility for public service delivery (water exploitation and development), the Ministry of Natural Resources and Environment (MONRE) was established in 2002 (GoV Joint Task Force 2003:2). Since, MONRE is the country's leading authority on the management of land, water and the environment. With regard to the water sector, MONRE is in charge of water resources assessment, water allocation and the regulatory management of surface water, groundwater and water quality.[6]

[6] Decree No. 25/2008/ND-CP promulgated by the Government on March 04, 2008 defining the functions, tasks, powers and organisational structure of the Ministry of Natural Resources and Environment.

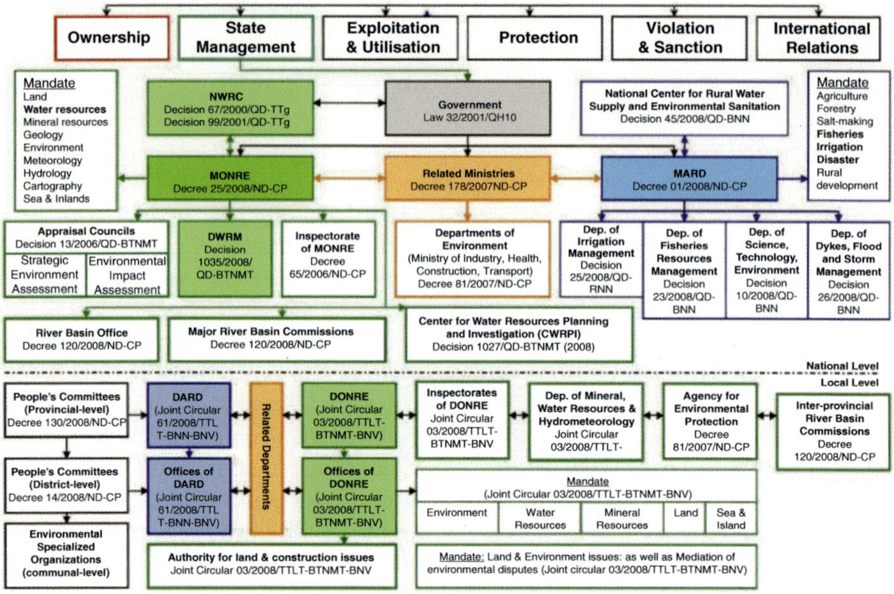


**Fig. 6.1** Institutional setup of water resources management in Vietnam (Nguyen Thi Phuong Loan 2010a:111)

Moreover, a number of other ministries, such as the Ministry of Health and the Ministry of Construction, have a stake in water resources management, which illustrates that the multiple sector reforms rather increased institutional fragmentation. The differentiation of policy areas has also generated a number of tailored and detailed legislations for the respective sub-sectors. These include the 'National Strategy for Natural Disaster Prevention, Response and Mitigation', environmental policies, a hydropower development plan, a set of guiding documents for dyke, wetland and coastal management, the 'National Rural Clean Water and Sanitation Strategy' and several other strategy papers and implementation plans. A few agencies such as the National Water Resources Council (NWRC) and the River Basin Office, have been set up to incorporate these different perspectives and concerns, yet their role remain an advisory and limited one (see below).

Ministries are organised in administrative and professional departments, and have research and planning institutes, as well as business agencies (e.g. state-owned enterprises), attached to them (Waibel 2010:18f.). Furthermore, state management agencies are divided into four vertical levels, namely the centre, the province, the district and the commune. Despite several decentralisation programmes, national policies are expected to be fully translated by the local departments and offices representing the ministries at province and district levels. However, local governments do not operate in a uniform way – as the following chapter will show.

## 6.2 Water Resources Management Practices

### *6.2.1 Inside the State and 'Everyday Politics'*[7] *in Vietnam*

State domination and the monopolisation of political power under the Vietnamese Communist Party remain part of a significant continuum in the post-renovation era of Vietnam (Fforde 2005:149–150). Although from the mid-1980s the economy gradually shifted towards market principles and social life dramatically changed, the political system, based on Leninist-Marxist ideas of statehood, remained rather untouched (Fforde 2008:5, 2010). Porter's (1993) analysis of the Vietnamese government system of the 1990s is still valid to date. According to him, "democratic centralism (*tập trung dân chủ*)" and "collective mastery" (*làm chủ tập thể*) build the core legal and organisational elements of the political and administrative systems. The first relates to the unconditional obedience of lower scales to superiors in state and party structures, and creates strict hierarchies of top-down decision making and implementation. Bottom-up reporting, in contrast, (theoretically) enables the paternalistic party-state to know best about what needs to be done and justifies that decisions are taken on behalf of the collective (Porter 1993:120; Gillespie 2005:47–50). The slogan "*the party is the leader, the state is the manager and the people are the masters*"[8] demonstrates the rationale of the Vietnamese Communist Party rule. It is therefore the Communist Party which controls and provides the key directions for policymaking, including those of the water sector.

Moreover, and despite economic liberalisation, the legacy of centralist-bureaucratic planning is still practiced to date. National policies are based on 10-, 5-year and yearly plans, which are developed by the central government. To define short-, medium- or long-term planning targets, the state apparatus collects detailed numbers about socio-economic development in all parts of the country. Local authorities submit data to the higher authorities, for example the number of households with access to wells, public water supply stations and hygienic latrines. Provincial and national authorities base the definition of planning targets on these numbers. Once the targets are set, local level administrative bodies are required to implement the

---

[7] The concept of 'everyday politics' is borrowed from, and refers to, the work of Kerkvliet (2003, 2005). It considers the relationship and interactions between people (society) and state (at local level). According to the author, and despite the prevalence of top-down approaches, people have the capacity and means to influence policy practices substantially through their daily actions. 'Everyday politics' is an arena where people and the state meet to negotiate and bargain over policy implementation, and where people silently and subliminally can oppose and change policies by their collective actions. Since those who represent the state (state officials/cadres) are also part of the society, it is difficult to distinguish clearly between state and society. Consequently, this concept even matters for grasping local-central government relations.

[8] Đảng lãnh đạo, Nhà nước quản lý, Nhân dân làm chủ.

plans. This system is based on the strong belief that the entire world is calculable and administrable through numbers and book-based omniscience, a belief which stems from the traditional east-Asian Mandarin rule (Woodside 2006:82).[9]

The current system excludes critical reviews and monitoring of state management activities by extra-bureaucratic forces. Besides administrative and multiple sector reforms, decentralisation and the overwhelming tendency of bureaucratisation in the post-*Doi Moi* era have led to an increased complexity and bloating of the state apparatus. This in turn has produced a confusing system of institutional fragmentation with gaps and overlaps in mandates and responsibilities, where, cynically speaking, everybody is in charge but nobody is responsible. Several Vietnamese scholars have demonstrated that the current conditions constitute a breeding ground for activities which undermine official policies and national decisions, but serve the individual interests of state bureaucrats and allied networks at various scales (Koh 2001a). Gainsborough (2005), for instance, argues that the contemporary Vietnamese state is divided into a public and a private sphere, and that in the wake of economic liberalisation and privatisation the private sphere has hollowed out its public counterpart.

Hence, the concept of 'everyday politics' (Kerkvliet 2005) has to be placed centre stage in policy analysis. Beyond the scope of formal institutions and national level politics one will find significant "*discrepancies between what state leaders have decided and what people in society actually do*" (Kerkvliet 2003:31), including those individuals and groups inside the state apparatus itself. The central government therefore appears as strong and weak simultaneously – strong in terms of a rigidly organised hierarchical structure and control over the public sphere, weak in terms of lack of enforcement of national provisions and lack of control over lower administrative levels (Koh 2001b:536, 2001c).

In this respect, Fforde (2008:7) points out that the relationship between the centre and local authorities may significantly differ across the country and empirical studies have to unravel the respective political structures of a specific locality under investigation. "*The emperor's law stops at the village gate*"[10] is a popular idiom that reflects well the antagonistic relationship between central and local state bureaucracy. Pike argues that in pre-colonial Vietnam, official politics would go through the royal court system, while informal politics lived at the local scale in hundreds of bamboo fence-surrounded villages, to where the central bureaucracy barely had access

[9] A managerial bureaucratic administration based on meritocratic ideals of statehood (mandarinate) and rule (rule of the talented) emerged early in Vietnamese history and derived from the Chinese model of administration and Confucian ethics. State officials (mandarins) were recruited through examination, in which candidates had to prove their ability to govern. Socio-economic deficiencies such as poverty and economic downturn were considered failures of state politics (Woodside: 24 f., 30, 59). These traditional Confucian ideals are still inherent in the Vietnamese conception of statehood and bureaucratic state management, which reflects that beneath the surface of Leninism/Socialism traditional-Confucian notions of rule and governance prevail (Dao Minh Chau 1996; Luttmer 2000).

[10] Phép vua thua lệ làng.

(Pike 2000:271–272). Moreover, a historical-induced, deep geographical regionalism has further driven particularism within state power and facilitated informal political practices (Pike 2000:277–278). There is ample evidence that these features have carried on to date. Beresford (1995:10) and Thayer (1995:55) both argue that socialist bureaucratic polity has always been highly decentralised in an informal manner, and even described provincial governments as independent kingdoms that allude to central decisions when it comes to the protection of their particularistic interests. This has created tensional relations and hidden struggles between central and local parts of the apparatus over the control of resources and functions.

The following case studies from the Mekong Delta present findings of the author's field research between 2008 and 2010 and addresses different aspects of water resources practices. By contrasting formal prescriptions in the form of policies and laws with the daily practices and implications, the authors come to similar conclusions about the ambivalent nature of state politics in Vietnam.

### *6.2.2 Establishing River Basin Organisations (RBOs) – A Long Way to Go*[11]

According to the Law on Water Resources (1998), the primary planning and management unit for water is the river basin. When the law was promulgated in Vietnam, no institution had the mandate to implement this directive, but it made the provisions for a non-business agency under MARD to fulfil the task of river basin planning (Article 64).[12] Subsequently, a legal document for the establishment of River Basin Organisations was produced and donor investments into river basin planning expanded. According to the legislation, the RBOs' main task was to enable the unified management of the catchment areas, across the administrative borders dividing them (Taylor and Wright 2001). In 2001, the first RBOs were set up: one in the Red River Delta and two in Ho Chi Minh City, namely the Office of the Dong Nai River Basin and the Office of the Cuu Long (Mekong Delta) River Basin. These three basins cover more than three provinces each and were selected as pilots, as they represent the major large basins of the country.

In the initial phase, the creation, funding and capacity building of the new agencies were dependent on donors,[13] and national investments were reduced to a minimum. This illustrates the low level of interest of Vietnamese water management officials, who "... *would [rather] refer to agencies or boards, seen as better reflecting their Vietnamese*

[11] This section of the article is basically taken from Waibel (2010:34–37).

[12] "The law does not, however, provide for the establishment of river basin organisations, as various donors stated in their documents" (Molle & Hoanh 2008:23, FN 28).

[13] While the ADB was engaged in the Red River and the Dong Nai Basin Development, the Australian government assisted the Mekong Basin. DANIDA and other bilateral cooperation agencies also played an important role.

*translations*" (Molle and Hoanh 2008:16, FN 18). As a consequence, they remained critical about whether this institution could fit their administrative and decision-making structures and culture. Molle and Hoanh (2008:2) thus argue that the concept of RBOs in Vietnam "*...has been quite disconnected from the existing institutional framework*". This argument is also supported by the research findings of the authors.

When the responsibility for RBOs was shifted from MARD to MONRE in 2006, the main idea was to concentrate all competencies for strategic planning in the water sector in one central agency. With this move, MONRE officially wrested charge of preparing River Basin plans, and in 2008 the ministry drafted a new decree on river basin management.[14]

Following this decree, each RBO is made up of a council and a secretariat. The council is composed of representatives from both the provinces and national-level ministries and is supposed to meet twice a year. The secretariat represents the operational unit but is practically understaffed. The RBOs of the Dong Nai and Mekong Delta river basins, for instance, have five staff each and a budget which clearly limits the office's activities to mainly administrative tasks (interview with an RBO officer 13.04.2008).

In September 2008, the Southern Institute of Water Resources Planning (a research institute under the authority of MARD, and the host of two RBOs) submitted a draft master plan for the Dong Nai River Basin, which was the first plan to be developed by one of the large RBOs initiated in 2001.[15] This can be considered as progress towards IWRM, although planning processes and outcomes definitely deserve critical attention (Pham Cong Huu et al. 2009). Similarly, there is no valid information on how such a plan would be implemented. Moreover, the emergence of parallel institutions and overlapping so-called environmental protection master plans raises new questions about the mandate and responsibilities of RBOs. These plans are drafted by new river basin organisations, established within the framework, and subsequent secondary regulations of the Law on Environmental Protection (2005). These organisations are called Environmental Protection Committees and deal with environmental management at a basin scale. They are set up under the authority of MONRE, are independent from RBOs established (in 2001) under MARD and report directly to the prime minister (Grothe 2009:53ff.). Such parallel structures potentially increase institutional fragmentation rather than help to reconcile sector-specific and diverse local interests.

[14] Decree No. 120/2008/ND-CP, issued by the Government on December 01, 2008, regulating river basin management. This decree was urgently needed because river basin management was omitted in two decrees issued in 2008, revising and clarifying the functions, tasks and responsibilities of MARD (Decree No. 01/2008/ND-CP of January 03, 2008) and MONRE (Decree No. 28/2008/ND-CP of March 04, 2008).

[15] A first evaluation (water sector review) in 1996 drew the following picture (World Bank et al. 1996:45 f.): River basin master plans for the national segments of the Mekong Delta (1993) and the Red River Delta (1995) were completed and three more were in preparation. All existing plans had been prepared by different ministries, and inconsistencies with other plans, affecting water allocation and use, were detected. With regard to water resources management at a sub-basin or basin-scale, no formal agreements among provinces existed (World Bank et al. 1996:43).

Recent history illustrates the continuous struggle over water resources management functions at ministerial level. In fact, the establishment of RBOs touches on questions of power and finance and adds to the already existing conflicts between MONRE and MARD (Molle and Hoanh 2008:29):

> Since RBOs are largely promoted by foreign partners and thus likely to be associated with the future delivery of loans and projects, they may also 'attract' more investments, which make their control even more desirable. In other words RBOs' legitimacy as 'registration chambers' for projects – rubber-stamped with the seal of IWRM – can be attractive for the departments traditionally involved in structural interventions (and perhaps for investment banks alike).

As generally stated, RBOs have never worked effectively, and in all the policy fields investigated in this paper, the Mekong Delta RBO played no role at all. In addition to the already mentioned constraints, the challenge of coordinating national and provincial interests and concerns appears to be a crucial point. Prior to the establishment of RBOs the only coordinating, cross-province mechanisms were those within MARD, but "*... any major issues affecting more than one province ... [were] ... usually handled by separate discussion with each province*" (AUSAID 2003:68). From a local perspective, Provincial People's Committees were reluctant to support the establishment and operation of RBOs, because their existence would entail a potential transfer of decision-making powers from the provinces to an inter-provincial body (CRDE and IESD 2006). Since out of all 15 river basins in Vietnam a vast majority covers more than two or more provinces, the question of inter-provincial coordination remains a critical concern. To date, no legal provisions concerning regional management have been developed for Vietnam (Scott and Chuyen 2004:98), which constitutes another constraining factor in this regard.

To sum up, RBOs appear as empty shells, lacking the resources – but more importantly the decision-making power – for strengthening IWRM in Vietnam's river basins. More precisely, the government has not demonstrated a real and profound interest in making the RBOs work, which again supports the argument of Molle and Hoanh (2008:2) that IWRM is, by and large, incompatible with Vietnamese water resources management institutions.

### *6.2.3 Technical Progress with Side-Effects: Dyke Planning and Implementation*

The Mekong Delta's climate and topography are characterised and greatly affected by clearly distinct dry and rainy seasons, as well as annual flooding. Floods constitute both a valuable resource for local production and, in some places and points in time, a risk and potential disaster for local livelihoods. For a long time, people living in the delta have adapted to the flooding regime and peculiar climatic conditions (Ehlert 2012). Similarly, technical solutions for the specific challenges of the water environment were continuously developed and favored by the different political regimes (Biggs et al. 2009:212ff.).

Up to the 1990s, flood control measures were not systematically planned in the sense that a comprehensive regional approach to flood protection was missing. However, shortly after the country's reunification in 1975, the irrigation system was identified as a historical solution to mitigate floods and thereby allocate water for agricultural production. Thereafter, local agencies and farmers built dykes to protect their production areas, but no central state flood control system existed. Between 1975 and 1990, both low and high dykes, in combination with the irrigation system, were constructed. The low dyke (so-called August dyke) was built mainly to protect the summer-autumn rice crop at the beginning of floods from July to August, while high dyke construction was used to control floods year round. However, this system could not fully prevent flood damage, which irregularly but frequently occurred.

During the same period, the central government started to investigate how national and local flood management policies could best address the risks and danger of flood disasters. Within this context, the NEDECO Master Plan was developed. In order to find relevant measures for different floodplains in the delta, the NEDECO Master Plan (1993) suggested that low dykes should be built to control floods in the low flooded areas, and high dykes constructed in the deep flooded plains. However, delta flooding is very complex and depends on topography, rain regime and tide. In order to control floods, the Vietnamese flood planners divided the floodplains into different sub-zones and selected relevant flood control measures for each of the floodplains. Although the Vietnamese government officially promoted an integrated and diverse flood control approach, where flood avoidance, flood adaptation and flood control measures were pursued, the dyke control system remained the dominant activity (To Van Truong 2000). In the 1990s, the government adopted a 5-year plan (1996–2000) and issued Decision 99TTg/1996 to deal with the specific flood risks of the Mekong Delta; MARD was assigned to organise and implement this decision through a corresponding programme.

Within this context, MARD commissioned the SIWRP, which operates under the ministry's authority, to prepare a flood control plan for Can Tho City. Similar plans already existed for a number of other provinces, but Can Tho City was the first province in the Mekong Delta that would develop a flood control scheme for an average level of flooding. A first result of the process was that dykes were identified as the priority flood control measure for the delta (SIWRP 1998, 2005). The plan itself was developed over several years and MARD invited a number of ministerial-level agencies to make their contributions. Similarly, as empirical research has shown (Pham Cong Huu 2012), local-level organisations such as DARD and other departments from Can Tho City were involved in the consultation process. However, as several representatives argued, they could only contribute their opinion to already drafted documents and were not considered as partners in the proper sense. The plan itself was developed at the office desk of SIWRP. Local planning and research institutes were not invited to participate in any capacity whatsoever, and the incorporation of local communities was very limited. The whole process therefore illustrates the usual top-down approach in decision-making, where local expertise and concerns are broadly ignored. This produced adverse effects, as investigations at province and commune levels have shown.

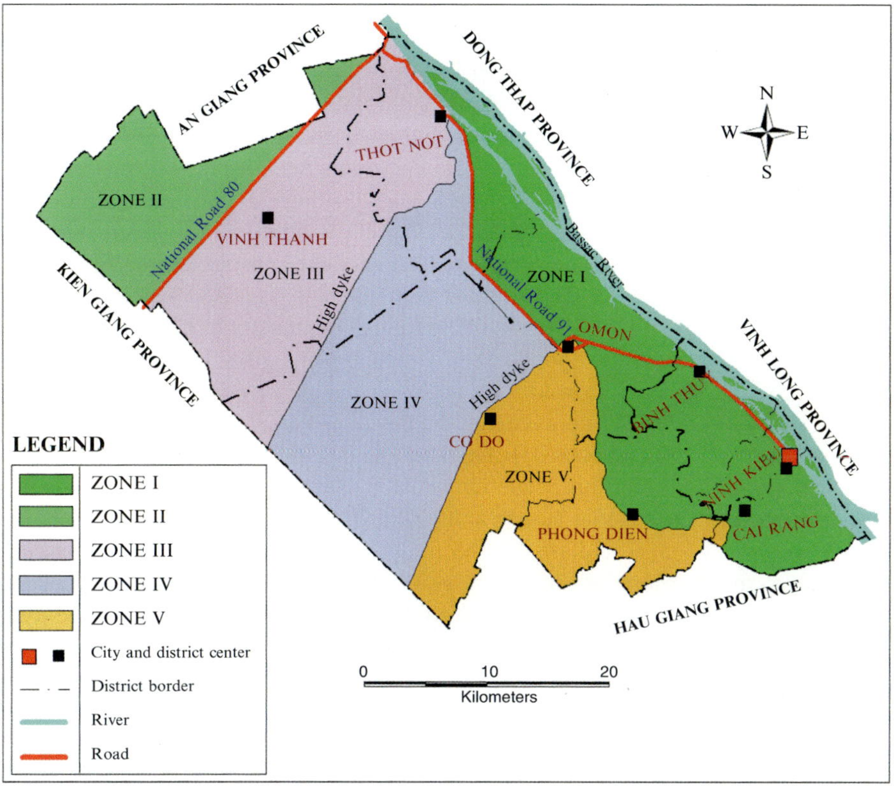


**Map 6.1** Dyke system planning in Can Tho city (Source: Pham Cong Huu 2012, based on Data from SIWRP collected in 2008)

Indeed, the undoubtedly positive effects of dyke construction measures such as the protection of farms are counterbalanced by a wide range of potential and real negative impacts on people's livelihoods and local ecosystems. Previous studies have already showed that the Mekong Delta's flood protection system leads to erosion, plant diseases and soil fertility decline in the protected areas (Tran Nhu Hoi 2005; Duong Van Nha 2006; Sarkkula et al. 2008). Other negative impacts are changes in flow velocity and rising annual flooding levels (Le Thi Viet Hoa et al. 2007; Käkönen 2008), as well as natural fish resources and biodiversity degradation (Hirsch et al. 2006).

With regard to Can Tho City, and according to the plan, the province was divided into five different flooding zones (see Map 6.1), where the topography, soil quality, irrigation and other infrastructure were taken into consideration. The implementation of the plan started in 2004, and during the past years dyke construction has proceeded in Zones II and V. Due to financial constraints, works in the other zones have not yet materialised. In practice, the dyke systems were not only built in the rural floodplains, but also in urban areas where residential zones should be protected and transport systems developed. In addition, the dykes were supposed to control floodwater flows from the Long Xuyen Quadrangle into Can Tho City, as well as its discharge into the Western and Eastern Seas.

As mentioned above, local communities did not participate in the design of the new system and their specific needs were basically ignored. As a consequence, the implementation of the dyke system created conflicts between central planners and the targeted communities. Local organisations and authorities constructed low dykes in almost all of the floodplains, while the central planning design anticipated high dykes as being appropriate: one survey in 2008 on dyke construction in Vinh Thanh District actually showed that 21,816 ha of farmland was protected by low dykes and only 5,029 ha lined with high dykes. These figures demonstrate that farmers clearly prefer the low dykes and do not necessarily follow plans made by central agencies.

The new dyke system showed both negative and positive impacts. The benefits notably include rural road improvement, the development of aquaculture production zones and the reduction of flood damage risk. The negative impacts encompass natural fish exhaustion, soil fertility reduction, erosion as well as changing and prolonged inundation levels. Le Thi Viet Hoa et al. (2008) found that inundations last approximately 5–10 days longer and are 0.2–0.3 m deeper in some places near or between high embankments. Moreover, the majority of farmers perceived water pollution as a severe problem in the protected floodplains. This is also closely linked to the dyke system, because the dykes block the flow-out of polluted waters. Moreover, intensive farming and the increasing use of agrochemicals are encouraged in the flood-protected zones. Centralist dyke system planning has dramatically changed the natural flooding regime of Can Tho City and produced a series of unintended, negative effects.

An analysis of the complex planning process clearly shows that the governmental perspective merely focused on protecting and improving agricultural production and preventing flood disasters. It must also be emphasised that most information and data taken into consideration by the planning agencies were secondary data from the city and district departments, while primary data and information from local communities were hardly sought. As a result, the experience and flood adaptation strategies of local people in the floodplains did not receive the appropriate attention. In addition, no studies on the expectations, attitudes and perceptions of the beneficiaries were carried out. Thus, traditional technocratic and engineering aspects emerged as dominant features in the new dyke system, while the 'living with floods' approach was relegated.

### *6.2.4 Water Quality Management: Urgent Problems, Temporary Solutions*

In Can Tho City, most economic sectors depend on or are related to water, yet the main freshwater sources, namely groundwater, (collected) rainwater and surface water (DONRE 2009), are increasingly threatened by overexploitation and pollution (Dung 2003). In particular, the Hau River (the last part of the Mekong River before emptying into the East Sea) is heavily polluted by organic, chemical and microbiological substances (Nuber et al. 2008), and the related drainage canals suffer from untreated wastewater discharges. According to the results of environmental monitoring conducted in 2007 by the Environmental Protection Department of the Southwest Region (belonging to MONRE), the amount of suspended solid waste was 12.760 mg/l, Nitrogen 171 mg/l and Sulphur 9.8 mg/l (*Can Tho Newspaper* 14.12.2008). These indices clearly exceeded the allowed national water quality standards by more than 100 times. The percentage of Coliform (2.400.000 MNP/100 ml) especially has now exceeded nearly 27 times the safe legal limit (ibid) (see also Chap. 13).

The sources of water pollution are manifold and include both domestic and production-related pollutants. At the household level, poverty reduction and higher income levels lead to changing consumption patterns, which in turn result in growing waste production (O'Rourke 2004). Domestic sewage treatment plants are non-existent, and most rural households still use traditional, open sanitation infrastructure (Herbst et al. 2009). Sewage from approximately 1.2 million inhabitants is discharged daily, directly or indirectly, into open waters. A population growth of 4.52% (2004–2009; General Statistics Office of Vietnam (GSO) 2011a) and creeping urbanisation further contribute to increasing water stresses.

With regard to the productive sectors, industries, agriculture and aquaculture constitute the main water polluters in the delta. In Can Tho City, four industrial parks accommodate mostly agriculture-related industries such as food processing and feedstuff, as well as pesticide and fertilizer production (WISDOM 2009). The non-existence of common wastewater treatment plants constitutes a major challenge (Anh Phuong 2009). In fact, private treatment plants of "rather poor condition" (Herbst et al. 2009:700) exist, but entrepreneurs are reluctant to pay for the waste treatment fee and consequently discharge solid and liquid wastes frequently into open waters in order to save costs.

Another main pollution source for surface water is agriculture, especially the intensive rice cultivation sector, as rice farmers almost exclusively draw on agrochemicals (Toan and Sebesvari 2010, Unpublished data, personal communication; cited in Reis 2012; see also Chap. 13). As intensification mounts, farmers use much more than the required level of fertilizers and pesticides (MONRE 2008), which are then washed into the canals and rivers.

Similarly, aquaculture farming constitutes a serious threat to the local environment. In 2008, fish farms, representing almost half of all farms in Can Tho City (159 out of 315 in 2008; GSO 2011b), had 12,900 ha of water surface under cultivation (GSO 2011c). Critical factors causing water pollution are fish feed, fish waste, pesticides and veterinary drug use, especially antibiotics (Khoi and Van 2008). The uncontrolled digging of ponds, wastewater discharge[16] and the continuous need for renewing the pond water seriously affect water quality (Vo Thi Lang et al. 2009:21; Nguyen Thi Phuong Loan 2010b; DONRE 2009).

[16] For example, according to the results of a survey analysing the environmental consequences of fish breeding systems, 60.2% of respondents answered that they discharge wastewater into public water sources, while the rest (39.8%) dump it on private land (Vo Thi Lang et al. 2009: 21).

According to the legal framework, the Department of Natural Resources and Environment (DONRE)[17] of Can Tho City shall coordinate with other related departments to set up and implement plans for the effective utilisation and sustainable development of water resources, as well as apply measures to prevent and control the degradation and depletion of water sources. Such key water-related departments embrace the Department of Agriculture and Rural Development (agriculture, forestry, aquaculture, irrigation and rural development), the Department of Finance (price of clean water), the Department of Construction (water supply to urban areas), the Department of Industry and Trade (technical standards for industrial products, environmental sanitation and safety), the Department of Health (clean water standards) the Department of Transport (inland waterway navigation) and the Department of Justice (examination and reviewing of legal documents issued by the PC).

The institutional complexity and fragmentation of state management functions with regard to water quality management are also reflected in the corresponding legal framework developed at the province level. In fact, more than 100 legal documents were issued by the People's Committee of Can Tho City between 2002 and 2010 in order to elaborate on and amend decisions on water resources protection and water pollution management. Nevertheless, the current legal framework does not meet real-life requirements. As the research on the problems of law enforcement on wastewater management in Can Tho City revealed, legal pluralism, in the sense of multiple secondary and subordinate legislations at provincial level, exists. For some of the relevant issues, local regulations are missing, for others, contradictions between local and national regulations emerge (e.g. on which basis natural resources taxes are being calculated). As a consequence, there is confusion about the applicability of legal regulations (Nguyen Thi Phuong Loan 2010b), which is further exacerbated by the fact that many local state authorities in charge of water management lack not only specialised knowledge but also legal awareness on water management.

In addition, water quality management practices do not correspond to the regulations expressed on paper. For example, research at commune level found that regulations regarding wastewater treatment and discharge in the aquaculture sector are not enforced at all. The rationale behind this has different facets: poor farmers are not punished by local authorities because financially punitive measures would seriously endanger their livelihoods, while more affluent farming households are not punished because local authorities have to maintain good relationships in order to, for example, make use of their position by asking for donations to improve

[17] DONRE operates at three administrative levels. It was established as a specialised organisation under the People's Committee (PC), assisting the PC to implement its state management tasks on water and mineral resources, land, geology, environment, hydrometeorology and cartography. DONRE's offices are accountable for state management of the environment, water resources, mineral resources, land and sea, as well as island issues in urban and rural districts. At communal level, civil servants in charge of land and construction issues shall also be responsible for environmental issues, as well as mediation in environmental disputes (Joint Circular No. 03/2008/TTLT-BTNMT-BNV, PC's Decision No. 30/2008/QD-UBND, and PC's Decision No. 08/2009/QD-UBND).

community development.[18] Furthermore, many large enterprises, factories and manufacturers that are well informed about water laws bypass the existing regulations and thereby benefit from the weak regulatory framework used for water and environmental state management. In fact, the factories actually choose to pay the "dirt-cheap" fines instead of having wastewater treatment systems installed, because the latter are a prohibitively more expensive option. In some cases, wastewater treatment plants are being established in order to respond to the demands of the Environmental Impact Assessment (EIA), but their actual use and operation is still limited. Financial punishment is thus considered a "temporary solution" (*giải pháp tạm thời*), which must be replaced by a more effective sanctioning system (interview with the DONRE of Can Tho City 23.03.2009).

Local state agencies demonstrate that they still have other priorities. In many provinces, the "red carpet" (thảm đỏ) is laid out to attract foreign investments, while the industries' technical capacity to treat wastewater and therefore minimise its environmental pollution, is often ignored. Additionally, companies collaborate with local state officers and environmental police, e.g. when they agree on the dates of upcoming "unscheduled" inspections.[19] Such arrangements apparently bring benefits to the involved parties but definitely do not help to improve upon water quality management.

Despite recent efforts made to improve judicial competences, the right to initiate lawsuits against Decisions/Acts issued by administrative agencies has remained unused. The People's Court in Can Tho City has not yet exercised its new competence of dispute settlement on environmental issues (since July 1996). At commune level, a few small dispute cases have been solved by mediation between communal cadres in charge of land and construction (interview with the Department of Justice of Can Tho City 23.03.2009). The broader picture shows that the legal framework and current law enforcement practices do not meet the ever growing challenges of water pollution. It further appears as if environmental degradation is accepted as a cost of economic growth and rising living standards, and that the future rather than the present will be compromised. Fundamentally, state cadres and entrepreneurs are the immediate beneficiaries, while (in particular the poorer segments of) the population suffers from water quality degradation. Nevertheless, most water users are also polluters, and awareness about water pollution in farming systems and aquaculture production is just starting to rise (see also Chap. 13). More severe consequences will definitely emerge in the future.

---

[18] Content is based on preliminary research findings of the WISDOM project; forthcoming in 2012.

[19] Such practices were revealed by the head of the Environmental Police, Can Tho City in his presentation "Enforcement mechanisms for environmental regulations in Vietnam", held at the "Workshop on Waste Water Management in Industrial Zones – Challenges and Solutions", Can Tho City, 2nd March 2011 and by a German expert on industrial waste water management, who works in a project in Tra Noc Industrial Zone, Can Tho City (personal communication, 04.11.2011).

### *6.2.5 Rural Water Supply in Can Tho – Serving the People?*

Domestic water supply in the rural areas of the Mekong Delta – a region where water is naturally abundant – is characterised by diverse water sources. Until recently, rural households almost exclusively used rain and river water for their domestic purposes. Rain water, however, is not available during the dry season, which lasts for around 4–5 months (December to April). The central role of river water for domestic water supply goes back to the fact that settlements are located along the dense network of rivers and canals. Commonly, river water is filled into buckets to transport it to the house and then stored in jars. Due to the increasing pollution of surface water since the 1980s, though, it has become more and more popular to drill private wells during the last years. However, as research has shown, this is a rather expensive practice due to groundwater depths of around 60–100 m, and thus only relatively wealthy households can afford to drill wells.

With the 'National Rural Clean Water and Sanitation Strategy'[20] (NRWSS), Vietnam adopted its first sector policy for Rural Water Supply and Sanitation (RWSS) in 2000. The national goal is to provide all rural people with sufficient clean water and hygienic latrines by 2020 (MoC and MARD 2000:11). The basic principle of the NRWSS is *sustainability*, yet it does not specify the meaning of the term in detail, or what is intended to be sustainable. Moreover, the policy follows the approach of 'demand responsiveness', whereby, in contrast to the former supply-oriented approach, users are now supposed to decide for themselves what kinds of facilities they want, how they will organise construction, operation and maintenance and how they will pay for these elements. Construction is normally carried out by users, although alternatively they can hire a contractor. The operation and maintenance of facilities are managed by the users.

For the period 2006–2010, the 'Rural Water Supply and Sanitation National Target Programme II' (RWSS NTPII) was the main programme used for implementing the NRWSS. NTPII is supported by a consortium of three international donors – Denmark, Australia and the Netherlands – and reaffirms the basic principles of the NRWSS, but develops in more detail how these principles are to be put into practice. In particular, the policy considers 'community participation' as a critical 'success factor' (SRV 2006:16) and thus modifies formal planning mechanisms. Planning is to be carried out in a decentralised way, meaning that communes ("with participation of the community") should develop commune RWSS work plans, which then should be submitted to the districts; the districts collect and summarise all plans and prepare district plans. Provincial plans are to be prepared on the basis of district plans (SRV 2006:36).

Research in Can Tho City has shown that the national policy on RWSS is of limited relevance to policy practices on the ground. Although decision-making power on the allocation of funds for the NTPII lies foremost with MARD and the Ministry of Planning and Investment (MPI) on the national level, influence on the

[20] Approved by the Prime Minister's Decision No. 237/1998/QD-TTg of December 03, 1998.

implementation of RWSS policy is strongly in the hands of the provincial agency CERWASS.[21] A provincial NTP committee, which (in line with national policy) is supposed to decide on RWSS issues, exists on paper but has never met in practice. Although many different agencies are formally involved in the decision-making procedure, CERWASS is the agency that finally chooses policy instruments. All other involved actors in fact only execute formal administrative procedures. Moreover, 10 years after the release of the NRWSS, and contrary to its aim, the management and operation of water supply facilities remain completely in the hands of the government agency.

Planning for RWSS is not executed based on bottom-up developed work plans, as intended by NTPII, but instead follows the 'old-style' Leninist way. 'Community participation' is thereby understood in terms of one of the core doctrines of the prevailing "socialist political-legal canon" (Gillespie 2005:47 ff.), 'democratic centralism' (tập trung dân chủ). From this perspective, the democratic rights of the working class are secured by 'proletarian dictatorship' (Gillespie 2005:48), and the party state thus represents and acts according to the will of the people (Minh Nhut Duong 2004:5). Local and district authorities can report their demands to the higher levels within the scope of the policy given by CERWASS – the implementation of new piped water schemes. In this sense, policymaking is 'demand-oriented'.

> The local governments know the problems of the local areas. They travel around the households and talk to the people. They know what the people want, it is not necessary to ask the households (interview with Vice Director of CERWASS, 06.04.2008).

According to CERWASS officials, the "most urgent areas" are then chosen for new constructions, based on the statistics and suggestions submitted by local authorities. In recent years, CERWASS has implemented hundreds of groundwater abstracting, small-scale piped water supply schemes for supplying the rural population with clean water.

However, contrary to the official discourse, the research revealed that people's demands – communicated through local authorities – are not necessarily the criteria according to which the location of new water supply stations is decided upon. In contrast, a series of statements leads to the conclusion that other considerations play a major role in the decision-making process. As authorities pointed out, local areas and hamlets have to fulfil certain conditions in order to qualify for piped water supply from CERWASS. Most important in this respect is potential economic efficiency. Piped water supply in the rural areas of the Mekong Delta is an economically tight business, mainly due to the 'competition' of piped water with other water sources. Moreover, CERWASS has a very low budget, which does not allow the permanent subsidising of a loss-making business. To gain sufficient fees from a water supply station, a potential location has to possess enough households that are willing to connect to the station, and at the same time these households must not be located too far from each other, in order to limit the costs of constructing tubes. According to local

[21] CERWASS = Centre for Rural Water Supply and Environmental Sanitation. CERWASS is an agency under the Department of Agriculture and Rural Development (DARD) and MARD for the national scale.

authorities, 'suitable' locations must have more than 100 households willing to pay for piped water supply within a distance of 1,500 m. In many areas that do not yet have piped water supply, it is not feasible to construct water supply stations because too many households already have a private well. The poorer households living in these areas therefore still rely on using polluted water sources, in particular during the dry season, when rainwater is not available. The piped water schemes approach leaves behind areas in which it is economically unfeasible to construct a network, because households are too poor to pay, the population density is too low or too many households have a well and do not demand piped water. Hence, the "demands of the people", i.e. the data submitted to CERWASS by local authorities, do not necessarily play a role when it comes to the decision about new water supply stations. To date, year-round clean water supply is still problematic for one-third to half of the rural households.

Further, the sustainability of the current policy approach is questionable. On the one hand, this is related to the critical economic situation of the CERWASS system, which averts the comprehensive coverage of all areas with water supply stations and the connection of all households to the networks. A low ability and willingness to pay for piped water, a very low water tariff, the separation and privatisation of urban water supply and rural settlement patterns all contribute to the financial instability of the system. The other critical factor for the future sustainability of the system is the use of groundwater for such piped schemes, which is ecologically unsustainable due to the fast depletion of the resource. According to various assessments (Nuber et al. 2008; Nuber and Stolpe 2008; MONRE 2009), it is very likely that the stations will start suffering water shortages within the next 5–10 years.

### *6.2.6 Water Engineering Liberalisation: Old Wine in New Bottles*

The recent success story of developing the Mekong Delta into Vietnam's 'rice bowl' in only three decades is linked closely to new technical and scientific innovations in hydraulic engineering and water control. Progress made in salinity intrusion control, irrigation and drainage, as well as flood management, contributed significantly to the delta's agro-economic upswing, together with economic liberalisation, after *Doi Moi* (Biggs et al. 2009; Käkönen 2008) and transformed the delta into a hydraulic landscape (Evers and Benedikter 2009b). Following the green revolution in the 1970s, and supported by *Doi Moi* reforms, the hydraulic infrastructure was continuously expanded and provided for steadily increasing agricultural growth rates (Evers and Benedikter 2009a; Yasuyuki 2001; Le Meur et al. 2005). To date, large amounts of public funds[22] generated from ODA, revenues

[22] In the period from 2000 to 2010, the Vietnamese government invested a total of 2.8 billion USD into the preservation and construction of its hydraulic infrastructure (including hydropower plants) and, according to the next 5-year plan, investment is expected to grow (Benedikter 2010).

from the sale of government bonds and allocations from the national state budget have been invested in water infrastructure development (Benedikter 2010).

In default of a new river basin plan for the Mekong Delta, most hydraulic schemes currently under construction still refer to the NEDECO Master Plan (1993), which focuses on water control schemes for boosting agricultural production. Current practices, however, take recent developments into account and tend to adapt the planning framework – while formerly, hydraulic engineering focused on single constructions (canals, dykes and pumping stations), today, large-scale irrigation and drainage schemes are being developed. These schemes include road infrastructure projects, solid embankments and sluices for full control over water in- and outflows[23] (White 2002:33–36; Miller 2003:219–223). Moreover, contemporary water resources planning takes water demands for agriculture, domestic use, industries and aquaculture into account and is more complex.

Hydraulic works, which constitute the infrastructural backbone of water resources management and exploitation in the Vietnamese sense of the term, are defined as public property managed by the state. In accordance with the administrative division of the state apparatus, a complex bureaucratic management system comprising state agencies and semi-private companies has been developed to manage hydraulic infrastructure. This system, however, strictly follows administrative boundaries, instead of water flow regimes and hydrologic systems. The insistence on administrative boundaries in water flow management vividly illustrates the absence of "IWRM" principles in hydraulic infrastructure management. In this respect, the lack of cooperation and communication between provinces negatively affects cross-boundary devices and operations.

With regard to the decentralisation of water management functions, responsibilities are shared according to the size and scale of the construction. In principle, large hydraulic works such as inter-provincial schemes come under the auspices of central government, more precisely MARD. Smaller primary and secondary canals, sluice gates and dykes are under the management of provincial governments authorities, in particular DARD (Doan The Loi no date:2). Each of the administrative units receives a budget for maintenance and construction, following their respective annual plan. Smaller hydraulic works in the context on-field irrigation are managed by farmers' groups, which provide labour and funds to sustain their activities (Benedikter et al. forthcoming 2012).

According to the national regulations,[24] state authorities are in charge of monitoring and planning, while hydraulic works operation and maintenance fall under the responsibility of irrigation and drainage management companies (IDMCs) (Doan The Loi no date:9–12). Following MARD regulations, such companies are to be set up in every province. Nonetheless, investigations into 10 out of the 13 Mekong

[23] Examples are Nam Măng Thít, Ô Môn – Xà No, Bắc Vàm Nao and Quản Lộ – Phụng Hiệp.

[24] Circular 43/1984/TC-TTLB issued jointly by the Ministry of Water and the Ministry of Finance (1984); Decree No. 143/2003/ND-CP issued by the Government on November 28, 2003 detailing the implementation of a number of articles of the Ordinance on the exploitation and protection of irrigation works.

Delta provinces in 2008/09 revealed a distinctive administrative pluralism. For instance, water service delivery for rural production comes under the responsibility of state management agencies (DARD) in some of the provinces (Can Tho City, Soc Trang, Vinh Long), whereas others (Bac Lieu, Tien Giang) charge IDMCs with these tasks. This shows that provincial governments pursue different institutional practices, while local conditions may be more decisive for water resources management than centrally determined standards. Similarly, the division of responsibilities between district and provincial authorities differs significantly from province to province. While in Can Tho City sluices are under the responsibility of the provincial sub-department of water resources under DARD, in Vinh Long sluice gate management has been delegated to district local authorities (offices of agriculture and rural development). This large distinction in the administrative setup and allocation of responsibilities within different provinces, which is not in line with national standards, makes inter-provincial cooperation even more complicated.

Despite the announcement made by the government to foster decentralisation and promote local governance, research on the *Ô Môn – Xà No*[25] water control scheme showed that top-down decisions are still rigorously enforced. *Ô Môn – Xà No* is one of the three sub-projects of the national Water Resources Development Project for the Mekong Delta, based on the NEDECO Master Plan. The project was initiated by MARD and the World Bank and aimed to improve flood control for rural production growth in Can Tho and Kien Giang provinces (World Bank 1999:27) – ironically an area almost untouched by floods. The planned intervention consists of a closed ring dyke combined with culverts and sluice gates to control water in- and outflow.

During the planning phase in the late 1990s, local authorities and scientific experts in the project region raised concerns about the feasibility and suitability of the project. According to the critics, the construction plan was inspired by similar projects implemented in the Red River Delta previously (Northern Vietnam), but did not match with the distinct water regime of the project region. Nevertheless, MARD and the World Bank started construction works and failed to provide opportunities for local government participation in the decision-making and implementation processes of the project. In the first phase (1999–2009), over US$20 million were invested in the construction of sluice gates, culvers, embankments, canal dredging and other water engineering activities (World Bank 1999:27).

[25] Ô Môn – Xà No was appraised as inappropriate and environmentally problematic by local authorities in Can Tho City and elsewhere in the project region (interviews 2009). According to the authorities, the sluice gates were too narrowly constructed and therefore water inflow into the scheme was constrained. This causes severe water pollution problems, accelerates sedimentation, creates higher levels of acidity and severs water scarcity during the dry season (no flushing possible).When sluices are closed, boat traffic is hampered, which causes additional costs for agricultural good transportation. In a similar way, local newspapers argued, when they cynically assessed the schemes as a "Red River design project" imposed to the Mekong Delta. Between the lines, critics targeted both the wrong design by northern engineers and central government's domination in the implementation process (Tuoi Tre Newspaper 14.04.2008, Tuoi Tre Newspaper 27.02.2008, Can Tho Newspaper 27.02.2008).

In the wake of *Doi Moi*, market-based biddings were introduced under the Law on Tendering (National Assembly of Vietnam 2005) as a new means of allocating investment funds in public-funded infrastructure projects. Moreover, state-owned enterprises were (partly) converted into joint-stock companies (equitisation policy).[26] The enactment of the Law on Private Enterprises (1990) finally allowed the establishment of private hydraulic engineering firms and their involvement in public-invested infrastructure projects.

However, as the case of *Ô Môn – Xà No* shows, procedures have barely changed since sector liberalisation came into force. Funds for large-scale infrastructure projects are controlled by MARD, which issues contracts for project implementation to planning and research institutes, semi-private consulting firms and equitised, state-owned construction companies. Most of the financially profiting contractors, namely research and planning institutes under MARD and equitised water engineering and construction firms, maintain close relationships with the national agency. Notably, engineering firms formerly under MARD (and partly the Ministry of Construction) are headed by North Vietnamese engineers who are related to MARD officers via colleagueship, friendship (university), regional provenance or even kinship. After 10 years of project implementation, the scheme was still not completed and the government released another US$50 million for a second project phase (2009–2012), despite remaining local critiques with respect to environmental and socio-economic impacts (MARD 2009). This time, no ODA funds were provided. According to a statement issued by MARD's hydraulic construction project management board in charge of the Mekong Delta region, the planning institutes and firms involved would remain the same as in phase I. Although Vietnam claims to be a market-oriented economy, obviously the introduction of market-based allocation mechanisms in public-funded projects – such as tendering – seems to remain rather ineffective, as long as strategically operating networks allude to them in order to safeguard their monopolistic access to public resources.

Similar resources networks can be found at the local scale of the water engineering sector, where funds are controlled by local governments. As sector liberalisation began no more than 10 years ago, private hydraulic engineering and construction companies have begun to develop only recently. Due to their rather small scale, their focus is on local construction and dredging projects. Empirical research conducted in the delta has sought to capture the background of this newly emerging entrepreneurship. The environment under which such private businesses emerged was minted by strong inter-linkages between business and local bureaucracies. In many investigated cases, the actual ownership in private firms was either through (former) local politicians, a local cadre of the water bureaucracy or a staff member of a provincial state-owned enterprise. Private firms with no such ties put a lot of effort into establishing and maintaining patron-client relations and resources networks with local bureaucrats, who promote and protect their businesses.

[26] Equitisation (cổ phần hóa) describes the Vietnamese way of privatising state-owned enterprises by converting them into joint-stock companies.

At first glance, one might see the whole sector undergoing profound transition towards market-economy principles, where all kinds of businesses can be involved and contractors are selected following competitiveness and performance criteria. Beneath the superficial guise of institutional change, however, what strikes the observer is continuity rather than change. As already argued by Gainsborough (2009), and Greenfield (1993) even earlier, SOE reforms, or the transformation of state assets into private property, and the appearance of private business over the past decades have been taking place with the strong involvement of the party-state bureaucracy, incumbent directors of SOEs and (partly) the military. Hence, privatisation promoted by the socialist regime itself has certainly changed institutional arrangements, but in a way that it offers even better opportunities of self-enrichment for the bureaucratic polity (water bureaucracy) and their allies (emerging business) by securing a share of the pie. State assets have been transformed into private property – a step towards market principles indeed – but therewith have fallen into the hands of the old water bureaucracy for satisfying their rent-seeking aspirations. The elimination of 'old socialist practices', and approaches of strictly separating state functions from business, have created new linkages, often informally placed, between bureaucracy and business. This indicates two different processes: (1) the hybridisation of the bureaucrats into the corporate sector and (2) emerging alliances between bureaucrats and businessmen.

## 6.3 Conclusion

Extensive field research on water-related policy practices shows that IWRM in the Mekong Delta exists on paper, but has not been transformed into practice. First of all, the formal existence of the Mekong Delta River Basin Organisation has, in fact, not changed the conventional sector-specific and technocratic approach to water governance. Water policy planning continues to target single sub-sectors, like flooding and irrigation, water quality and water supply, separately from each other. There is no integrative approach to water governance, which appreciates the complexity of water-related problems in the delta. The sector-specific view on water issues has a substantially negative impact on the appropriateness of current policy approaches for solving the problems the population is facing. For instance, the lack of access to clean drinking and domestic water in rural areas is closely interrelated with the pollution of surface water resources, but water quality management does not play any role in RWSS policy.

In spite of existing policies which emphasise IWRM principles, virtually none of these principles is applied when it comes to actual practices. 'Community participation', as understood in the global water policy discourse, is a concept rather 'alien' to Vietnamese cultural and political history. Concepts of participation and grassroots democracy are adapted to fit the Vietnamese socio-cultural value system, where collective interest reigns supreme over individual interests and rights (Minh Nhat Duong 2004:21–22; Pham Duy Nghia 2005:80).

Historically, the Vietnamese political system has been based on the model of the Confucian paternalistic family, where relationships and hierarchies between people and socio-political institutions are strictly defined and kept in 'natural' harmony. Social order is maintained by political elite who act on behalf of the people (principle of *chính nghĩa* – exclusive righteousness). This ideological underpinning was incorporated in the Marxist-Leninist post-colonial regime, and legitimises the contemporary system, where the party state holds the central authority which represents the will of the Vietnamese people (Minh Nhut Duong 2004:5; Gillespie 2005:47 ff.; Gainsborough 2005:37). Elitism and the concept of monolithically-organised political power are therefore intrinsic characteristics of the Vietnamese political culture. According to this culture, citizens are sufficiently represented by local authorities and thus 'participate' in decision-making (democratic centralism). However, empirical research has revealed that the 'participation' of local communities in water governance does not translate into practice, as local authorities are, more often than not, excluded from policymaking and planning.

Furthermore, water resources management still follows a strict technical bias that places engineering activities and related infrastructural interference centre stage; the idea of exerting human domination over nature by technological innovation still overwhelms the whole water sector. 'Water resources development for effective exploitation' to stimulate growth still reflects the traditional paradigm of water management (*thủy lợi*). Against this background, water resources management is less integrative and neither ecologically nor environmentally sustainable. The case of RWSS has shown evidentially that the current approach to solving the 'domestic water crisis' in the Mekong Delta is based on groundwater exploitation on a large scale. This strategy cannot be sustainable in the long run, as it only provides a (temporary) solution that will create even more severe problems in the future. In a similar way, water quality management lacks effective instruments, although the reduction of water pollution is imperative – clean water is becoming scarce, and changes in water resources management are urgently needed. Here, short-term thinking outstrips long-term-oriented ideas of sustainable economic growth. Provincial governments compete over investments in local industrialisation, including foreign direct investments, and environmental protection standards and law enforcement are relegated to play a negligible role.

Decentralisation exists to a certain extent, but does not follow the principle of subsidiarity. Firstly, the devolution of decision-making powers and management responsibilities stops at the provincial level, hindering the development of solutions appropriate for local conditions at district and commune level. Local authorities can submit reports and petitions to the provincial-level agencies, but in fact do not exert any influence on decision-making. Secondly, it is evident that decentralisation is non-existent with regard to far-reaching, large-scale infrastructure projects that involve large amounts of government and/or donor funds. The case of the *Ô Môn – Xà No* water control project illustrates the prevalence of strict top-down decision enforcement, even where local authorities as well as citizens have shown a lot of dissatisfaction with the project. While the local stakeholders opt for more decentralisation and participation in planning, their central counterparts are unwilling to accept devolution, since this would diminish their power and sovereignty over administrative functions and – respectively – monetary resources.

Principles of economic efficiency have been formally introduced into water management, but functioning markets are levered out in practice due to the coalition, or even hybridisation, of bureaucracy and private entrepreneurs (Evers and Benedikter 2009a). Making use of public offices for gaining additional income is a well-known phenomenon in Vietnam (Gainsborough 2005:27), and was illustrated by the case of water engineering sector liberalisation. As further shown for the case of water quality management, those in charge of environmental protection and the polluters (companies, farms, etc.) often build arrangements for mutual benefit at the expense of water quality. This elucidates that economic issues, whether serving public welfare in the sense of economic development or the individual interests of resources networks, excel at environmental and ecological concerns of sustainability and even outsmart formal regulations and standards that have been formulated by those not complying with them.

At commune level, people live in relatively closed village communities, where local cadres are strongly embedded in a social environment that is minted by strict moral commitments to the lives of their fellows and community welfare. Under such conditions, local cadres rather follow their sentiments *(tình cảm)* with the community and single individuals, instead of rigorously enforcing what the law actually requests them to do (Koh 2004:221). By promoting commune welfare, they gain people's respect and social rewards as good cadres, or in a more traditional sense as ethically correct-acting mandarins. As shown for aquaculture, the rigorous national water quality standards would endanger people's livelihoods, since the majority of fish farmers do not comply with wastewater management regulations. However, a commune cadre turns a blind eye to such violations to keep good relations to their commune fellows and promote community welfare. At the same time, farmers might reward him/her for his/her 'ethically proper behaviour' in one of a number of ways for mutual benefit. This underlines that water governance at local level is not based on rules that are the same for everyone and every jurisdiction (the rule of law), but that the enforcement of laws and policies is embedded in a traditional moral code system and informal arrangements of 'everyday politics' at the local scale.

While on the one hand the managerial state bureaucracy is a closed unit that acts as a strategic group, there are on the other hand fierce struggles within the state bureaucracy over the division of functions and responsibilities. As shown in the case of RBOs and the relationship between MARD and MONRE, struggles over administrative responsibilities and corresponding financial resources within the bureaucratic apparatus fuel bureaucratisation and institutional complexity. This process follows the rationale that resources networks within the bureaucracy compete with each other by maintaining and constantly expanding their access options to government funds and ODA. Such internal struggles can be observed in the vertical direction between ministries and other central state agencies, as well as between central and local bureaucratic units, as shown with the *Ô Môn – Xà No* water control scheme. Although national directives and legal frameworks are bounding for all state management agencies, they often remain vague and require local specification.

Nevertheless, and as the case of water quality management standards has shown, provincial bureaucracies define their own institutional arrangements and organisational setups and hereby even ignore national regulations and policies somehow they create their own 'kingdoms'. Mechanisms to identify, sanction or control such contradictions are apparently not effective or even completely missing.

It is clear that the 'IWRM discourse' serves the interests of the state bureaucracy as a collectively acting group striving for power and resources. In 2008, Vietnam was the number one recipient of World Bank funds[27] (Cling et al. 2009:4) and ranked fifth among all ODA recipients (OECD 2010:2). Policies and institutional frameworks fulfilling international standards of 'good water governance' are required for obtaining access to donor loans and grants. However, while donors provide substantial funding and influence policy formulation, their influence over policy practices, which follow the rationality of the Vietnamese bureaucratic apparatus, remains very limited (Reis 2012). The fact that IWRM, a principle developed by a Western philosophy-dominated donor community, currently cannot function as an effective guideline for water policy practices in Vietnam is, in essence, not caused by a lack of capacity, inadequate institutional arrangements or the shortage of resources; rather, it is the result of the peculiar structural features of the contemporary state in Vietnam.

**Acknowledgements** The findings presented herein result from the joint effort of a number of researchers involved in the WISDOM (Water Related Information System for the Sustainable Development of the Mekong Delta, Vietnam) project. It strongly draws on empirical research mainly conducted in 2008–2010 in Can Tho City. The authors would like to thank its scientific partner in the Mekong Delta, the Mekong Delta Development Research Institute (MDI) of Can Tho University, for its strong academic and administrative support. Furthermore, a number of staff from local state agencies and other relevant organisations in Can Tho City and beyond offered their time and patience and responded to the many questions which emerged during the field work. We also owe our gratitude to the many farmers, entrepreneurs and household representatives who participated in personal interviews and surveys and shared their knowledge and individual perceptions with the researchers. At ZEF, Serge Birtel helped in meeting the editor's demands for correct citation and layout.

## References

Anh Phuong (2009) Can Tho – improving the investment environment is a strategic goal. The Local Economy 2, Overview (No. 6). http://dautumekong.vn/index.php/en/news-a-event/125-cn-th-ci-thin-moi-trng-u-t-la-mc-tieu-chin-lc?format=pdf. Accessed 1 Mar 2011

AUSAID (2003) Water resources management assistance project. Component 3. Management Study No 3: Inter-Institutional Sharing and Processing of Water Quality Data. Canberra

Bach Tan Sinh (2002) Government and NGO partnership in managing community-based water resources in Vietnam: a case study of Thai Long Dam project. Business Strategy and the Environment 11(2):119–129

[27] Granted by the International Development Association (IDA) in the form of project aid and budget support.

Benedikter S (2010) Hydraulic engineering and strategic group formation in Vietnam and the Mekong Delta. http://www.wisdom.caf.dlr.de/en/content/hydraulic-engineering-and-strategic-group-formation-vietnam-and-mekong-delta. Accessed 1 Mar 2011

Benedikter S, Waibel G (2012) Collective economy and local water management interfaces in the Mekong Delta of Vietnam, (forthcoming)

Beresford M (1995) Interpretation of the Vietnamese economic reforms, 1979–85. In: Fforde A (ed) Researching the Vietnamese economic reforms, 1979–86: Vietnam's transition to a market economy: an introduction to theoretical and empirical issues. Macquarie University, School of Economic & Financial Studies, Sydney, pp 1–16

Biggs D (2004) Between the rivers and the tides: a hydraulic history of the Mekong Delta 1820–1975. Dissertation, University of Washington, Washington DC

Biggs D, Miller F, Hoanh CT, Molle F (2009) The delta machine: water management in the Vietnamese Mekong Delta in historical and contemporary perspectives. In: Molle F, Foran T, Käkönen M (eds) Contested waterscapes in the Mekong Region: hydropower, livelihoods and governance. Earthscan, London, pp 203–225

Buhmann K (2007) Building blocks for the rule of law? Legal reforms and public administration in Vietnam. In: Balme S, Sidel M (eds) Vietnam's new order: International perspectives on the state and reform in Vietnam. Palgrave Macmillan, New York, pp 237–253

Can Tho Newspaper [Bao Can Tho] (27.02.2008). Dự án thủy lợi Ô Môn – Xà No. Ì ạch, nhiều bất cập

Can Tho Newspaper [Bao Can Tho] (14.12.2008). Cần Thơ trước vấn nạn ô nhiễm nước – Bài 1: Sông Hậu kêu cứu! [Can Tho in the face of water pollution – Part 1: Hau river calls for help!]. http://www.baocantho.com.vn/?mod=detnews&catid=184&p=&id=26154. Accessed 1 Mar 2011

CCFSC [Central Committee for Flood and Storm Control] and MARD [Ministry for Agriculture and Rural Development] (2001) Second national strategy and action plan for disaster mitigation and management in Vietnam – 2001 to 2020. Hanoi

Cling J-P, Razafindrakoto M, Roubaud F (2009) Desperately seeking model countries: the World Bank in Vietnam. DIAL document de travail. http://www.dial.prd.fr/dial_publications/PDF/Doc_travail/2009-04.pdf. Accessed July 2010

Conca K (2006) Governing water: contentious transnational politics and global institution building. MIT Press, Cambridge, MA

CRDE [Center for Resource Development and Environment] and IESD [Institute for Environment and Sustainable Development] (2006) Integrated management of river basins in Vietnam for sustainable development. Ministry of Plan and Investment, Vietnam Agenda 21

Dao Minh Chau (1996) Administrative concepts in confucianism and their influence on development in Confucian countries. Asian Journal of Public Administration 18(1):45–69

Doan The Loi (no date) Hiện trạng tổ chức quản lý và giải pháp nâng cao hiệu quả các hệ thống thủy lợi. http://www.vncold.vn/Modules/CMS/Upload/10/YKien_BinhLuan/HientrangTLW_P.pdf Accessed 1 Mar 2011

DONRE [Department of Natural Resources and the Environment of Can Tho City] (2009) Report on development of environmental quality in Can Tho City from 1999 to 2008, Can Tho City

Dung NH (2003) Economic and health consequences of pesticide use in paddy production in the Mekong Delta, Vietnam. Research report. http://203.116.43.77/publications/research1/ACF124.html. Accessed 1 July 2009

Duong Van Nha (2006) Research on the impacts of dyke on socio – economic issues and environment at the floodplains in An Giang Province. Report for Viet Nam – Netherlands Research Programme (VNRP)

Ehlert J (2012) Beautiful Floods. Environmental Knowledge and Agrarian Change in the Mekong Delta, Vietnam. ZEF Development Studies. LIT Verlag, Berlin

Evers H-D, Benedikter S (2009a) Hydraulic bureaucracy in a modern hydraulic society – Strategic group formation in the Mekong Delta. Water Alternatives 2(3):416–439

Evers H-D, Benedikter S (2009b) Strategic group formation in the Mekong Delta – The development of a modern hydraulic society. ZEF working paper, No. 35. Center for Development Research, University of Bonn, Bonn. http://www.zef.de/fileadmin/webfiles/downloads/zef_wp/wp35.pdf. Accessed 1 Mar 2011

Fforde A (2005) Vietnam in 2004: popular authority seeking power? Asian Survey 45(1):146–152

Fforde A (2008) Political processes in a Leninist State: the politics of rural development policy in contemporary Vietnam. Paper presented at the EUROVIET conference, Hamburg, 5 June 2008

Fforde A (2010) Vietnam: water policy dynamics under a Post-Cold War communism. Water Alternatives 3(3):552–574

Fontenelle J-P (2001) Vietnam Red River Delta irrigation management. GRET, Nogent-sur-Marne

Fritzen S (2006) Probing system limits: decentralisation and local political accountability in Vietnam. The Asia Pacific Journal of Public Administration 28(1):1–23

Gainsborough M (2005) Rethinking Vietnamese politics: will the real state please stand up? Bristol-Mekong Project working papers. University of Bristol, Bristol

Gainsborough M (2009) Privatization as state advance: private indirect government in Vietnam. New Political Economy 14(2)

Gillespie J (2005) Changing concepts of socialist law in Vietnam. In: Gillespie J, Nicholson P (eds) Asian socialism & legal change: the dynamics of Vietnamese and Chinese reform. ANAU E Press and Asia Pacific Press, Canberra, pp 45–75

Global Water Partnership (2000) Integrated water resources management, vol 4, TAC background papers. Global Water Partnership, Stockholm

GoV [Government of Vietnam] Joint Task Force (2003) Transition of water resources management from MARD to MONRE. Final report. Hanoi

Gray ML (2003) NGOs and highland development: a case study in crafting new roles. In: Kerkvliet BJT, Koh DWH, Heng RH-K (eds) Getting organised in Vietnam: moving in and around the socialist state. Institute of Southeast Asian Studies, Singapore, pp 110–125

Greenfield G (1993) The development of capitalism in Vietnam. In: Milisand R, Panitch L (eds) Between globalism and nationalism: socialist register 1994. Merlin Press, London, pp 203–234

Grothe S (2009) Integrated water resources management and institutional change in Vietnam and Poland – a comparative analysis with special consideration of the river basin level. Inaugural dissertation, Ernst-Moritz-Arndt-Universität Greifswald, Greifswald

GSO [General Statistics Office of Vietnam] (2011a) Average population by province. http://www.gso.gov.vn/default_en.aspx?tabid=467&idmid=3&ItemID=9880. Accessed 1 Jan 2011

GSO [General Statistics Office of Vietnam] (2011b) Number of farms in 2008 by kind of activity and by province. http://www.gso.gov.vn/default_en.aspx?tabid=469&idmid=3&ItemID=8892. Accessed 1 Jan 2011

GSO [General Statistics Office of Vietnam] (2011c) Area of water surface for the aquaculture by province (2008). http://www.gso.gov.vn/default_en.aspx?tabid=469&idmid=3&ItemID=10074. Accessed 1 Mar 2011

Herbst S, Benedikter S, Koester U, Phan N, Berger C, Rechenburg A, Kistemann T (2009) Perceptions of water, sanitation and health – a case study from the Mekong Delta, Vietnam. Water, Science and Technology 60(3):699–707

Hirsch P, Carrard N, Miller F, Wyatt A (2006) Water governance in context: lessons for development assistance. Australian Water Research Facility, AusAID and Australian Mekong Resource Center, University of Sydney

Käkönen M (2008) Mekong Delta at the crossroad: more control or adaptation? Ambio 37(3):205–212

Kerkvliet BJT (2003) Authorities and the people: an analysis of state-society relations in Vietnam. In: Luong HV (ed) Post-war Vietnam: dynamics of a transforming society. Rowman & Littlefield, Oxford, pp 27–53

Kerkvliet BJT (2005) The power of everyday politics: how Vietnamese peasants transformed national policy. Cornell University Press, Ithaca/London

Khoi D, Van H (2008) Bài 1: Sông Hậu kêu cứu. http://www.baocantho.com.vn/?mod=detnews&catid=184&p=&id=26154. Accessed 1 Jan 2011

Koh D (2001a) Negotiating the socialist state in Vietnam through local administrators: the case of karaoke shops. SOJOURN 16(2):279–305

Koh D (2001b) The politics of a divided party and Parkinson's State in Vietnam. Contemporary Southeast Asia 23(3):533–551

Koh D (2001c) State-society relations in Vietnam: strong or weak? In: Singh D (ed) Southeast Asian Affairs 2001. ISEAS, Singapore, pp 369–387

Koh D (2004) Urban government: ward-level administration in Hanoi. In: Kerkvliet BJT, Marr DG (eds) Beyond Hanoi: local government in Vietnam. Institute of Southeast Asian Studies, Singapore, pp 197–228

Le Meur P-Y, Hauswirth D, Leurent T, Lienhard P (2005) The local politics of land and water: case studies from the Mekong Delta. GRET, Paris

Le Thi Viet Hoa, Nguyen Huu Nhan, Wolanski E, Tran Thanh Cong, Shigeko H (2007) The combined impact on the flooding in Vietnam's Mekong River Delta of local man-made structures, sea level rise, and dams upstream in the river catchment. Estuar Coast Shelf Sci 71(1–2):110–116

Le Thi Viet Hoa, Shigeko H, Nguyen Huu Nhan, Tran Thanh Cong (2008) Infrastructure effects on floods in the Mekong River Delta in Vietnam. Hydrol Process 22(9):1359–1372

Luttmer N (2000) Konfuzianismus und Kommunismus in Vietnam. ASIEN 75:61–80

MARD [Ministry Of Agriculture and Rural Development] (2009) 593/QĐ-BNN-XD – Quyết Đinh: Phê duyệt dự án đầu tư xây dựng công trình Ô Môn – Xà No giai đoạn 2 tỉnh Hậu Giang, Kiên Giang và thành phố Cần Thơ

Miller F (2003) Society-water relations in the Mekong Delta: a political ecology of risk. PhD, University of Sydney, Sydney

Minh Nhut Duong (2004) Grassroots democracy in Vietnamese communes. The Centre for Democratic Institutions, Research School of Social Sciences, The Australian National University, Canberra

MOC [Ministry Of Commerce] and MARD [Ministry of Agriculture and Rural Development] (2000) National rural clean water supply and sanitation strategy up to year 2020. http://www.cerwass.org.vn/English/index_e.htm. Accessed 1 Feb 2008

Molle F, Chu Thai Hoanh (2008) Implementing integrated river basin management: lessons from the Red River Basin, Vietnam. M-Power working paper, IDR, M-Power and IWMI, Colombo

Molle F, Mollinga PP, Wester P (2009) Hydraulic bureaucracies and the hydraulic mission: flows of water, flows of power. Water Alternatives 2(3):328–349

MONRE [Ministry Of Natural Resources and Environment] (2006) National water resources strategy – towards the year 2020. Culture-Information Publishing House, Hanoi

MONRE [Ministry Of Natural Resources and Environment] (2008) Mekong Delta's environment degrading: 20.09.2008. http://www.monre.gov.vn/monreNet/Default.aspx?tabid=255. Accessed 1 Oct 2008

MONRE [Ministry Of Natural Resources and Environment] (2009) Experts warn that southern Vietnam aquifers seriously depleted. 01.07.2009. Online news, water. http://www.monre.gov.vn/monreNet/Default.aspx?tabid=255&ItemID=67935. Accessed 1 July 2009

National Assembly of Vietnam (2005) Luật Đấu Thầu số 61/2005/QH11

NEDECO (1993). Master plan for the Mekong Delta in Viet Nam. A perspective for sustainable development of land and water resources. A report to government of Vietnam, State Planning Committee, World Bank, Mekong Secretariat, UNDP. Netherlands Engineering Consultants, Arheim, the Netherlands

Nguyen Thi Phuong Loan (2010a) Legal framework of the water sector in Vietnam. ZEF working paper No. 52, Center for Development Research, University of Bonn, Bonn. http://www.zef.de/fileadmin/webfiles/downloads/zef_wp/wp52.pdf. Accessed 1 Jan 2011

Nguyen Thi Phuong Loan (2010b) Problems of law enforcement in Vietnam: the case of wastewater management in Can Tho City. ZEF working paper no. 53, Center for Development Research, University of Bonn, Bonn. http://www.zef.de/fileadmin/webfiles/downloads/zef_wp/wp53.pdf. Accessed 1 Jan 2011

Nguyen Xuan Tiep (2008) Pim models in Vietnam – experience lessons. In: Nguyen Xuan Tiep (ed) Nông dân Tham Gia Quản Lý Công Trình Thủy Lội (Participatory Irrigation Management and Emerging Issues). Nhà Xuất Bản Nông Nghiệp, Hanoi, pp 297–336

Nuber T, Stolpe H (2008) Challenges of the groundwater management in Can Tho City, Vietnam. In: Presentation at the BGR symposium on sanitation and groundwater protection, 14 Oct –17 Oct 2008, Hannover. http://www.geopotenzial-nordsee.de/nn_324952/EN/Themen/Wasser/Veranstaltungen/symp__sanitat-gwprotect/present__nuber__pdf,templateId=raw,property=publicationFile.pdf/present_nuber_pdf.pdf. Accessed 1 July 2009

Nuber T, Stolpe H, Ky Quang Vinh, Vu Van Nam (2008) Modelling the groundwater dynamics of Can Tho City – challenges, approaches, solutions. In: Proceedings of the international SANSED workshop – decentralized water treatment systems and beneficial reuse of generated substrates, April 15–18, 2008, Can Tho University, Vietnam, pp 188–195

O'Rourke D (2004) Community-driven regulation: balancing development and the environment in Vietnam. Massachusetts Institute of Technology, Cambridge, MA

OECD [Organisation for Economic Co-operation and Development] (2010) Development aid at a glance. Statistics by region. http://www.oecd.org/dataoecd/59/5/42139479.pdf. Accessed 1 Aug 2010

Perret S (2006) New paradigms, policies and governance in the water sector. In: Perret S, Farolfi S, Hassan R (eds) Water governance for sustainable development: approaches and lessons from developing and transitional countries. Earthscan, London

Pham Cong Huu (2012). Floods and Farmers. Politics, Economics and Environmental Impacts of Dyke Construction in the Mekong Delta, Vietnam. ZEF Development Series. Lit Verlag, Berlin

Pham Cong Huu, Ehlers E, Subramanian SV (2009) Dyke system planning: theory and practice in Can Tho City, Vietnam. ZEF working paper no. 47. Center for development Research, University of Bonn, Bonn. http://www.zef.de/fileadmin/webfiles/downloads/zef_wp/wp47.pdf. Accessed 2 Mar 2011

Pham Duy Nghia (2005) Confucianism and the conception of the law in Vietnam. In: Gillespie J, Nicholson P (eds) Asian socialism & legal change: the dynamics of Vietnamese and Chinese reform. ANAU E Press and Asia Pacific Press, Canberra

Pike D (2000) Informal politics in Vietnam. In: Dittmer L, Fukui H, Lee PNS (eds) Informal politics in East Asia. Cambridge University Press, Cambridge, UK, pp 269–289

Porter G (1993) Vietnam: the politics of bureaucratic socialism, Politics and International Relations of Southeast Asia. Cornell University Press, Ithaca/London

Reis N (2012) Tracing and making the state. Policy practices and domestic water supply in the Mekong Delta. ZEF Development Series. Lit Verlag, Berlin

Reis N, Mollinga PP (2009) Microcredit for rural water supply and sanitation in the Mekong Delta – policy implementation between the needs for clean water and 'beautiful latrines'. ZEF working paper no. 49, Center for Development Research. University of Bonn, Bonn. www.zef.de/module/register/media/a0a4_WP49_Reis-Mollinga.pdf. Accessed 1 Mar 2011

Rogers P, Hall AW (2003) Effective water governance. TEC Background Paper, 7. Global Water Partnership. Stockholm, Sweden

Sarkkula J, Koponen J, Keskinen M, Kummu M, Inkala A, Käkönen M, Lauri H, Varis O, Veijalainen N, Virtanen M, Rasphone A, Souksakoun K, Savuth Y, Rath H, Vongsa TS, Toan TQ, Nghia PD, Kiem LV (2008) Hydrological, environmental and socio-economic modelling tools for the Lower Mekong Basin impact assessment. WUP-FIN Phase 2: impact assessment report

Scott S, Truong Thi Kim Chuyen (2004) Behind the numbers: social mobility, regional disparities, and new trajectories of development in rural Vietnam. In: Taylor P (ed) Social inequality in Vietnam and the challenges to reform. Institute of Southeast Asian Studies, Singapore, pp 90–122

SIWRP [Southern Institute for Water Resources Planning] (1998) Mekong Delta flood planning. The main report of the Southern Institute for Water Resources Planning (SIWRP), the Ministry of Agriculture and Rural Development (MARD) (Vietnamese)

SIWRP [Southern Institute for Water Resources Planning] (2005) Mekong Delta general planning: Flood control dyke system planning section in the Mekong Delta. The main report of the Southern Institute of Water Resources Planning, Ministry of Agriculture and Rural Development, vol 2 (Vietnamese)
SRV [Socialist Republic of Vietnam] (2006) National target program for rural water supply and sanitation (2006–2010). Final draft, Hanoi
Taylor P, Wright G (2001) Establishing river basin organisations in Vietnam: Red River, Dong Ni River and Lower Mekong Delta. Water Science and Technology 43(9):273–281
Thayer CA (1995) Mono-organizational socialism and the state. In: Kerkvliet BJT, Porter DJ (eds) Vietnam's rural transformation. Institute of Southeast Asian Studies (ISEAS), Singapore, pp 39–64
To Van Truong (2000) Referring to the historic flood in the year 2000 and flood control strategies and measures in the MD. SIWRP, Ho Chi Minh City
To Van Truong, Nguyen Ngoc Anh (2002) Water resources development for Socio-economic stability and development strategy in the Mekong Delta. Paper presented at the National Workshop on Water, Food and Environment, Hanoi, 24–25 Sept 2002
Tran Nhu Hoi (2005) Dykes for the flooding areas in the Mekong Delta. Agriculture Publishing House (Vietnamese), Ho Chi Minh City
Tuoi Tre Newspaper (27.02.2008) Dân khổ vì công trình tiền tỉ
Tuoi Tre Newspaper (14.04.2008) Hết hạn vay vốn nhưng vẫn chưa hoàn thành
UNDP [United Nations Development Programme] (2001) Modernizing governance in Vietnam. UNDP, Hanoi
Vo Thi Lang et al (2009) Environmental consequences of and pollution control options for pond "Tra" fish production in Thotnot District, Cantho City, Vietnam. EEPSEA Research Report 2009 No. 2009-RR3
Waibel G (2010) State management in transition: understanding water resources management in Vietnam. ZEF working paper no. 55, Center for Development Research, University of Bonn, Bonn. http://www.zef.de/fileadmin/webfiles/downloads/zef_wp/wp55.pdf. Accessed 1 Mar 2011
White I (2002) Water management in the Mekong Delta: changes, conflicts and opportunities, Centre for Resource and Environmental Studies. The Australian National University, Canberra
Wischermann J (1999) 'Keine Modernisierung der Gesellschaft ohne aktive Teilhabe der Bevölkerung!' – 'Civic Organisations' und ihr Verhältnis zu 'Governmental Organisations' im vietnamesischen Transformationsprozeß. In: Duy Tu Vu, Will G (eds) Vietnams neue Position in Südostasien. Verbund Stiftung Deutsches Übersee-Institut, Hamburg, pp 121–156
WISDOM (2009) Industrial parks and investment capital in the Mekong Delta. http://www.wisdom.caf.dlr.de/en/content/industrial-parks-and-investment-capital-mekong-delta. Accessed 1 Jan 2011
Woodside A (2006) Lost modernities: China, Vietnam, Korea and the hazards of world history. Harvard University Press, London
World Bank (1999) Project appraisal document on a proposed credit of SDR 72.8 million to the Socialist Republic of Vietnam for the Mekong Delta Water Resources Project
World Bank (2009) Vietnam development report 2010. Modern institutions. World Bank, Hanoi
World Bank, ADB [Asian Development Bank], FAO [Food and Agricultural Organisation], UNDP [United Nations Development Programme], NGO Water Resources Group, and Institute of Water Resources Planning (eds) (1996) Vietnam water resources sector review. Selected working papers. Hanoi
World Bank, DANIDA, MONRE (2003) Vietnam environment monitor 2003. Water, Hanoi (http://siteresources.worldbank.org/INTEASTASIAPACIFIC/Resources/Vietnam-Environment-Monitor-03.pdf)
Yasuyuki K (2001) Canal development and intensification of rice cultivation in the Mekong Delta: a case study in Can Tho Province, Vietnam. Southeast Asian Studies 39(1):70–85
Zingerli C (2004) Politics in mountain communes: exploring Vietnamese grassroots democracy. In: McCargo D (ed) Rethinking Vietnam. RoutledgeCurzon, London, pp 53–66

# Part III
# Water Resources and Related Risks

# Chapter 7
# Groundwater Resources in the Mekong Delta: Availability, Utilization and Risks

**Frank Wagner, Vuong Bui Tran, and Fabrice G. Renaud**

**Abstract** Surface water resources in the Mekong Delta (MD) are under increasing strain due to unplanned extraction, pollution, salinization and climate change effects. As a consequence, groundwater has become an increasingly important resource since the 1990s. Nevertheless, in some regions the ongoing exploitation has reached the capacity limits of the aquifers, challenging scientists and decision makers in order to satisfy the increasing water demand in the MD. New groundwater exploration studies are often hindered by the complexity of the subsurface structure and the hydrogeological system of the MD, which still needs further investigations. Therefore, this chapter provides an overview of the hydrogeology in the MD in Vietnam based on recently published and unpublished studies. It interlinks Late Pleistocene and Holocene sea level change with modern groundwater resources and concludes with possibilities and obstacles regarding current exploitation and future exploration studies. Furthermore, the final section focuses on the role of groundwater for domestic, industrial and agricultural water supply in the MD and addresses implications for the future exploration and management strategies.

F. Wagner (✉)
Sub-Department of Groundwater Resources – Quality and Dynamics, Federal Institute for Geosciences and Natural Resources (BGR), Stilleweg 2, 30655 Hannover, Germany

Center for Water Resources Planning and Investigation (CWRPI), No 93, Lane 95, Vu Xuan Thieu Street, Sai Dong Ward, Long Bien District, Hanoi, Vietnam
e-mail: frank.wagner@bgr.de

V.B. Tran
Technical Department, Division for Water Resources Planning and Investigation for the South (DWRPIS), No. 59, Street No. 2, Binh An Ward, District No. 2, Ho Chi Minh City, Vietnam
e-mail: buitranvuong@gmail.com

F.G. Renaud
United Nations University - Institute for Environment and Human Security (UNU-EHS), UN Campus, Hermann-Ehlers-Str. 10, Bonn 53113, Germany
e-mail: renaud@ehs.unu.edu

F.G. Renaud and C. Kuenzer (eds.), *The Mekong Delta System: Interdisciplinary Analyses of a River Delta,* Springer Environmental Science and Engineering,
DOI 10.1007/978-94-007-3962-8_7, © Springer Science+Business Media Dordrecht 2012

## 7.1 Geologic Evolution of the Mekong Delta

The rapid uplift of the Proto-Himalaya mountain chain initiated the subsidence of adjacent tectonic basins in Southeast Asia. The pre-Cenozoic basement of the NW[1] trending Mekong basin began to subside in late Neogene time. High erosion rates have resulted in the mobilization of large amounts of material, which was collected by the tributaries of the major receiving river systems and transported to the South China Sea; recent erosion rates have reached approximately 160 million tons of sediments per year (Xue et al. 2010). During millions of years, deposition and accumulation at the river mouth in combination with ongoing subsidence of the Mekong basin finally resulted in the formation of the extensive Mekong Delta (MD) complex with a maximum thickness of more than 600 m along the basin axis (Anderson 1978). Offshore, the basin passes eastward into the narrow, ENE trending Vung Tau basin filled with up to 7,000 m Cenozoic deposits (Fontaine and Workman 1997).

NE-SW directed faults divide the igneous, metamorphic and sedimentary bedrock into blocks of increasing depths towards the coastline. The secondary tectonic directions are NW-SE and N-S (Division for Geological Mapping for the South of Vietnam (DGMS) 2004). The crosscutting of faults with NE-SW and NW-SE directions resulted in blocking of the bedrock near the mouth of the Bassac River with maximum depth of about 3,000 m below surface level in the SE of Tra Vinh province. Minor faults in N-S direction have been detected in the West of the MD.

Plate-tectonics and faulting have established the frame for further development of the MD. Alternating periods of alluvial and marine sedimentation during late Cenozoic time resulted in the accumulation of heterogeneous strata. Thus, evolution and architecture of the MD subsurface is governing availability as well as quality of groundwater. The following sub-section describes sea-level fluctuation during late Quaternary time and its relevance for the availability of modern groundwater resources in the MD.

### *7.1.1 Impact of Paleo-Sea Level Change on Modern Groundwater*

The geologic and geomorphologic development of the MD was controlled by the glaci-eustatic sea-level change and the ongoing tectonic subsidence of the Mekong Basin. Continuous accumulation of the sediment load was interrupted by repetitive erosion events during the glacial periods due to the associated marine regression in the late Pleistocene. Especially the last glacial event hugely impacted both distribution and interconnection of groundwater bearing strata in the subsurface of the MD as well as on the salinity of pore waters as demonstrated below.

During the last interglacials sea-level high stand about 125,000 years BP, extended coastal areas of today's MD were flooded. Hence, the area was exposed to seawater intrusion and regional salinization of pore water within permeable

---

[1] N = North, W = West, S = South and E = East

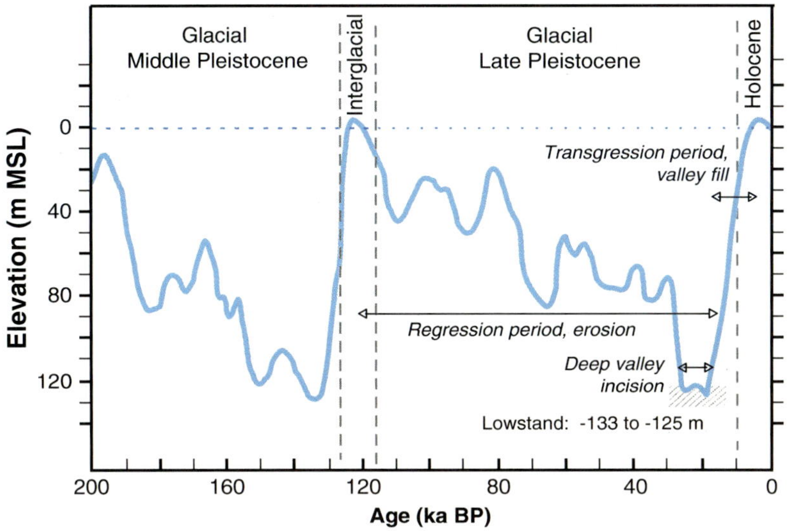


**Fig. 7.1** Sea level fluctuation relative to modern sea-level (*MSL*) during the last 200,000 years, reconstructed from the Pacific $\delta^{18}O$ record (redrawn after Waelbroeck et al. 2002). Maximum low stand during last glacial was about −133 to −125 m (Hanebuth et al. 2009)

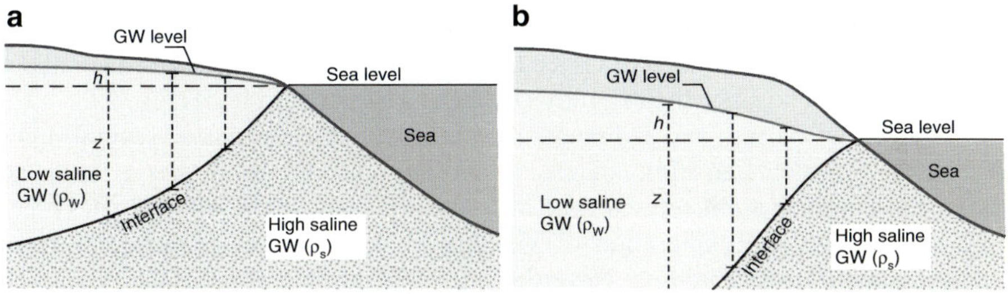


**Fig. 7.2** Relationship of low saline groundwater (*GW*) head (*h*) and depth of transition to high saline groundwater (*z*) according to the *Ghyben-Herzberg principle*. Its application for sea-level high stand (**a**) and sea-level low stand (**b**) indicates the origin of deep fresh groundwater lenses in the MD

Pleistocene and underlying sediments. Subsequently, the last glacial period began about 120,000 years BC. It was accompanied by a general trend of sea-level lowering of several tens of meters and interrupted by temporary transgressive events (Fig. 7.1). Marine regression implies a deeper discharging system, increasing the relief and the erosion energy of the receiving Mekong River and its tributaries.

During the maximum sea-level low stand up to 125–133 m below modern sea-level (Hanebuth et al. 2009), erosion of earlier Pleistocene sediments culminated in the incision of deep river valleys (Xue et al. 2010). Thus, compared to modern times, a much steeper hydraulic gradient between recharge areas in the inland and the ocean was developed. This directly impacts the depth of the interface between fresh and saline groundwater in coastal areas as shown by the *Ghyben-Herzberg principle* (Fig. 7.2).

$$z = h\frac{\rho_w}{\rho_s - \rho_w}$$

z depth of the salt-water interface below sea-level
h elevation of the water table above sea-level
$\rho_w$ density of fresh low saline groundwater
$\rho_s$ density of high saline groundwater

Based on the difference in density, the depth to which fresh water extends below sea-level (z) is approximately 40 times higher than the water table above the sea-level (h) assuming that stationary and unconfined conditions are given (Fetter 1972, 2001). Currently, the seaward directed natural hydraulic gradient of groundwater in the Mekong area is generally very low (Fig. 7.2a). During periods of mean low sea-level, a much steeper gradient (h) of more than 10 m potentially results in flushing of pore water down to hundreds of meters depth.

During the transgression starting from 19,000 to 20,000 years BP, the paleo river valleys were rapidly filled with estuarine, tidal, and fluvial sediments (Xue et al. 2010). Accumulation peaked from 13,000 to 9,500 years BP when sea-level rose rather constantly at a rate of approximately 10 mm/year and the valley was almost filled (Tjallingii et al. 2010). Accumulation stopped in the late Holocene with fine grained, low permeable flood sediments and peat layers coming apart from the active river channels. These beds represent protective layers preventing regional resalinization of the deeper aquifers[2] during sea-level rise. Thus, fresh water lenses were locally preserved at hundreds of meter depths. Local sea-level reached its maximum about 5,500 years BP, and since then slightly declined to the modern level (Fig. 7.1).

The cycle of erosion and sedimentation described above occurred several times during the repeated transitions of glacial and interglacial periods. As a result, an extremely complex architecture of the MD subsurface as described in the next section complicates the exploration of yielding fresh groundwater resources from Neogene and Pleistocene time. Nevertheless, salinization and pollution of surface water and shallow groundwater have led to increasing exploitation of low saline groundwater from aquifers up to 400–500 m depths.

## 7.2 Groundwater Resources in the Mekong Delta

### *7.2.1 Hydrostratigraphy and Aquifer System*

As demonstrated above, the subsurface structure and, thus, the hydrogeology setting of the MD is somewhat complex. Active faulting and repetitive transgression and regression events during deposition of the sediments have resulted in a very

[2] In this text, the term *aquifer* represents a water saturated and relatively high permeable geological formation, transmitting and yielding significant quantities of water; an *aquiclude* is a low to non-permeable formation and represents a barrier for groundwater transport.

heterogeneous structure of intersecting aquifers and aquicludes. Sedimentary strata of the Mekong complex with relevance for groundwater supply last from the late Neogene (Miocene, Pliocene) up to recent Holocene time. Table 7.1 presents a stratigraphic overview based on the geological studies by DGMS.

According to the Vietnamese nomenclature (DGMS 2004), the Cenozoic Mekong complex can be distinguished into eight hydro geological units, namely Holocene (*qh*), Upper Pleistocene ($qp_3$), Upper- Middle Pleistocene ($qp_{2-3}$), Lower Pleistocene ($qp_1$), Middle Pliocene ($n_2^2$), Lower Pliocene ($n_2^1$), Upper Miocene ($n_1^3$) and Upper-Middle Miocene ($n_1^{2-3}$). Generally, each hydrogeological unit has been divided into two parts. The upper part is composed of a low permeable silt, clay or silty clay. A lower rather permeable part consists of fine to coarse sand, gravel, and pebble with medium to high water yield. Generally, the yield of all aquifers varies from medium to high (1 to >5 l/s), exclusively the Holocene aquifer due to its more complex composition (see below and Table 7.1).

The two cross sections illustrated in Figs. 7.4 and 7.5 provide an overview of the spatial distribution and interconnection of the hydrogeological units within the complex architecture of the Delta's subsurface. Basically, the aquifer system in the MD has an artesian basin structure. The deepest area of the basement is located below the Tien and Hau Rivers and rises to the NE, N and NW borders. A brief characterization of the hydrogeological units and their composition is summarized below (DGMS 2004).

#### 7.2.1.1 Holocene (*qh*)

Recent Holocene sediments outcrop almost the whole MD with an area of 40,000 km$^2$. The Holocene hydrogeological unit comprises various sediments from mainly three facies types: Lower to Middle Holocene sediments ($qh_{1-2}$) of alluvial and marine origin consist mainly of clayey silt and fine sand and are rich in organic matter. Alluvial, marine and eolian sediments ($qh_{2-3}$) include 5–10 m thick remnants of sand dunes from paleo-sea shores which can be found in Mo Cay, Ba Tri, Tra Cu, Long Toan districts of Ben Tre and Tra Vinh provinces. Upper Holocene sediments ($qh_3$), accumulated in river valleys and flood plains, consist of clayey silt and fine sand.

The permeable layers of the Holocene unit represent the *qh* aquifer. The shallow wells screened in the *qh* aquifer have depths up to 30 m and vary in yield from low to medium (0.1–2.0 l/s). The groundwater levels are generally between 0.5 and 3.0 m above modern sea-level (msl, see next subsection). Whereas Holocene aquifers mostly carry salty or brackish water, sand dune sediments provide limited recharge of fresh groundwater which is locally used for domestic water supply.

#### 7.2.1.2 Upper Pleistocene ($qp_3$)

The Upper Pleistocene unit is widely distributed over the whole MD, mainly overlaying by Holocene sediments. Only in NE part of the MD $qp_3$ sediments are cropping out. The unit consists of the Cu Chi and Moc Hoa formations, both showing alluvial and marine-alluvial characteristics.

**Table 7.1** Time scale of Late Cenozoic Strata separated in geological and hydrogeological units of the MD according to DGMS (2004)

| Era | System | Series | Subseries | Geological unit | Hydrogeol. unit | Facies type[a] | Base age (Mio a BP) |
|---|---|---|---|---|---|---|---|
| Cenozoic | Quaternary | Holocene | Upper | $Q_{II}^{3}$ | $qh_3$ | mv, b, mb, m, ab, am, a | 0.002 |
| | | | Middle | $Q_{II}^{2-3}$ | $qh_{2-3}$ | bm, mb, am, ab | 0.006 |
| | | | Lower | $Q_{II}^{1-2}$ | $qh_1$ | m, am, a | 0.012 |
| | | Pleistocene | Upper | $Q_{I}^{3}$ | $qp_3$ | m, am, a | 0.126 |
| | | | Middle | $Q_{I}^{2-3}$ | $qp_{2-3}$ | m, am, a | 0.781 |
| | | | Lower | $Q_{I}^{1}$ | $qp_1$ | m, am, a | 1.806[b] |
| | | | | $N_{2}^{3}$ | – | m, am, a | 2.588[b] |
| | Neogene | Pliocene | Middle | $N_{2}^{2}$ | $n_{2}^{2}$ | m, am, a | 3.600 |
| | | | Lower | $N_{2}^{1}$ | $n_{2}^{1}$ | m, am, a | 5.332 |
| | | Miocene | Upper | $N_{1}^{3}$ | $n_{1}^{3}$ | m | 11.608 |
| | | | Middle | $N_{1}^{2-3}$ | $n^{2-3}$ | am | 15.97 |
| | | | Lower | $N_{1}^{1}$ | – | – | 23.03 |

Associated base ages are modified according to IUGS-ICS ratification in 2009 (Gibbard et al. 2010)

[a](a) – alluvial; (m) – marine; (b) – swamp, bog; (v) – aeolian; according to DGMS Vietnam

[b]Quaternary base has been redefined by IUGS-ICS in 2009 to 2.58 Mio a BP, including the former upper Pliocene

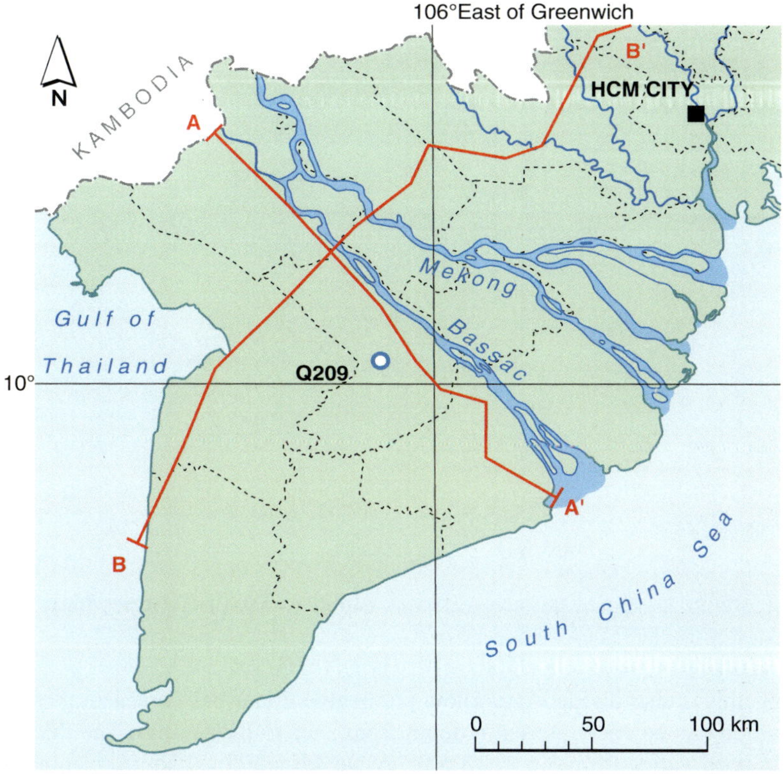


**Fig. 7.3** Sketch map of the Vietnamese part of the MD with locations of cross sections (Figs. 7.4 and 7.5) and National Monitoring Station Q209 (see next sub-section)

From the hydrogeological point of view, *qp3* strata can be divided into two parts: A lower high permeable aquifer is covered by an upper low permeable aquitard layer, generally consisting of silt to clay size fraction. The top of the aquitard can be encountered in depths of 10.1–37.3 m below ground level (m bgl) and consists of a thickness from 12.1 up to 20.6 m. The underlying aquifer is composing of fine to coarse sand and has a thickness from 9.4 to 22.4 m. The *qp3* aquifer is weakly confined.

Despite the complicated structure, the upper Pleistocene aquifer provides low saline fresh groundwater in an area of about 8,500 $km^2$, located in Vinh Hung (Long An province); My Tho (Tien Giang province), and Tieu Can, Cau Ngang (Tra Vinh province). In these areas, its exploitation for domestic water supply is very common.

#### 7.2.1.3 Upper-Middle Pleistocene ($qp_{2-3}$)

Outcrops of $qp_{2-3}$ can only be found in An Giang province. In other areas of the MD they are overlaid by the upper Pleistocene aquifer. The lithology is dominated by sediments of alluvial, marine alluvial, and marine origin, and hence, is rather diverse.

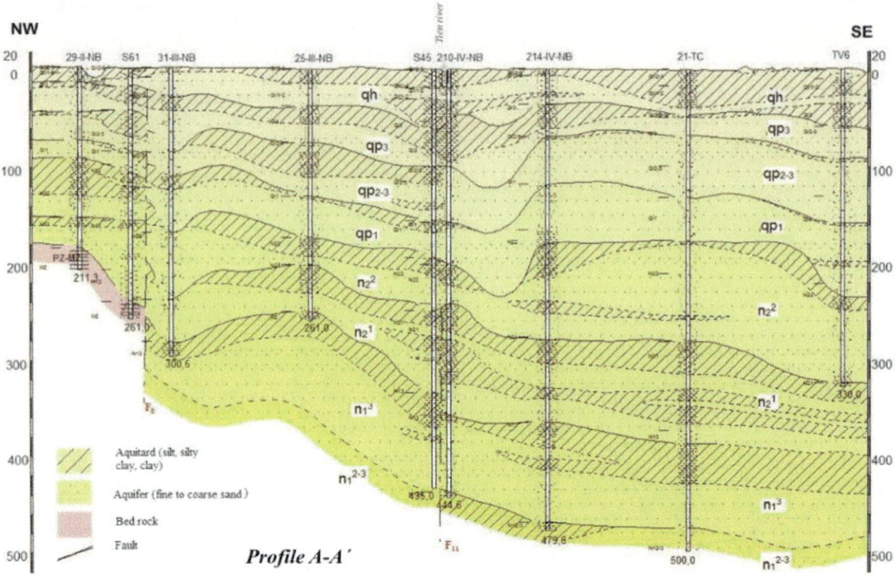


**Fig. 7.4** NW-SE directed hydrogeological profile cutting the MD plain (DGMS 2004)

This unit is also divided into a low permeable upper part, consisting of silt and clay which can be encountered in depth from 31.6 to 81.7 m bgl. The thickness of this aquitard varies from 3.6 to 13.5 m. A weakly confined aquifer represents the lower part, consisting of well sorted and high permeable fine to coarse sand with a total thickness of 17.3–56.2 m. There are several hydraulic windows connecting the $qp_{2-3}$ aquifer with the overlaying aquifer of $qp_3$ unit.

Low saline groundwater extends to the areas of more than 20,000 km², located in Tra Vinh, Bac Lieu, and Ca Mau provinces. In these areas the $qp_{2-3}$ aquifer has a high yield and provides groundwater of generally good quality.

#### 7.2.1.4 Lower Pleistocene ($qp_1$)

Sediments of the Lower Pleistocene unit $qp_1$ are generally of alluvial origin. Only in the area of the Ca Mau peninsula the $qp_1$ unit shows a marine facies. The top part consists of an aquitard with 8–16 m thickness. The lower part is an aquifer with permeable fine to coarse sand and a thickness varying from 14.2 to 43.9 m. The depth to the top of the aquifer inclines strongly from NW to SE.

There are several hydraulic windows where this aquifer has direct hydraulic contact to the overlaying Middle Pleistocene $qp_{2-3}$ aquifer. The yielding lower Pleistocene aquifer is commonly exploited for water supply in Can Tho, Kien Giang, Bac Lieu, and Ca Mau provinces.

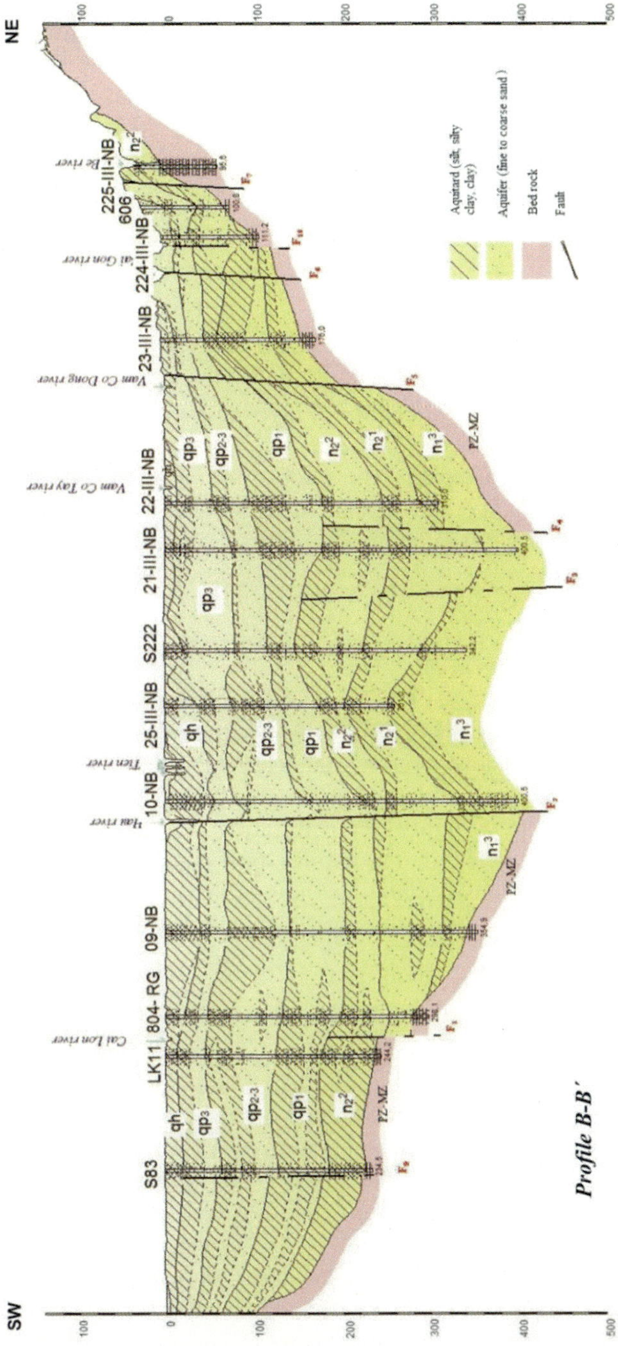


**Fig. 7.5** NE-SW directed hydrogeological profile cutting the MD plain (DGMS 2004)

#### 7.2.1.5 Middle Pliocene ($n_2^2$)

The alluvial to marine sediments of the Middle Pleistocene are divided into an upper aquitard and a lower aquifer. Depth of the top of the Pliocene has been detected from 80 up to 200 m bgl. The upper aquitard has an average thickness of 5–20 m, whereas the lower aquifer consists of 30–57 m strata of fine to coarse sand. The middle Pleistocene aquifer contains low saline groundwater in two areas with total 19,000 km$^2$. One is located around the northern part of the Tien River, and the other is located within Can Tho, Bac Lieu, and Ca Mau provinces.

#### 7.2.1.6 Lower Pliocene ($n_2^1$)

The uppermost low permeable aquitard layer of the lower Pliocene unit can be encountered at depths between 196 and 280 m bgl with an average thickness of 12–20 m. This aquitard covers the underlying aquifer consisting of fine to coarse sand with an average thickness of 41–46 m. Fresh groundwater is extracted from the $n_2^1$ unit in Dong Thap, Long An, Can Tho, Bac Lieu, and Ca Mau provinces.

#### 7.2.1.7 Upper Miocene ($n_1^3$)

Up to now, there are very few wells providing information from the pre-Pliocene units, thus, the alluvial and alluvial-marine sediments of the upper Miocene $n_1^3$ have not yet been studied in detail. Based on the available data, the $n_1^3$ unit starts from 258 to 364 m depth bgl. A highly confined aquifer is covered by 12–24 m thick silt and weathered silty clay layers. The aquifer itself consists of semi-compacted fine to coarse sand with a considerable thickness of 40 up to 72 m. In the vicinity of HCMC and Dong Thap province, the upper Miocene provides low saline groundwater of high quality.

#### 7.2.1.8 Upper-Middle Miocene ($n_1^{2-3}$)

The hydrogeological unit of Upper-Middle Miocene represents the deepest confined aquifer in the Mekong river basin and it is directly overlying Mesozoic bed-rocks. Hydrogeological characteristics and groundwater potential of this unit are still under-researched. The limited information is based on rather few drillings. The upper aquitard's thickness varies from 100 to 200 m and consists of clay and sandy silt. The top of the underlying aquifer has been encountered at depths of 600 and 900 m, respectively. It consists of compacted fine to medium sand with intercalated gravel layers and has up to 200 m thickness.

### 7.2.2 *Groundwater Dynamics*

Operated by the Division for Water Resources Planning and Investigation for the South (DWRPIS) since 1991, the national groundwater-monitoring network represents an important observation source for variations groundwater quantity and quality with time. Today the national monitoring network in the Mekong river plain (including HCMC area) comprises 62 stations, where 204 monitoring wells are screened in eight unconsolidated aquifers and two hard rock aquifers. Facing the large and diverse MD area, these numbers clearly illustrate that a much higher monitoring density is needed to get a clearer picture about groundwater dynamics, which is crucial for sustainable groundwater management. Long-term monitoring in the Mekong area shows that groundwater level fluctuation is caused by natural factors (climate and hydrological factors) as well as anthropogenic impacts such as groundwater extraction, lake and canal construction, irrigation and drainage systems.

The impact of these factors can be illustrated by the national monitoring station Q209, which is located in Vinh Long province close to the Mekong river (Fig. 7.3), consisting of five wells screened at different depth. According to the dynamics of level fluctuation, the screened aquifers can be allocated to three hydraulic groups (Fig. 7.6). The shallow Holocene aquifer shows a stable trend since the beginning of data collection in 1993. Deficits in the Holocene groundwater budget are balanced by recharge through repetitive phases of irrigation and inundation. The remarkable seasonal amplitude of up to 1 m indicates a close hydraulic connection with the close-by Mekong River. The increasing exploitation of groundwater since the 1990s results in a decreasing trend of groundwater levels in both Pleistocene aquifers of up to 25 cm/year. Overexploitation is even more distinct in the deeper Neogene aquifers (at Q209 > 400 m bgl) with a decreasing trend of more than 40 cm/year during the last 10 years.

The illustrated time series of groundwater level fluctuation in Q209 can be taken as a typical example. They shed some light on groundwater dynamics in the central part of the MD. Recharge in shallow groundwater is significant due to close connections with the river, tributaries and irrigation channels. Whereas low permeable aquicludes separating $qh - qp$ and $qp - n$ (see previous section), these hydraulic units seem to behave as relatively isolated systems. On a regional level, this need to be further researched and, thus confirmed. The amount of recharge replenishing *qp* and *n* units caused by side flow from hard rocks in the basement and at the borders of the Mekong Basin is still not well-understood. Based on hydraulic parameters, Benner et al. (2008) calculate groundwater residence times of 100–1,000 years in the upper Pleistocene aquifer. Consequently, resident times in deeper Pleistocene and Neogene aquifers are from several to tens of thousands of years and recharge is probably very limited, if there is any.

In the whole MD, groundwater levels in the Holocene aquifer are generally less than 5 msl (m above modern sea level). Previous to their extensive exploitation, deeper Pleistocene and Neogene aquifers were reported to be artesian with natural water levels up to 30 msl (Anderson 1978). Owing to low hydraulic gradients,

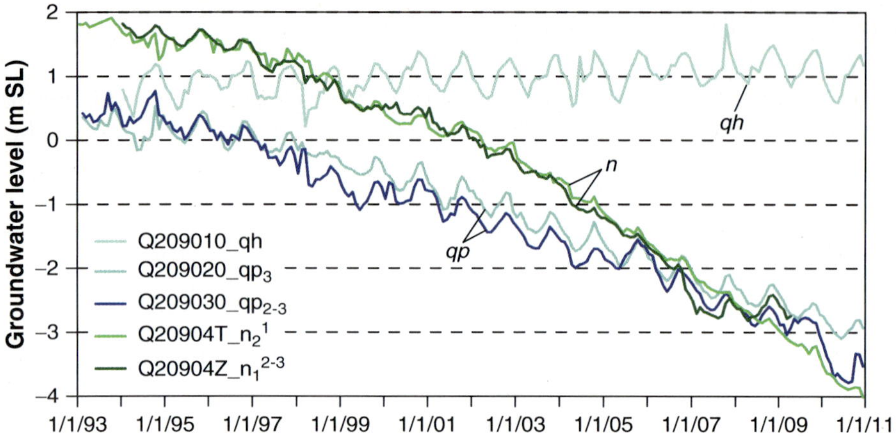


**Fig. 7.6** Time series of groundwater level fluctuation in National Monitoring Station Q209 (elevation 2.1 msl, location see Fig. 7.3) since 1993. Five wells are screened in hydrogeological units *qh*, $qp_2$, $qp_{2-3}$, $n_2^1$, $n_1^{2-3}$ (Data source: CWRPI, Vietnam)

horizontal groundwater flow velocities in the permeable layers are generally rather low (in the Pleistocene aquifer 1–13 m/year, Benner et al. 2008). However, because of abstraction cones, the natural groundwater flow is at least locally disturbed and may cause horizontal and vertical influx from adjacent saline ground and surface water bodies.

Figure 7.7 presents data of registered new extraction wells and the authorized amount of groundwater extracted in the most important aquifers in the MD from 2001 to 2004, based on water supply data from the DWRPIS. The general trend to a decreasing number of new extraction wells, registered each year, goes along with an increasing trend of extracted groundwater for each well. Although, the $qp_{2-3}$ unit became less important in 2004, the uppermost Neogene aquifer $n_2^2$ represents the favorite target for new groundwater exploration projects since 2003. This clearly explains the strong groundwater level drawdown in Neogene aquifers, not only observed in Q209 monitoring station (see Fig. 7.6), but also in large areas of the MD.

### 7.2.3 Groundwater Quality and Salinization

Hydrogeochemistry of pore water in alluvial-deltaic sediments is very heterogeneous and naturally reflects the impact of numerous lithological, biogeochemical, (paleo-) hydrogeologic, and hydrological factors. Moreover, groundwater composition is superimposed by widespread groundwater utilization and waste water infiltration. Nevertheless, this sub-section describes some general quality characteristics

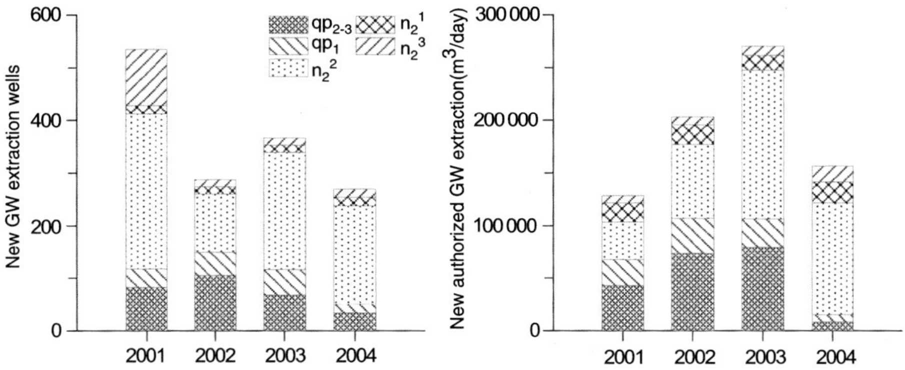


**Fig. 7.7** Number of registered new extraction wells (*left*) and the amount of authorized groundwater (*GW*) extraction in the MD (*right*) in the five most relevant hydrogeological units during period 2001–2004 (Data source: DWRPIS, Vietnam)

of Mekong groundwater, always keeping in mind that even within a small spatial scale, considerable quality changes can be observed.

The terms fresh or low saline, brackish and saline groundwater are represented by total dissolved solids (TDS) content of <1 g/L, 1–3 g/L and >3 g/L, respectively.

#### 7.2.3.1 Holocene Aquifer

The shallow Holocene aquifer generally contains non-potable brackish or saline water. Salt-water intrudes the Mekong River just about 50 km upstream during high water slack and low river discharge (Nguyen and Savenije 2006; Tamura et al. 2009). Nevertheless, Holocene aquifer is mostly brackish or saline even in the inland area where surface water is fresh. Only shallow wells located upstream towards the Cambodian border are tapping low saline groundwater between Bassac and Mekong river channels. Furthermore, high quality low saline water is locally accumulating during the rainy seasons in ancient beach and dune ridges, located near paleo-coastlines.

Besides the high salinity, relatively high sulfate content and lower pH values are quite common in shallow groundwater of the western MD due to pyrite dissolution from acid sulfate soils. Soils rich in pyrite (iron sulfate) are abundant in the tide-dominated areas of the Mekong (Husson et al. 2000). Similarly, soils high in aluminum and magnesium sulfate reported in Dong Thap Muoi ("Plain of Reeds") north of Mekong River have low pH pore water containing toxic levels of the dissolved metals (Anderson 1978). Furthermore, fecal pollution of the shallow groundwater has been reported (Isobe et al. 2004), probably caused by infiltrating sewage and domestic waste waters. The occurrence of dissolved trace metals such as arsenic in shallow groundwater is discussed in the following sub-section.

#### 7.2.3.2 Pleistocene and Pliocene Aquifer

Pleistocene and Pliocene aquifers are the principal sources of freshwater supply in the coastal part and to a large extend also in the inland part of the MD. However, all these aquifers contain brackish or salty water surrounding the isolated fresh water lenses. Groundwater in the deep aquifers $n_2^2$, $n_2^1$, $n_1^3$ can have elevated temperatures from 32°C to 39°C. Geothermal water has been found in Long Dien, Nhon Trach, Tra Vinh and Vinh Long provinces, probably originating from deep faults in the underlying bedrock (DGMS 2004).

In Dong Thap Muoi area ("Plain of Reeds", north of Mekong River) brackish to saline groundwater predominates the Pleistocene layers. In contrast, the underlying Pliocene aquifers contain a significant amount of low saline groundwater of Na-Mg-$HCO_3$ type. Saline water of the Na-Cl type is found in the East and West of the area and increases in salinity with depth. In addition to that, an enrichment of Ca originates from ion exchange in frame of salinization processes.

In the area of the Ca Mau peninsula, south of the Bassac River, the Pleistocene and Pliocene aquifers contain low saline groundwater of the type Na-Cl-$HCO_3$ and Na-Mg-Cl-$HCO_3$. The TDS of the fresh groundwater is usually higher in comparison to the eastern MD. Some traces of nitrite are locally found. The deeper Pliocene aquifers inhibit only few isolated lenses with fresh groundwater in a generally brackish to saline environment. The probably fossil groundwater is subject to extensive mining in Ca Mau and Soc Trang provinces.

#### 7.2.3.3 Arsenic and Trace Metals in Groundwater of the Mekong

The occurrence of high dissolved arsenic (As) levels mobilized from geogene sources in pore water has evolved to be a global issue, complicating groundwater usage worldwide (e.g. Welch and Stollenwerk 2003; Ravenscroft et al. 2009). In particular shallow aquifers in relatively young river plains and delta systems are affected due to their high content of immature and organic rich sediments. In these environments, the main release controlling drivers are biogeochemical redox processes and ion exchange. However, infiltration of organic rich waste water and pumping activities potentially accelerate As mobilisation and transfer. Well-known examples in South East Asia are the Ganges-Brahamaputra River Plain and Bengal Delta, the Red River Delta and the MD.

Some international studies about As in groundwater of the MD shed light on an alarming picture. Stanger et al. (2005) has analyzed 932 samples in the Vietnamese MD and found that 13% exceed 10 µg/L, which is the international valid threshold for As in drinking water. In a cross-national study (South Vietnam, Cambodia) Buschmann et al. (2008) found As levels exceeding 10 µg/L in 37% of the studied wells used for drinking water supply. Affected areas are generally restricted to the flood plains of the major rivers channels. In these areas, 1.2 million people are estimated to drink untreated well water of which 40% is exceeding 50 µg/L As, the

Vietnamese threshold for As in drinking water (Berg et al. 2007; Buschmann et al. 2007). While generally As levels decrease with increasing depth, some noticeable values (25 μg/L up to 60 μg/L) have locally been detected in the $n_2^2$ and $n_2^1$ aquifers at 200–300 m depth (DGMS 2001).

So far, no symptoms of chronic As poisoning have been indicated in the MD, possibly because of the relatively short-term usage of high As groundwater in less than 10 years (Berg et al. 2007). In comparison to other areas such as the North of Vietnam or Bangladesh, the As threat in the Mekong area seems to be lower. However, increased groundwater pumping is considered to accelerate As mobilization and transfer processes (Stute et al. 2007) and long-term exploitation may result in a drawdown of dissolved As into deeper aquifers (Winkel et al. 2011). Therefore, future systematic monitoring of As levels in groundwater is essential.

Beside As, other dissolved metals may have health effects when consumed continuously. High manganese levels are commonly found due to anoxic conditions in groundwater of the Mekong area (50% exceed 0.4 mg/L, maximum levels up to 34 mg/L, Buschmann et al. 2008). Other trace metals of minor concern are barium, cadmium, nickel, selenium, lead and uranium.

## 7.3 Groundwater Utilization in the Mekong Delta

The Mekong Delta in Vietnam is crisscrossed by an elaborate array of rivers belonging to the Mekong system and canals which were dug to facilitate irrigation and transport. Water dominates the landscape yet the region faces severe problems of freshwater supplies in some localities. These problems are linked to increasing periods of low flows during the dry season, pollution by the extensive use of agrochemicals and the release of industrial and household contaminants and by salinity intrusion from the coastal areas. Surface water is therefore costly to clean in order to bring to acceptable drinking standards and many people and communities still drink polluted surface water throughout the region. In rural areas, only 8–12% of the MD population has access to piped water, while 42–47% of the households use unprotected surface water (Netherlands Development Organisation (SNV) 2010). It is for these reasons that many see an increasing role for groundwater in terms of freshwater supply for the population and irrigation in the MD (Vo Thanh Danh 2008; Nguyen Van Sanh 2012) even though, as described above, the quality and availability of groundwater resources vary greatly within the MD.

Nationwide, hand dug wells remain the main source of water supply in rural areas (but not in the MD as surface aquifers are polluted and some areas flooded annually), followed by drilled wells which are affordable only to wealthier households (SNV 2010). Traditionally, surface water and canals were the main source of freshwater supply in the MD. These sources have been supplemented by rainwater cisterns and other forms of surface water storage, particularly in areas where surface water is at least seasonally brackish. Due to the general high salinity of the shallow groundwater, dug wells traditionally tap shallow fresh water lenses lying beneath dune and beach ridges of the MD. Initially, French technicians introduced drilling techniques in the 1930s and the first water supply wells at about 100 m depth were successfully installed in Bac Lieu and Ca Mau provinces (Anderson 1978). As drilling capabilities improved during the 1960s, some wells with depths greater than 500 m have been successfully completed in various parts of the delta.

Despite these early attempts at tapping groundwater in the MD, it is only recently (1990s) that groundwater exploitation has become a major source of freshwater for domestic use, in particular following interventions by UNICEF at the household level and the creation of Center for Rural Water Supply and Sanitation (CERWASS) which set up small water supply station tapping groundwater in the MD (see Reis 2012). According to Vo Thanh Danh (2008), groundwater in the region is currently accessed via unregulated private shallow tube-wells (more than one million) reaching depths of 80–120 m and by regulated groundwater plants accessing water in the deeper aquifers, at depths of 100–250 m. In Can Tho City, Nuber et al. (2008) noted that there were 400 medium sized groundwater supply stations with a capacity of up to 20 $m^3$/h-as well as 20 large wells. The volume of water extracted by households owning shallow tube-wells in Can Tho Province was shown to be significantly higher than the volume supplied to households via the groundwater plants, the reason for this being that the latter charge a price for supplying the water whereas water is of free access in the case of privately owned wells (Vo Thanh Danh 2008). In Can Tho City, falling water tables have already been noted due to a combination of low recharge rates and increased abstraction (Nuber et al. 2008). Falling water tables and pollution of the resource is in part explained by a lack of clear system of groundwater resources management. For example, in Tra Vinh province, Nguyen Van Sanh (2012) reported that mismanagement of groundwater resources is partly due to the fact there is un-clarity with respect to legislation and institutions governing the resource. A similar nationwide diagnostic is made by Dang Dinh Phuc (2008).

Nowadays, high quality groundwater is extensively used not only for domestic water supply, but also for irrigation, aquaculture and industrial purposes. Especially in the coastal areas of the MD groundwater represents an exclusive source for low saline water even though in some areas such as Ben Tre province the water is not consumed because it is too saline. Dang Dinh Phuc (2008) reported rapid increases in groundwater use in Vietnam with, for example, exploitation of 100 $m^3$/day in 1990 compared to 700 $m^3$/day in 2006 in Ho Chi Ming City. Similar trends are observed in main towns and industrial areas in Vietnam. In the Mekong Delta, ca. 350,000 $m^3$/day are exploited by water supply plants, industries and other institutions, the region ranking in third place in terms of exploited volume, far behind the South Central Coast (ca. 1.5 million $m^3$/day) and the Red River Delta (ca. 960,000 $m^3$/day) (Dang Dinh Phuc 2008).

In addition to the high arsenic levels reported in previous sections, groundwater resources are also degraded by direct and indirect human action: pollution by the same agrochemicals and other contaminants that affect surface waters, incompetent drilling methods, salinity intrusion that will be aggravated during this century by sea-level rise and over-abstraction, the latter induced by the fact that a majority of households in given localities prefer to access groundwater as opposed to other freshwater resources. As described previously, the shallowest aquifer in the region is already polluted and people avoid it when drilling tube-wells (see Vo Thanh Danh 2007).

In contrast, Reis (2012) showed that many people in Can Tho City remain cautious with respect to groundwater use because of the high iron concentrations in some localities and the fact that groundwater levels in the region are dropping rapidly (up to 0.5 m/year). This was already observed by Herbst et al. (2009) who, in a household survey carried out in a suburban area of Can Tho City, found that groundwater came only in third position with respect preference as drinking water, behind purchased purified water and rainwater, the latter being obviously more prevalent during the rainy season.

Drilling wells has a cost and Reis (2012) found through interviews in Can Tho City that this cost varies between €40 and €100 depending on depth. The implication is that only relatively wealthy households can afford the investment, a fact also supported by Nguyen Van Sanh (2012). Nguyen Van Sanh (2012) showed another type of social differentiation in Tra Vinh province where Khmer communities settle preferentially on sand banks where they cultivate upland crops and are therefore increasingly relying on groundwater for irrigation whereas Kinh communities settle preferentially in the lowlands where they cultivate rice and shrimps and rely on groundwater resources principally in the dry season.

## 7.4 Conclusions: Mitigating Groundwater Overexploitation and Salinization

Increasing overexploitation and salinization of deeper aquifers raised the question about origin and recharge of the deep fresh groundwater and new exploration strategies. The relationship between low mean sea-level and the origin of deep fresh groundwater which has been demonstrated previously, suggests that the recharge of this finite groundwater resource is limited. Detailed case studies need to quantitatively assess the groundwater side flows feeding the low saline aquifers of the MD, if there are any. However, it is obvious that the current exploitation status already exceeds the aquifer recharge capacities. Further exploration of other fresh groundwater lenses may only temporarily improve the situation.

Obviously, proper management and mitigation strategies aiming at more sustainable solutions are urgently needed. Thus, close connection between water policy, science and engineering, and a strong cooperation between different groundwater users such as water supply companies, agriculture, aquaculture and other industries as well as domestic households is crucial. Just to name some challenges potentially leading to a more sustainable groundwater usage in the MD:

- Reducing extraction: High quality groundwater is a finite resource and should be kept as national reserve. It should be used for domestic drinking water supply only and whenever possible replaced by surface water usage. Therefore:
  - Identification of sources for groundwater loss or misuse, together with the design and approval of mitigation strategies are necessary. Leaking wells, pipes, tubes and taps are common causes for wasting groundwater.
  - Alternatives need to be found for groundwater usage with lower priority, such as (i.) application of saline tolerant crops in agriculture, and (ii.) treatment and usage of surface water for industry, aqua- and agriculture.
- Optimizing extraction: Salinization of remaining fresh groundwater lenses can be minimized by applying appropriate exploitation strategies. This can only be achieved by having detailed knowledge about the local subsurface structure and their hydraulic characteristics.
- Increasing recharge: Potential areas need to identified, where technologies can be applied to increase the groundwater recharge with artificially infiltrated precipitation and low saline surface water. To minimize long-term risks feasibility studies are needed to assess carefully the hydrogeological and hydrogeochemical impact on the subsurface environment as well as to develop appropriate geotechnical and monitoring strategies.
- Conjunctive usage: Mixing of high quality water with poorer quality water may extend the available amount of water with still acceptable quality for water supply. This should be understood as an intermediate action while realizing the recommendations above.

**Acknowledgements** The authors thank the Center for Water Resources Planning and Investigation for provision of data from the National Groundwater Monitoring Network. Furthermore, gratitude is dedicated to the reviewer for the constructive comments contributing to the quality of this chapter. Many thanks are also submitted to Mr. Ulrich Gersdorf (BGR) for his kind support in designing Figs. 7.2 and 7.3.

## References

Anderson HR (1978) Hydrogeologic reconnaissance of the Mekong Delta in South Vietnam and Cambodia. Water supply paper, 1608-R, Washington, DC

Benner SG, Polizzotto ML, Kocar BD, Ganguly S, Phan K, Ouch K, Sampson M, Fendorf S (2008) Benner groundwater flow in an arsenic-contaminated aquifer, Mekong Delta, Cambodia. Appl Geochem 23:3072–3087. doi:10.1016/j.apgeochem.2008.06.013

Berg M, Stengel C, Pham TKT, Pham HV, Sampson ML, Leng M, Samreth S, Fredericks D (2007) Magnitude of arsenic pollution in the Mekong and Red River Deltas – Cambodia and Vietnam. Sci Total Environ 372:413–425. doi:10.1016/j.scitotenv.2006.09.010

Buschmann J, Berg M, Stengel C, Sampson ML (2007) Arsenic and manganese contamination of drinking water resources in Cambodia: coincidence of risk areas with low relief topography. Environ Sci Technol 41:2146–2152

Buschmann J, Berg M, Stengel C, Winkel L, Sampson ML, Trang PTK, Viet PH (2008) Contamination of drinking water resources in the Mekong delta floodplains: arsenic and other trace metals pose serious health risks to population. Environ Int 34(6):756–764. doi:10.1016/j.envint.2007.12.025

Dang Dinh Phuc (2008) General on groundwater resources. Department of water resources management, Water sector review project. Project ADB-TA- 4903 VIE

DGMS [Division for Geological Mapping for the South of Vietnam] (2001) Groundwater study of the Mekong Delta – Final report. Division of Geology and Minerals of the South of Viet Nam (unpublished)

DGMS [Division for Geological Mapping for the South of Vietnam] (2004) Research of geological structure and classification of N-Q sediments in Mekong Delta. In: Nguyen HD (ed) Division of Geology and Minerals of the South of Viet Nam (unpublished)

DWRPIS [Division for Water Resources Planning and Investigation for the South] (2010) Report on the results of the National Groundwater Monitoring Network for Nam Bo Plain. Division of Water Resources and Planning Investigation of Viet Nam (unpublished)

Fetter CW (1972) Position of the saline water interface beneath oceanic islands. Water Resour Res 8:1307–1314

Fetter CW (2001) Applied hydrogeology, 4th edn. Prentice Hall, Upper Saddle River

Fontaine H, Workman DR (1997) Vietnam. In: Moores EM, Fairbridge RM (eds) Encyclopedia of European and Asian regional geology: 774–782, Springer, Stuttgart

Gibbard PL, Head MJ, Walker MJC, The Subcommission on Quaternary Stratigraphy (2010) Formal ratification of the Quaternary System/Period and the Pleistocene Series/Epoch with a base at 2.58 Ma. J Quat Sci 25:96–102

Hanebuth TJJ, Stattegger K, Bojanowski K (2009) Termination of the Last Glacial Maximum sea-level lowstand: the Sunda-Shelf data revisited. Glob Planet Chang 66:76–84. doi:10.1016/j.gloplacha.2008.03.011

Herbst S, Benedikter S, Koester U, Phan N, Berger C, Rechenburg A, Kistemann T (2009) Perceptions of water, sanitation and health: a case study from the Mekong Delta, Vietnam. Water Sci Technol 60:699–707

Husson O, Verburg PH, Phung MT, Van Mensvoort MEF (2000) Spatial variability of acid sulphate soils in the Plain of Reeds, Mekong delta, Vietnam. Geoderma 97:1–19

Isobe KO, Tarao M, Nguyen HC, Le YM, Takada H (2004) Effect of environmental factors on the relationship between concentrations of coprostanol and fecal indicator bacteria in tropical (Mekong Delta) and temperate (Tokyo) freshwaters. Appl Environ Microbiol 70:814–821

Nguyen Van Sanh (2012) Tra Vinh case study – Vietnam. In: Renaud FG, Bigas H, Thomas R (eds) Sustaining people from the ground up: vulnerability and resilience in groundwater-dependent communities. UNESCO, Paris (in preparation)

Nguyen HD, Savenije HGG (2006) Salt intrusion in multi-channel estuaries: a case study in the Mekong Delta, Vietnam. Hydrol Earth Syst Sci 10:743–754

Nuber T et al (2008) Modelling the groundwater dynamics of Can Tho city – challenges, approaches, solutions. In: SANSED project (ed) Decentralised water treatment systems and beneficial reuse of generated substrates. Proceedings of the international SANSED workshop, 15–18 Apr 2008, Can Tho University

Ravenscroft P, Brammer H, Richards K (2009) Arsenic pollution – a global synthesis. Wiley-Blackwell, Hoboken

Reis N (2012) Tracing and making the state. Policy practices and domestic water supply in the Mekong Delta. ZEF Development Series. Lit Verlag, Berlin

SNV [Netherlands Development Organisation] (2010) Study of rural water supply service delivery models in Vietnam. Report from the Deutsch Technical Cooperation. http://www.snvworld.org/en/regions/asia/ourwork/Documents/Rural%20Water%20Supply%20in%20Vietnam%20-%20Final%20Report.pdf. Accessed 5 May 2011

Stanger G, Truong TV, Ngoc LTM, Luyen TV, Thanh TT (2005) Arsenic in groundwater of the Lower Mekong. Environ Geochem Health 27:341–357

Stute M, Zheng Y, Schlosser P, Horneman A, Dhar RK, Hoque MA, Seddique AA, Shamsudduha M, Ahmed KM, van Geen A (2007) Hydrological control of As concentrations in Bangladesh groundwater. Water Resour Res 43:1–11. doi:10.1029/2005WR00449W09417

Tamura T, Saito Y, Sieng S, Ben B, Kong M, Sim I, Choup S, Akiba F (2009) Initiation of the Mekong River delta at 8 ka: evidence from the sedimentary succession in the Cambodian lowland. Quat Sci Rev 28:327–344

Tjallingii R, Stategger K, Wetzel A, Phach PV (2010) Infilling and flooding of the Mekong River incised valley during deglacial sea-level rise. Quat Sci Rev 29:1432–1444

Vo Thanh Danh (2007) Economic value of groundwater protection in the Mekong Delta. Report prepared for the CAULES-DANIDA project

Vo Thanh Danh (2008) Household switching behavior in the use of groundwater in the Mekong Delta, Vietnam. Research report 2008-RR3, Economy and Environment Program for Southeast Asia (EEPSEA), Singapore

Waelbroeck C, Labeyrie L, Michel E, Duplessy JC, McManus JF, Lambeck K, Balbon E, Labracherie M (2002) Sea-level and deep water temperature changes derived from benthic foraminifera isotopic records. Quat Sci Rev 21:295–305

Welch AH, Stollenwerk KG (2003) Arsenic in ground water – geochemistry and occurrence. Kluwer, Dordrecht

Winkel LHE, Pham Thi KT, Vi ML, Stengel C, Amini M, Nguyen TH, Pham HV, Berg M (2011) Arsenic pollution of groundwater in Vietnam exacerbated by deep aquifer exploitation for more than a century. Proc Natl Acad Sci USA 108(4):1246–1251

Xue Z, Liua JP, DeMastera D, Nguyen LV, Ta TKO (2010) Late Holocene evolution of the Mekong Subaqueous Delta, Southern Vietnam. Mar Geol 269:46–60. doi:10.1016/j.margeo.2009.12.005

# Chapter 8
# Flood Hydraulics and Suspended Sediment Transport in the Plain of Reeds, Mekong Delta

**Heiko Apel, Nguyen Nghia Hung, Trinh Thi Long, and Vo Khac Tri**

**Abstract** The Mekong Delta is one of the largest and most intensively used estuaries in the world. It witnesses widespread flooding annually, which is both the basis of the livelihood for more than 17 million people, but also constitutes the major natural hazard in the region. Therefore, a thorough understanding of the hydrologic and hydraulic features as well as the associated sediment dynamics is urgently required for various planning purposes. In order to gain quantitative support for the existing knowledge of the floodplain inundation processes, a comprehensive monitoring scheme of the inundation hydraulics and sediment dynamics in channels and floodplains was established in a study area in the Plain of Reeds in the north-eastern part of the Vietnamese delta. With the collected data the inundation and sediment transport in the floodplains and the interaction between channels and floodplains

H. Apel (✉)
Section 5.4 Hydrology, German Research Centre for Geoscience (GFZ),
Telegrafenberg, 14473 Potsdam, Germany
e-mail: heiko.apel@gfz-potsdam.de

N.N. Hung
Research Centre for Rural Technical Infrastructure Development,
Southern Institute for Water Resources Research (SIWRR),
Ho Chi Minh City, Vietnam
e-mail: nguyennghiahung@yahoo.com

T.T. Long
Center for Environment Science Technolgy and Ecology,
Southern Institute for Water Resources Research (SIWRR),
Ho Chi Minh City, Vietnam
e-mail: ttlongvn@gmail.com

V.K. Tri
Science and International Cooperation Department,
Southern Institute for Water Resources Research (SIWRR),
Ho Chi Minh City, Vietnam
e-mail: vokhac.tri7@gmail.com

F.G. Renaud and C. Kuenzer (eds.), *The Mekong Delta System: Interdisciplinary Analyses of a River Delta*, Springer Environmental Science and Engineering,
DOI 10.1007/978-94-007-3962-8_8, © Springer Science+Business Media Dordrecht 2012

could be observed and quantified for the first time. Differences between the two major dike systems and management strategies in terms of inundation and sediment transport in the floodplains became obvious. Combining this finding with the observed reduction of sediment loads in the secondary channels with distance from the main river, a recommendation for the operation of the sluice gates in the area is given in order to optimize sediment input to the floodplains.

## 8.1 Introduction

The Vietnamese part of the Mekong delta is characterized by a high level of management through channels, dikes, and control structures such as sluice gates, culverts, and pumps. The man-made channel and dike system has greatly altered the natural hydrodynamic conditions in this part of the Delta. Today, the canal network covers more than 50,000 km, and it is under continuous development (Truong 2006). Along with the development of the canal network the construction of high level flood protection dikes has been enforced in the last decades, especially after the devastating flood in 2000, cutting off more and more floodplains from the channel network and thus natural inundation. In wide areas in the Delta the floodplain inundation is controlled by sluice gates which are managed by the responsible communities. In the areas completely enclosed by flood protection dikes, a trend towards growing three crops per year and blocking the natural inundation completely is observed. However, besides the obvious economic benefits of such a system, the adverse impacts manifest themselves in terms of water quality degradation (MRC 2007), riverbank erosion (Hung 2004), salinity intrusion (Sam et al. 2004; Nguyen and Savenije 2006), and a general increasing pollution of the environment.

The floodplains play an important role in the sustainability of the agro-ecosystems as well as the socio-economy of the Mekong Delta. In particular, they provide natural flood retention, regulating flood and tide levels during the wet season. This natural flood retention has been strongly altered by flood mitigation structures and the intensification of agriculture. Traditionally, the in- and outflow of water during the rising and falling stage, i.e. in July and December, of the flood period is controlled by the demands of rice cultivation. This management allows for two rice crops per year and is achieved by low dikes protecting the floodplains from the first flood waters until the rice is harvested in July. In December the surplus water is pumped out of the floodplain for plantation of the winter crop. The low dikes also ensure that substantial amounts of sediment and thus nutrients are deposited in the floodplain. These sediments are the thriving source of the very productive agricultural system in the Delta. Complete flood protection cuts this annual natural support of nutrients and is thus object of controversial debates. However, these debates are not based on quantitative information up to now.

This study, therefore, aims at the quantification of typical floodplain processes in the Mekong Delta with a focus on the Northern area of the Vietnamese part of the Delta. The research is performed in a representative study area of about 15 $km^2$, which witnesses the deepest annual inundations in the Delta. The area in Tam Nong District of the Dong Thap province covers all aspects of topography, man-made structures such as dikes, sluice gates, pumps, multiple land use (paddy rice, shrimp farming, fish farms, and vegetable) typical for intensively used floodplains in Vietnam. Additionally, the hydraulic conditions are typical for the northern part of the Vietnamese delta, where deep and regular inundations during the flood period dominate. The area is thus representative for most of the Northern part of the Vietnamese Delta, i.e. the Plain of Reeds and the Long Xuyen Quadrangle. The hydrology of the floodplains is controlled by the annual floods of the Mekong River and two different tidal systems, despite being about 180 km away from the shores.

## 8.2 Monitoring Network for Floodplain Processes

The monitoring network for the study area in Tam Nong consists of 21 pressure probes for water level monitoring and 7 stations monitoring water quality parameters (total suspended sediment concentration TSS, electrical conductivity, pH and temperature). The stations were distributed to strategic locations in the area (Fig. 8.1). By placing the probes in channels, on some low dikes, and in the floodplains, the different inundation controlling processes (dike overflow, sluice gate and pump operation) can be monitored, either directly or indirectly.

The water level stations (H-stations) are absolute pressure probes facilitating deployment at any location and can be submerged completely over long periods. The pressure readings were offline corrected for atmospheric pressure after retrieval of the probes. The water quality stations (T-stations) operate autonomously by solar power supply. Because the power supply and data loggers must not be inundated, the T-stations consist of a weather- and lightning-proof steel box hosting the solar panels, battery, and the data loggers. The box is mounted either on a bridge or on a 4 m high pylon in the floodplains. Three T-stations are located in channels, the remaining four in the floodplains. All four floodplain T-stations were supplemented with sediment traps in order to quantify the overall sedimentation over the flood season.

**Fig. 8.1** Study area in Tam Nong district in the Plain of Reeds, the North-eastern part of the Vietnamese part of the Mekong Delta with major dike lines, topography and monitoring network

This unique combination of sediment traps and continuous monitoring of suspended sediments facilitate detailed insights into the sediment dynamics on the floodplains. The monitoring network was installed in 2008. Two T-stations in the channels operated continuously, while the remaining stations were operational over the annual flood seasons from July to December. Details about the monitoring network and the sensors used can be found in Hung et al. (2012).

## 8.3 General Flood Hydrology and Hydraulics

In order to discuss the channel and floodplain inundation dynamics in a more structured way, the typical flood hydrograph in the Mekong Delta is divided into three distinct periods: "rising", "high", and "falling".

"Rising stage": In this stage the flood regime is controlled by the inflow into the Tonle Sap and the tidal influence. Flood control in the Vietnamese part of the Delta is mainly for crop protection. During this period the tidal influence is still strong. The tides may cause short term floodplain inundation in case of coincidence of high tides and early small flood peaks. The higher the general flood level, the more attenuated is the tidal amplitude. This is visible in the comparison of the different hydrographs of 2008 and 2009 at station T1 in Fig. 8.2. Also, the tidal influence is dampened with distance from the main river. This illustrated by the comparison of the hydrographs of 2009 for stations T1 and H11 in Fig. 8.1. H11 is located in a secondary channel at about 11 km distance from gauge T11 near the Mekong River.

"High stage": With rising water levels the hydrodynamics of the system changes. Floodplain inundation is initiated by overland flow or sluice gate operation in case of closed dike rings and the influence of the tides on the inundation process are diminishing. In the Plain of Reeds in the Delta the flood wave caused by the overland flow from the North-East from the Cambodian border meets the flood wave from the Mekong. The additional hydraulic head created by the overland flood changes the stage-discharge relationship in the secondary channels: at the same

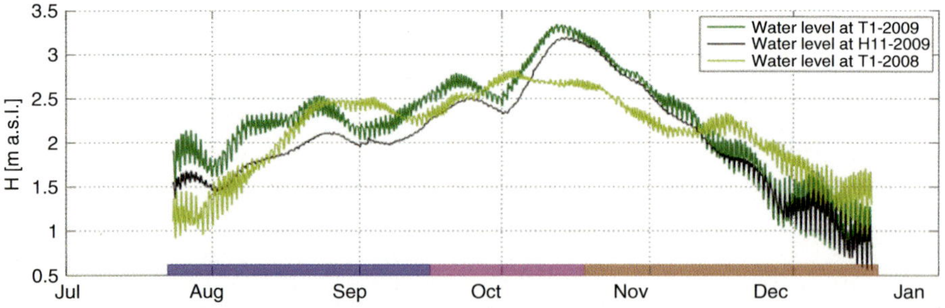


**Fig. 8.2** Water elevations at channel stations T1 (2008, 2009) and H11 (2009). The *colored boxes* in the *lower panel* indicate the stage of the *flood: blue*=rising, *purple*=high, and *brown*=falling

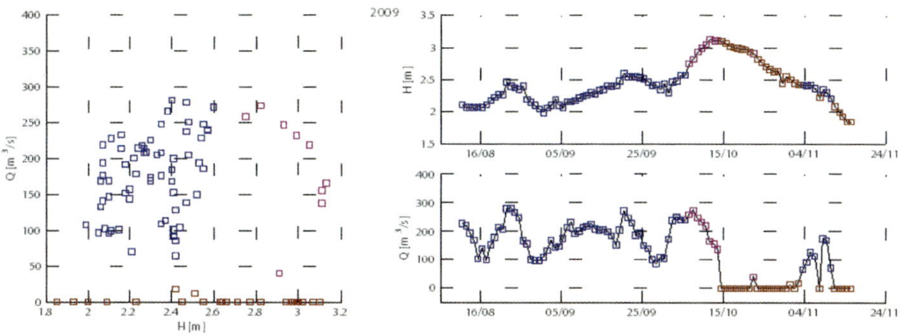


**Fig. 8.3** Stage-discharge relationships at station G1 (cf. Fig. 8.3) for flood season 2009 (*blue* = rising stage, resp. normal H-Q relation, *purple* = high stage or flood peak h-Q relation, and *brown* = falling stage or stagnant H-Q relation)

water elevation less water flows from the Mekong into the channels (cf. Fig. 8.3, purple markers).

"Falling stage": From mid-October the water levels generally fall and no further flood peaks caused by tropical typhoons occur. This stage is characterized by rising influence of the tides and widespread human activities to manage the remaining inundation waters in the floodplains. Traditionally, the waters are pumped from the floodplains when the water level in the channels falls below the elevation of the low dikes. This facilitates the timely planting of the first paddy crop of the new cropping season while controlling the required irrigation water for planting at the same time. A particular feature at the beginning of this stage, i.e. between mid-October and mid-November, are stagnant and even reversal flow conditions, mainly in east-west leading channels (cf. Fig. 8.3, brown markers). This is caused by the lower propagation velocity of the second flood wave coming overland from the Cambodian -Vietnamese border compared to the flood wave in the Mekong. Hence, a hydraulic head equal or higher than in the Mekong is created causing the stagnation or even reversal of the flow in the secondary channels.

Figure 8.3 illustrates the different stages through changing stage-discharge relation over the flood season 2009. The blue marked data represent the "normal" relation during low flow and the rising stage of the flood, which is dominated by the hydraulic head of the Mekong. The scatter in the data has to be attributed to the tidal influence. The purple data indicate the high stage of the flood, where the flood wave of the Mekong and the overland flood wave from the north-east coincide. During this phase the hydraulic head of the Mekong is reduced by the hydraulic head of the overland flood wave, causing a significant change in the H-Q relation. The brown data indicate stagnant of even reverse flow conditions in the first phase of the falling flood stage, where the hydraulic head of the Mekong is falling below or equal to the head of the overland flood wave. Slightly rising water levels in the Mekong by minor flood peaks alleviate this situation, which is well observable during the end of the flood season in the right panels of Fig. 8.3.

## 8.4 Tidal Influence

In order to gain more insights into the characteristics of the tidal influence, the tidal signal was separated from the hydrograph through a high-pass Butterworth digital frequency filter. An analysis of the amplitudes of the tidal signals for all channel stations reveals three distinct features:

1. The amplitude during low flow is in the range of 1 m, while it is reduced to 15 cm and below during peak flow.
2. The tidal amplitude is dampened with distance to the Mekong River, especially during high flows: While T1 close to the main river of the Mekong? shows an amplitude of still 20 cm during high flow, H11 at 11 km orthogonal distance from the Mekong shows hardly any tidal influence during this period (<3 cm). This is caused by the travel time in the channel, but also due to the additional hydraulic head imposed by the overland flood from the Vietnamese-Cambodian border.
3. The tidal amplitudes are reduced with lower order of secondary channels, i.e. the smaller the channels and the larger the distance from the Mekong the lower is the tidal amplitude.

This means in consequence that the tidal influence has little impact during high flows, i.e. on the inundation in the study area. However in other parts of the Delta closer to the coast the tidal influence on the flood characteristics is significant also during the flood season.

## 8.5 Floodplain Inundation Dynamics

The particular features of the channel-floodplain interaction, i.e. hydraulic linkage and anthropogenic interference, can be quantified in terms of timing and water levels recorded by the monitoring network. Fig. 8.4 shows the hydrographs of gauging stations in channels and the linked floodplains exemplarily for a compartment completely enclosed by high dikes (Fig. 8.4a) and a floodplain with low dikes (Fig. 8.4b) for 2009. In case of compartments with high dikes, the floodplain inundation is completely controlled by sluice gate operation. In Fig. 8.4a this is illustrated by the hydrographs of the nearest channels station and H6 in the floodplain along with the mean ground and dike levels of the floodplain compartment. The flood level is generally below the dike levels, thus no overbank flow occurred. The first minor flood peak does not cause inundation in the floodplain due to closed sluice gates. The opening of the sluice gates is recorded by station H6 within the dike ring. After opening of the sluice gates, the inundation of the floodplain follows the dynamics of the channel, but with a slight delay. This is partly due to the distance between the stations, but to a larger extend caused by the limited flow capacity of the sluice gates dampening the hydraulic link between channel and floodplain. This weak hydraulic connection is also responsible for the low tidal influence visible in the floodplain

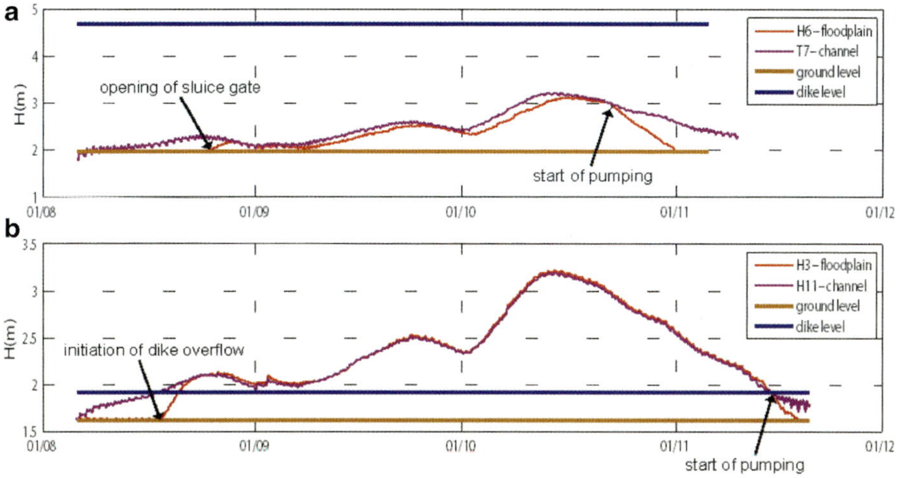


**Fig. 8.4** Comparison of channel and floodplain hydrographs in a dike ring with high flood protection dikes (**a**) and a floodplain compartment with low dikes (**b**) for the flood season 2009

hydrograph compared to the channel. At the end of the flood season, the closing of the sluice gates and the operation of the pumps is clearly detectable in the time series by the divergence of the hydrographs in floodplain and channel. The water level in the floodplain drops linearly without any visible tidal influence indicating the pumping operation and capacity.

In contrast to the closed compartments, the inundation of the low dike floodplain compartments is controlled by dike overflow. This is shown in Fig. 8.4b plotting hydrographs of station H11 in the channel and H3 in the floodplain. The onset of the floodplain inundation is clearly related to the lowest dike elevation in the vicinity of station H3: as soon as this level is exceeded, the inundation starts and quickly rises to the level in the channel forming a strong hydraulic link extending over the whole length of the dike. This is illustrated in Fig. 8.4b by the close match between floodplain and channel hydrograph including the tidal influence, which is identical in both channel and floodplain. As mentioned earlier, the pumping in these compartments cannot start before the water level drops below the dike level. This is also observable in Fig. 8.4b by the divergence of the floodplain and channel water levels at the end of the flood season. As in Fig. 8.4a the reduction of the water level in the floodplain is almost linear.

The analysis shows that human activities influence the floodplain inundation in two ways. First, the construction of dikes of different protection levels and purposes influence the flood propagation in general. Second, the operation of sluice gates and pumps delays the onset of the floodplain inundation, or causes an earlier drop of the water levels compared to the channels, respectively. The sluice gates differ in sizes and their operation schemes. However, they are a key factor in the hydraulic linkage between channel and floodplain compartment and have a significant impact on the inundation dynamics, thus also on the sediment dynamics on the floodplains.

The development of additional closed high level protected compartments in future has to take this aspect into account, especially regarding the sediment input into the floodplains. This particular aspect will be elaborated in the next section.

## 8.6 Sediment Transport in Channels and Floodplains

In addition to the continuous monitoring, water samples were taken in several campaigns. The purpose of the sampling was two-fold: (a) determination of the grain size distribution of the suspended sediments, and (b) the calibration of the turbidity sensors. The samples were taken at several depths in the channels with an automatic water sampler. The analysis of the samples revealed that the suspended particle size is very fine with a mean grain size (d50) smaller than 2 μm. The stratified sampling in the channels also revealed no vertical distribution of suspended sediment concentrations, which has to be attributed to the fine nature and floating capacity of the sediments.

The total suspended sediment concentration TSS in the channels varies over the flood season, as shown in Fig. 8.5. It indicates that the variation in suspended sediment is controlled by two mechanisms: a low frequency modulation by the annual flood waves generally bringing higher TSS concentrations, and a higher frequency variation correlated to the tidal influence. During low-flow, although tidal influence is stronger than under high-flow conditions there is no increasing suspended sediment supply from upstream. Therefore the variation of suspended sediment in this period is relatively small, and its concentration is often lower than 100 mg/l. During high flows, especially, between June and July the concentration increases gradually,

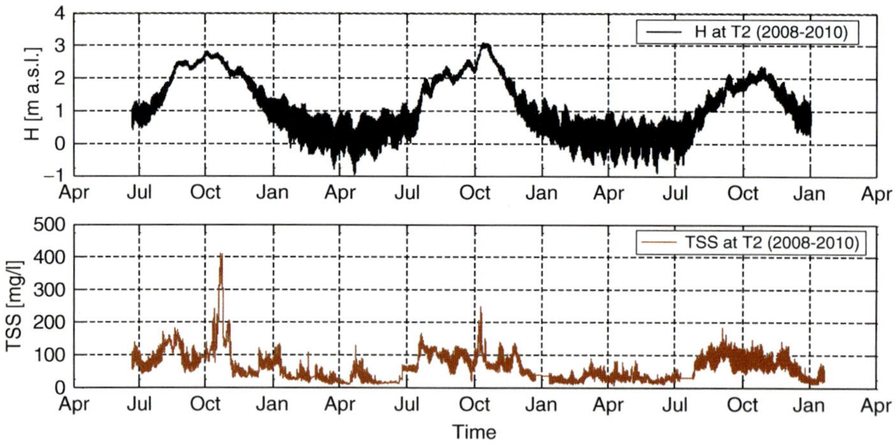


**Fig. 8.5** Water level and TSS concentration for three consecutive flood seasons (2008–2010) at channel station T2

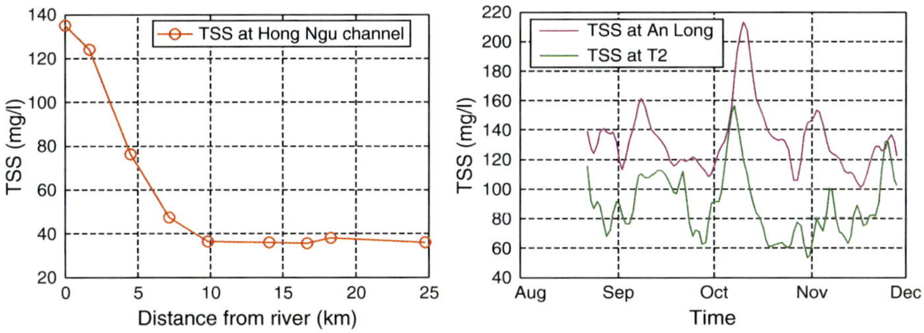


**Fig. 8.6** Reduction of suspended concentration along secondary channels. *Left panel*: results of a targeted monitoring campaign along a major secondary channel in September 2008; *right panel*: TSS at G1 (An Long, at the junction to the Mekong) and T2 of flood season 2009

reaching a peak in September or October. During high flow the tidal influence on suspended sediment loads is suppressed. The actual peak of TSS is controlled by the flood magnitude and duration and the intensity of the rainfall events in the Mekong basin.

TSS is also decreasing with increasing distance from the Mekong. This is quantitatively illustrated by Fig. 8.6. It shows an exponential reduction along a major secondary channel leading eastwards from the Mekong at the northern border of the investigation area (Fig. 8.6, left panel). The data presented was collected in a sampling campaign during peak flood in September 2008. The right panel in Fig. 8.6 shows the records of regular daily manual samples at An Long (G1 in Fig. 8.1) compared to the automatic records of station T2. The comparison reveals an almost constant reduction of TSS at T2 compared to G1 over the whole flood period. From both data sets shown in Fig. 8.6 it is calculated that a 50% reduction of TSS is reached at about 6.5 km distance from the main river.

As shown in the previous section, the particular hydraulic features of the inundation process in the floodplains of the Mekong Delta are the different hydraulic links between channels and floodplains. Sediments are therefore also transferred to the floodplains by different pathways, i.e. either through sluice gates or overbank flow. The floodplain compartments function as sinks of suspended sediment, with high concentrations of sediment entering especially during the onset of the inundation (Fig. 8.7). Most of the sediment will stay deposited in the compartments at the end of the flood season, except in locations near to sluice gates where erosion is dominant due to the currents induced by the outflow through the gates. The deposition of sediments mainly occurs during the first 2 weeks of floodplain inundation, when suspended sediment concentrations in both channels and floodplains are high. After that period the TSS load in the floodplains remains stable at a low level, despite changes in the channels indicating a reduced exchange of TSS once the hydraulic connection between channel and floodplain is closed (Fig. 8.7). Peaks of TSS in the channels do no longer flush into the floodplains. On the floodplains itself the TSS

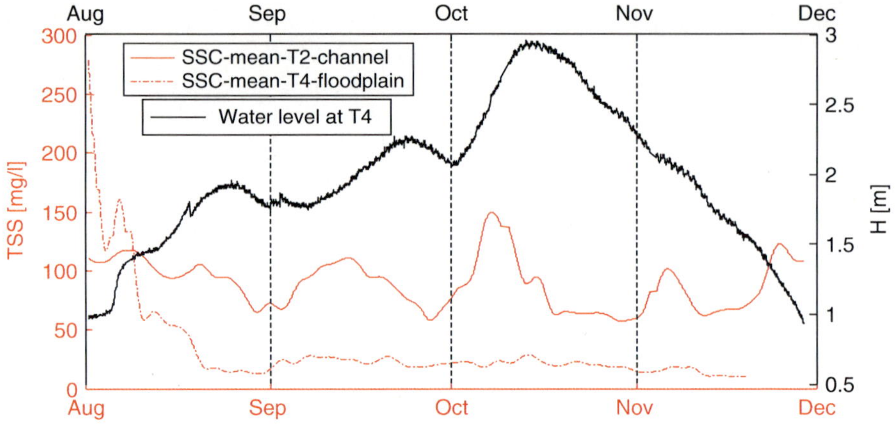


**Fig. 8.7** TSS concentration (moving window mean) at T2 in the channel and T4 in the flood plain during flood season 2009

concentration remains in a state of equilibrium where the near bed concentration, as measured by the monitoring system, changes mainly due to temperature, i.e. density changes in the water column. This causes a diurnal perturbation of the stratification of the water column without any significant sediment deposition.

## 8.7 Implications on Floodplain Sediment Management

Sedimentation in the floodplains is the major natural source of nutrient input in the Mekong Delta. Therefore the presented differences in the hydraulic link between channel and floodplain and the dynamics of the suspended sediments in the floodplains have strong implications on the sediment and thus nutrient input in the floodplains. The following three main findings are of relevance in this respect:

- Reduction of the suspended sediment concentrations with distance from the main river
- Highest TSS concentrations in floodplains in the first days of inundation
- Reduced link of channels and floodplains in areas protected by high dikes

In general this means that the first days of floodplain inundation are of major importance for the floodplain sediment deposition, in particular for the areas enclosed by high dikes. In these areas the hydraulic link to the channels providing the sediment input is reduced by the capacity of the sluice gates. Peaks in suspended sediment concentration in the channels occurring after the initial filling of the floodplains can hardly increase the sediment input in these areas. On the contrary, areas enclosed by low dikes can potentially receive additional sediments during the remaining flood period, although the data collected show that peaks in TSS in the channels do not cause peaks in the TSS in the floodplain after the initial inundation. However, it has to be noted that the flood levels in the observed period were very low. During more severe floods a closer link of TSS in channels and floodplains can be expected in low dike areas.

Combining these findings with the fact that TSS concentrations reduce with distance from the main river and the option of controlling floodplain inundation in high dike areas, a recommendation for the management of floodplain sediment input can be derived: Sluice gates located further away from the main rivers and channels should be open before those located in close proximity to the major rivers/canals in order to distribute the sediments more uniformly in the floodplains. However, this management requires a close coordination among the communes, which are responsible for the sluice gate control. This recommendation for the study area can be generalized for wide parts of the Plain of Reeds and the Long Xuyen Quadrangle, because similar hydraulic conditions, dike systems and land use exist throughout these areas.

## 8.8 Conclusions

The floodplain inundation dynamics in the Mekong Delta show a complex and differentiated pattern resulting from the interacting natural flood causes (flood waves from the Mekong and overland and tidal influence) and the anthropogenic interference on the natural system (dikes, sluice gates and pumps). In particular the floodplain compartments with high and closed dike rings exhibit a significant change from only partly influenced low dike compartments. Here the timing and duration of the inundation is completely controlled thus having a significant impact on the inundation dynamics and especially the sediment input into the floodplains. The data and findings presented here will be used to quantify the sediment input into the floodplains to substantiate the discussion of positive and negative impacts of high dike rings with quantitative information.

In addition, the presented findings may be used to optimize operation of sluice gates. The exponential decrease of suspended concentration in the secondary channels and the importance of the first flush of sediment into the floodplain suggest that sluice gates located further from the river should be opened first in order to capture more sediment in the floodplain compartments situated at larger distances to the main rivers/canals.

## References

Hung LM (2004) Research on the causes and solutions to prevent river bank erosion and deposition for the lower Mekong River system, Vietnam, Ministry of Science and Technology, Code KC08-15 (in Vietnamese)

Hung NN, Delgado JM, Vo Khac Tri, Le Manh Hung, Merz B, Bárdossy A, Apel H (2012) Floodplain hydrology of the Mekong Delta, Vietnam. Hydrol Process 26(5):674–686

MRC [Mekong River Commission] (2007) Diagnostic study of water quality in the Lower Mekong Basin, Technical paper no. 15, Mekong River Commission, Vientiane

Nguyen AD, Savenije HHG (2006) Salt intrusion in multi-channel estuaries: a case study in the Mekong Delta, Vietnam. Hydrol Earth Syst Sci 10:743–754

Sam L et al (2004) Study on the salinity intrusion to serve the economic – social development in the coastal zone of the Mekong Delta, July 2004 (in Vietnamese). The national program of the environmental protection and the natural; disaster prevention, code KC.08-18

Truong TV (2006) Flood identification, forecast and control in Cuu Long River Delta. Agriculture publication; 459 (in Vietnamese)

# Chapter 9
# Monsoon Variability and the Mekong Flood Regime

**Jose Miguel Delgado, Bruno Merz, and Heiko Apel**

**Abstract** The Mekong River in Southeast Asia owes its tidal annual flood regime to the monsoon. But the monsoon is a spatially and temporally variable circulation, with different annual to millennial variability for different regions. We focus on the separation between the Indian and the Western North-Pacific monsoon component to shed light on the interannual flood variability of the Mekong River.

The variance of flood discharge at four stations on the Mekong River is analyzed, as well as two well-known monsoon indices that represent annual monsoon intensity for the Indian Ocean monsoon circulation and for the Western North-Pacific monsoon circulation. An effort is made to identify the coarser temporal resolution that contains most of the interannual variability.

A close connection between the Western North-Pacific monsoon and the flood discharge in Kratie is found. This correlation is stronger for periods with enhanced variance. We find that the Western North-Pacific monsoon variance is a good predictor of flood variance. This dependence influences the probability of occurrence of a flood in the Mekong Delta.

## 9.1 Introduction

The variability of monsoon circulation of the lower latitudes is of extreme importance to the social and-economic development of South, Southeast and East Asia (Fig. 9.1). This fact is not only acknowledged within the scientific community, but also by a wider popular audience (Fagan 2009; Davis 2001). The scientific basis of such discussions is widely documented (Cook et al. 2010; Buckley et al. 2010). The impact of

J.M. Delgado (✉) • B. Merz • H. Apel
Section 5.4 Hydrology, German Research Centre for Geoscience (GFZ), Telegrafenberg, 14473 Potsdam, Germany
e-mail: jdelgado@gfz-potsdam.de; bruno.merz@gfz-potsdam.de; heiko.apel@gfz-potsdam.de

F.G. Renaud and C. Kuenzer (eds.), *The Mekong Delta System: Interdisciplinary Analyses of a River Delta,* Springer Environmental Science and Engineering,
DOI 10.1007/978-94-007-3962-8_9, 

monsoon variability on the flood regime of the Mekong River is not yet known, mainly due to a series of linkages that are hard to quantify. The El Niño-Southern Oscillation (ENSO) is a major interannual perturbation: the 2–7 year long periodic cooling and warming cycles of the tropical Pacific has an acknowledged effect on monsoon variability (Lau and Wang 2006) and on the intensification of the typhoon season in the Western Pacific (Elsner and Liu 2003).

The Asian monsoon can easily be understood as a unique entity, whose intensity would be described by its large scale circulation over the Indian Ocean (Webster and Yang 1992). Conceptually, it is a good model for understanding the common physical processes affecting this region: in fact, there is an Asian-wide disposition of the inter-tropical convergence zone (ITCZ) to exceptionally high latitudes during spring due to ascending air over heated land masses, causing a high-altitude air flow to comparably cooler areas over the ocean towards South (the Indian Ocean and the Western North-Pacific). This pressure gradient generates surface level flows that evaporate water from the warm pools in the Western Indian Ocean (shores of east Africa, Arabian Sea and Red sea) and Western Pacific. These flows finally cause convectional instability over the Asian continent and produce large amounts of rainfall.

However, the hypothesis that the whole monsoon domain reacts to common forcing on the whole range of time scales has been proved flawed by Wang et al. (2001), who clearly showed different variability and geographical extent for the East Asian component, the Indian component and the Western North-Pacific component. The differences are mainly due to the different geographic boundary conditions (Wang et al. 2005): the Indian Ocean is totally land-bounded to the north, whereas East Asia has an open ocean to the east, and land to the north and south. These boundary conditions translate into distinct internal feedback mechanism (Maher and Hu 2006) and may lead to a relationship with ENSO and the Australian monsoon.

The Mekong Delta is located in a transition zone between the Indian monsoon and the Western North-Pacific monsoon as shown in Fig. 9.1. Discharge into the Delta is a mixture of these two components: (a) precipitation from the Indian monsoon is mainly forced by the convective heat source over the Bay of Bengal (Wang et al. 2001) and arrives earlier than (b) precipitation from the Western North-Pacific monsoon, forced by a convective heat source over the South China Sea. The variability of this component will be discussed later in this chapter.

From a hydrological point of view, the inflow to the delta is the result of a non-linear aggregation of atmospheric, surface and subsurface processes, which span over the entire Mekong River basin. The configuration of the basin, elongated in the north-south direction, covers tropical and sub-tropical latitudes and tropical monsoon to humid continental climate according to the Köppen classification (Peel et al. 2007). Topography is also an important factor to understand the origins of runoff over the basin. In the upper basin, much of the spring discharge is due to snow melt from the Himalaya mountains; in the lower basin the mountains along the border between Laos, Cambodia and Vietnam cause high rates of orographic precipitation, making these sub-catchments the most productive of the basin in terms of runoff. Consequently, the interannual flood variability is not the same for the whole of the basin. Delgado et al. (2010) showed that a changing flood variability pattern could be identified along the river; more recently,

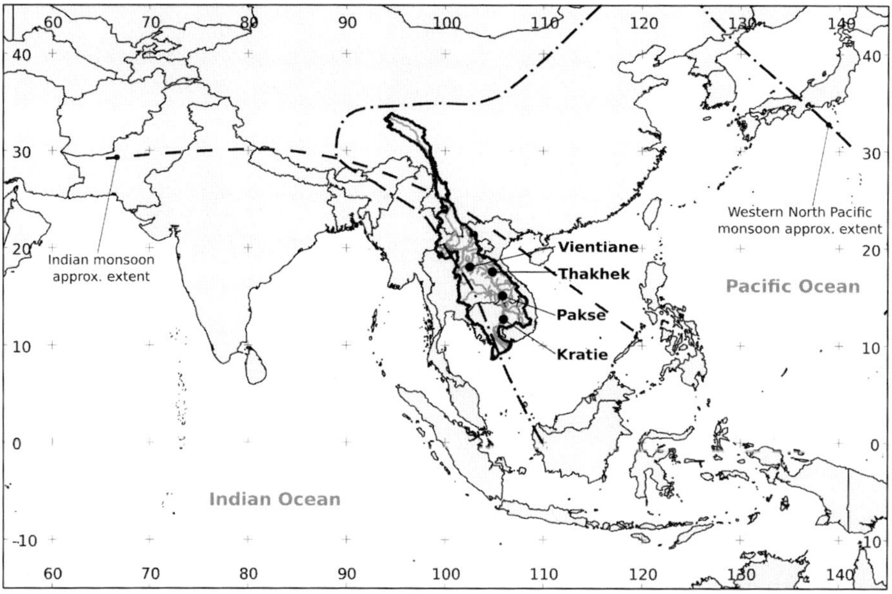


**Fig. 9.1** Asian monsoon regions. Mekong River, its basin and the location of the gauging stations used in this study are shown. The approximate Western Pacific monsoon and Indian monsoon extents were taken from Holmes et al. (2009)

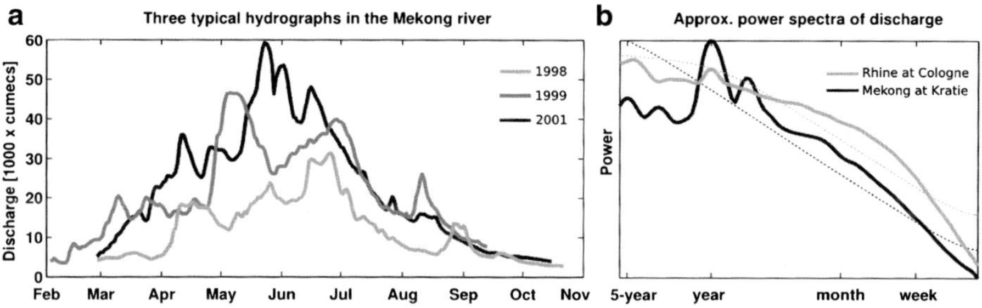


**Fig. 9.2** Flow variability in Kratie. Discharges during the flood season of 1998, 1999 and 2000 are depicted in (**a**). The estimated power spectrum of the Mekong at Kratie and of River Rhine at Cologne are compared in (**b**). 95% significance levels of a background red-noise spectrum were drawn in *thinner lines*. The annual cycle in the Mekong is stronger than in the Rhine, whereas the inter- and intra-annual variability is comparably weaker

Xue et al. (2011) added a spatial dimension to the problem, relating discharge in the frequency space with atmospheric circulation.

However, interannual variability is only a measure of the variation of the severity of a flood through time. The severity of a flood in a given year can be evaluated by the extreme water level, the duration of the inundation and the number of flood peaks. This is shown in Fig. 9.2a by the discharge recorded on what might be considered an

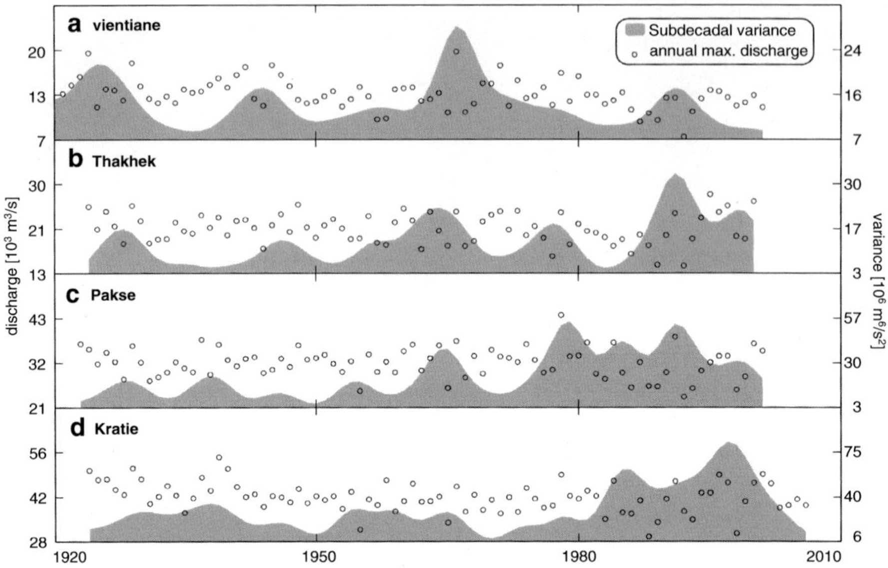


**Fig. 9.3** *Dots* represent annual maximum discharges (*left axis*). The *shaded areas* show the subdecadal variability estimated with the Morlet wavelet (the Morlet wavelet is one of a set of orthogonal functions that, just like a sine or a cosine function, allow the transformation of a periodic function between its time domain and frequency domain. The advantage of using a complex wavelet and not a sine and cosine is that it is well defined both in the frequency and time domain) (*right axis*). Lower stations show a later enhancement in variance

*average* (1999), a *dry* (1998) and a *flood* (2001) year. There, the typical features of the flood wave passing a gauging station are evident, although the length of the flood event is almost 1 year, much longer than in rivers under a temperate climate. The possibility of a two peak hydrograph is apparent as well, but both peaks belong to the same flood wave that arrives in the monsoon months. Due to the monsoon cycle the annual flood regime is very regular. With a record of roughly 80 years on a one flood per year basis, it is possible to construct a probabilistic framework of representative hydrograph types (Apel et al. 2006). A flood hazard assessment for the delta will have to take that into account in order to represent well the hydraulic boundary conditions.

In Fig. 9.2b, the power spectrum of the daily discharge at Kratie is compared with a higher latitude European river. It is clear that the annual cycle is more pronounced in the Mekong than in colder latitudes (Adamson et al. 2009), mainly because of the cyclic circulation of the monsoon. The more extreme-prone behavior of temperate rivers is here shown by its higher interannual variability. This chapter focuses mainly on interannual variability and therefore uses representative values for the intensity of the flood season: the annual maximum discharges. These time series are plotted in Fig. 9.3 for Kratie, Pakse, Thakhek and Vientiane, along with their moving variance.

In the time domain, the limited length of climate datasets constrains the analysis of temporal variability. Fortunately, discharges in the Mekong basin have been recorded for almost one century in at least four gauging stations (with notable interruptions in measurements, like for example during the late 1970s for Kratie in Cambodia, where data was interpolated from upstream gages). Other sources of information regarding the monsoon are available from proxy datasets from dendrochronology (Cook et al. 2010), marine geology (Jian et al. 2001) and speleology (Zhang et al. 2008). These datasets show that the variability of the monsoon has different forcing mechanisms for different time-scales. On geological time-scales, the movement of tectonic plates (by changing the land mass configuration) dominates; the Milankovich cycles (a model for earth harmonics) act on the millennial scale, as well as solar radiation; sun spots have centennial to decadal variations (Schwabe-Hale cycles); finally ENSO is responsible for interannual variability.

The objective of this chapter is to quantify changes in monsoon intensity and variability that affect the Mekong Delta. Further, the linkages between the two monsoon components that affect the region and the variability of floods in the delta will be examined. The main upstream boundary condition for explaining floods in the Mekong Delta is the inflow of water from the basin. Due to its topography the delta is also significantly affected by tides, storm surges and, on a larger time-scale, by sea level rise and the sinking of the Delta. However, due to the complex dynamics of flood propagation in floodplains which are highly affected by human interventions, this chapter will focus only on hydrology immediately upstream of the delta, at Kratie and Pakse. The impacts in the delta will be the subject of further study.

## 9.2 Changes in Flood Variance

The justification for focusing on variance changes and inhomogeneity is twofold: firstly, with respect to flood hazards, variability is more important than averages (Katz and Brown 1992), due to the fact that small changes in variance can alter the configuration of the tails of the frequency distribution and have a greater effect on the probability of extremes than the same change in the averages; secondly flood variances in the downstream reaches of the Mekong River have been changing over time in the last century, with consequences for flood hazard estimation (Delgado et al. 2010).

Other studies, e.g. Campbell (2007) and Xue et al. (2011), mentioned trends of averages, either by applying the Mann-Kendal test[1] or by simple linear regression. Three points summarize the results from these publications that focus on change in the Mekong River: *(a)* generally, a downward trend in averages was detected in floods in the Mekong river; *(b)* there is an upward trend in variance in the lower part of the Mekong basin and *(c)* the increases in variance outweigh the effect of the decreases in averages on the probability of an extreme event. These studies

[1]The Mann-Kendall test is a non-parametric (does not assume that the data follow a certain distribution) statistical test. It is used to reject or accept the hypothesis that the time series of interest is trend free, by evaluating a rank-based statistic.

assumed a continuous, monotonic variation either on averages or variance over time. They assume an *a priori* linear, polynomial or rank-based variation. This is appropriate for detection of changes, but fails to describe the dynamics of the change, for which a better adapted model is necessary. In the present study, wavelet[2] theory was used to identify changes in variance (Torrence and Compo 1998). It offers a very flexible decomposition of a time-series, allowing for an adaptive time and frequency definition.

Time series for annual maximum discharge in four gauges in the Mekong are given in Fig. 9.3. In the background of each panel, the average sub-decadal variance computed from the wavelet power spectrum is shown. The average sub-decadal variance is defined as the variance of all scales lower than 10 years, averaged over the frequency domain. For more details see Torrence and Compo (1998). Several features in the data are evident: *(a)* the enhancement in variance during the two last decades of the twentieth century in Thakhek, Pakse and Kratie; *(b)* the enhancement in variance during the 1960s in Vientiane; *(c)* the relationship of (b) with one extreme event in 1966; *(d)* the longer term enhancement of variance in the two downstream stations of Kratie and Pakse. This indicates that a north-south spatial pattern in interannual variability within the basin exists, which is mainly due to the analogous spatial variability of the monsoon. Dam construction can also be of importance, but its signal is not as clear as the monsoon: most changes due to dam construction were only detected in upstream gauges and did not affect the flow at Pakse or Kratie (Lu and Siew 2006). Massive deforestation and reservoir building affected the Korat Plateau in Thailand, but this region contributes with only 5% of the total discharge at the river mouth (MRC 2005).

To validate these observations, Fig. 9.6a shows the variance of the annual maximum discharge in Pakse for the period before and after 1976, decomposed into its different discrete modes, with uncertainty bounds. We see that the 4-year variance before 1976 is lower than after this year. The fact that the 95% uncertainty bounds do not intersect each other, indicates a significant increase in variance. Although not shown in the figure, the same increase is detected for Kratie and Thakhek, although statistically insignificant for the latter. For Vientiane, instead of an increase in variance, a statistically insignificant decrease after 1976 can be witnessed. This is in line with the variance presented in Fig. 9.3.

## 9.3 Changes in Monsoon Variance

The quantification of the Asian monsoon at the continental scale was mainly accomplished within the last two decades (Wang et al. 2001). Before that, studies focused mostly on the local/national scale. A simple way of quantifying and predicting the

[2]The Morlet wavelet is one of a set of orthogonal functions that, just like a sine or a cosine function, allow the transformation of a periodic function between its time domain and frequency domain. The advantage of using a complex wavelet and not a sine and cosine is that it is well defined both in the frequency and time domain.

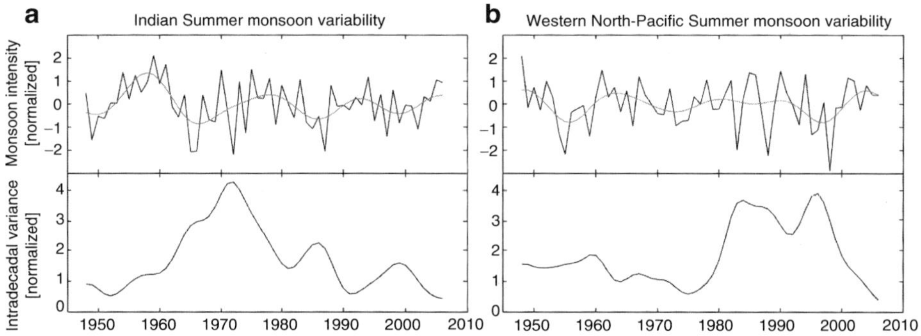


**Fig. 9.4** Two normalized monsoon indexes (*upper panels*) and their sub-decadal variance, estimated with the Morlet wavelet (*lower panels*). The indexes have different periods of enhanced variance

interannual variability of the Asian monsoon at a continental scale is to derive indices based on representative measurable processes. Wang and Fan (1999) reviewed the most important measures of monsoon intensity. More recently, many authors used these indexes for different purposes, such as characterizing the monsoon variability (Goswami 2006), evaluating and validating climate model predictions (Paeth et al. 2008; Cherchi and Navarra 2003) or investigating trends in monsoon intensity (Chase et al. 2003).

Separate indices for the Indian Summer monsoon and the Western North-Pacific Summer monsoon were used, following results from Wang et al. (2001). The intensity of the monsoon is represented by the difference between 850 hPa zonal wind velocities averaged over key regions for the monsoon circulation during June, July, August and September of each year. These indices can be seen in Fig. 9.4 together with their sub-decadal variance, obtained following Torrence and Compo (1998). This range of variance encompasses significant scales of variability and is in phase with the flood variability in the Mekong River, as it will be seen in the next section. It is also limited by the length of the time series available in order to avoid edge effects.

The most relevant features in Fig. 9.4 are: a period of enhanced variance of the Indian monsoon in the 1960s, which evens out after 1970, and a similar, but longer period of enhancement in the Western North-Pacific Summer monsoon during the last quarter of the twentieth century. These results were also obtained with different methods by Wang et al. (2001). The reason for different periods of enhancement of the two indices is the greater response of the Western North-Pacific monsoon to thermal conditions in the Pacific, for which ENSO plays a crucial role. Wang et al. (2003) have shown that El-Niño tends to enhance the Western Pacific monsoon circulation to a greater extent than the Indian monsoon (IM) circulation. Its variance also suffers an increase in the last quarter of last century (Torrence and Compo 1998).

Although both components of the monsoon deliver great amounts of precipitation to the basin, they have a distinct spatial and temporal variability. The Indian monsoon affects the whole of the basin, although with greater prevalence in the northern part.

The rainfall over the southern part of the basin is affected by the general monsoon circulation, but is forced mostly by the intensity of the Western North-Pacific circulation, including the coupling with the Australian monsoon. The moist air masses travel across the equator, curve to the right (due to the Coriolis acceleration) and enter the Mekong region (McGregor and Nieuwolt 1998). By then the flow has a strong zonal component (along the earth's parallels), therefore converging to the highlands along the Vietnamese border, which leads to intense orographic precipitation. Later in the season, typhoons develop and may make landfall on the coast of Vietnam, prolonging the flood season or contributing to the development of an additional flood peak. Some of the flood peaks in the record are due to a combination of landfall of typhoons and monsoon rainfall. The 2000 event is a famous example of such a mixture of effects.

The Western Pacific circulation is responsible for forcing rainfall over the area that most contributes to the discharge to the Mekong Delta. This area, identified in MRC (2005) as consisting of the sub-basins draining to tributaries between Pakse and Kratie, has an area of about 13% of the whole Mekong basin. It is responsible for about 23% of the flood volume in the delta. Enhancements in either the intensity of the circulation or in its interannual variance are thus likely to produce consequences in flood hazard.

## 9.4 Quantifying the Relationship Between Monsoon and Floods

Ideally, proving a cause effect relationship between two phenomena would require either a controlled repetition of the experiment in a laboratory, or the mathematical modelling of the underlying physical processes. The former is impossible, due to the scale of the phenomenon. The latter is underway, but until now, has yielded inconclusive results. However, the correlation between monsoon and flood frequency can be quantified statistically, by plotting the cross wavelet spectrum of annual maximum discharge and a monsoon index. If there is a consistent phase relationship between the two, it is likely that the monsoon forces the variability of floods in the Mekong. That is what is shown in Fig. 9.5a, where the upper panel represents the cross wavelet spectrum and the lower panel represents the average sub-decadal variance of the Western North-Pacific monsoon index and the annual maximum discharge measured in Kratie.

The WNPM causes positive enhancements in floods in the Mekong, even on interannual time scales. As is shown in Fig. 9.5a, the sub-decadal variance of the monsoon modulates the variance of the annual maximum discharge. Although not shown, the same applies to flood volume. Further, on the upper panel, it is shown that the two signals are significantly in phase (arrows pointing to the right) for most parts of the time domain.

By separating times of enhanced variance from times of lower variance (the threshold being the median of the sub-decadal variance) we could improve the correlation to significant values (Fig. 9.5b). This means that for periods of greater monsoon variance, the Western Pacific monsoon is a strong forcing mechanism for

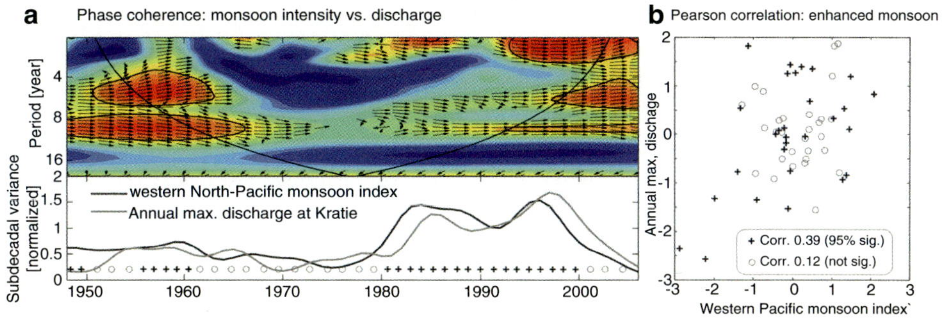


**Fig. 9.5** Correlation between annual maximum discharge in Kratie and Western North-Pacific monsoon index. The *upper panel* of (**a**) shows the correlation in phase coherence; its *lower panel* shows the sub-decadal variance. The *scatter plot* (**b**) distinguishes between above-average variance and below-average variance (also marked in **a**). The Spearman correlation coefficient shows an increased agreement during enhancement of variance

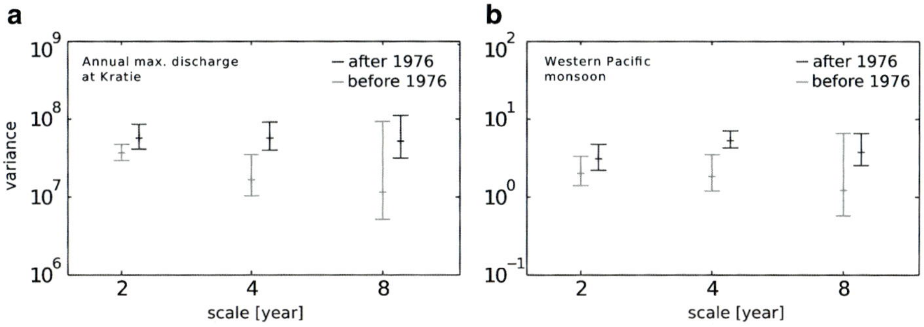


**Fig. 9.6** Discrete wavelet variance estimated for a monsoon index and annual maximum discharge in Pakse before and after 1976. There is a significant difference between the two periods for both variables

floods in the downstream reaches of the Mekong River. The same was not achieved by considering the Indian monsoon index.

To test if the changes detected in the 1970s were statistically significant, a step change test was performed in the year 1976 for the station Pakse, upstream of Kratie. Pakse was chosen to avoid possible errors introduced by the interpolation of the data in Kratie in the late 1970s. A regime shift in the Pacific Ocean has been detected around this year by several authors (e.g. Kerr 1992; Overland et al. 2008). The method used was suggested by Percival (1995) and discretizes the variance of the signal into different scales, which is consistent with the approach used in this study. The results are presented in Fig. 9.6 for scales of 2, 4 and 8 years. The 4-year variance increases after 1976 in both monsoon intensity and annual maximum discharge. The 2-year and 8-year scales also present a parallel behavior. This suggests a close linkage between both time-series with regard to variance. It is likely that the regime shifts in the Pacific Ocean significantly changed the probability of occurrence of an extreme flood in the Mekong. The relationship between monsoon variability and flood variability has implications for flood risk assessment: a mechanism by which the flood frequency distribution is forced by a measurable circulation index is suggested. Additionally, the construction of simple and effective flood scenarios under climate change is at hand just by deriving a monsoon index from a coupled general circulation model.

## 9.5 Implications for the Mekong Delta

The measure of climate change has commonly been understood as the anomaly around an estimated mean. This chapter points out that changes in variance of the monsoon were detected in the Western Pacific region that affected the flood variability of the Mekong River, especially in its lower part. These changes have a stronger impact in the probability of occurrence of extreme floods, than the anomalies in the mean value of the time series (Katz and Brown 1992; Delgado et al. 2010). Although monsoon climates are characterized by comparably low interannual variability, in the case of the Mekong this variability had a tendency to increase in the southern areas of the basin.

The Mekong Delta is directly affected by flood waves and typhoons formed in the southeastern parts of lower Mekong basin and its vicinity. The changes detected in this study affected the Delta, not because of the year by year increase of the incoming flood waves, but due to a growing variability. The massive man-made morphological changes of the Delta – the construction of embankments and channels, as well as the adaptation of flow regimes to human needs – introduced changes in the flow conditions that are still difficult to quantify (Dung et al. 2011). By introducing a new perspective in the study of flood variability, this chapter hopes to advance an explanation to the changes occurring in the upstream hydraulic boundary conditions of the Delta.

## 9.6 Conclusions

This chapter presented an overview of the main aspects regarding interannual variability of floods in the Mekong River and how it is forced by the monsoon variability. The general monsoon circulation was separated in two components: the Indian monsoon and Western North-Pacific monsoon. These proved to have different regions of influence in the Mekong River basin. The northern part is dominated by a surface level flow of moist air entering the basin through the Bay of Bengal forced by the Indian monsoon. The southern part is under the effect of the Western North-Pacific monsoon, receiving a moist surface level flow from the Indian Ocean that is forced by the adjacent Western North-Pacific circulation.

The two components of the monsoon have different periods of enhancement. The Western North-Pacific monsoon, with enhanced variance after the late 1970s, coincides with the variance of floods in Kratie, Pakse and Thakhek. Further upstream, the relationship with this component is not evident, which is consistent with the monsoon regions defined earlier.

Changing variance in the severity of floods is a defining aspect of flood hazard in the Mekong Delta (Delgado et al. 2010). By proving that a regime shift in flood variance took place in the past, we render stationarity in flood hazard assessment obsolete. A climate-driven non-stationary flood frequency model is therefore necessary.

Longer measured time series are necessary to better define the linkages between the Western North-Pacific monsoon and flood variability. In the absence of enough data, the method of choice is to model the physical processes and from long model runs understand the interactions between the large scale processes. An alternative is to use available proxy datasets that represent past climates. Both directions will in the future certainly open new perspectives for the hydrology of this region.

## References


Adamson P, Rutherfurd I, Peel M, Conlan I (2009) The hydrology of the Mekong River. In: Campbell I (ed) The Mekong: biophysical environment of an international river basin. Elsevier, New York, pp 53–76

Apel H, Thieken A, Merz B, Blöschl G (2006) A probabilistic modelling system for assessing flood risks. Nat Hazards 38(1):79–100. doi:10.1007/s11069-005-8603-7

Buckley BM, Kevin JA, Penny D, Fletcher R, Cook ER, Sano M, Nam LC, Wichienkeeo A, Minh TT, Hong TM (2010) Climate as a contributing factor in the demise of Angkor, Cambodia. Proc Natl Acad Sci 107(15):6748–6752. doi:10.1073/pnas.0910827107

Campbell IC (2007) Perceptions, data, and river management: lessons from the Mekong River. Water Resour Res 43. doi:10.1029/2006WR005130

Chase TN, Knaff JA, Kalnay E (2003) Changes in global monsoon circulations since 1950. Nat Hazards 29(2):229–254. doi:10.1023/A:1023638030885

Cherchi A, Navarra A (2003) Reproducibility and predictability of the Asian summer monsoon in the ECHAM4-GCM. Clim Dyn 20(4):365–379. doi:10.1007/s00382-002-0280-6

Cook ER, Anchukaitis KJ, Buckley BM, D'Arrigo RD, Jacoby GC, Wright WE (2010) Asian monsoon failure and megadrought during the last millennium. Science 328(5977):486–489. doi:10.1126/science.1185188

Davis M (2001) Late Victorian holocausts: El Niño famines and the making of the third world. Verso, London

Delgado JM, Apel H, Merz B (2010) Flood trends and variability in the Mekong River. Hydrol Earth Syst Sci 14(3):407–418. doi:10.5194/hess-14-407-2010

Dung NV, Merz B, Bárdossy A, Thang TD, Apel H (2011) Multi-objective automatic calibration of hydrodynamic models utilizing inundation maps and gauge data. Hydrol Earth Syst Sci 15:1339–1354. doi:10.5194/hess-15-1339-2011

Elsner JB, Liu K (2003) Examining the ENSO–typhoon hypothesis. Clim Res 25:43–54. doi:10.3354/cr025043

Fagan BM (2009) Floods, famines, and emperors: El Niño and the fate of civilizations. Basic Books, New York

Goswami BN (2006) The Asian monsoon: interdecadal variability. In: Wang B (ed) The Asian monsoon. Springer, Berlin/Heidelberg, pp 295–327

Holmes JA, Cook ER, Yang B (2009) Climate change over the past 2000 years in Western China. Quat Int 194:91–107. doi:10.1016/j.quaint.2007.10.013
Jian Z, Huang B, Kuhnt W, Lin HL (2001) Late quaternary upwelling intensity and East Asian monsoon forcing in the South China sea. Quat Res 55(3):363–370. doi:10.1006/qres.2001.2231
Katz RW, Brown BG (1992) Extreme events in a changing climate: variability is more important than averages. Clim Chang 21:289–302. doi:10.1007/BF00139728
Kerr RA (1992) Unmasking a shifty climate system. Science 255(5051):1508–1510. doi:10.1126/science.255.5051.1508
Lau NC, Wang B (2006) Interactions between the Asian monsoon and the El Niño/Southern Oscillation. In: Wang B (ed) The Asian monsoon. Springer, Berlin/Heidelberg, pp 479–512
Lu XX, Siew RY (2006) Water discharge and sediment flux changes over the past decades in the Lower Mekong River: possible impacts of the Chinese dams. Hydrol Earth Syst Sci 10:181–195. doi:10.5194/hess-10-181-2006
Maher BA, Hu M (2006) A high-resolution record of Holocene rainfall variations from the western Chinese Loess Plateau: antiphase behaviour of the African/Indian and East Asian summer monsoons. Holocene 16(3):309–319. doi:10.1191/0959683606hl929rp
McGregor GR, Nieuwolt S (1998) Tropical climatology: an introduction to the climates of the low latitudes. Wiley, New York
MRC [Mekong River Commission] (2005) Overview of the hydrology of the Mekong Basin. MRC, Vientiane
Overland J, Rodionov S, Minobe S, Bond N (2008) North Pacific regime shifts: definitions, issues and recent transitions. Prog Oceanogr 77(2–3):92–102. doi:10.1016/j.pocean.2008.03.016
Paeth H, Scholten A, Friedrichs P, Hense A (2008) Uncertainties in climate change prediction: El Niño-Southern Oscillation and monsoons. Glob Planet Chang 60:265–288. doi:10.1016/j.gloplacha.2007.03.002
Peel MC, Finlayson BL, McMahon TA (2007) Updated world map of the Köppen-Geiger climate classification. Hydrol Earth Syst Sci 11:1633–1644. doi:10.5194/hess-11-1633-2007
Percival D (1995) On estimation of the wavelet variance. Biometrika 3:619–631. doi:10.1093/biomet/82.3.619
Torrence C, Compo GP (1998) A practical guide to wavelet analysis. Bull Am Meteorol Soc 79:61–78. doi:10.1175/1520-0477(1998)079<0061:APGTWA>2.0.CO;2
Wang B, Fan Z (1999) Choice of South Asian summer monsoon indices. Bull Am Meteorol Soc 80:629–638. doi:10.1175/1520-0477(1999)080<0629:COSASM>2.0.CO;2
Wang B, Wu R, Lau KM (2001) Interannual variability of the Asian summer monsoon: contrasts between the Indian and the Western North Pacific-East Asian monsoons. J Clim 14(20):4073–4090. doi:10.1175/1520-0442(2001)014<4073:IVOTAS>2.0.CO;2
Wang B, Clemens SC, Liu P (2003) Contrasting the Indian and East Asian monsoons: implications on geologic timescales. Mar Geol 201:5–21. doi:10.1016/S0025-3227(03)00196-8
Wang P, Clemens SC, Beaufort L, Braconnot P, Ganssen G, Jian Z, Kershaw P, Sarnthein M (2005) Evolution and variability of the Asian monsoon system: state of the art and outstanding issues. Quat Sci Rev 24(5–6):595–629. doi:10.1016/j.quascirev.2004.10.002
Webster PJ, Yang S (1992) Monsoon and ENSO: selectively interactive systems. Q J R Meteorol Soc 877–926. doi:10.1002/qj.49711850705
Xue Z, Liu JP, Ge Q (2011) Changes in hydrology and sediment delivery of the Mekong River in the last 50 years: connection to damming, monsoon, and ENSO. Earth Surf Process Landf 36(3):296–308. doi:10.1002/esp.2036
Zhang P, Cheng H, Edwards RL, Chen F, Wang Y et al (2008) A test of climate, sun, and culture relationships from an 1810-year Chinese Cave record. Science 322:940–942. doi:10.1126/science.1163965

# Chapter 10
# Vulnerability, Coping and Adaptation to Water Related Hazards in the Vietnamese Mekong Delta

**Jörn Birkmann, Matthias Garschagen, Vo Van Tuan, and Nguyen Thanh Binh**

**Abstract** This chapter deals with the conceptualization, identification and assessment of the vulnerability of different social groups to water related hazards in the Vietnamese Mekong Delta. The Mekong Delta is globally seen as one of the key hotspots of climate change related risks due to its exposure to floods, salinization and potential sea level rise. In order to underline the multifaceted nature of vulnerability to natural hazards and climate change the paper outlines vulnerability profiles of different households and socio-economic groups in selected hazard prone areas, notably in rural communities exposed to high floods, coastal communities exposed to saline intrusion and urban communities exposed to urban and tidal flooding. The different locations selected for the assessment of vulnerability allow comparing how different local context situations and hazard phenomena might influence specific coping and adaptation strategies. The socio-economic transformation processes and policy reforms that have affected all three locations are examined in terms of their influence on vulnerability and capacities. The chapter provides a contribution to a further enhancement of methods, data bases and quality criteria for moving from an impact oriented risk assessment to a forward-looking vulnerability assessment

J. Birkmann (✉) • M. Garschagen
United Nations University - Institute for Environment and Human Security (UNU-EHS), UN Campus, Hermann-Ehlers-Str. 10, Bonn 53113, Germany
e-mail: birkmann@ehs.unu.edu; garschagen@ehs.unu.edu

V. Van Tuan • N.T. Binh
United Nations University - Institute for Environment and Human Security (UNU- EHS), UN Campus, Hermann-Ehlers-Str. 10, Bonn 53113, Germany

Mekong Delta Development Research Institute (MDI), Can Tho University, Can Tho, Vietnam
e-mail: vvtuan@ctu.edu.vn; ntbinh02@ctu.edu.vn

F.G. Renaud and C. Kuenzer (eds.), *The Mekong Delta System: Interdisciplinary Analyses of a River Delta,* Springer Environmental Science and Engineering, DOI 10.1007/978-94-007-3962-8_10, that can inform future adaptation strategies. In this regard the chapter makes a contribution for linking disaster risk reduction (DRR) and climate change adaptation (CCA) discourses. Particularly, the analysis of vulnerabilities to creeping-changes has often not been sufficiently addresses and incorporated in DRR strategies in Vietnam.

## 10.1 Introduction

Delta regions such as the Mekong Delta are increasingly seen as hotspots of risks in the light of growing frequency and magnitude of extreme events and climate change (see also Chap. 2). Especially, the Mekong Delta in Vietnam is seen as a key region where the capacities for climate change adaptation and disaster risk reduction have to be enhanced and effective strategies implemented urgently. The paradigm in both discourses in Vietnam, however, is still predominantly focusing on the adjustment of physical structures and the development of structural adaptation strategies, such as new dyke systems, to deal with actual and potential consequences of climate change. At the same time, complementary approaches for reducing the vulnerabilities in the social and ecological domains remain under-emphasised (Birkmann 2011a; Garschagen 2010a, c; Garschagen et al. 2009; Birkmann et al. 2010).

While structural measures for risk reduction and climate change adaptation are needed, there is little information on how these structural measures increase or decrease the vulnerabilities of different communities. Particularly in Delta regions the interconnectivity of coastal zones and the linkages between up- and downstream communities in flood plains underscore that such adaptation and risk mitigation measures are not solely influencing the communities where such measures are implemented, but also imply changes, for example, for downstream communities when dealing with dyke constructions in upstream areas. The implications of such intervention measures on medium- and long-term vulnerability reduction have not sufficiently been evaluated. The vulnerability profiles outlined in this chapter therefore provide important methodological toolkits and examples on how vulnerability and risk can be identified and how modifications due to such intervention measures, such as dyke systems, relocation and sluice gates, might reduce or increase the vulnerability of different social groups and communities in flood plains and coastal zones. Particularly, when examining these structural and technical adaptation measures and their potential consequences more in-depth, it becomes evident that these measures require additional adaptation processes for people and social-ecological systems to deal with the changes introduced. The success and appropriateness of the adaptation measures might even be determined by the ability or willingness of people to move forward with a second order adaptation process (see Birkmann 2011a). In this regard, the different vulnerability profiles are major factors that determine whether such strategies can enhance resilience or whether these measures are obstacles for further resilience building.

The identification and assessment of vulnerability are key to improving development strategies and integrated concepts within the framework of Integrated Water Resource Management (IWRM). While flood risk management has long been considered a central component within IWRM (UNESCO 2009), particularly with respect to risk mapping, most of the information gathered and used in such approaches is still focusing solely on the hazard and exposure component of risk. That means vulnerability and capacities of societies exposed to such flood hazards is often not sufficiently captured.

### *10.1.1 Vulnerability and Risk Research in Vietnam*

Despite some recent work on the social dimension of risks and vulnerability, the overall approach and focus on flood and coastal risk management strategies in Vietnam has to date been dominated by a strong emphasise on engineering-based technical solutions. Examples include larger dyke constructions after the major floods in the north of the Vietnamese Mekong Delta in 2000 and 2001 as well as the construction of sluice gates to protect saline intrusion in coastal provinces and the various relocation projects conducted in flood prone areas in An Giang and Dong Thap. The strategies outlined above are linked to a rather technocratic paradigm of disaster risk reduction drawing on the notion that natural hazards have to be controlled or fought against and that natural hazards are an external threat to human society resulting in the need for taming this hazard or nature as a whole (Birkmann 2006a; Casimir 2008; Käkönen 2008).

In contrast to a purely hazard oriented risk management, the international scientific discourse underscored in the last decades that the potential damage of natural hazards does not solely depend on the hazard as such, rather its shows that the vulnerabilities in a societal system play a key role in influencing the actual damage or the damage potential. Vulnerability can in this context be understood as the internal pre-conditions in a society or a social-ecological system that influence the potential to experience harm due to a given hazard (Thywissen 2006). Compared to vulnerability, risk can be understood as the product of a given hazard and a vulnerable societal system that is exposed to this hazard. Risk often also encompasses the probability of the hazard occurrence. Hence, effective risk reduction and risk management strategies – for example linked to IWRM – require an improved understanding of the vulnerabilities of communities at risk.

The vulnerability assessment within the project of developing a Water-Related Information System for the Sustainable Development of the Mekong Delta in Vietnam (WISDOM) is taking up this challenge and derives vulnerability profiles for different social groups and communities exposed to flooding and salt water intrusion. The case study focus is on households coping with high floods in the province for Dong Thap in the northern part of the Vietnamese Mekong Delta, on urban households dealing with tidal flooding and river bank erosion in urban and peri-urban areas in Can Tho and finally on rural households that face saltwater intrusion in the coastal province of Tra Vinh.

## 10.2 Conceptual Approaches to Risk, Vulnerability and Adaptation

### *10.2.1 The Social Construction of Risk*

In contrast to approaches that mainly associate risk with the occurrence of a natural phenomenon such as flooding, there is growing evidence that disasters in the context of natural hazards are not primarily determined by the physical event but rather by the vulnerability of the community or social-ecological system exposed to such events. That means the impact of a hazard event on a society is strongly shaped by the human action and by the internal pre-conditions of social systems (Hewitt 1983; Burton et al. 1993; Oliver-Smith and Hoffman 1999; Wisner et al. 2004; Müller-Mahn 2005; Birkmann 2006a). These vulnerabilities vary between different social groups leading to socially stratified susceptibility patterns and disaster pathways (Chambers 1989). While in the 1980s and 1990s this discussion was mainly informed by researchers working in the field of Disaster Risk Reduction (DRR) and development research, the discussion today has also involved the climate change community. Most notably, the IPCC special report SREX (IPCC 2011, 2012) addresses the new challenges of dealing with extreme weather events through the lens of vulnerability and risk.

### *10.2.2 Livelihoods, Risk and Vulnerability – A Conceptual Frameworks*

Vulnerability approaches have been shaped in strong interaction with other approaches in the field of poverty, natural hazard research and development studies. One of the most prominent examples is the Sustainable Livelihood framework (DFID 1999) which puts emphasis on the acquisition, substitution and utilisation of different types of assets that an individual or household has for pursuing livelihood strategies. The availability of, access to, and operational effectiveness of these asset portfolios are thereby regulated by transforming structures and processes that shape livelihood outcomes and the internal dimension of vulnerabilities (Chambers1989; Bohle 2001). Although the framework is used in a wide range of studies, it is still relatively fuzzy in terms of what transforming structures are and how livelihood outcomes influence also the vulnerability context (see Birkmann 2006a).

Another framework that has been developed in the context of risk research is based on the notion of coupled social-ecological systems (Turner et al. 2003). The authors of this framework argue that vulnerabilities consist of exposure, sensitivity and resilience whereby the latter comprise coping response, impact response and adaptation response (including adjustments to system components) (Turner et al. 2003). Their framework places the human-environmental system in the centre of its analysis. Focusing on the dynamic processes that drive vulnerability and eventually

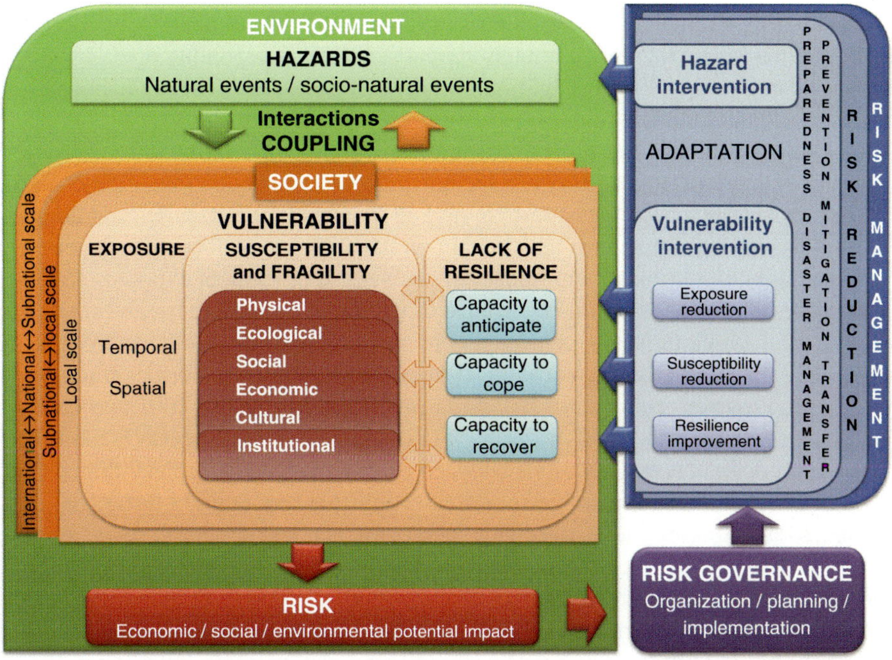


**Fig. 10.1** The MOVE framework (Own figure, based on concepts of Cardona 1999: 65, 2001; Turner et al. 2003; Bogardi and Birkmann 2004; IDEA 2005; Birkmann 2006b; Carreño et al. 2007)

can lead to disaster, Wisner et al. differentiate within their Pressure and Release Model (PAR) three components which together shape the progress of vulnerability. Systemic and institutional root causes can comprise limited access to resources, exploitive structures and lack of power. Dynamic pressures comprise destabilising processes at various scales that are inadequately governed and regulated like, for example, rapid urbanisation or ecosystem degradation. In combination, root causes and dynamic pressures can lead to unsafe conditions and an increase in vulnerability.

A final framework that will be introduced here is based on work dealing with the development of a holistic approach to vulnerability assessment linked to a system-science perspective, including reference to the concept of feedback loops and the theory of cybernetics. At its core, the MOVE framework (see Fig. 10.1) differentiates key factors of vulnerability and shows various thematic dimensions such as social, economic, environmental and institutional vulnerability. In terms of the key factors of vulnerability it is important to note that exposure is a contested aspect. In this framework exposure describes the extent to which a unit of assessment falls within the geographical range of a hazard event. Exposure extends to fixed physical attributes of social systems (infrastructure) but also human systems (livelihoods, economies, cultures) that are spatially bound to specific resources and practices which may also be exposed. Exposure then is qualified in terms of spatial and temporal patterns. The second key factor of vulnerability in the framework is susceptibility.

That means even if a system or object is exposed, it does not necessarily mean that the system is vulnerable, since it might have a very low susceptibility and a high coping or adaptive capacity. Susceptibility (or fragility) describes the predisposition of exposed elements (social and ecological) to suffer harm. The third factor that determines vulnerability is the societal response capacity of the community or system exposed. This response capacity can include pre-event risk reduction, in-time coping and post-event response measures. Compared to adaptation processes and adaptive capacities, coping capacities focus mainly on the ability to maintain the system as it is in the light of a hazard event impacting the system or element exposed. The term hazard describes the potential occurrence of natural, socio-natural or anthropogenic events, such as earthquakes or tsunamis as geo-hazards and floods or droughts as well as storms as hydro-meteorological hazards. Compared to former frameworks such as the holistic approach of risk assessment (Cardona 1999, 2001 and Cardona and Barbat 2000) or the BBC-framework (Birkmann 2006a), the MOVE framework also emphasises coupling relations between environmental and societal dimensions. Coupling emphasizes the framework's assertion that any defined hazard is given form and meaning by interaction with social systems, and similarly, social systems are influenced by their actual and perceived hazard context and environmental conditions. Hence, the concept of a coupled social-ecological system as outlined in the resilience research community (e.g. Walker et al. 2004, 2006; Berkes et al. 2003; Folke 2006) and further developed in the disaster risk research (Birkmann 2011b) offers an important perspective. Finally, the framework also underscores that risk governance (see e.g. Renn 2008) is a process that could span the whole diagram, but that was placed on the intersection between risk and adaptation in order to underline that risk management is linked to decisions and actions performed by both governmental stakeholders or institutions and non-governmental actors (e.g. individual households). Risk management undertaken by these stakeholders include tasks on risk reduction, through prevention, mitigation, preparedness or transfer. The framework stresses the fact that hazards might not only be of natural origin. Many hazards that reveal societal vulnerability, such as climate change related hazards, can be classified as socio-natural since the actual hazards are influenced by humans, such as anthropogenic climate change. In addition, purely anthropogenic hazards would include technical accidents, such as accidents in nuclear power plants.

### *10.2.3 Coping and Adaptation*

The concepts of coping and adaptation have gained substantial prominence in scientific and political discourses revolving around natural hazards and climate change. The terms are thereby often mixed without clearly differentiating their specific notions and implications; doing so, however, is necessary and of great help for guiding decision making and analytical approaches. We understand coping as a short-term, often reactive, response to deal with the impacts of a hazard during or after the hazards strikes and, by doing so, to minimize the effects of a disaster.

Adaptation, in contrast, implies a longer time frame and a notion of planned, strategic, target-oriented and coordinated action. While coping measures are mainly undertaken within the existing frame of processes and structures in a system and are generally not aimed at altering the principles of operation of that system as such, adaptation often implies an adjustment of system components, processes and structures in response to experienced or anticipated hazards or climate change impacts in order to moderate harm or even exploit beneficial opportunities (strongly mollified definition on the basis of IPCC 2007: 869). Some activities assessed within the following vulnerability assessment showed elements of both coping and adaptation. We take this, however, not as an argument for discarding those two concepts. Rather we take these observed hybrid forms as motivation for analysing the interactions of rather spontaneous short-term response that remains within the semantics of a given system or worldview (coping) with actions and measures that follow a more strategic approach and that implies changes of systems (adaptation). The latter involves a longer time horizon and the possibility of changing the ontological and operational landscape of an existing system altogether (see e.g. Adger et al. 2005). Questions that arise in this context include, for example, the interplay of generic and specific adaptive capacities (Handmer 2003), interlinkages between erosive coping capacities that degrade livelihood assets (e.g. selling of livestock during crisis situation), or the question, how coping mechanisms have to be adapted in view of climate change.

Overall, the MOVE framework and the sustainable livelihood framework are two guiding concepts that informed the case study research in Vietnam. The presentation of the different vulnerability profiles in the three case study locations presented below is therefore systematized according to the three key factors of vulnerability (a) exposure, (b) susceptibility and (c) societal response capacities in terms of coping and adaptation processes. Since the case study work is still ongoing some of the case studies can be presented more in-depth than others, however, the guidance the frameworks have provided for the analysis will become evident.

## 10.3 Vietnam: Formal Disaster Risk Management and Climate Change Adaptation

### *10.3.1 Disaster Risk Management*

Owing to the country's long experience with natural hazards and resulting disasters, Vietnam has a comprehensive system of formal disaster risk management. The coordination and implementation of disaster risk management falls within the remit of the Ministry of Agriculture and Rural Development (MARD). The mandate of the Ministry defines that the section of water resources has to "unify the management of dike construction and protection, headwork for prevention of floods and typhoons and efforts to prevent and combat flash flooding, floods, typhoons, drought, and landslides along riversides and coastal areas" (SRV 2009). In order to supervise the implementation and maintenance of disaster prevention measures (such as dykes and shelters) and to organize disaster response (e.g. on-the-spot dyke repair units), MARD and its subsidiary departments are in charge of coordinating committees of flood and storm control of which one is in place at the national level and at every province, district and commune/ward levels, respectively. These committees are headed by high ranking officials of the Ministry or by leading officers of the People's Committee of the respective level and include representatives from the different planning agencies and other relevant institutions such as the Red Cross or the Women's Union. In November 2007, the Prime Minister approved the new long-term National Strategy for Natural Disaster Prevention, Response and Mitigation to 2020. This strategy confirms the role of MARD as focal agency for disaster risk management and sets out guiding principles and objectives for the next years. These comprise the improvement of early warning systems, the improvement of planning and building codes in view of natural hazards, the fostering of capacity building at all levels, the relocation of people in disaster-prone areas and the upgrade of structural protection measures such as sea dykes, flood resistant embankments or storm shelters (SRV 2007).

### *10.3.2 Climate Change Adaptation*

Similar to developments in the international community and in most countries, the formulation of official climate change adaptation policies is a rather recent phenomenon in Vietnam. Earlier initiatives had been mainly focusing on climate change mitigation strategies[1] (e.g. the first drafts for the correspondence with the UNFCCC in the 1990s). However, owing to the exceptional risk that climate change implies for Vietnam, the Government as well as other actors within the ministries and from the international cooperation community in Vietnam pushed for a speedy process of institutionalising climate change adaptation efforts. In December 2008, the Prime Minister approved the National Target Program to Respond to Climate Change (SRV 2008). This program officially acknowledges the importance of climate adaptation for the overall sustainable development of the country. It, therefore, calls for adaptation efforts of all levels and sectors and for the mainstreaming of

[1] The authors follow the commonly agreed terminology in which climate change response comprises the two domains of climate change adaptation and climate change mitigation. Adaptation in this notion refers to the "adjustment in natural or human systems in response to actual or expected climatic stimuli or their effects, which moderates harm or exploits beneficial opportunities" (IPCC 2007). Mitigation comprises all measures that reduce green house gas emissions, limit their growth or enhance sinks. Due to the focus of this paper, the authors have to limit the discussion to adaptation measures here, however, fully acknowledge that measures from both domains are absolutely necessary and that "mitigation is the best adaptation" in the long run.

**Table 10.1** Objectives and targets of the National Target Program to Respond to Climate Change

| Objective[a] | Targets to be achieved by 2010 (selection) | Targets to be achieved by 2015 (selection) |
|---|---|---|
| Assessment of climate change impacts in Vietnam | Scenarios based on existing data<br>Pilot projects for assessment | Update/completion |
| Identification of response measures | Implement pilot/test projects in different sectors and locations | Wide scale implementation based on lessons learned |
| Establishment of scientific and practical basis for response measures | Development of a national science and technology program on climate change | Completion and updating of climate change database<br>Update research and implement results |
| Consolidation of organisational structure and capacity building amongst relevant institutions | Development of framework for legal documents and mechanisms<br>Coordination amongst ministries, sectors and localities | Mechanisms to prioritize climate change activities |
| Raising public awareness and human resources development | Over 10% of population and over 65% of Government staff with basic knowledge on CC | Over 80% of population and 100% of Government staff with basic knowledge on CC |
| Promotion of international cooperation and support | Establish bilateral and multilateral cooperation mechanisms for implementing the national target plan | Completion and effective implementation |
| Mainstreaming into development planning (socio-economic, sectoral, local) | Guidance documents and classification of measures | Mainstreaming into future planning<br>Assess implementation for period 2010–2015 |
| Development and implementation of action plans and (pilot) projects | Ministries and local authorities complete action plans | Action plans being implemented |

Source: Own table based on SRV (2008)
[a]For the official wording of the objectives and tasks refer to SRV (2008)

climate change adaptation into general planning processes. In order to achieve this goal the plan appoints responsibilities and tasks to government sectors and institutions. The plan specifies that the Ministry for Natural Resources and the Environment (MoNRE) will act as the focal agency for all climate change response activities in Vietnam. The ministry and its subsidiary departments at the lower levels, therefore, have to coordinate response measures and facilitate the communication of other ministries, sectors and the localities with the national Government (which has the ultimate responsibility). Table 10.1 summarizes the objectives of the National Target Program as well as the most important tasks including the envisaged timetable for completion.

### *10.3.3 Frictions, Tensions and Limits for Implementation*

The legal and institutional framework for climate change adaptation and disaster risk management laid out above might create the impression of a neatly organised and effective system with clear competencies assigned and procedures defined. However, the implementation of these target programmes and plans is in fact hampered by a number of factors. The programmes sketch out the broad visions of future directions in disaster risk management and climate change adaptation. However, there is little concrete guidance provided on how to implement or achieve these goals – particularly in terms of cross-scale transfer from the national to the local level (Garschagen 2010a). Institutional responsibilities are in many cases ambiguous, particularly with respect to the synchronisation between disaster risk reduction and climate change adaptation tasks and funds. The fact that these two domains are coordinated by two different line ministries hampers the effective integration of activities and political initiatives and rather leads to competition for resources and responsibilities (see Garschagen et al. 2009). The latter point is of particular importance as MARD and MoNRE have similar mandates and areas of interests, causing competition over resources, competencies and power.

## 10.4 Projected Future Climate Change Impacts

Recently, a number studies set out to analyse potential climate change impacts on Vietnam and the Mekong Delta building on climatological and hydrological modelling. The International Centre for Environmental Management (ICEM) concluded in its 2008 assessment, that based on current population and land use patterns, a sea level rise of 1 m would directly affect around one third the Delta's area (including roughly 10,000 km$^2$ of agricultural land) and over one quarter of its roughly 17 million inhabitants (Carew-Reid 2008). Sea level rise also implies severe challenges in terms of soil and water salinization. In addition, climate change is predicted to alter precipitation patterns leading to an even stronger concentration in the rainy season with higher precipitation levels during that time and the risk of increasing numbers of short-term heavy precipitation events throughout the entire year (MoNRE 2009; Hoang and Tran 2006; Chaudhry and Ruysschaert 2007). At the same time, the risk of prolonged drought periods is expected to rise during the dry season (Tuan and Chinvanno 2011). An overall rise in air and sea surface temperature is further expected to enhance the typhoon activity in the South China Sea (Elsner et al. 2008), leading to intensified typhoon occurrence in the Mekong Delta (CFSC 2004; see also Chap. 2).

## 10.5 Vulnerability Profiles in Selected Case Study Areas

The vulnerability assessment within the WISDOM project comprises three different case study areas along the main branches of the Mekong River, i.e. Dong Thap on the Tien Giang River; Can Tho and Tra Vinh on the Hau River. This enables us to draw comparisons with respect to vulnerabilities at community and household levels for different hazard patterns and different socio-economic conditions – representing the main bio-physical and socio-economic zones in the Delta (Chaps. 3 and 4). Dong Thap in the northern Mekong Delta comprises mainly rural communes involved principally in agricultural production, mainly highly intensified rice production with two or three cropping cycles per year which heavily relies on hydraulic infrastructures and water management. At the same time, aquatic common property resources provide an important livelihood supplement for a considerable part of the population. Given its location in the Plain of Reeds, Dong Thap is exposed to severe flooding with long durations and high flooding depths during the rainy season. The province therefore belongs to the Delta's provinces suffering the highest flood damages and losses over the last decades. The City of Can Tho, in contrast, constitutes the main socio-economic centre of the Delta. While the administrative entity of Can Tho City comprises nine districts including rural areas, the research underlying this chapter focused mainly on the three main urban and peri-urban districts, i.e. Ninh Kieu, Binh Thuy and Cai Rang. The urban centre constituted by these three districts is growing and spreading rapidly. While Can Tho is not exposed to extended flooding like in Dong Thap, it is exposed to urban flooding and is projected to be exposed to increased flooding in the future. Tra Vinh Province is located downstream of the Hau River next to the coast. It is therefore exposed to different hazards with salinity intrusion and freshwater shortages being the main challenges. Consisting mostly of rural population, Tra Vinh hosts a large variety of agriculture and aquaculture production systems with shrimps, rice and sugar cane being the most prominent outputs. Figure 10.2 shows the location of the different case study areas in the Delta as well as the differences in predominant hazard exposure, socio-economic conditions and vulnerabilities, and predominant coping and adaptation strategies. More detailed descriptions and analysis of each of the locations are provided below.

## 10.6 Communities Exposed to High Floods – Case Study Dong Thap

In the Mekong Delta (MD), annual floods have existed for thousands of years along the main stream of the Mekong River, which originates from the Tibetan Plateau in China and passes through Myanmar, Thailand, Laos, Cambodia and Vietnam. The flood season lasts from late July through December. The flood peak is reached

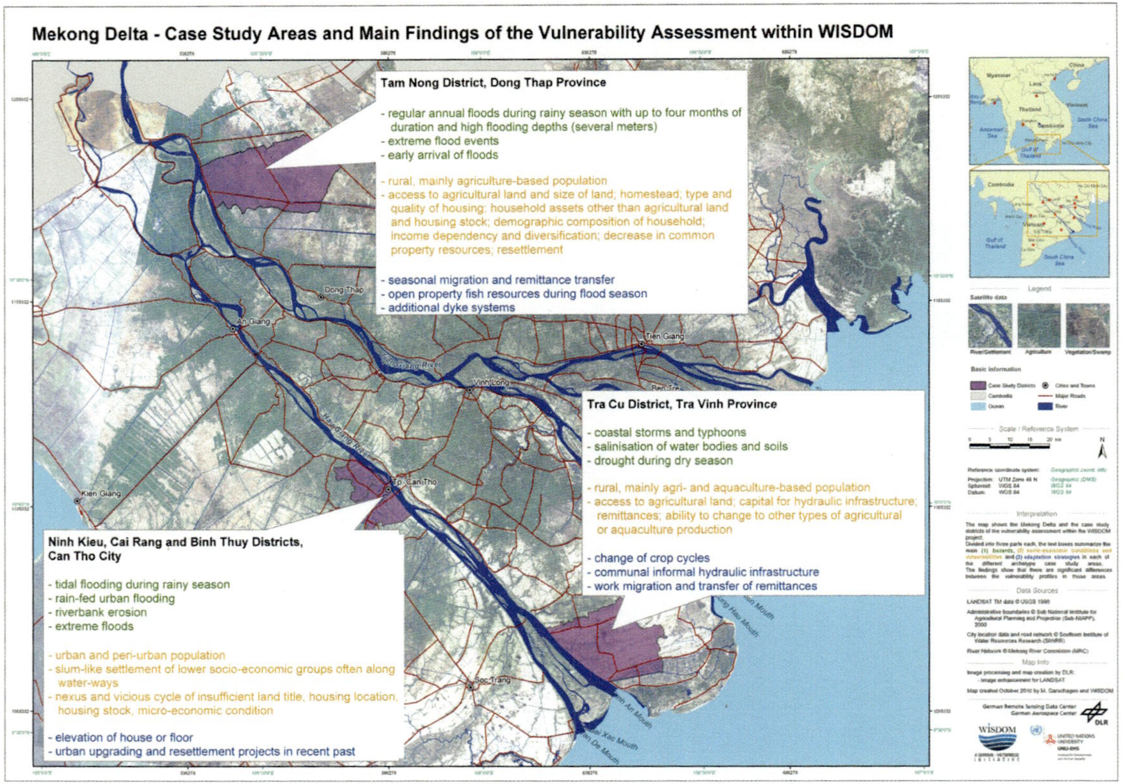


**Fig. 10.2** The three study sites in Vietnam. The map shows the case study districts and lists the main findings in the text boxes. According to the MOVE framework presented above, *green text* stands for the main hazards, *orange* for socio-economic characteristics and key vulnerabilities and *blue* for the most important adaptation strategies identified (Source: Own figure Garschagen 2010b)

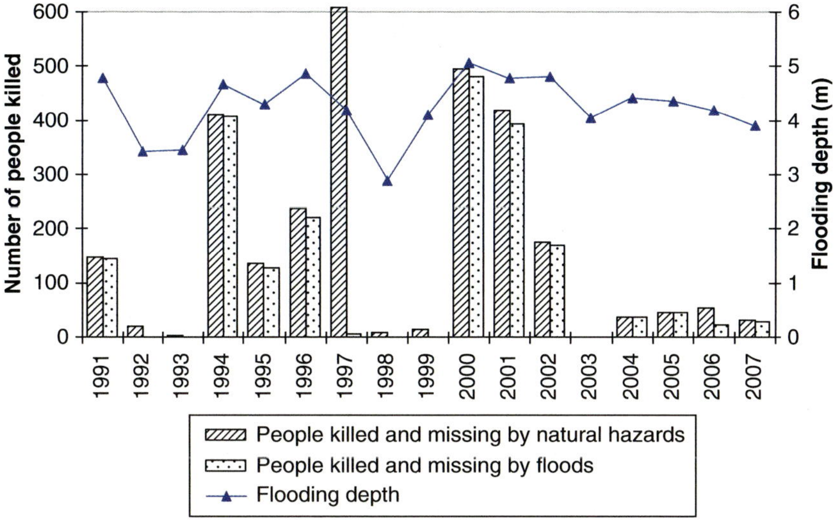


**Fig. 10.3** People killed and missing by natural hazards and floods in the Mekong Delta (Source: Vo Van Tuan, data from CFSC, Dong Thap CFSC 1990–2006). (The number of people killed and missing by natural hazards in 1997 was 2,244)

between late September and mid October. The flooded area accounts for 53.3% of the natural area and over 50% of the delta's population live within its extent. The two regions most affected by flooding are the Plain of Reeds and the Long Xuyen Quadrangle (see Chap. 3). The Plain of Reeds is characterized by its shallow topography, extended wetland and severe acid sulphate soils which have been exploited particularly to cultivate floating rice since the 1960s and high-yielding rice varieties since the early 1990s.

Flooding levels vary from 0.5 to 4 m in the upper and middle parts of the delta. Daily flooding depth varies between 5 and 7 cm during normal flood years and 10–20 cm during years with strong flooding (Hoi 2005). Flood levels in the Dong Thap case study area are categorized according to the water levels of the Tien River at Tan Chau gauging station located in the upper Mekong Delta. According to this categorization, flood years with water levels below 4 m are called low floods (lũ nhỏ). Water levels between 4 and 4.5 m constitute normal floods (lũ thường), water levels of more than 4.5 m are considered high floods (lũ lớn). These flood levels coincide to a certain extent with the damage caused by flooding (Figs. 10.3 and 10.4).

The research sites are located in both inland and river bank areas in Dong Thap, one of the three provinces most prone to floods in the upper MD. This province is severely affected by acid sulphate soils which are leached by flood water (Minh et al. 1997; Chap. 14). Rice is the main crop, accounting for 94.3% of planted area of the annual crops (Dong Thap Statistical Office 2008). Double rice cropping is the most popular production mode, and triple rice cropping has become the standard in the fully-protected areas in the province. The case study of Dong Thap indicates that

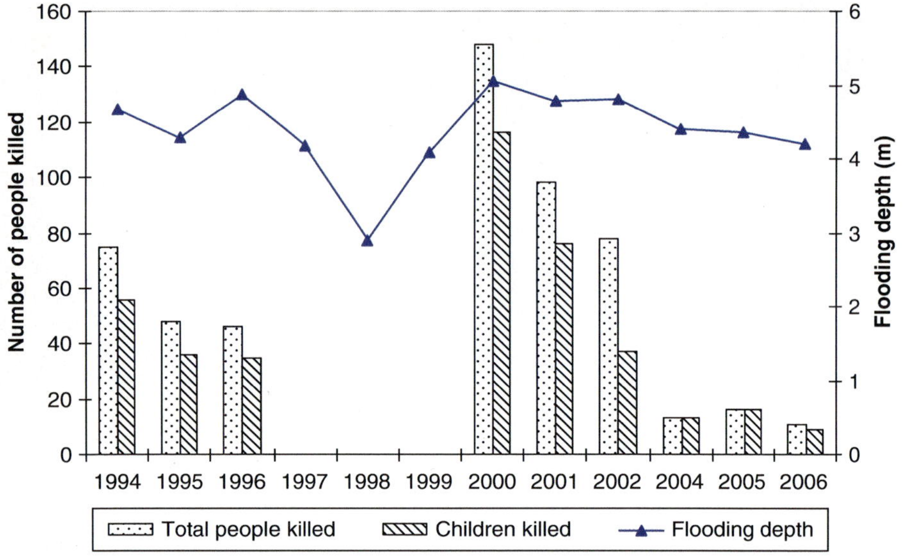


**Fig. 10.4** People killed by floods in Dong Thap (Source: Vo Van Tuan, data from Dong Thap CFSC 1994–2006)

formal interventions (relocation and embankments) have influenced rural livelihoods of different socio-economic groups. In order to analyse these influences and household level vulnerability in detail, 370 households were investigated through standardized questionnaires, with the number of households located in the floodplains (rice fields), along local roads, in the residential clusters (relocation sites), and in the residential dyke being 81, 169, 41 and 79, respectively.

## *10.6.1 Hazard Profile*

Floods are natural phenomena, but very high floods cause serious problems to human lives, infrastructures and crops. High floods are caused simultaneously by the combination of large upstream discharge affected by tropical typhoons or low atmospheric pressures, long and heavy rainfall in the delta itself and high tidal levels in the canals and rivers reducing their drainage capacity (Be et al. 2007). The recorded high floods in the last four decades seem to be increasing in frequency, particularly, the years 1961, 1966, 1978, 1984, 1991, 1994, 1995, 1996, 2000, 2001, 2002 and 2011 are to be mentioned. From 2003 to 2010, flooding depths have decreased; however, high floods may cause high damages after a series of low floods because of a decrease in flood preparedness due the decrease in "normal" flooding and related experience (see Fig. 10.4). More than 2,000 ha of Autumn-Winter rice crop area in Dong Thap were destroyed due to dyke failure during the 2011 floods. Given the low and flat topography of the Mekong Delta, relatively small changes in flooding depths can cause great differences in flood impact and losses (see Figs. 10.3 and 10.4). According to statistical analysis of the past flooding events, the most significant economic problems and human fatalities are associated with flooding depths exceeding 4.5 m.

Vietnam has for a long time been exposed to many different natural hazards including typhoons, flash floods, landslides and others. Flooding alone contributes around one third of all the disasters caused by natural hazards. Recently, while the number of people missing and killed by hazards has remained more or less constant, economic losses have significantly increased. In the upper MD, flooding is considered as a major natural hazard so that almost all hazard victims and damages were caused by slow-onset floods. Within the upper MD, Dong Thap can be considered a major hotspot of flood vulnerability and flood losses. Between 1994 and 2006 the impact on people and children registered in Dong Thap accounted for 21% and 27% of the impact in the entire Mekong Delta, respectively. In the extreme flood of 2000, Dong Thap alone accounted for 31% of all the casualties in the Mekong Delta and 35% of all the children drowning in the entire Delta. Economic losses in Dong Thap accounted for 21% of all the economic losses experienced in the entire Delta.

In terms of stratified vulnerability, children under six suffered the highest fatalities, accounting for 74% of total fatalities in Dong Thap (Fig. 10.4). According to Dong Thap CFSC[2] staff, most children drowning occurred in poor[3] households and at night time when the children crawled and fell into the water while their parents were sleeping. During flooding, poor children were also not sufficiently protected by adults since their parents and other adult members usually worked in floodplains.

The main economic losses due to floods in the upper MD are related to crop loss, and destruction of houses and infrastructure. Double rice cropping is the major farming system in the flood-prone areas. The second crop, Summer-Autumn rice, is harvested at the beginning of the flooding season, usually in May or June. This explains why in the past there have been large areas of Summer-Autumn rice destroyed or severely affected due to the early onset of floods. In order to protect Summer-Autumn rice and develop Autumn-Winter rice, semi- and fully-protected dyke systems have been constructed by the government in most areas.[4] Since a third rice crop has been introduced in the fully-protected area, the Autumn-Winter season is characterized by high risk of being affected (Fig. 10.5).

In the rural delta, temporary houses are common, accounting for approximately 60% of the poor households. These houses are easily destroyed or damaged

[2] Committee for Floods and Storm Control.

[3] Poverty in this context refers to a participatory wealth ranking conducted within the framework of the research. The main criteria in this ranking were: land title, housing quality, income related to differentiated livelihood activities, other physical capital assets.

[4] Only in the newly cultivated areas in the Long Xuyen Quadrangle the vulnerability to early floods remains a major problem, given that these areas are not yet protected by dyke systems.

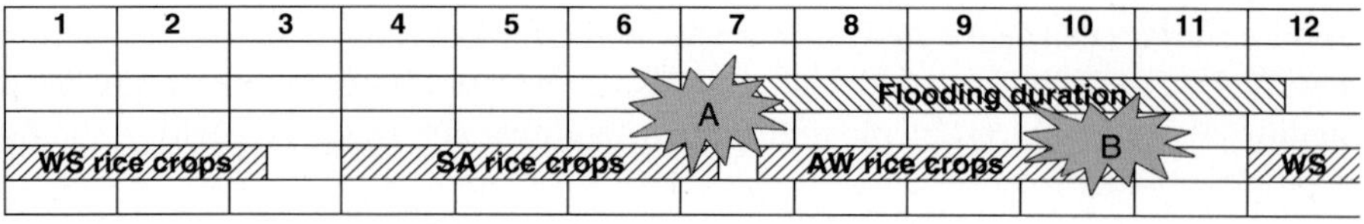


**Fig. 10.5** Relation between flood season, crop cycles and vulnerability in Dong Thap Province. (*A*) Late summer-autumn (*SA*) rice crops were/are at risk of being affected if the flood arrives early. (*B*) During the autumn-winter (*AW*), rice crops are exposed to dyke breakage due to high floods (Source: own focus group discussion with key informants in Phu Hiep commune 2009)

particularly due to long duration of floods and high flood peaks. Flood damages have decreased due to low flooding; however, elements exposed and susceptible to floods may increase because of an acceleration of economic activities and physical interventions. In particular the planted areas of Autumn-Winter rice grown during the flooding season in Dong Thap have increased rapidly over the last decade (Dong Thap Statistical Year Book 1998–2008), and embankments have been continuously constructed in floodplains.

## *10.6.2 Exposure to Floods*

### 10.6.2.1 Exposure to Different Flooding Levels

In the rural Plain of Reeds, people usually live in or close to the floodplains, either along low dykes, along high dykes or in elevated places as residential clusters (Fig. 10.6). These patterns of settlements depend on their in-migration periods, wealth, land ownership or resettlement policies. In the densely populated areas, other facilities such as electricity lines, tap-water, elevated local roads, schools and markets are constructed by local governments. In Phu Hiep commune, the pioneers, who settled in the area in the 1960s, moved to the elevated places and have been protected by trees surrounding their houses and high dykes (Fig. 10.7).

People who live in floodplains are strongly exposed to floods. Their houses are directly exposed to flooding right from the early phase of the flooding season. They live far from elevated roads and densely populated areas so that they often lack access to basic infrastructure (e.g. electricity, tap-water, elevated roads, schools and markets). In the flooding season, they travel by boats which constrains them and their children in their access to basic services like schooling and healthcare services. It is difficult for them to evacuate to high places or to receive external support since they live far from densely populated areas and lack transportation means. People who live along low dykes are affected by medium flooding. Their houses are easily affected by floods and strong winds. In contrast, people who live along elevated roads can escape directly from floods. Their houses are protected by high dykes and trees. They easily move to elevated roads if their houses are destroyed by strong

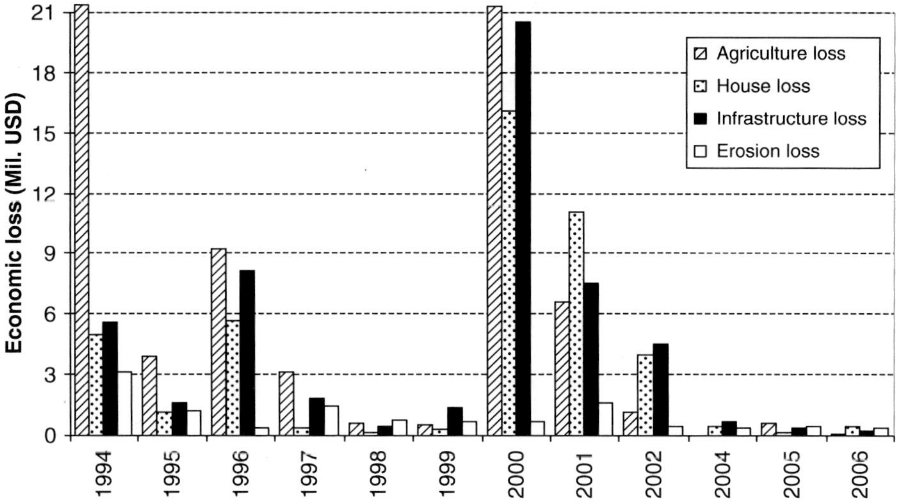


**Fig. 10.6** Total estimated economic loss by floods in Dong Thap (Source: Vo Van Tuan, data from Dong Thap CFSC 1994–2006). Economic loss was converted to the value in the year 2000, USD/VND = 14,177

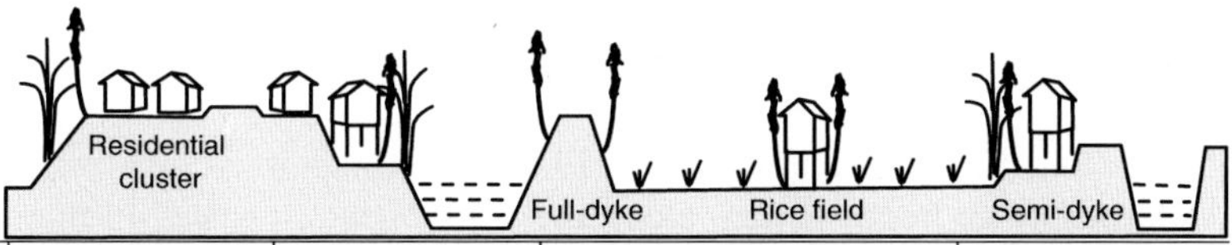


| | | | | |
|---|---|---|---|---|
| Settlement periods | Relocated in 2003 from the flood-plains, shaped by the relocation policy | Resettled in the 1960s, forced by the former government | Informally settled in the 1980s-1990s, attracted by opportunities to obtaining agricultural land and flood-related resources | Informally settled in the 1980s-1990s, attracted by opportunities to obtaining agricultural land |
| Basic infrastructure | Electricity line, tap-water, schools, market, transportation | Electricity line, tap-water, schools, market, transportation | Lacks of access to basic infrastructure | Electricity line, schools |

**Fig. 10.7** Transect map of the inland site, Phu Hiep Commune (west-east direction) (Source: Figure by Vo Van Tuan, transect walk in Phu Hiep commune 2008)

winds. However, according to focus group discussions in Phu Hiep in 2009, these people struggle with transportation problems as their house are linked with elevated roads and dense vehicle movements. People who live in residential clusters or on residential dykes are not affected directly by floods, but they have encountered other constraints regarding the changes in their lifestyles and livelihood activities. Before resettlement, most of them were poor people with no residential land and high exposure to flooding. These relocated households were offered subsidized house foundations and housing frames, to be paid back through installments over a period of 10 years (see in detail Birkmann et al. 2012).

In addition to the place of living, the exposure of people to floods depends on their cropping patterns which are changing due to both formal and informal measures. Major crops exposed to floods have been changing from floating rice before the 1980s to Summer-Autumn rice in 1999 and Autumn-Winter rice after the 2000s due to the changes in rice-based farming systems and embankments. In reality, large planted areas of Summer-Autumn and Autumn-Winter rice crops were destroyed or reduced in yield by big floods in the past.

#### 10.6.2.2 Landownership and Exposure to Floods

Land ownership is an important factor enabling inhabitants to achieve rural livelihood outcomes. From the socio-economic crises of the early 1980s, the 6th Party Congress of the Vietnam Communist Party adopted an economic reform policy "Doi Moi" in order to restructure Vietnam's legal, regulatory, administrative, investment and policies from the centrally-planned economy into a market-oriented economy with "socialist characteristics" (Bryant 1998). Households were considered as autonomous and independent economic units, and they were provided agricultural land. When the collective system was dismantled in 1988, and the land law was reformed in 1993 (subsequently modified in 1998 and 2003), farm households had the right to use their land in the long-term and could transfer, exchange, lease, inherit or mortgage this land. However, the rapid economic changes also implied for many farmers severe economic risks. In this regard some households had to cope with a high debt by transferring or leasing their plot of land (see Chap. 4).

Exposure to floods differs along the lines of land ownership that shapes rural livelihood activities. There is significant high correlation between wealth and agricultural land ownership at the research site.[5] The major reasons for landlessness in Phu Hiep Commune are that immigrants failed in the initial production of the floating rice, high-yielding rice, intensive vegetables or intensive snake head fish. The landless households earn their main income from off-farm activities such as fishing and off-farm wage activities which normally only cover their daily costs of living. Therefore, they must partly rely on flood-based activities in the risky flooding condition. Moreover, during flooding events, children in landless households are inadequately supervised by others given that their parents and other adult members are usually working in the floodplains. Recently, landless people in rural areas have tried to shift into other income-earning activities (e.g. non-farm jobs in urban areas) and have to revert to seasonal off-farm activities since flood-related resources decline quickly. Remittances have therefore become an important part of these household's income base (own household survey 2009).

[5]The significant correlation is shown by the high value of the Pearson correlation coefficient (0.688**).

### 10.6.3 Susceptibility

The susceptibility of people to slow-onset floods in the upper MD is influenced by natural, socio-economic and governance factors. As a large rural population relies on agriculture, particularly rice cultivation, the lack of access to agricultural land has an essential influence on rural livelihoods or their susceptibility. Landless households are constrained in accessing formal loans and implementing livelihood diversification. They are dependent on the susceptible flood-related resources, open assets, which decline quickly through the increase in embankment, agro-chemical use and over-exploitation. Recently, inhabitants have gradually shifted into other livelihood options like urban non-farm activities; however, it is difficult for the poor to access other livelihood activities due to their low educational schooling or vocational training.

### 10.6.4 Coping and Adaptation

#### 10.6.4.1 Coping

People apply a series of coping activities just before and during flooding in order to mitigate flood impacts. They increase the stability of the house through installing supportive wires just before floods. In addition, they build grass fences to break waves and remove several planks of the floor in order to prevent them being destroyed by flood waves, and then gradually lift up their house floor in response to the increase in flooding depth. Secondly, flood-related livelihoods are an important issue for flood-affected people, particularly the poor. Floods provide both advantages and disadvantages; therefore, people engage in flood-based activities, especially fishing. Flood-related products provide food not only during flooding but also during the dry season. This means that fish caught during the flooding season is often processed into dry fish or salted fish that can also be consumed in the dry season. Thirdly, people try to pay increased attention to the protection of dependent persons, especially children during high floods. People who engage in fishing usually assign adult members to take care of their children and the elderly, or children stay with them in the floodplains. They improve wooden fence to protect children from floods and prepare home-made lifebuoys while fishing. Women in poor households often try to access and prepare food and medicine during flooding. Normally they borrow money from private money lenders at high interest rates. Besides coping at the individual household level, a series of formal coping activities are implemented by the Committee for Flood and Storm Control (CSFC). This institution is organized at all administrative levels and includes various departments. During high floods, children can be protected at daycare centers which are organized at local people's houses in densely populated areas. Table 10.2 provides an overview of different coping and adaptation strategies identified in the upper MD to deal with floods.

**Table 10.2** Coping and adaptation strategies in Dong Thap

| Coping strategies | Adaptation strategies |
|---|---|
| **1. Informal/individual coping strategies** | **1. Informal/individual adaptation strategies** |
| (1) Adjusting housing condition | (1) Improving housing condition |
| Support house with wires just before floods | Gradually elevate homestead |
| Build grass fence at low floods | Grow trees or bushes surrounding homestead |
| Remove several planks of house at low floods | Build wooden stilt houses |
| Lift up house floor | Improve good wooden/concrete stilt houses |
| Evacuate when houses are strongly flooded | |
| (2) Coping with livelihood disruption | (2) Improving flood-related livelihoods |
| Small-scale fishing | Large-scale fishing |
| Collect flood-related vegetables | Cultivate flood-related products (snakehead fish, freshwater prawn and vegetables) |
| Sell livestock or give up agriculture products just before floods | Build high or solid cages |
| Seasonally migrate for remittance | Adopt flood-related knowledge & experiences |
| (3) Protecting dependents or people | (3) Living with floods |
| Prepare children protection facilities | Improve housing condition & physical assets |
| Take children along for fishing or travelling | Assign adults to take care of children |
| Prepare man-made life saving devices | Informally relocate along elevated roads |
| | Train children to swim |
| **2. Formal/governmental coping strategies** | **2. Formal/governmental adaptation strategies** |
| (1) Mitigating agricultural damage | (1) Protecting agriculture |
| Army to help to harvest crops & protect dykes | Construct semi- and full embankments |
| Subsidize equipment | |
| (2) Protecting people's life | (2) Implementing flood risk management |
| Organize evacuation of affected households | Relocate flood-affected households |
| Organize child daycare centers | Provide swimming training for pupils |
| Adjust schooling time | |
| (3) Coping with livelihood disruption | (3) Improving rural livelihoods |
| Provide basic needs (foods, medicine, clothes) | Provide credits for flood-related agriculture |
| Provide fishing tools (boats, nets, hooks) | Train people for other occupations |

Source: Vo Van Tuan, KIP, focus group discussions, in-depth interviews, Dong Thap 2008–2010

#### 10.6.4.2 Adaptation

Adaptation to floods is considered as a long-term strategy, associated with learning either before or after an extreme event occurs. Both informal (individual adaptation at household level) and formal adaptation (initiated by governmental agencies) strategies are used to adapt to slow-onset floods. Informal adaptation strategies include improvement of housing condition, flood-related livelihood adaptation and flood-related knowledge accumulation. Wealthier people, who have had relatively stable livelihood resources and access to livelihoods assets, usually apply adaptation strategies. In contrast, the poor often implement a series of coping activities. Overall, research indicates that people have gradually improved their conditions for living regarding homestead elevation, planting (Nam 1992), and concrete housing construction and physical household asset enhancement. Moreover, people have implemented their flood-based livelihoods (e.g. flood-related agriculture, crop calendar adjustment, migration) in order to adapt to livelihood disruptions.

Several formal adaptation strategies are applied in the flood-prone areas, in which the embankment and relocation are predominant. Embankment has been popularly implemented in the 1990s and 2000s. It aims at protecting Summer-Autumn crops and develops Autumn-Winter crops within the fully-protected dykes. Regarding dyke systems, farmers have adjusted their seasonal crop calendars which protect Summer-Autumn crops from early floods. However, embankments have created the strong seasonality of crop cultivation and constrained the exploitation of flood-related resources which previously used to be of high importance for poor households in order to complement income and livelihood assets. Embankments therefore have contributed to the decrease of off-farm income and flood-related activities. Resulting from this, there are different perceptions relating to dyke systems regarding varying land ownership and flood-related income dependency. Landless people have a preference to take flood-related benefits without embankments while land-owning farmers have a preference for embankments which allow for improved and more secure rice cropping.

Relocating landless households prone to floods has been implemented in the MD after the historic 2000 floods through a series of official documents issued by the government. In the rural Plain of Reeds, normally one or two residential clusters or dykes were constructed in each commune. Regarding this, the resettled households escape from direct flood impacts; however, after initial resettlement, these households struggled with lack of basic infrastructure and the disruption of social networks and livelihood activities. These problems have gradually been addressed by informal and formal measures.

### *10.6.5 Summary: Main Drivers of Vulnerability in Dong Thap*

In the context of rural floodplains, vulnerability to floods of local residents is shaped by various factors. Poverty constrains them to improve their housing condition as well as flood-responsive assets. Landless residents who rely on flood-based resources are severely exposed to floods and more susceptible since these resources decline rapidly due to dyke systems and the intensive use of agro-chemicals. Moreover, due to low educational levels, young landless laborers have usually taken manual non-farm jobs in urban areas, often implying temporal migration. In contrast, farmers who own agricultural land could access formal financial institutions, diversify their livelihoods and gain benefits from embankment that protect and enable the development of rice production.

## 10.7 Can Tho City: Vulnerability Profiles to Future Floods

### *10.7.1 Role of the Study Area Within the Wider Vulnerability Assessment*

Climate change adaptation and responding to natural hazards within cities is not an entirely new field of research but should – for good reasons – receive increased attention (Birkmann et al. 2010). Living in a heavily urbanising world (in particular with heavily urbanising countries like Vietnam), cities increasingly become places of accumulated risk to natural hazards, while at the same time creating specifically urban vulnerabilities and development challenges (Pelling 2003; Birkmann et al. 2010). In Vietnam, the strong urbanisation of the past decade and the expected intensification in future urbanisation processes, coupled with challenges resulting from wider socio-economic and political transformation affecting the resilience particularly of poor urban dwellers, calls for in-depth case study research to explore the details of vulnerability pathways and adaptation potentials of an increasing share of the Vietnamese society.

Can Tho City is the centre of the Mekong Delta in terms of population, economy, transport and other key functions. Administratively, the City comprises nine districts with altogether almost 1.2 million inhabitants. However, the majority of these districts comprise mainly rural hamlets based on agricultural activities. The urban core of Can Tho City is made of Ninh Kieu District and parts of Cai Rang as well as Binh Thuy districts, being home to around 350,000 inhabitants. Can Tho City provides central services reaching out to the entire Mekong Delta, e.g. in education, services or trade. Moreover, Can Tho City has a higher industrial output than any other province in the Mekong Delta.

Owing to the rapid urbanisation and transformation processes that can be observed in Can Tho over the last decades and that are expected to intensify in future, the city is increasingly characterised by social stratification and fragmentation resulting, for instance, from forced resettlement, rocketing land prices, modifications to social security networks or labour migration. Disparities with respect to short-term coping as well as long-term adaptation capacities to water-related hazards and in particular to future hazards expected in the context of climate change are therefore on the rise between different population groups in the city.

**Fig. 10.8** Tidal Flooding in Can Tho City (Source: Own pictures Garschagen 2009)

In summary, the case study of Can Tho City contributes to the general goal of the vulnerability assessment within the WISDOM project to compare vulnerability, coping and adaptation patterns along the lines of different socio-economic groups as well as different hazard profiles represented in the Delta. The Can Tho City case study region adds the urban – and often neglected peri-urban (Garschagen et al. 2011) – focus to this exercise and, in terms of the hazard pattern, constitutes an area that has not yet experienced drastic hazards in the past but will be affected heavily in future due to climate change.

The research in Can Tho draws from in-depth household interviews followed by a standardised household survey campaign conducted in late 2009 covering 588 households in different parts of Ninh Kieu, Cai Rang and Binh Thuy. In addition, expert interviews at province, district, ward and residential group level were conducted.

## *10.7.2 Hazards*

Compared to Dong Thap and its flooding situation the urban areas of Can Tho have in the past experienced rather few and mild natural hazards. While the northern, rural districts of Can Tho are affected regularly by flooding during the rainy season, the urban districts in the south are beyond the extent of common seasonal flooding. This means that the urban areas of Can Tho are (so far) not facing extended flooding with substantial flooding depths during the rainy season. However, many parts of the city are subject to tidal flooding during those months, whereby streets and houses can be flooded up to 1 m for several hours – particularly during the monthly tide peaks towards the end and middle of the lunar month. This tidal flooding, however, often implies substantial problems for the affected households, businesses or infrastructure elements (Fig. 10.8).

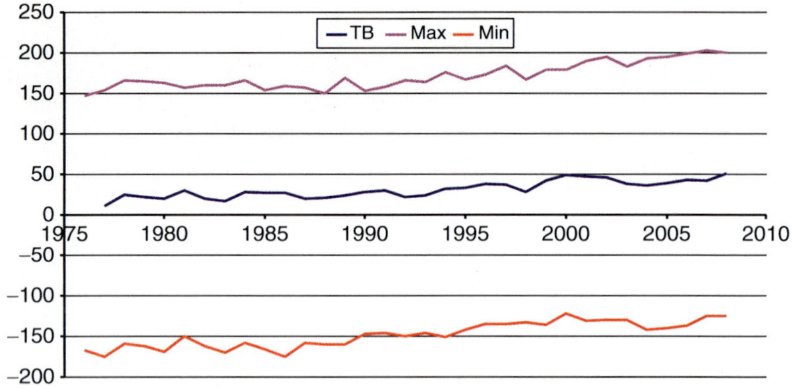


**Fig. 10.9** Annual minimum, maximum and average water levels in Hau River at Can Tho City in relation to the statistical mean sea level defined for Southern Vietnam (Source: DoNRE Can Tho City 2009)

### *10.7.3 Hazard Exposure*

Apart from the existing hazard of tidal flooding, Can Tho City is projected to be highly exposed to several climate change related hazards in future, making it an interesting case study for exploring how a dynamically growing city is dealing with the challenge of adapting to future (and new) hazards. These climate change related hazards comprise slow-onset progressive hazards such as sea level rise, an expected increase in annual river-based flooding, changes in temperature patterns as well as rather irregular hazards such as typhoons and smaller storms, heavy precipitations, rising variability in precipitation and increased tidal amplitudes in relation to storm surges at the coast. Importantly, these hazards are likely to occur in combination under conditions of climate change, leading to multi-hazard situations in which impacts can cause feedbacks in multiple ways. Resulting from this combination of hazards and the risk of concurrent or cascading multi-hazard phenomena, is a complex mix of potential secondary hazards, i.e. hazards which emerge out of the change in primary hazard patterns. The range of such secondary hazards is large, including, for example, the increased spread of water- or vector-borne diseases in slum areas with insufficient sanitation, draining and sewerage infrastructure. Other hazards are likely to include increased river bank erosion and the consecutive collapse of buildings and other infrastructure, but also soil and water salinisation, the lack of freshwater resources or heat-related health problems in quarters with poor ventilation.

Interestingly, the frequency and intensity of tidal flooding in the city has been increasing over the last years which has to be seen in relation to a rising trend in maximum levels in Can Tho's Hau River (one of the main branches of the Mekong) (see Fig. 10.9).

The reasons for this increase in the river's water levels are not yet entirely clear and are contested in the scientific and particularly in the political arena. Amongst

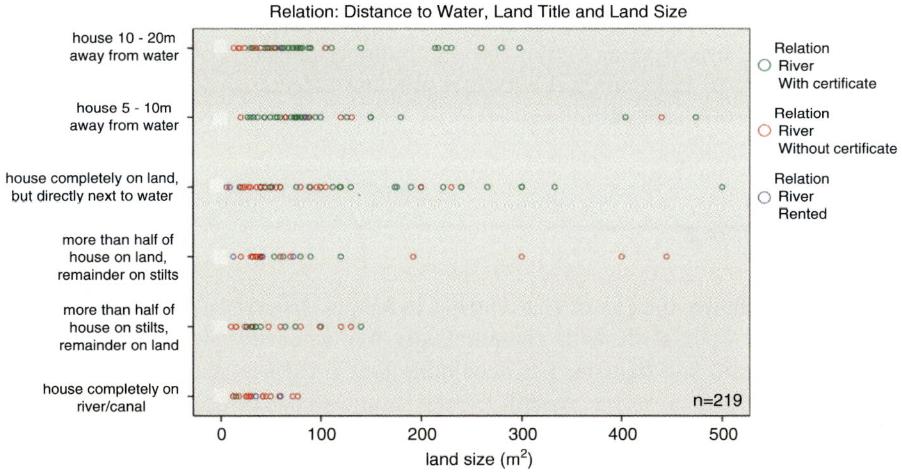


**Fig. 10.10** Relationship between physical exposure, land title and land size (Source: Own draft Garschagen 2010b)

the reasons most prominently stated are climate change, land subsidence in Can Tho City, but most importantly hydraulic infrastructure development upstream in the Mekong Delta (in particular embankments and other dyke constructions in Dong Thap and An Giang that block off former retention areas and channel more water downstream) as well as dams in upstream riparian countries. As precise assessment exercises, e.g. on the basis of hydrological modelling, are lacking so far, the specific contribution of each of these factors is hard to assess. However, according to prevalent expert opinion in Vietnam, the two latter reasons are by far the most important to describe the trends of rising water levels in the last decades. It can be assumed that adaptation to climate change related hazards will to a considerable part draw on strategies and mechanisms that have earlier been applied in other context settings, e.g. elevation of house floors. Therefore, the situation of rising water levels in the Hau River over the last decades and the analysis of adaptation patterns to this change, allows to generate "lessons learned" with respect to the adaptive capacities to certain climate change hazards such as sea level rise or the increase in urban flooding.

Within Can Tho, the nexus of physical exposure, socio-economic status and land title could be identified as one of the major criteria for regulating socially stratified exposure to tidal flooding and river bank erosion, also affecting the susceptibility of the respective households. Figure 10.10 indicates that poorer socio-economic groups often have to live in slum-like housing (with limited space) which is built close to or even on the water due to the fact that the households lack sufficient resources for acquiring land title and houses in less exposed areas further away from the water. These households experience an increased exposure not only to flooding and its secondary hazards but also to river bank erosion.

In addition, such settlements on or close to the banks of rivers and canals are usually characterised by an extremely high building density, coupled with poor or lacking drainage infrastructure. As a result, the effects of flooding and heavy precipitation events are amplified.

## *10.7.4 Coping*

As indicated above, the city of Can Tho has in the past been only exposed to a rather limited range of hazards with comparatively low intensity. Hence, the need for deploying coping mechanisms has been quite low in the past and experiences with coping are limited. However, specific coping responses can be observed with respect to the currently predominant hazards. These are detailed below.

### 10.7.4.1 Tidal Flooding

Households affected by tidal flooding (mainly during the rainy season) in Can Tho apply several coping measures in response to this hazard. In the absence of physical flood protection measures like dykes within the city, the most prominent strategy to prevent flood damage to furniture or appliances in the house is to elevate those assets for the time of expected or actual flooding, using bricks or other materials. In addition, many households have small-scale flood-prevention devices installed in their house, for example, in their door frames in order to protect at least some parts of the house from flooding. In the peri-urban areas, small-scale dyke systems (i.e. earth dykes of up to 1 m height) are quite common which protect a number of households that are located close together. The maintenance of these dyke systems is often supervised by the local People's Committee at ward level in the sense that the local officer or group leader has to arrange the workforce required for the maintenance of the dyke or for necessary extensions. This work is usually implemented before the advent of the rainy season and is conducted by household members (usually men) of the households located within the embanked area.

In some cases, interviewees reported that they evacuate household members when they expect the highest flood peaks (in the middle and end of the lunar calendar). This is especially the case if the flood peak is expected to be reached at night time, meaning that infants and small children who have to sleep on the floor are at particular risk of being affected. In those cases, children and infants have been reported to be sent to relatives in less exposed, i.e. less flood-prone, areas.

The number of floors a household has available has great influence on the range of possible coping measures in urban areas with respect to vertical evacuation of people and physical assets. In general, poorer and lower middle-class households – often living in slum-like housing conditions without land certificate – in most cases do not have sufficient resources to build houses with more than a ground floor and can, hence, not revert to vertical evacuation of persons or assets within their own house. But even for those households that do have several floors, vertical evacuation is only of limited feasibility as many assets cannot be easily moved to the first floor (kitchen appliances, gas cookers, motorbikes etc.). However, these households usually have a much smaller physical exposure and, hence, experience less flooding, even in the ground-floor.

#### 10.7.4.2 Storms

Many households reported to evacuate their children when strong storms are expected. The most prominent places for evacuation in Can Tho are communal buildings like pagodas, temples, churches or schools. The evacuation does sometimes follow the warning issued by the local People's Committee officers; however, in most cases it rather seems to be triggered by storm warnings through the television. Evacuation durations are usually quite short and have not been observed to last longer than 24 h. In addition, many households – in particular those with a poor or provisional housing stock – do prepare their homes when storms are expected in order to prevent damage. In particular provisional roofs made of banana leaves or corrugated metal sheets are in such cases secured with additional wooden elements, tightened with ropes or supported with additional ballast such as sand bags or old motorbike tires.

#### 10.7.4.3 Committee for Flood and Storm Control

Like all other provinces and districts in Vietnam, Can Tho and the case study districts of Ninh Kieu, Binh Thuy and Cai Rang have committees for flood and storm control in place (CFSC) which are in charge or organising the disaster preparation, response and recovery. However, given the low disaster profile to date in these districts, the action required by this committee and the implementing bodies of disaster management (mainly the local People's Committees, Fatherland's Front, military, police) was comparatively minimal. Nevertheless, in cases of smaller disasters, the committee organises the support for affected households. For the case study areas those are mainly households whose house has been affected or destroyed by river bank erosion. In addition, the CSFC organises trainings for local officers of the People's Committees and other organs in order to prepare, teach and practise response measures for the case of larger disasters.

### *10.7.5 Adaptation*

#### 10.7.5.1 Tidal Flooding

With respect to longer-term structural adaptation measures, adjustments to the housing stock are the most important measures that can be observed. Due to the overall rise in water levels and intensity as well as frequency in flooding events,

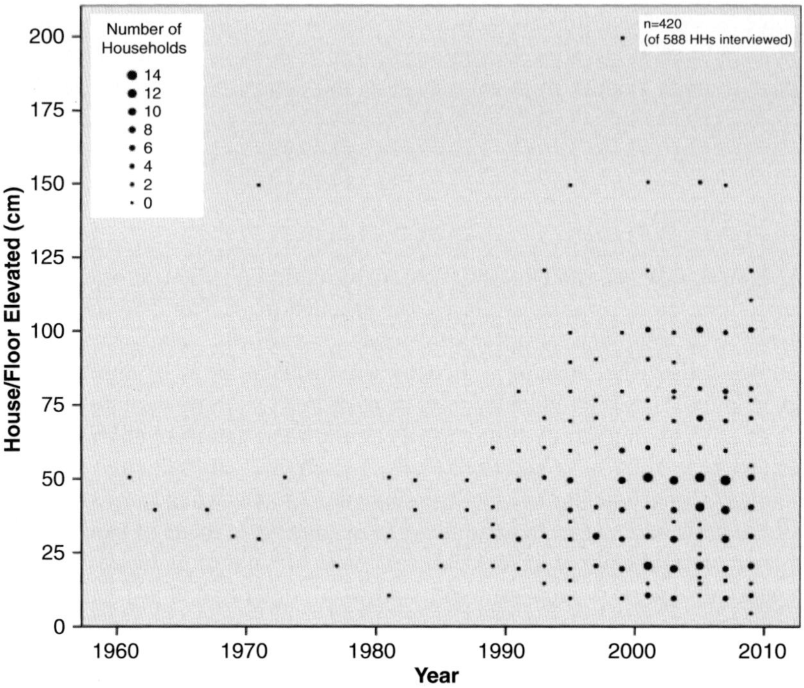


**Fig. 10.11** House and/or floor elevation as an adaptation measure (Source: Garschagen 2010b)

many households have been observed to elevate their house or parts of their house in order to mitigate flooding exposure. Of the 588 households interviewed in flood-exposed areas within Ninh Kieu, Cai Rang and Binh Thuy, 71% reported that they have elevated their floor or parts of their floor at least once over the last 50 years. In general, two trends could be identified in this context: First, the number of households elevating their floor or house has been increasing over the last years. Second, the amplitude of elevation is increasing (see Fig. 10.11). Besides the trend of increasing flood levels and frequencies, this development is surely also influenced by the fact that the absolute number of residents in Can Tho is increasing and that the general socio-economic development over the last years enables more households to accumulate the necessary resources to implement such adaptation measures. Figures. 10.12 and 10.13 indicate that the respective household members are by far the most important actors with respect to financing and implementing the adaptation measures.

#### 10.7.5.2 River Bank Erosion

Households located directly along rivers or canals in Can Tho are often affected by river bank erosion, eroding also the foundations of the house or the sediment layers at the river banks that stilts are build into. As a result, the affected households have

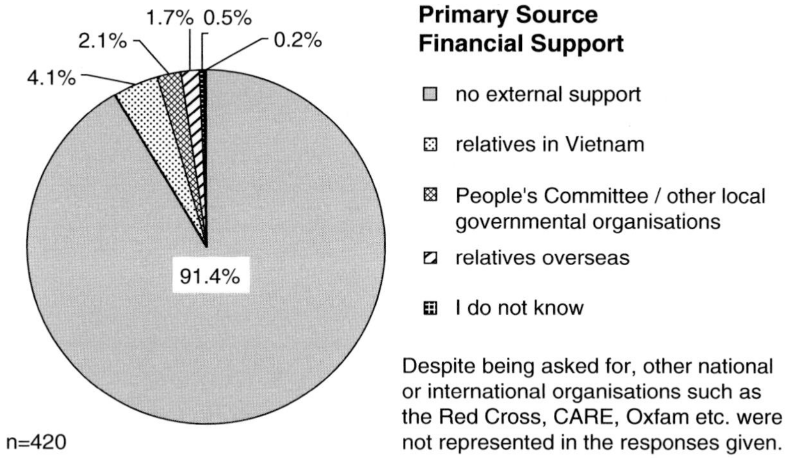


**Fig. 10.12** Financing informal adaptation (Source: Birkmann et al. 2010)

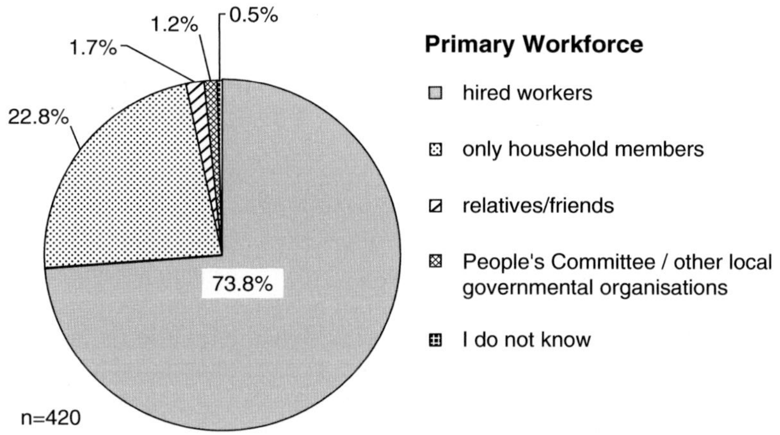


**Fig. 10.13** Implementing informal adaptation (Source: Birkmann et al. 2010)

been observed to be under constant pressure to maintain or rebuild parts of their house, meaning in particular erosion protection walls and/or stilts. Figure 10.14 illustrates some rebuilding efforts triggered by river bank erosion (here in combination with a vertical elevation due to rising water levels). Depending on the available resources of the households, the rebuilding is either done by hired workforce or mainly by household members with the support of relatives, neighbours and friends who share this type of labour force, support in a reciprocal manner. The building material used was reported to be mainly of low quality such as second-choice wooden poles or sand and gravel which is collected from construction sites (Fig. 10.14).

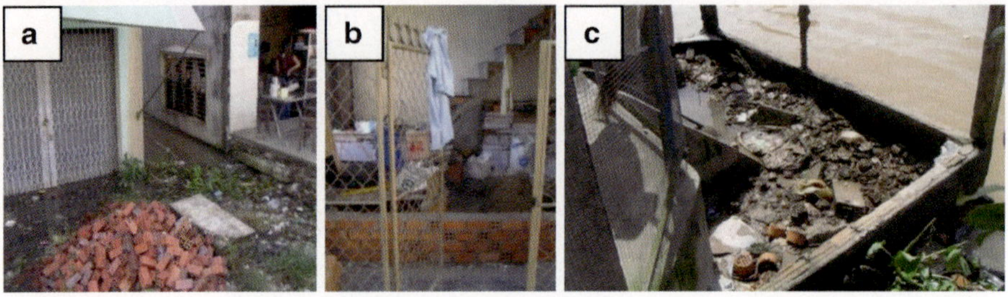


**Fig. 10.14** (**a**) Floor elevation, (**b**) floor elevation, and (**c**) river bank erosion protection and floor elevation (Source: own pictures Garschagen 2009)

### *10.7.6 Summary: Main Drivers of Vulnerability in the Urban and Peri-urban Cluster*

In summary, the vulnerability and adaptive capacity in the urban and peri-urban areas of Can Tho is shaped by a number of interacting factors. Poverty and a low socio-economic status often force households to live in locations with high hazard exposure. The poor quality of makeshift dwellings leads to a high physical susceptibility and limits the options for coping in flooding situations. On larger scale, the morphology of such makeshift settlements increases the risk of, for example, disease spreading during floods due to insufficient sanitation and drainage infrastructure. The capacity for long term adaptation – and hence vulnerability reduction – is limited not only through poverty but also through institutional factors. The majority of highly exposed households in makeshift settlements lacks official land title, making it hard to impossible to acquire bank loans for implementing structural improvements or fostering economic activities. Social networks within the community are hence important not only for coping with acute flooding but also for implementing provisional in situ adaptation. The main burden of such endeavors has to be borne by the affected population with rather limited support from governmental entities.

## 10.8 Tra Vinh: Vulnerability to Salinization in Coastal Areas

### *10.8.1 Role of the Study Area Within the Wider Vulnerability Assessment*

Salinity intrusion is a normal phenomenon in the coastal provinces of the Mekong Delta of Vietnam. In the dry season, seawater can penetrate into the Delta to a distance of about 40–60 km inland and about 2.1 million ha (about 55% of total Delta area) are affected by salinity (Miller 2003; Sam 2006). It causes many problems for agricultural development and people's livelihoods (Hashimoto 2001; Hossain et al. 2006; Hanh and Furukawa 2007). An analysis of rice productivity in the coastal area of the Mekong

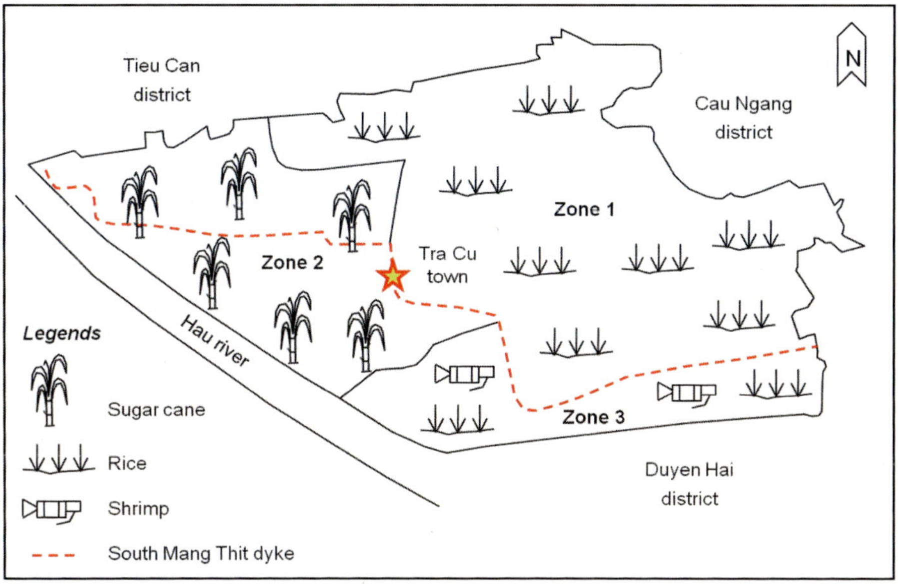


**Fig. 10.15** The Tra Cu map showing three different zones (Source: Mapping exercise with DARD staff in Tra Cu district in 2009)

Delta has shown that rice cropping intensities decreased with increasing salinity level in canal water (Kotera et al. 2008). Potentially, within the context of global climate change – especially sea level rise – the Mekong coastal region would be among the most affected regions globally in terms of impacts on population, biodiversity degradation, land loss and water-related hazards including flooding and storms (Dasgupta et al. 2007; Hanh and Furukawa 2007; Carew-Reid 2008). With around 65 km of coast line, Tra Vinh province would be affected the most by sea level rise. Carew-Reid (2008) projected that around 45.7% of the province area would be inundated by 1 m sea level rise (the third highest percentage per province in Vietnam).

Tra Cu district which covers an area of 370 $km^2$ in Tra Vinh province was selected for the research (Fig. 10.15), as: (1) the district has been affected by salinity problems and drought; (2) the district encompasses various socio-economic groups and different ethnicities (i.e. Vietnamese and Khmer ethnic, high poverty rate); (3) economic activities are diversified due to different ecological zones (i.e. freshwater zone for intensified rice farming; brackish water zone for aquaculture and brackish water zone for sugar-cane farming; see Table 10.3). Therefore, data and information was collected and compared in these three zones as well as between different socio-economic groups. The results from Tra Cu are not only useful for the district itself but also for other coastal regions with similar conditions. The household survey targeted 512 households and was conducted in May 2010. The interviewed households were equally distributed in three agro-ecological zones in Tra Cu District (Table 10.3). Within each zone, a stratified sampling was applied. In total, 46% of the respondents were Kinh and 54% Khmer.

**Table 10.3** Characteristics of three different zones in the study site

| | Zone 1 (rice zone) | Zone 2 (sugar-cane zone) | Zone 3 (aquaculture zone) |
|---|---|---|---|
| Irrigation system | Freshwater whole year<br>Good irrigation system due to SMTS (100% inside SMTS) | 1/3 freshwater whole year (inside SMTS) and 2/3 affected by brackish water in the dry season (outside SMTS) | Brackish water in the dry season (100% outside SMTS) |
| Main economic activities | 2–3 rice crops per year<br>Upland crops (maize, peanut, etc) | Sugar-cane<br>1–2 rice crops per year | Shrimp culture<br>1 rice crop in the wet season and 1 shrimp crop in the dry season |
| | Cattle, pigs, poultry | Upland crops (maize, peanut, etc)<br>Less animal husbandry | Less animal husbandry |
| Major hazards and problems | Freshwater scarcity in the dry season (especially from February to April) | Brackish water intrusion can destroy or reduce crop production | Shrimp diseases<br>Brackish water can destroy or reduce crop production |
| | Drought, especially along sand ridges | Affected by tidal influence (flooding) | Freshwater scarcity |
| | Brackish water leakage | Whirlwind | Affected by tidal influence (flooding) |
| | Whirlwind | Storm (seldom) | Whirlwind Storm (seldom) |

Source: Focus Group Discussion and Key-Informant interviews 2009

The case study area is situated down-stream at the Mekong River with much differences in term of socio-economic conditions from the other two case study areas in Dong Thap and Can Tho. Thus, combining the results from three study sites will provide a full picture of vulnerability in the whole Mekong delta.

## *10.8.2 Past Hazards*

The economic losses caused by the effects of salt water intrusion in 2005 was estimated at USD 45 million or 1.5% of annual rice production in the Mekong Delta (MARD 2005). Apart from economic losses, salinity intrusion also implies ecological degradations and health problems. It degrades water quality, damaged aquatic ecosystems and threatened bio-diversity in coastal areas (White 2002; Hanh and Furukawa 2007). As a result, livelihoods of downstream communities are extremely affected, particularly for poor people who rely mainly on natural resources (Hossain et al. 2006; Nhan et al. 2007). Salinity intrusion and drought in the dry season

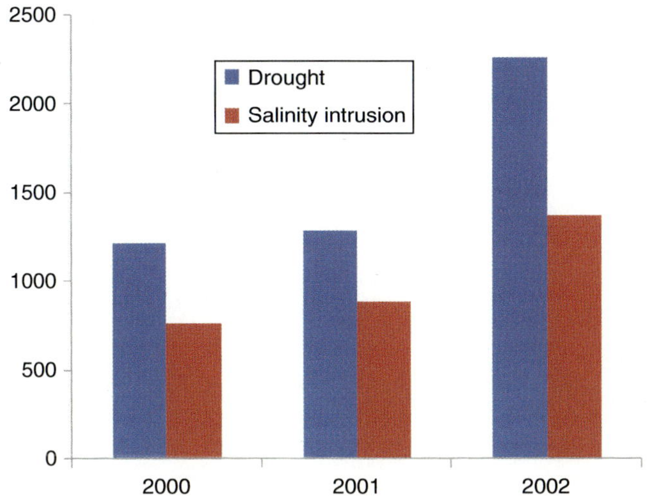


**Fig. 10.16** Affected upland crop areas by salinity intrusion and drought in Tra Cu district, Tra Vinh province (Source: Nguyen Thanh Binh, based on DARD-TC 2003)

are – according to the data available – increasing in the coastal region of the Mekong Delta. An example from Tra Cu district shows that upland crop areas affected by (i) saline water, increased from 763 to 1,375 ha between 2000 and 2002 and by (ii) drought, from 1,221 to 2,266 ha in the same period (Fig. 10.16). Given data restrictions with respect to salinity measurements, this analysis is based on a very short time frame; however, these numbers may function as an alarm signal for the likelihood of a longer trend.

Typhoons have rarely happened in the past but are expected to increase in intensity and frequency in the future. Since 1997, there have been two harmful typhoons in the Mekong Delta namely typhoon Linda in 1997 and typhoon Durian in 2006. In Tra Vinh province, losses were estimated at around 82.76 billion VND for typhoon Durian. In terms of housing, the typhoon caused 830 fully and 2,352 partly destroyed houses and in agricultural sector, it destroyed 3,033 ha of rice, 524 ha of fruit trees, 135 ha of sugar cane and 128 ha of upland crops in the province in 2006 (own expert interviews 2009).

### *10.8.3 Exposure to Natural Hazards*

Before 1995, the district was strongly affected by salinity intrusion which caused freshwater scarcity in the dry season (November to May) due to low flow from Hau River and tidal influence from the East Sea. Since 1995, a series of embankments and sluice gates (Fig. 10.17) have been built in the district to prevent sea water intrusion under the framework of the South Mang Thit Sub-project (SMTS) supported by the World Bank.

**Fig. 10.17** Sluice gate and dyke system under the framework of South Mang Thit sub-project (Source: own picture N.T. Binh 2010)

Based on hydrologic regimes, topographic conditions, and irrigation systems, Tra Cu is divided into three different zones (see Fig. 10.15 and Table 10.3). Each zone has particular activities and faces varying constraints which are summarized in Table 10.3.

In zone 1, brackish water intrusion has been controlled thanks to the SMTS which allowed for agricultural development as farmers can grow two or three rice crops per year. However, from February to April, low discharge from the Hau River and less local rainfall cause freshwater scarcity for crop irrigation and animal husbandry. Additionally, brackish water leakage during closed-gate period is also recorded. The household survey revealed that rice yields were on average 41% lower in the areas affected by such leakage.

In zone 2, sugar-cane farming is the main agricultural production type. This zone is strongly affected by brackish water intrusion and tidal influence, especially in areas of low elevation and located outside the SMTS (Fig. 10.18). Salinity intrusion occurs between December and May while tidal influence (causing flooding) becomes very important from November to January. During the last 15 years, the salinity level was highest in 1998 and lowest in 2000 (on expert interviews 2009). The salinity levels for these 2 years at Vam Buon station (around 30 km from the sea), for example, where 13.9 g/l in April 1998 and 4.4 g/l in March 2000 (Sam 2007). Both brackish water intrusion and tidal influence constrain the agricultural production. Salinity intrusion affects more than 500 ha of sugar-cane and 300 ha upland crops annually (own expert interviews 2009). 45% of the interviewed sugar cane farmers reported to have been affected in the past by salinity intrusion and tidal flooding which caused

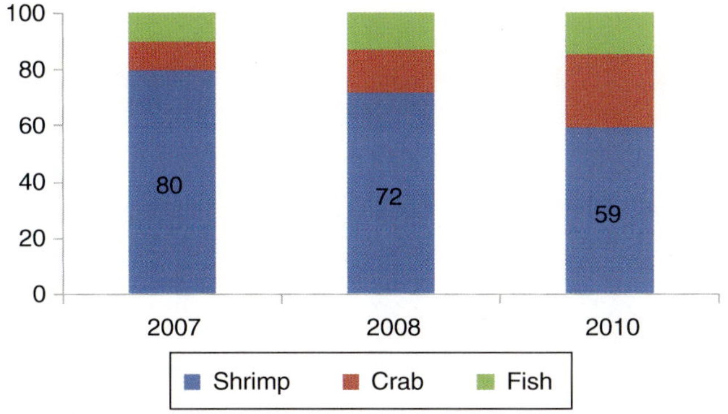


**Fig. 10.18** Shifts in aquaculture production in the coastal areas (outside of SMTS) of Tra Vinh (Source: Nguyen Thanh Binh, based on DARD-TV 2010)

not only a reduction in the amount of production but also in quality. Resulting from this are increased costs of production and decreasing prices for the produces.

In zone 3, salinity intrusion and tidal influence are greater than zone 2. The maximum salinity concentration was recorded at 25 g/l at La Bang station (around 15 km from the sea) in April 1998. The rice-shrimp integrated farming system is popular in this zone. Farmers grow rainfed rice in the wet season and cultivate tiger shrimp in the dry season. If the rain stops early and salinity levels become higher the rice production is at risk of being lost. Currently, salinity intrusion, tidal influence and freshwater scarcity in the dry season create severe problems for agricultural development in the study site.

Sea level rise is likely to severly affect the whole area, even inside the SMTS area due to its low topography. In combination with an increased likelihood of severe low flows in the Mekong River, this could imply that freshwater intake points for the SMTS will be increasingly influenced by brackish conditions threatening rice production and other crops within the original SMTS freshwater zone. It was identified that if the sea level increases 20–30 cm the SMTS dyke systems will be ineffective (Key-informant interview 2009). Additionaly, under climate change the occurrence of storms and whirlwinds (which are rare now) will increase in future. Moreover, upstream interventions (i.e. deforestation, irrigation and hydropower development) will change river flows and could increase the exposure of downstream communities to floods as well as low-flows.

### *10.8.4 Coping and Adaptation*

To cope with and adapt to salinity intrusion and related problems, the government and the local population have applied various strategies. Governmental responses include e.g. dyke constructions or the development of sluice-gates, while individual farmers

**Fig. 10.19** (**a**) Individual earth walls to cope with tide; (**b**) integrated rice-shrimp farming (Source: own picture N.T. Binh 2010)

often cope and adapt through crop calendar adjustments, crop or varieties changes, water storage and migration.

#### 10.8.4.1 Dyke Construction

Besides large projects such as the SMTS which was planned and built by the central government, many smaller projects have been implemented to prevent brackish water intrusion and tidal influence in Tra Cu. These are funded by the province and/or district. In high risk areas, farmers have also protected themselves by building small earth walls around their fields (Fig. 10.19a). These investments are very costly and suitable for the wealthier groups while the poor groups cannot afford to implement such measures. Generally, both types of dyke systems have shown many advantages. Referring to the effects of the SMTS project, 44% of the respondents have reported to benefit from increased crop intensification, 23% from agricultural diversification, 19% from increased freshwater availability for domestic usage, and 14% from other advantages. However, dyke systems like the SMTS project have also caused negative impacts including a reduction in natural fish resources and increasing water level outside the dyke areas. During the focus group discussions and KI expert interviews in 2009, local people reported that before the construction of sluice gates and dyke systems natural fish resources were abundant and contributed a significant part to the total household income, especially for landless and poor people. An estimation from Long Truong hamlet leader in zone 1 showed that in 1990, natural fish contributed around 15% of total household income; the contribution is now only 1% (own expert interviews 2009). When sluice gates are closed, the crops inside the dykes are protected from salinity intrusion but it causes serious inundation on the other side of the dykes. Particularly in 2005, the outside areas could not cope with the high water levels and the sluice gates had to be opened to avoid catastrophic failure.

#### 10.8.4.2 Crop Calendar Adjustments

Based on experiences and information gathered from the Department of Agriculture and Rural Development, farmers have adjusted their crop calendar due to salinization problems. For example, if the rain comes later they will seed later and vice versa. But this sometimes can put people at risk due to abnormal weather (i.e. shorter rainy period, earlier salinity intrusion). Thus, it is necessary to improve the weather forecast system by using both modern technologies and indigenous knowledge.

#### 10.8.4.3 Changes in Crops or Varieties

This option consists of choosing other crops than rice which need less water, such as maize and water-melon. In aquaculture areas (zone 3), before the 1990s, farmers grew only one traditional rice crop in the wet season but later on they introduced shrimp in the dry season. In recent years, shrimp farming faced diseases and environmental pollution. To cope with the situation, farmers culture crabs or other fish species instead of shrimp (Fig. 10.18). The integrated rice-shrimp farming system is a suitable system in coastal areas (Binh et al. 2009). Therefore, it is important to do more research with respect to these options in order to diversify agricultural activities and utilize land and water resources in saline affected areas.

#### 10.8.4.4 Water Storage and Groundwater Exploitation

In the wet season, farmers harvest and store rain water in jars or small tanks in order to use it in the dry season, mostly for drinking and cooking. For other types of household consumption, people use groundwater from individual drilled wells or rural tap water supply systems (these systems are newly developed). Along the sand ridge areas, groundwater is also exploited for watering upland crops. In the 1990s, many handle wells were drilled under a UNICEF (United Nations International Children's Emergency Fund) program. According to the Department of Natural Resource and Environment, there are more than 14,000 drilled wells in Tra Cu today. Currently, the use of groundwater is free of charge which could cause excessive exploitation in the near future (Chap. 7). Thus, research on groundwater markets is necessary for better management of this resource in the region.

#### 10.8.4.5 Migration

Before the SMTS was built, natural fish resources were considered as a source of income for local people, especially the poor. However, after the construction of the SMTS, natural fish stocks were reduced which affects mainly the poor people who have previously relied on open access, common pool, natural fish resources. Crop failures due to water related hazards have caused many difficulties to people's livelihoods. On the other hand, local industrial activities have not much developed but rural labour activities are increasing. Therefore, a number of young people have moved to cities (e.g. Ho Chi Minh City, industrial zones in South East Vietnam) to find new jobs since 1995. It is estimated that around 10% of total population have migrated out the district (own expert interview 2009). Most of them are poor and unskilled.

#### 10.8.4.6 Adaptation of Vulnerable Social Groups

The Khmer population in the region shows relatively high poverty rates when compared to other groups (TCSO 2009; Chap. 4). Household livelihood activities differ between wealth groups. The poor rely much on unskilled off-farm and/or non-farm wage labour due to their low education levels. With respect to the households interviewed in the survey, household income of Kinh populations was on average 1.3 times higher than that of Khmer households and the income of wealthier famers was 2.3 times higher than the income of households with a poverty certificate: this has detrimental implications in terms of adaptive capacities. The government has many policies to reduce poverty among such population (i.e. the 135 program for improving infrastructure and living conditions in disadvantaged communities, The 134 Program for supporting land, houses and tap-water in minority ethnic population, etc). However, they do not seem to be very effective and stable due to their single disciplinary and top-down approach that did not include to stimulate community participation and empowerment. People just above the poverty line can easily fall below that line if they suffer from shocks such as diseases or crop failure. It is necessary to develop more effective measures and increase investment for rural poor areas (i.e. extension, education, micro-credit, job creation, health care programs). The way to set up such programs should change from the current "top-down" approaches towards an inclusion of participatory and multi-disciplinary components in order to make them more useful and stable.

### *10.8.5 Summary: Main Drivers of Vulnerability in Tra Vinh*

Vulnerability to water related hazards in Tra Cu results from many different factors, not only in the terms of socio-economic factors but also due to close coupling processes within social-ecological systems (e.g. farming systems). A high poverty rate, especially amongst the Khmer population is considered as the most important constraint for enhancing adaptive capacity. The income and livelihoods of people are highly dependent on fluctuating price levels for agricultural products. In particular zones 2 and 3 are more susceptible to market integration as sugarcane and shrimp prices are unstable compared to rice prices. Additionally, the dyke systems cause socially differentiated effects as people inside the system benefit (mostly rice farmers) while people outside the system are affected by increased occurrence of flooding.

## 10.9 Comparison and Synthesis

The three case studies and the different profiles of vulnerability show that vulnerability is much more complex than the spatial exposure to hazards. The research underscores that particularly in the context of the Mekong Delta, exposure is only one of the three main dimensions constituting vulnerability besides resilience and sensitivity. In particular the examples of access to land, housing types and the number of floors as well as the different abilities to cope and adapt are key determinants of vulnerablity.

### *10.9.1 Hazard Specific Versus Hazard Non-specific Factors and Drivers of Vulnerability*

The comparison of vulnerabilities to the three different hazard profiles reveals not only the key drivers of vulnerability but also the fact that a differentiation can be made between vulnerability factors that are rather hazard independent (i.e. occur in all case study areas) and those that are hazard dependent (i.e. are only of importance in a specific hazard context). This finding leads to a number of implications for policy making and Integrated Water Resources Management. While the hazard independent vulnerability calls for general policies tackling the related problems in a broad scale (e.g. improving the institutional access to land and land titles), the hazard specific drivers imply the need for more specific and localised responses (e.g. supporting structural housing improvement of the most exposed people). Figure 10.20 provides an overview of both types of factors. For example, it is evident that particularly for marginalized populations exposed to floods in rural and urban areas, the access to land is a major difficulty, since it also means that in case of floods their access to compensation regimes for losses is limited, as they have no official land certificate.

### *10.9.2 Dynamics in Vulnerability to Natural Hazards in Vietnam*

Overall, the research in the three locations highlights that vulnerability is much more dynamic and socially differentiated even within micro-scale geographical areas than is often thought. The analysis and comparison of two rural case studies with the urban case study area have shown that vulnerability configurations at household level may be much more heterogeneous – even within a homogeneous region – than often assumed within the rather centralised top-down planning structure in Vietnam. Hence, disaster risk management as well as climate change adaptation projects have to acknowledge the need for small-scale assessment and for developing water management and adaptation strategies that accommodate differentiated

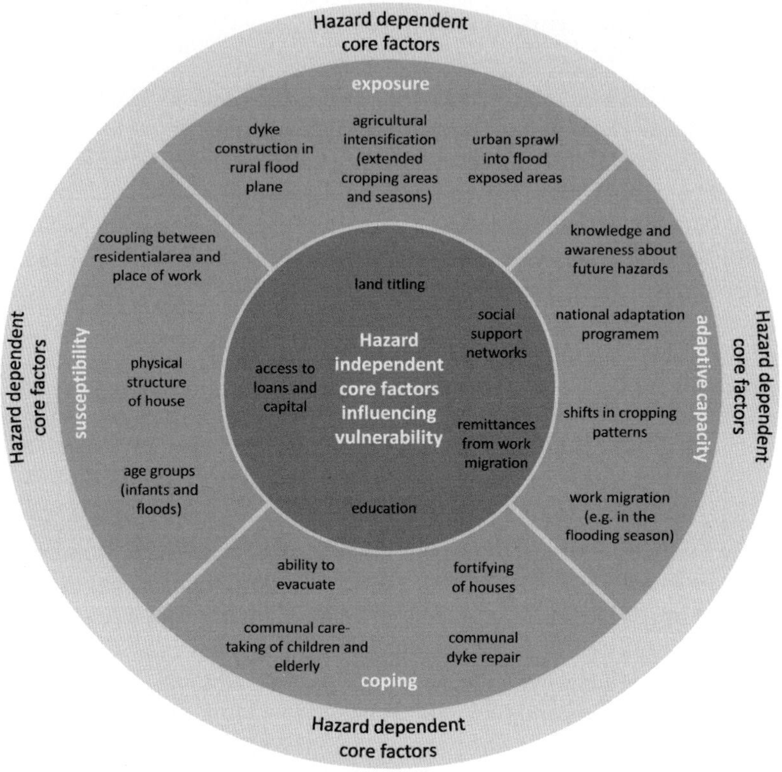


**Fig. 10.20** Selected factors and drivers of vulnerability differentiated into hazard dependent and hazard independent aspects (Source: Own figure)

solutions at small-scales while also providing efficient measures at larger scales. This requires a serious re-consideration of the set of adaptation and disaster mitigation measures that are predominantly discussed at the moment and usually apply a macro-focus (e.g. large-scale dyke systems for protecting certain districts or city quarters or large scale resettlement projects; for a review on Ho Chi Minh City and other cities, see Birkmann et al. 2010)).

### *10.9.3 Lessons for Adaptation to Extreme Events and Risk Reduction*

Additionally, the case studies show that adaptation options and measures in one region might impact strongly the exposure to hazards and coping capacity in other regions. The dyke systems in Dong Thap for example are constructed by water resources planners in order to reduce exposure but have been proven to deteriorate the livelihood base of specific population groups, in turn increasing their vulnerability towards future hazard events and might also increase the flood risk of downstream communities, such as the city of Can Tho.

### 10.9.4 Socio-Economic Tipping Points

The case studies also provide first evidence that different households and different socio-ecological systems have different potential tipping points, meaning that phases of partial collapse and crises can sometimes be observed at the individual household level. For example in years of increased salinity levels in Tra Vinh, income losses force the household to send members to urban areas for seasonal off-farm occupation. Although temporal migration could also be seen as a coping or adaptation strategy, it is evident that in some cases the regional livelihood strategies might reach tipping points that require renewal or might cause a collapse of the livelihood activities. Furthermore, the notion of thresholds and tipping points can also be transferred to adaptation processes. The gradual up-lift of houses or housing-floors observed in Can Tho may be working for a certain time – i.e. every time the rising water levels necessitate a new up-lift and/or the household has acquired sufficient resources for carrying this out. However, after a while the costs for the next up-lift will raise exponentially as also the roof, door, windows, pipe-works etc. would have to be adjusted, potentially pushing the costs beyond the feasible limits for the respective household.

### 10.9.5 Vulnerability and Development Pathways

Finally, all research findings suggest indirectly that vulnerability reduction in the Mekong Delta has to be considered as integral part of development processes. The vulnerability profiles underline that the determinants of vulnerability are shaped by both: (a) the changing hazard context that poses new challenges for people's coping and adaptive capacities as well as (b) the socio-economic changes and development pathways in different regions – rural or urban – that heavily modify people's susceptibility, exposure as well as coping and adaptive capacities. In particular, the influences of local decisions on transforming processes and structures need to be better understood at the micro level. The vulnerability profiles provide first insights into the variety of factors that need to be understood when aiming to develop sustainable and effective risk reduction and climate change adaptation strategies to hazards, such as floods or salinization. Solely on the basis of the detailed vulnerability profiles of different social groups policy makers can judge how potential interventions and risk reduction measures will influence these groups differently. The overall expectation that economic growth in general will lead to a vulnerability reduction per se must be highly questioned looking at the results of the vulnerability profiles examined.

## References


Adger N, Arnel NW, Tompkins EL (2005) Successful adaptation to climate change across scales. Glob Environ Chang 15:77–86

Be TT, Sinh BT, Miller F (2007) Literature analysis challenges to sustainable development in the Mekong Delta: regional and national policy issues and research needs. The Sustainable Mekong Research Network (Sumernet), Bangkok

Berkes F, Colding J, Folke C (2003) Navigating social-ecological systems: building resilience for complexity and change. Cambridge University Press, Cambridge

Binh NT, Tam NTT, Can ND (2009) Effective factors to change of farming systems in saline affected area of My Xuyen district, Soc Trang province. In: Du PV (ed) The first conference on development of sustainable rice-shrimp farming systems in coastal areas of the Mekong delta. Department of Crop Production, Ministry of Agriculture and Rural Development, pp 37–48

Birkmann J (2006a) Measuring vulnerability to natural hazards. Towards disaster resilient societies. United Nations University Press, Tokyo/New York/Paris

Birkmann J (2006b) Indicators and criteria for measuring vulnerability: theoretical bases and requirements. In: Birkmann J (ed) Measuring vulnerability to natural hazards – towards disaster resilient societies. United Nations University Press, Tokyo/New York/Paris, pp 55–77

Birkmann J (2011a) First- and second-order adaptation to natural hazards and extreme events in the context of climate change. Nat Hazards 58:811–840

Birkmann J (2011b) Regulation and coupling of society and nature in the context of natural hazards. In: Brauch HG, Oswald-Spring U, Mesjasz C, Grin J, Kameri-Mbote P, Chourou B, Birkmann J (eds) Coping with global environmental change, disasters and security. Springer, Berlin, pp 1103–1127

Birkmann J, Garschagen M, Kraas F, Quang N (2010) Adaptive urban governance: new challenges for the second generation of urban adaptation strategies to climate change. Sustain Sci 5(2):185–206

Birkmann J, Garschagen M, Fernando N, Tuan VV, Oliver-Smith A, Hettige S (2012) Dynamics of vulnerability – Relocation in the context of natural hazards and disasters. In: Birkmann J (eds) Measuring Vulnerability to Natural Hazards – Towards disaster resilient societies. UNU Press, Tokyo. (in press)

Bogardi J, Birkmann J (2004) Vulnerability assessment: the first step towards sustainable risk reduction. In: Malzahn D, Plapp T (eds) Disasters and society – from hazard assessment to risk reduction. Logos Verlag, Berlin, pp 75–82

Bohle H-G (2001) Vulnerability and criticality: perspectives from social geography. IHDP Updat 2(2001):1–7

Bryant J (1998) Communism, poverty, and demographic change in north Vietnam. Popul Dev Rev 24(2):235–69

Burton I, Kates RW, White GF (1993) The environment as hazard. Oxford University Press, New York

Cardona OD (1999) Diagnóstico local de riesgos naturales en Santa Fe de Bogotá para la planificación y medidas de mitigación. Panamericana-Secretaría de Salud, Bogotá

Cardona OD (2001) Estimación Holística del Riesgo Sísmico utilizando Sistemas Dinámicos Complejos. Doctoral dissertation, Technical University of Catalonia, Department of Terrain Engineering. http://www.desenredando.org/public/varios/2001/ehrisusd/index.html. Accessed 4 July 2010

Cardona OD, Barbat HA (2000) El Riesgo Sísmico y su Prevención, Cuaderno Técnico 5. Calidad Siderúrgica, Madrid

Carew-Reid J (2008) Rapid assessment of the extent and impact of sea level rise in Viet Nam, climate change discussion paper 1, ICEM – International Centre for Environmental Management, Brisbane, Australia

Carreño ML, Cardona OD, Barbat AH (2007) Urban seismic risk evaluation: a holistic approach. Nat Hazards 40(1):137–172

Casimir M (2008) The mutual dynamics of cultural and environmental change: an introductory essay. In: Casimir M (ed) Culture and the changing environment: uncertainty, cognition and risk management in cross-cultural perspective. Berghahn, Oxford, pp 1–58

CFSC [Committee for Flood and Storm Control, Socialist Republic of Vietnam] (2004) National report on disaster reduction in Vietnam. For the world conference on disaster reduction, Kobe-Hyogo, Japan, 18–22 Jan 2005. Hanoi

Chambers R (1989) Editorial introduction: vulnerability, coping and policy. IDS Bull 20(2):1–7

Chaudhry P, Ruysschaert G (2007) Climate change and human development in Viet Nam. Human development report 2007/2008. UNDP

DARD-TC [Department of Agriculture and Rural Development in Tra Cu] (2003) Report on upland crop survey. No 10/BC-NN, Tra Cu district, Tra Vinh province, Vietnam

DARD-TV [Department of Agriculture and Rural Development in Tra Vinh] (2010) Report on aquaculture results in coastal areas in 2010, Tra Vinh Province, Vietnam [Report in Vietnamese Language]

Dasgupta S, Laplante B, Meisner C, Wheeler D, Jianping Y (2007) The impact of sea level rise on developing countries: a comparative analysis. World Bank, Washington, DC

DFID (1999) Sustainable livelihood guidance sheets. http://www.nssd.net/pdf/sectiono.pdf. Accessed 19 July 2010

Dong Thap Statistical Office (1998–2008) Statistical yearbook of Dong Thap

DONRE Can Tho City [Department of Environment and Natural Resources of Can Tho City] (2009) Adapting Can Tho City to climate change impacts and the work of the committee for climate change adaptation.- Presentation given at the expert meeting on vulnerability, climate change risk and adaptation in Can Tho City, Can Tho City, 5–7 Oct 2009

Elsner J, Kossin J, Jagger T (2008) The increasing intensity of the strongest tropical cyclones. Nature 455:92–95

Folke C (2006) Resilience: the emergence of a perspective for social-ecological systems analyses. Glob Environ Chang 16(3):253–267

Garschagen M (2010a) Challenges and opportunities of climate change adaptation in high risk areas using the example of the Mekong Delta, Vietnam. In: Dölemeyer A, Zimmer J, Tetzlaff G (eds) Risk and planet earth. Vulnerability, natural hazards, integrated adaptation strategies. Schweizerbart, Stuttgart, pp 56–65

Garschagen M (2010b) Assessing present and future vulnerabilities to water-related hazards in the Mekong Delta's urban areas: the challenge to factor-in dynamic trends under conditions of transformation and climate change. Presentation at the third WISDOM scientific seminar, Bonn, 27–29 Jan 2010

Garschagen M (2010c) Potential humanitarian crises and climate change adaptation in the coupled social-ecological systems of the Mekong Delta, Vietnam. In: Shen X, Downing TE, Hamza M (eds) Tipping points in humanitarian crises: from hot spots to hot systems. United Nations University, Institute for Environment and Human Security, Bonn, pp 45–55 Source, 13:45–55

Garschagen M (2011) Resilience and organisational institutionalism from a cross-cultural perspective – an exploration based on urban climate change adaptation in Vietnam. Nat Hazards. doi:10.1007/s11069-011-9753-4

Garschagen M, Renaud F, Birkmann J (2011) Dynamic resilience of peri-urban agriculturalists in the Mekong Delta under pressures of climate change and socio-economic transformation. In: Steward M, Coclanis P (eds) Environmental change and agricultural sustainability in the Mekong Delta. Springer, Dordrecht

Garschagen M, Binh NT, Thach LN (2009) Vietnam: the challenge of integrating disaster risk reduction and climate change adaptation. In: Birkmann J, Tetzlaff G, Zentel K (eds) Addressing the challenge: recommendations and quality criteria for linking disaster risk reduction and adaptation to climate change, vol 38, DKKV publication series., pp 18–20

Handmer J (2003) Adaptive capacity: what does it mean in the context of natural hazards? In: Smith JB, Klein RJT, Huq S (eds) Climate change, adaptive capacity, and development. Imperial College Press, London, pp 51–70

Hanh PTT, Furukawa F (2007) Impacts of sea level rise on coastal zone of Vietnam. Bull Fac Sci Univ Ryukyus 84:45–59

Hashimoto TR (2001) Environmental issues and recent infrastructure development in the Mekong Delta: review, analysis and recommendations with particular reference to large-scale water control projects and the development of coastal areas. Australian Mekong Resource Centre, University of Sydney, Working paper No. 4

Hewitt K (1983) Interpretations of calamity from the viewpoint of human ecology. Allen & Unwin, London

Hoang DC, Tran VL (2006) Developing various climate change scenarios of 21 century for regions of Viet Nam. Sci Tech Hydro-Meteorol J 541

Hoi TN (2005) Dyke systems in the inundated areas in the Mekong Delta. Agricultural Publishing House, Ho Chi Minh City

Hossain M, Ut TT, Bose ML (2006) Livelihood systems and dynamics of poverty in a coastal province of Vietnam. In: Hoanh CT, Tuong TP, Gowing JW, Hardy B (eds) Environment and livelihoods in tropical coastal zones: managing agriculture-aquaculture-fishery conflicts. CABI, Wallingford, pp 30–47

IDEA [Inter-American Development Bank] (2005) Indicators of disaster risk and risk management – main technical report. English and Spanish edn, National University of Colombia/Manizales, Institute of Environmental Studies/IDEA, Inter-American Development Bank, Washington, DC. http://idea.unalmzl.edu.co. Accessed 4 July 2010

IPCC [Intergovernmental Panel on Climate Change] (2007) Appendix I: glossary. In: Parry M et al (eds) Contribution of working group II to the fourth assessment report of the intergovernmental panel on climate change. Cambridge University Press, Cambridge

IPCC [Intergovernmental Panel on Climate Change] (2011) Managing the risks of extreme events and disasters to advance climate change adaptation – a special report of Working Group I and Working Group II of the intergovernmental panel on climate change – summary for policy makers, Geneva. http://www.ipcc.ch/news_and_events/docs/ipcc34/SREX_FD_SPM_final.pdf. Accessed 28 Nov 2011

IPPC [Intergovernmental Panel on Climate Change] (2012) Special report on managing the risk of extreme events and disasters to advance climate change adaptation (SREX). Cambridge University Press, Cambridge

Käkönen M (2008) Mekong delta at the crossroads: more control or adaptation. AVISO 37(3): 205–212

Kotera A, Sakamoto T, Nguyen DK, Yokozawa M (2008) Regional consequences of seawater intrusion on rice productivity and land use in coastal area of the Mekong river delta. Jpn Agric Res Q (JARQ) 42(4):267–274

MARD [Ministry of Agriculture and Rural Development] (2005) Vietnam News Agency, 25 May 2005

Miller F (2003) Society-water relations in the Mekong Delta: a political ecology of risk. PhD dissertation. University of Sydney, Department of Geography

Minh LQ, Tuong TP, van Mensvoort MEF, Bouma J (1997) Contamination of surface water as affected by land use in acid sulphate soils in the Mekong River Delta, Vietnam. Agric Ecosyst Environ 61(1):19–27

MONRE [Ministry of Natural Resources and Environment of Vietnam] (2009) Climate change and sea level rise scenarios for Vietnam. Hanoi

Müller-Mahn D (2005) Von "Naturkatastrophen"zu "Complex Emergencies". In: Müller-Mahn D, Wardenga U (eds) Die Entwicklung integrativer Forschungsansätze im Dialog mit der Praxis. Integrative Forschungsansätze, Leipzig, pp 69–78

Nam N (1992) Culture in the fruit garden (in the Mekong Delta). Cultural Publishing House, Ho Chi Minh City

Nhan DK, Be NV, Trung NH (2007) Water use and competition in the Mekong Delta. In: Be TT, Sinh BT, Miller F (eds) Challenges to sustainable development in the Mekong delta: regional and national policy issues and research needs. Sumernet, Bangkok, pp 143–188

Oliver-Smith A, Hoffman SM (1999) The angry earth: disaster in anthropological perspective. Routledge, New York

Pelling M (2003) The vulnerability of cities – natural disasters and social resilience. Earthscan, London

Renn O (2008) Risk governance – coping with uncertainty in a complex world. Earthscan, London

Sam L (2006) Salinity intrusion in the Mekong delta. Agricultural publication house, Ho Chi Minh City

Sam L (2007) Research on salinity intrusion in the dry season for socio-economic development in coastal districts of Tra Vinh province. Department of Science and Technology, Tra Vinh

SRV [Socialist Republic of Vietnam] (2007) Decision to approve the National Strategy for Natural Disaster Prevention, Response and Mitigation to 2020. No: 172/2007/QĐ-TTg. Hanoi

SRV [Socialist Republic of Vietnam] (2008) Decision on the approval of the National Target Program to Respond to Climate Change, Decision No. 158/2008/QD-TTg. Hanoi

SRV [Socialist Republic of Vietnam] (2009) Ministry of Agriculture and Rural Development. Tasks and Authorities. Hanoi. http://xttmnew.agroviet.gov.vn/TestE/AboutMARD/mandate.asp. Accessed 20 May 2009

TCSO [Tra Cu Statistical Office] (2009) Statistical book 2005–2008. Tra Cu Statistical Office, Tra Vinh Province

Thywissen K (2006) Components of risk: a comparative glossary. UNU-EHS SOURCE, no. 2, Bonn

Tuan LA, Chinvanno S (2011) Climate change adaptation in the Mekong River Delta and key concerns on future climate threats. In: Stewart M, Coclanis P (eds) Environmental change and agricultural sustainability in the Mekong Delta, vol 45, Advances in global change research., pp 207–217

Turner BL II, Kasperson RE, Matson PA, McCarthy JJ, Corell RW, Christensen L, Eckley N, Kasperson JX, Luerse A, Martello ML, Polsky C, Pulsipher A, Schiller A (2003) A framework for vulnerability analysis in sustainability science. Proc Natl Acad Sci (PNAS) 100(14): 8074–8079

UNESCO (2009) IWRM Guidelines at River Basin Level Part 2.2 – The Guidelines for Flood Management. http://www.unesco.org/water/news/pdf/Part_2-2_Guidelines_for_Flood_Management.pdf. Accessed 13 July 2010

Walker B, Holling CS, Carpenter SR, Kinzig A (2004) Resilience, adaptability and transformability in social–ecological systems. Ecol Soc 9(2):5

Walker B, Gunderson L, Kinzig A, Folke C, Carpenter S, Schultz L (2006) A handful of heuristics for some propositions for understanding resilience in social-ecological systems. Ecol Soc 11(1):13

White I (2002) Water management in the Mekong delta: changes, conflicts and opportunities, vol 61, Technical documents in hydrology IHP-VI. UNESCO, Paris

Wisner B, Blaikie P, Cannon T, Davis I (2004) At Risk: natural hazards, people's vulnerability and disasters, 2nd edn. Routledge, London

# Part IV
# Water and Environment

# Chapter 11
# Biodiversity of the Mekong Delta

**Ian C. Campbell**

**Abstract** Although the Mekong Delta has been substantially modified by humans, primarily to support rice-based agriculture, it still supports a diverse biota, albeit a small fraction of what must have been there prior to human settlement. While there only two mammals of conservation significance, the Hairy-nosed Otter and the Dugong, remaining in the delta there are at least 37 species of birds of conservation significance and 470 species of fish have been recorded, of which 28 are endemic to the Mekong and 4 are known only from the delta. The delta also contains a number of distinct vegetation communities although most are now reduced to small remnants. The aquatic invertebrate fauna of the delta has been substantially impacted by humans, with the littoral fauna of the main channels particularly species poor, probably reflecting the intensity of human impact in those locations. National Parks and conservation reserves in the Delta are very small, but they play an extremely important role in maintaining biodiversity. Threats to the biodiversity of the delta include growing human populations and intensification of agriculture, with increasing use of fertilizers and pesticides and poldering altering flood water levels during the high flow season. Water quality is poor and will probably decline further, almost entirely due to human activities in Viet Nam rather than impacts from upstream countries. Fishing is intense in the Delta, as it is elsewhere in the Mekong Basin and the pressure will grow in future. Dams on the Mekong will alter flow patterns within the delta, increasing dry season flows and decreasing wet season flows, blocking fish passage and probably decreasing sediment delivery, which in turn may lead to shoreline erosion. Finally, the greatest long term threat to the Mekong Delta is undoubtedly climate change. An increase in sea level of 65 cm would lead to a loss of about 5,200 km$^2$ or 13% of land area within the Delta, with very serious consequences not only for the biodiversity of the Delta, but also for the economy of Viet Nam.

I.C. Campbell (✉)
School of Biological Sciences, Monash University and Principal River Scientist (GHD), Lonsdale Street, Melbourne 3000, Australia
e-mail: I.C.Campbell@bigpond.com

F.G. Renaud and C. Kuenzer (eds.), *The Mekong Delta System: Interdisciplinary Analyses of a River Delta*, Springer Environmental Science and Engineering,
DOI 10.1007/978-94-007-3962-8_11, 

## 11.1 Introduction

The Mekong Delta is relatively young in geological terms (Gupta 2009; Carling 2009a, b). The sediment has been described as having a Holocene carapace over deeper Quaternary sediments (Carling 2009a, b; Chap. 7). The upper reaches of the delta have been developed as a result of fluvial activity by the river, while further downstream the influence of marine factors, including changes in sea level, are far more evident.

As with any estuary, the Mekong delta forms an ecotone – a boundary between two or more ecosystems. In the Mekong Delta there are the boundaries between the river and the sea, the land and the water, and between the mountains and the coastal plain. Ecotones always support a wide variety of life forms because they support some organisms characteristic of each of the ecosystem types fringing them, as well as species which pass through migrating from one to another. Whenever wetlands are involved, long distance migratory birds are potential temporary residents, and the Mekong Delta is no exception.

The Mekong Delta includes three broad landforms (Buckton and Safford 2004). In the northern and central part of the delta is floodplain, which includes the area known as the "Plain of Reeds". This area was historically seasonally flooded by the waters of the Mekong and contained large areas of wetland forest and diverse grasslands. There is also a coastal system of sand ridges, mangroves and mudflats in the eastern and southern parts of the delta. Mangroves occurred primarily in the Ca Mau Peninsula and around the river mouths. Thirdly there is a low lying area in the interior of the Ca Mau Peninsula which derives freshwater largely from rainfall, and which contains peat swamps as well as scattered spectacular limestone outcrops of the Kien Giang karsts.

Prior to human settlement the Delta must have been a site of very substantial biological diversity. However, deltas are also attractive sites for human occupation – having fertile flat land with abundant fresh water which makes them suitable for agriculture, abundant fish which makes them more attractive as a place to subsist, while also being close to the sea so simplifying trade.

It is not surprising, therefore, that deltas have often been densely settled by humans. Many civilizations have initially developed around deltas – the Nile, the Tigris and Euphrates and the Indus and Ganges are three large deltaic regions which have supported large human populations (Chap. 2).

The consequence of dense human populations and intensive agriculture is always a drop in biodiversity. In this respect the Mekong is not different from many of the world's other major deltas.

Biodiversity can be characterised in a number of different ways. Many people use the term almost synonymously with charismatic megafauna. That is large, spectacular and (usually) rare animals. These include marine and terrestrial mammals and birds, which are important, often as flagship or icon species that can be used in biodiversity conservation campaigns because of their high level of public recognition. In order to conserve such species the ecosystems on which they depend must also be conserved, so many other species will be conserved incidentally. However the large charismatic species comprise only a small proportion of the species on earth.

Most species on earth are invertebrates. Wilson (1988) estimated that over 70% of all described species of organisms (including microorganisms, plants and animals) were arthropods – the group of invertebrates which includes insects, crustaceans and spiders and their relatives. The rate at which new arthropod species are described and named is also very high compared with most other groups of organisms, so that proportion has no doubt increased since Wilson's estimate was published.

Unfortunately, like most other places on earth, the invertebrate fauna of the Mekong Delta is still poorly known. It is the rarer and more charismatic larger species about which there is the most information. Plants also have been relatively poorly documented in any comprehensive sense in the Delta. This Chapter is a review of the accumulated information on the biota under a number of different groupings: mammals and birds, plants, fish and aquatic invertebrates, before discussing some of the factors that threaten the Delta's biodiversity.

## 11.2 Terrestrial Biota

### *11.2.1 Mammals*

In contrast to other parts of Vietnam, or the Mekong Basin, there are relatively few native vertebrate animals in the Mekong Delta other than the fish. That is basically a reflection of the great changes that have been wrought by several centuries of human occupation. Many of the vertebrates associated with deltas are aquatic or semi- aquatic, because the environment is usually wet, often dissected by numerous river channels that animals must be able to cross. So it is an ideal habitat for species which are strong swimmers, such as elephants and tigers, as well as the more aquatic species such as otters. However the large terrestrial species such as tigers and elephants that undoubtedly once lived in the Mekong Delta are now long gone – eliminated from the area by hunting and competition for space by humans.

Notable mammals include the rare Hairy-nosed Otter (*Lutra sumatrana*) the Small Clawed Otter (*Aonyx cinerea*) and the Dugong (*Dugong dugon*). Otters are small carnivorous mammals associated with freshwater environments in the Northern Hemisphere. The otters have always been rare in southern Vietnam, at least in historical times (Sterling et al. 2006), and the Hairy-nosed Otter was feared extinct in the Mekong Delta area until camera trapping confirmed that both the Hairy-nosed and Small Clawed Otter species were present in U Minh National Park in 2000.

Dugongs are large relatively sluggish marine end estuarine herbivores which feed on seagrass. The species extends from Africa, through India and Southeast Asia, the Malay Peninsula, Indonesia and Australia. Dugongs are now restricted in Vietnam to a small population in Con Dao National Park, which consists of islands off the Mekong Delta. While the habitat may be classified as marine, the freshwater plume from the Mekong extends several kilometres out to sea (Hortle 2009) and the islands certainly lie within it. As a consequence, changes to the flow regime of the Mekong River would certainly impact the habitat of the dugong.

### 11.2.2 Birds

Buckton and Safford (2004) provide a detailed review of the birds of the Mekong Delta and their status. They recorded 247 species of birds from the Mekong Delta in the period since 1988. There are earlier records, dating back to the 1800s but a sufficient number of the early records are doubtful either because the species are considered unlikely (Delacour 1970) or because the locality records are not sufficiently clear or specific to be useful (Buckton and Safford 2004).

Of the 247 species about half are considered wetland dependant and about a third are considered to be migrants using the delta area. Twenty bird species recorded from the Delta are listed as globally threatened or Near Threatened (Table 11.1) (Buckton and Safford 2004). Twenty one species of water birds occur in the Delta in internationally important numbers, where this is defined as likely to be 1% or more of the species numbers globally present in the Delta.

Species of particular conservation importance occurring within the Delta include the South-East Asian Sarus Crane (*Grus antigone sharpii*) of which an estimated 40% of the population spends part of the year within the Delta. About 3% of the global population of the Indian Cormorant (*Phalacrocorax fuscicollis*) occurs at a single site within the Delta, and the Little Cormorant (*Phalacrocorax niger*), Little Egret (*Egretta garzetta*), Cattle Egret (*Bubulcus ibis*), and Glossy Ibis (*Plegadis falcinellus*) are each represented by between 10% and 20% of their biogeographic populations.

A number of other species are said to have occurred within the Delta previously but are no longer present. These include the Great-billed Heron (*Ardea sumatrana*), Giant Ibis (*Pseudibis gigantea*), Milky Stork (*Mycteria cinerea*), White-winged Duck (*Cairina scutulata*) and Indian Skimmer (*Rhynchops albicollis*). Other species, such as the Greater Adjutant and the Black-necked Stork have been reported to have bred in the Delta in the past, but are now only rare occasional visitors.

Clearly the Delta constitutes a globally significant area for bird conservation. It not only supports a large number of bird species overall, but it supports a large proportion of the global populations of many species and provides a habitat for many species that are in real danger of global extinction.

### 11.2.3 Reptiles and Amphibians

The reptile and amphibian fauna of the Mekong Delta has not been thoroughly documented as have the birds and fish. There are accounts of the fauna of some localities, most notably Tram Chim National Park (Krohn 2009) and Lang Sam wetland reserve (Nguyen and Wyatt 2006). Krohn (2009) in a relatively superficial short term investigation recorded 19 species of snakes and lizards, one turtle and five amphibians from Tram Chim.

**Table 11.1** Species of birds recorded from the Mekong Delta since 1988 which are either of global conservation significance (threatened or near threatened) or for which the Delta is believed to host >1% of the global population

| English name | Scientific name | Significant population | Conservation status |
|---|---|---|---|
| Spot-billed Duck | *Anas poecilorhyncha* | Yes | |
| Sarus Crane | *Grus antigone* | Yes | Globally vulnerable |
| Eurasian Curlew | *Numenius arquata* | Yes | |
| Eastern Curlew | *Numenius madagascariensis* | No | Globally near threatened |
| Asian Dowitcher | *Limnodromus semipalmatus* | Yes | Globally near threatened |
| Kentish Plover | *Charadius alexandrinus* | Yes | |
| Greater Sand Plover | *Charadius leschenaultii* | Yes | |
| Malaysian Plover | *Charadrius peroni* | No | Globally near threatened |
| Oriental Pratincole | *Glareola maldivarum* | Yes | |
| Caspian Tern | *Sterna caspia* | Yes | |
| Little Cormorant | *Phalacrocorax niger* | Yes | |
| Indian Cormorant | *Phalacrocorax fuscicollis* | Yes | |
| Little Egret | *Egretta garzetta* | Yes | |
| Cattle Egret | *Bubulcus ibis* | Yes | |
| Chinese Egret | *Egretta eulophotes* | Yes | Globally vulnerable |
| Purple Heron | *Ardea purpurea* | Yes | |
| Great Egret | *Casmerodius alba* | Yes | |
| Chinese Pond Heron | *Ardeola bacchus* | Yes | |
| Javan Pond Heron | *Ardeola speciosa* | Yes | |
| Black Bittern | *Dupetor falvicollis* | Yes | |
| Glossy Ibis | *Plegadis falcinellus* | Yes | |
| Black-headed Ibis | *Threskionis melanocephalus* | Yes | Globally near threatened |
| White-shouldered Ibis | *Pseudibis davisoni* | No | Globally critical |
| Painted Stork | *Mycteria leucocephala* | Yes | Globally near threatened |
| Bengal florican | *Houbaropsis bengalensis* | No | Globally endangered |
| Black-tailed Godwit | *Limosa limosa* | Yes | |
| Spotted Greenshank | *Tringa guttifer* | No | Globally endangered |
| Spoon-billed Sandpiper | *Calidris pygmaeus* | No | Globally vulnerable |
| Grey-headed Fish Eagle | *Ichthyophaga ichthyaetus* | No | Globally near Threatened |
| Greater Spotted Eagle | *Aquila clanga* | No | Globally vulnerable |
| Oriental Darter | *Anhinga melanogaster* | No | Globally near threatened |
| Black-faced Spoonbill | *Platalea minor* | No | Globally endangered |
| Spot-billed Pelican | *Pelecanus philippensis* | No | Globally vulnerable |
| Black-necked Stork | *Ephippiorhynchus asiaticus* | No | Globally near threatened |
| Lesser Adjutant | *Leptoptilos javanicus* | No | Globally vulnerable |
| Greater Adjutant | *Leptopilos dubius* | No | Globally endangered |
| Asian Golden Weaver | *Ploceus hypoxanthus* | No | Globally near threatened |

Data from Buckton and Safford (2004)

### *11.2.4 Vegetation*

The major vegetation units of the Mekong delta have been recently reviewed by Rundel (2009). Remnant tree stumps indicate that extensive areas of the delta were once covered with *Melaleuca* forest, but relatively little of that original forest now remains (Rundel 2009; Tran Triet et al. 2000), and natural grasslands and sedgelands are reduced to small remnants. The major remaining semi-natural wetland areas of the Mekong Delta are the Plain of Reeds, the Ha Tien Plain and the U Minh wetlands (Sterling et al. 2006).

The freshwater swamp forests of *Melaleuca cajeputi* grow in seasonally inundated areas where the soil is permanently saturated and the salinity is low. The Melaleucas grow in relatively dense pure stands, so the vegetation diversity in these forests is low, and the most diverse plant communities in the delta are the sedgelands and grasslands, which may contain up to a 100 plant species.

Other notable plant communities within the delta are the peat swamps, most notable within the U Minh wetlands where the peat layers are 1–3 m deep, and the mangrove forests. The mangroves grow mainly in areas influenced by marine salinity, with bands of species which are ecologically differentiated by their tolerance to salinity and inundation. In the delta the white mangrove (*Avicennia alba*) dominates the stands closest to the sea, with red mangrove species (*Rhizophora* spp.) on the landward side and *Nypa* palms in brackish areas closest to fresh water (Sterling et al. 2006). However, the total area of mangroves has been reduced to 20% of the area of six decades ago (Thang Nam Do and Bennett 2008).

The Kien Karst outcrops in the Ca Mau Peninsula also support a number of plants endemic to the karstic environments. Several, including a newly described species of begonia (*Begonia bataiensis*) and a newly described orchid (*Calanthe kienluongensis*), are known from nowhere else (Lê Quang Khôi 2010)

## 11.3 Aquatic Biota

### *11.3.1 Fish*

The Mekong River basin is noted for its fish diversity. The actual number of fish species in the river may never be known, with some species potentially lost before they are even recorded. Determining the number of fish species in the basin is difficult because there are unsurveyed areas, there is material already collected still to be fully studied by taxonomists, and there are difficulties in deciding how to account for transient marine species in the lower river (Valbo-Jørgensen et al. 2009). Based on the Mekong Fish Database Valbo-Jørgensen et al. (2009), identified 924 named species as occurring in the Mekong Basin, of which 898 are indigenous.

Within the Mekong Delta the problem of transient species is most acute. The estuarine ichthyofauna includes a number of component fish categories based on occurrence within the delta:

- Freshwater species which may enter from the river upstream and whose major populations are in the river upstream. Vidthayanon (2008) identified over 160 species in this category from the delta, although many of the species for the freshwater Cambodian section of the Mekong could sometimes occasionally occur in the delta.
- Freshwater species which tolerate brackish water and may live within the estuary. In the Mekong Delta there are 89 species in this category.
- Estuarine species whose major populations are within estuaries although they may venture into the upstream riverine or the downstream marine environment. Within the Mekong Delta there are four known species in this group.
- Marine species which may occur more or less briefly in the estuary. Within the Mekong Delta there are 47 species in this category according to Valbo-Jørgensen et al. (2009), although Vidthayanon (2008) identified 80 species in the category.
- Migratory species which pass through the estuary on their way to breed in either the marine or freshwater environment. These include catadromous species which breed in the estuary or the sea and migrate into freshwater to live, which, in the Mekong includes the giant mottled eel (*Anguilla marmorata*) and the seas bass (*Lates calcarifer*). This group also includes anadromous fishes which breed in freshwater but then live most of their lives in the seas. In the Mekong this includes the pangassiid catfishes *Pangasius krempfi* and *P. mekongiensis.*
- Species which appear to live in water regardless of salinity and so occur in fresh, brackish and marine environments. In the Mekong there are 111 species known in this category.

Vidthayanon (2008) listed 460 fish species as recorded from the waters of the delta with over 360 species "commonly" occurring. The fauna is dominated by Cypriniformes (carps barbs and minnows) and Perciformes (gobies) which together comprise over 30% of the species.

Twenty-eight fish species endemic to the Mekong are found in the waters of the delta (Vidthayanon 2008). Four species occur only within the delta: dwarf catfish (*Akysis* filifer), Mekong sea catfish (*Hemipimelodus daugueti*) the Mekong goby (*Stenogobius mekongensis*) and the Mekong blind sole (*Typhlachirus elongates*). A further two species are restricted to the delta and the lower Mekong: Cambodian fighting fish (*Betta stiktos*) and grand lake sprat carp (*Thryssocypris tonlesapensis*).

The disparities in estimates of numbers of fish species and their categories reflect both the fish diversity and the relative lack of research effort. The large numbers of fish species mean that the effort required understanding and documenting their ecology and behaviour is substantial. Relative to the variety of fish, and the size and importance of the fishery fish, the research effort in the delta, as with elsewhere in the Mekong, is small. Consequently much of the information is anecdotal and scattered.

The Mekong River basin is recognised as having unusually high fish diversity (Valbo-Jørgensen et al. 2009) with only the Amazon and the Zaire rivers being comparable (Campbell 2009a; MRC 2003). Consequently it is not surprising that the fish diversity of the delta is also high, but data on diversity of comparable estuaries are generally lacking.

The delta supports an important fishery for both wild-caught fish and aquaculture (Hortle 2009). Both fisheries impact on the fish diversity of the delta, as will be discussed below. Nguyen Xuan Vinh and Wyatt (2006) suggest that five fish species have disappeared from the Plain of Reeds in recent times, although the only one that they fully identify is the Marbled Goby (*Oxyeleotris marmorata*). Vidthayanon (2008) considered the species common, so if the extinction in the Plain of Reeds is real it would appear to be local.

### *11.3.2 Aquatic Invertebrates*

The aquatic invertebrate fauna of the Mekong Delta is much less well known than the fish fauna. Pham Van Mien (2002) documented 102 zooplankton species in the Ca Mau Peninsula including 20 species of rotifers, 17 cladoceran species and 46 copepod species (Table 11.2). He also listed 55 species of molluscs and 26 species of non-planktonic (benthic and pelagic) crustacea (Table 11.3). The sources of the data are not cited, but were presumably Vietnamese Government agencies.

The Mekong River Commission operated a biomonitoring program from 2004 to 2008 which included six sites within the Vietnamese portion of the delta (MRC 2008a, 2009). The sampling included invertebrates from the edges (littoral) and from soft sediments as well as zooplankton, animals drifting in the water, and benthic diatoms, algae which grow attached to solid surfaces. Not all the sites included in the program were sampled in every year and sites in the delta were sampled in 2004, 2006 and 2008 (MRC 2010). The invertebrate data discussed here included that from 2004, when two sites in the Vietnamese section of the delta were sampled (MRC 2006), and 2006 when six Vietnamese delta sites were sampled including the two (Tan Chau and Chau Doc) that were sampled in 2004 (MRC 2008a).

From the zooplankton samples 36 species were identified in 2004 and 47 in 2006. In addition there were abundant unidentified larvae of copepod crustaceans and bivalve and gastropod molluscs on both sampling occasions. The zooplankton was dominated by rotifers in both sets of samples. Rotifers are characteristic of the zooplankton of large rivers (Hynes 1970), so the results suggest that the zooplankton at the sites sampled are predominantly riverine rather than marine in origin. The dominant taxa differed between the two sampling events, the rotifer genus *Polyarthra* was abundant on both occasions but in 2004 another rotifer genus, *Keratella* was abundant while in 2006 amoeboid protists from the genus *Difflugia* were the next most abundant group. This genus of protists was not recorded at all in 2004.

Littoral invertebrates, those living along the edges of the stream channels, have a very low diversity in the Mekong Delta. In 2004, 24 species were recorded, but in 2006

**Table 11.2** A list of zooplankton species recorded from the Ca Mau Peninsula

| | | |
|---|---|---|
| **ROTATORIA** | *Chydorus sphaericus* (O. F. Muller) | Oithonidae |
| Philodinidae | *Alonella excisa excisa* (Fischer) | *Oithona rigida* (Rosendorn) |
| *Rotaria neptunia* (Ehrenberg) | *Euryalona orientalis* (Daday) | *Oithona similis* (Claus) |
| *Rotaria rotaria* (Pallas) | *Dunhevedia crassa* (King) | *Oithona brevicornis* (Giesbrecht) |
| Notommatidae | *Leydigia acanthocercoides* (Fischer) | *Sapphirina nigromaculata* (Claus) |
| *Scaridium longicaudum* (Muller) | *Alona rectangula* (Sars) | *Limnoithona sinensis* (Burckhardt) |
| Asplanchnidae | **COPEPODA** | Corycaeidae |
| *Asplanchna sieboldi* (Leydig) | Eucalanidae | *Corycaeus speciosus* (Dana) |
| Lecanidae | *Eucalanus subcrassus* (Giesbrecht) | Cyclopidae |
| *Lecane (L.) leontina* (Turner) | *Eucalanus mucronatus* (Giesbrecht) | *Tropocyclops prasinus* (Fischer) |
| *Lecane (L.) luna* (Muller) | Paracalanidae | *Paracyclops fimbriatus* (Fischer) |
| Lecane (*L.) curvicornis* (Murray) | *Paracalanus parvus* (Claus) | *Ectocyclops phaleratus* (Koch) |
| *Lecane (Monostyla) bulla* (Gosse) | *Paracalanus aculeatus* (Giesbrecht) | *Mesocyclops leuckarti* (Claus) |
| Euchlanidae | *Calocalanus pavo* (Dana) | *Thermocyclops hyalinus* (Rehberg) |
| *Dipleuchlanis propatula* (Gosse) | *Acrocalanus gracilis* (Giesbrecht) | Ectinosomidae |
| Brachionidae | Euchaetidae | *Microsetella norvegica* (Boeck) |
| *Brachionus angularis* (Gosse) | *Euchaeta plana* (Mori) | *Microsetella rosea* (Dana) |
| *Brachionus calyciflorus* (Pallas) | Temoridae | Macrosetellidae |
| *Brachionus plicatilis* (Muller) | *Temora turbinata* (Dana) | *Macrosetella gracilis* (Dana) |
| *Brachionus urceus* (Linnaeus) | Scolecithricidae | Tachidiidae |
| *Brachionus quadridentatus* (Hermann) | *Scolecithrix danae* (Lubbock) | *Euterpina acutifrons* (Dana) |
| *Platyias quadricornis* (Ehrenberg) | Centropagidae | Canthocamptidae |
| *Platyias patulus* (Muller) | *Centropages furcatus* (Dana) | *Elaphoidella javaensis* (Chappuis) |
| *Keratella tropica* (Apstein) | *Centropages calaninus* (Dana) | **CNIDARIA** |
| Testudinellidae | Pseudodiaptomidae | Leptomedusae |
| *Pompholyx complanata* (Gosse) | *Pseudodiaptomus beieri* (Brehm) | *Obelia* sp. |
| *Trochosphaera solstitialis* (Thorpe) | *Pseudodiaptomus incisus* (Shen *et* Tai) | **EUPHAUSIACEA** |

(continued)

**Table 11.2** (continued)

| | | |
|---|---|---|
| Filiniidae | *Schmackeria dubia* (Kiefer) | Euphausidae |
| *Filinia longiseta* (Ehrenberg) | *Schmackeria speciosa* (Dang) | *Euphausia* sp. |
| *Tetramastix opoliensis* (Zacharias) | *Diaptomidae* | **MYSIDACEA** |
| **CLADOCERA** | *Vietodiaptomus hatinhensis* (Dang) | *Mesopodopsis slabberi* (P. J. Bened) |
| Polyhemidae | *Allodiaptomus mieni* (Dang) | **DECAPODA** |
| *Evadne tergestina* (Claus) | *Tropodiaptomus vicinus* (Kiefer) | Luciferidae |
| Sididae | *Eodiaptomus draconi-signivomi* (Brehm) | *Lucifer pennicillifer* (Hansen) |
| *Diaphanosoma excisum* (Sars) | *Neodiaptomus visnu* (Brehm) | *Sergestidae* |
| *Latonopsis australis* (Sars) | *Lucicutidae* | *Acetes chinensis* (Hansen) |
| Macrothricidae | *Lucicutia flavicornis* (Claus) | **CHAETOGNATHA** |
| *Macrothrix triserialis* (Brady) | Pontellidae | Sagittidae |
| *Ilyocryptus halyi* (Brady) | *Calanopia thompsoni* (A. Scott) | *Sagitta enflata* (Grassi) |
| Daphniidae | *Calanopia elliptica* (Dana) | **PROCHORDATA** |
| *Moina dubia* (de Guerne et Richard) | *Labidocera minuta* (Giesbrecht) | *Appendicularidae* |
| *Moinodaphnia macleayii* (King) | *Labidocera sinilobata* (Shen et Lee) | *Oikopleura longicauda* (Vogt) |
| *Simocephalus elizabethae* (King) | Acartiidae | *Oikopleura rufescens* Fol. |
| *Ceriodaphnia rigaudi* (Richard) | *Acartia clausi* (Giesbrecht) | **OSTRACODA** |
| *Ceriodaphnia laticaudata* (O. F. Muller) | *Acartia pacifica* (Steuer) | Cypridae |
| Chydoridae | *Acartia erythraea* (Giesbrecht) | *Cypris subglobosa* (Sowerby) |
| *Chydorus eurynotus* (Sars) | *Acartiella sinensis* (Shen et Lee) | *Stenocypris derupta* (Vavra) |
| | Tortaniadae | *Heterocypris anomala* (Klie) |
| | *Tortanus forcipatus* (Giesbrecht) | **PROTOZOA** |
| | *Chydorus sphaericus* (O. F. Muller) | Vorticellidae |
| | *Alonella excisa excisa* (Fischer) | *Vorticella* sp. |
| | *Euryalona orientalis* (Daday) | *Zoothamium arbuscula* (Ehrenberg) |

Unpublished data from Pham Van Mien (2002)

**Table 11.3** A list of aquatic invertebrates recorded from the Mekong Delta

| | | |
|---|---|---|
| **CRUSTACEA – COPEPODA** | *Macrobrachium lanchesteri* (De Man, 1911) | Lymnaeidae |
| Diaptomidae | *Macrobrachium mekongense* (Dang, 1998) | *Lymnaea swinhoei* (H. A. Adams, 1866) |
| *Allodiaptomus mieni* (Dang) | *Macrobrachium idae* (Heller, 1862) | *Lymnaea viridis* (Quoy et Galmard, 1832) |
| *Allodiaptomus raoi* (Kiefer, 1936) | *Macrobrachium mammillo-dactylus* (Thalwitz, 1892) | Planorbidae |
| *Tropodiaptomus oryzanus* (Kiefer, 1937) | *Machrachium sintangense* (De Man, 1892) | *Gyraulus convexiusculus* (Hutton, 1849) |
| *Tropodiaptomus vicinus* (Brehm, 1955) | Atyidae | *Hippeutis (Helicorbis) umbillicalis* (Benson, 1836) |
| *Viettodiaptomus hatinhensis* (Dang, 1977) | *Caridina weberi sumatrensis* (De man, 1892) | |
| *Viettodiaptomus tridentatus* (Dang et Ho, 1985) | *Caridina nilotica bengalensis* (De man, 1908) | *Segmentina (Polypilis) hemisphaerula* (Benson, 1842) |
| *Neodiaptomus botulifer* (Kiefer, 1974) | **MOLLUSCA** | |
| *Neodiaptomus malaindosinensis* (Lai and Fernando, 1978) | **GASTROPODA – PROSOBRANCHIA** | **MOLLUSCA – BIVALVIA** |
| *Eodiaptomus draconisignivomi* (Brehm, 1952) | Thiaridae | Arcidae |
| Pseudodiaptomidae | *Stenomalania reevei* (Brot) | Scaphula pinna (Benson, 1856) |
| *Pseudodiaptomus beieri* (Brehm, 1951) | *Thiara scabra* (Muller) | Mytilidae |
| *Pseudodiaptomus incisus* (Shen and Lee, 1963) | *Melanoides tuberculatus* (Muller) | *Limnoperna siamensis* (Morelet, 1875) |
| *Schmackeria bulbosa* (Shen and Tai, 1964) | *Sermyla tornatella* (Lea) | *Brachidontes arcuatulus* (Hanley, 1844) |
| *Schmackeria speciosa* (Dang) | *Tarebia granifera* (Lamarck) | Amblemidae |
| Schmackeria dubia Kiefer | Pilidae | *Pilsbryochoncha exilis exilis* (Lea, 1839) |
| **CRUSTACEA – DECAPODA** | *Pila conica* (Gray, 1856) | *Pseudodon inoscularis callifer* (Martens, 1860) |
| **Brachyura** | *Pila polita* (Decrayes, 1830) | *Pseudodon vondembus chianus ellipticus* (Conrad, 1865) |
| Parathelphusidae | *Pila ampullacea* (Linnaeus, 1758) | *Hyriopsis (Hyriopsis) bialatus* (Simpson, 1900) |
| *Siamthelphusa beauvoisi* (Rathbun, 1902) | Bithyniidae | *Chamberlainia hainesiana* (Lea, 1856) |
| *Somanniathelphusa dugasti* (Rathbun, 1902) | *Bithynia (Digoniostoma) siamensis siamensis* (Lea, 1858) | *Cristata bellua* (Monelet, 1866) |

(continued)

**Table 11.3** (continued)

| | | |
|---|---|---|
| *Somanniathelphusa sinensis* (H. Milne Edwards, 1853) | *Bithynia (Digoniostoma) siamensis goniomphalus* (Morelet, 1866) | |
| *Somanniathelphusa dangi* (Durren et Ng.) | *Wattebledia crosseana* (Wattebled, 1886) | *Ensidens ingallsianus ingallsianus* (Lea, 1852) |
| *Somanniathelphusa germaini* (Rathbun, 1902) | Stenothyridae | *Uniandra contradens tumidula* (Lea, 1856) |
| Potamidae | *Stenothyra monilifera* (Benson) | *Uniandra contradens crossei* (Deshayes, 1876) |
| *Ranguna cochinchinensisis* (De Man, 1898) | *Stenothyra glabrata* (A. Adams, 1851) | *Physunio superbus* (Lea, 1843) |
| | Iravadiidae | *Trapezoideus exolescens exolescens* (Gould, 1843) |
| **DECAPODA – MACRURA** | *Fairbankia cochinchinensis* (Bavay et Dautzenberg, 1912) | Dressenidae |
| Palaemonidae | Assimineidae | *Sinomytilus harmandi* (Rochebrune, 1881) |
| *Leandrites indicus* (Holthuis, 1950) | *Cyclotropis carinata* (Lea, 1836) | Corbiculidae |
| *Leptocarpus potamiscus* (Kemp, 1917) | Viviparidae | *Polymesoda (Geloina) proxima* (Prime, 1864) |
| *Exopalaemon vietnamicus* (Nguyeãn Vaên Xuaân, 1992) | *Mekongia rattei* (Crosse et Fischer, 1876) | *Corbicula blandiana* (Prime, 1864) |
| *Palaemon curvirostris* (Nguyeãn Vaên Xuaân, 1992) | *Filopalludina (Filopalludina) filosa* (Reeve, 1863) | *Corbicula bocourti* (Morelet, 1863) |
| *Palaemon concinnus* (Dana, 1852) | *Filopaludina (Filopaludina) doliaris* (Gould, 1844) | *Cobicula castanea* (Morelet, 1865) |
| *Palaemon serrifer* (Stimpson, 1860) | Potamididae | *Corbicula cyreniformis* (Prime, 1862) |
| *Palaemon pacificus* (Stimpson, 1860) | *Cerithidea (Cerithideopsilla) cingulata* (Gmelin, 1790) | *Corbicula tenuis* (Clessin, 1867) |
| *Palaemon semmelincki* (De Man, 1879) | *Cerithidea (Cerithidea) charbonnieri* (Petit, 1851) | *Corbicula moreletiana* (Prime, 1867) |
| *Macrobrachium rosenbergii* (De man, 1879) | *Terebralia sulcata* (Born, 1778) | *Corbicula baudoni* (Morlet, 1886) |
| *Macrobrachium mirabile* (Kemp, 1917) | Buccinidae | *Corbicula leviuscula* (Prime, 1864) |
| *Macrobrachium pilimanus* (De Man, 1879) | *Clea (Anenton) helena* (Philippi, 1847) | *Corbicula solidula* (Prime, 1861) |
| *Macrobrachium equidens* (Dana, 1852) | *Macrobrachium lanchesteri* (De Man, 1911) | Pisidiidae |
| | | *Pisidium (Afropisidium) clarkeanum* (G. et H. Neville, 1871) |

Unpublished data from Pham Van Mien (2002)

only 12 species were recorded at a larger number of sites. In 2006 only a freshwater shrimp, *Caridina* sp. and an air-breathing aquatic bug, *Micronecta*, from the family Corixidae, were reasonably abundant at any sites. All other species collected were represented by only a few individuals. In 2004, those two taxa were also abundant, but bivalve molluscs (small mussels), snails, chironomid midge larvae and oligochaet and polychaet worms were also abundant.

The littoral environment of the Mekong Delta appears to be in particularly poor condition, at least along the larger channels. It is heavily utilized often being overhung by houses and other buildings near settlements, with waste water directly discharged into it. In other areas it is impacted by jetties and other facilities for boats, aquaculture cages and engineering stabilization works. Erosion is active in a number of areas, as it usually is in deltas. Finally the physical habitat is impacted by boat wash, since the waters of the delta are intensively used for navigation by boats of all sizes.

With a poor condition habitat it is not surprising that the invertebrate diversity is low. Regardless of water quality few invertebrates can persist in such environments. Littoral invertebrates of more natural systems are generally adapted to living in macrophytes, the large plants that fringe these systems, but along the main channels of the delta few macrophytes persist.

Benthic invertebrates are those living within the mud, sand and other soft sediments in the river channel. They are sampled using a "grab" on a long rope which takes a sample of mud. The mud is sieved and the animals present are picked out and identified. Soft muddy habitats are abundant in deltas such as that of the Mekong, and the habitat is usually subject to less physical disturbance by human activity than the environments along the edges of the channel, unless there has been channel dredging, so the fauna would be expected to be relatively rich in species and abundant.

Within the Mekong Delta the benthic fauna is reasonably rich compared with the littoral fauna. In 2004, 20 species were collected (MRC 2006) and in 2006, 36 species (MRC 2008a). However, in 2004 only 2 species were collected at Chau Doc, neither abundant. This apparently reflected a difficulty in finding a soft enough habitat for the grab to sample rather than an indication of environmental stress. Sixteen species were collected at the same site in 2006.

Bivalves, particularly the genus *Corbicula*, were conspicuous members if the benthic fauna. The genus occurs in both fresh and brackish water, so is well suited to estuaries. It occurred in huge numbers at Chau Doc in 2004. Oligochaet worms were also abundant on both sampling occasions – notably species of *Limnodrilus* and *Branchyura* which are very tolerant of low oxygen. Insects, which tend to be characteristic of freshwater rather than brackish or saline water, were abundant at Tan Chau but not Chau Doc in 2004, but only a single species of caddisfly was collected at either site in 2006.

The MRC bioassessment data demonstrate that the sites in the lower delta are substantially impacted by human activity (Campbell et al. 2009). That is especially true of the biota in the littoral or edge habitats. The benthic, or bottom dwelling biota is in slightly better condition, but still characteristic of a fauna subject to low oxygen conditions which is characteristic of systems suffering from nutrient enrichment.

### *11.3.3 Aquatic Plants*

Far less has been documented about the floral biodiversity of Vietnam than about the faunal diversity. The MRC bioassessment project has documented diatoms, single celled algae which grow readily on solid surfaces, throughout the Mekong (MRC 2006, 2008a, 2009). The number of diatom taxa encountered in the Delta was generally low – 41 species in 2004 and only 32 in 2006. The genera *Navicula, Nitzschia* and *Luticula* were abundant at most sites on most occasions. The low number of species could be attributed both to the relative scarcity of solid surfaces which could be colonized and sampled, and the generally poor water quality at the sites.

## 11.4 Threats to Biodiversity

### *11.4.1 Population and Population Pressure*

Population growth is a fundamental driver of much environmental degradation. The Mekong Delta, like many other large river deltas, has conditions which make it a productive agricultural area. It has abundant freshwater, flat land which is easy to work, rich soils washed down by the river, and, as a tropical system warm weather and abundant sunlight. Consequently it is capable of supporting, and does support, a high human population density which has grown rapidly over the past 200 years, and continues to grow. Population densities within the Delta exceed 350 people per $km^2$ over much of the area (Hook et al. 2003, Chap. 4). As population grows, the demands of the population on the environment, and the stresses created increase. More land is cleared for farms, existing agriculture intensifies, more roads, houses, offices and factories are built.

### *11.4.2 Agriculture*

Agriculture within the Mekong Delta is more intense than elsewhere in the Mekong basin, or elsewhere in Vietnam. The delta produces about 50% percent of Vietnam's annual food requirements – rice, fish, fruit and vegetables, so it performs a vital role in feeding the country, but also, because Vietnam is a rice exporter, the Delta is also an important source of foreign income.

The clearing of forest and drainage of swamps and wetlands to permit agriculture has been historically one of the major contributors to the evident loss of biodiversity within the Delta. Thang Nam Do and Bennett (2008) reported that only 20% of the mangroves present in 1943 now remain. Many areas unsuitable for rice or crop production have been cleared to construct aquaculture ponds, mainly for shrimp. The rate of clearance has now dropped, but this is mainly because there are fewer uncleared areas remaining.

With the intense levels of agriculture there is also higher use of fertilisers and pesticides than elsewhere in the region. And consequently nutrient and pesticide runoff is an issue that impacts aquatic biodiversity in the Delta (Chap. 13).

One of the great changes that has occurred, and continues to occur in the Mekong Delta has been a huge increase in the area of poldering (Chap. 3). Poldering involves the construction of levee banks around areas containing rice paddies in order to exclude the flood waters. Instead of rice being planted at the commencement of the natural flood and growing as the flood waters naturally rise, the level at which the water rises is controlled by allowing flood water to enter through water control structures. The advantage for the farmers is that instead of rice being grown in seed beds and the seedlings transplanted, the seed can be directly planted in the fields. With the natural flood, rice had to be planted as seedlings because the taller seedlings were less liable to be "drowned" through a sudden rise in water level. But direct planting of seed is far less labour intensive, so is preferred by farmers. In addition, where water levels are closely controlled rice cropping has increased from one crop a year to two or even three crops.

The development of large scale polders has had a dramatic effect on the water flows within the delta. Essentially the floodplain now contained behind the levee is no longer connected to the river as it once was and is no longer available to store flood waters. Consequently the flood heights in non-poldered areas are substantially higher, because they now have to store greater volumes of water than previously. So those areas are now being inundated for longer periods and to greater depths than previously, disrupting their ecology, while the natural ecosystems within the levee are also completely modified.

### *11.4.3 Water Quality*

The water quality of the Mekong River delta suffers from two sets of circumstances. As with any estuary it suffers the consequences of any factors which impinge on water quality from upstream within the Mekong basin. Secondly, because the delta is a centre of intensive agriculture and population growth, it also suffers from local impacts. The waterways of the delta have the poorest quality water of any locations sampled by the Mekong River Commission within the Mekong basin (Campbell 2007; MRC 2008b). Nutrient levels are high and are increasing, presumably through runoff of fertilizers used in agriculture, as well as the food and waste materials derived from instream aquaculture activities and sewage and urban wastewater (Chaps. 13 and 14).

Investigations conducted by the Mekong River Commission failed to detect pesticides or other persistent organic pollutants in water and sediment of the delta (MRC 2007), however several more recent studies have detected these compounds. Minh et al. (2007) recorded high levels of DDT and PCBs particularly in sediment near urban areas. Furthermore their data indicated that the input of DDT to the system was recent. Carvalho et al. (2008) also detected relatively high concentrations of

DDT with an average of 6.3 ng/g dry weight in sediments and 38.6 ng/g in soft tissues of molluscs collected from the delta. They also detected PCBs, endosulphan and chlordane. Finally, Pham Van Toan et al. (2010) detected 13 of 15 target pesticides they analysed from within water samples collected from waterways within the delta. They found that the fungicide isoprothiolane and the insecticide buprofezin were the compounds present in the highest concentrations (up to 20.8 and 16.5 μg/L respectively), and in 70% of their 434 samples there were 4 or more pesticide residues detected.

Water quality may impact aquatic animals directly, but also has potential indirect effects. As nutrient levels rise, so do the risks of toxic algal blooms which could trigger fish kills. Slow water movement in the dry season makes this the time of year when risks are greatest.

Poor water quality in the Mekong Delta is almost entirely a consequence of human activities within Viet Nam. The contribution of pollutants from Cambodia is trivial. Poor water quality has certainly impacted on aquatic biodiversity in the Delta already, and unless Vietnam begins to seriously address the sources of contamination, the problems will become much more serious, and impact on industries like the export aquaculture industry as well as the biodiversity.

### *11.4.4 Fishing*

The Mekong River, including the Delta is intensively fished (Hortle 2009) using a wide variety of fishing techniques (e.g. see Deap Loeung et al. 2003). While there is no evidence of any decline in the overall fish catch as yet, there has clearly been a decline in large fish species (Mattson et al. 2002). Large fish species are more susceptible to fishing pressure because of their longer lives and the length of time they take to reach reproductive size. In addition many of the methods used to catch smaller species will also catch large species. Fishers catching large species as "bycatch" will usually retain the large specimens as a bonus. Thus, although they are now too rare to be specifically targeted by fishers, the "subsidy" provided by the more numerous smaller species allows fishers to also catch large species. There is an absence of data to document whether fish species have been lost to the Delta, and if so whether the cause has been fishing or other factors such as changing habitats, but there is no doubt that fishing has caused at least a reduction in abundance of some species.

### *11.4.5 Exotic Invasive Species*

Invasive species are a threat to biodiversity globally. Within the Mekong Delta there are a number of exotic invasive species, but the species that has attracted the most attention, and is of greatest concern is the giant mimosa plant, *Mimosa pigra*. This invasive bush, which grows rapidly in tropical wetlands, is extremely difficult to eradicate once it has become established. Tran Triet et al. (undated) reported that manual control methods were effective in U Minh Thuong where the invasion was detected early on, but the species has become a major problem in Tram Chim national park. The species has proven a problem in other tropical wetland systems around the world, including areas such as Kakadu in northern Australia (Skeat et al. 1987).

### *11.4.6 Mekong Dams*

In recent years there has been a resurgence in water resource development in the Mekong Basin, with a number of newly constructed dams in the Chinese section of the river and a number of others about to be constructed, or planned or under consideration in Laos and Cambodia (Campbell 2009a, c, 2011). This development constitutes the greatest short term threat to biodiversity in the Mekong Delta.

The developments will impact the Delta in three ways. They will alter the hydrology of the river, they will block fish migration pathways, and they will trap nutrients and sediment.

The intent of the dams is to bring about alterations in the hydrology of the river. While increases in irrigated areas are often discussed together with hydropower, the primary purpose or all the dams proposed or under construction is hydropower production. Consequently the major impact on hydrology will not be an overall reduction of streamflow, but rather an alteration in flow with a (relatively small) decrease in wet season flows and a fairly large relative increase in dry season flows (Campbell 2009b; Adamson et al. 2009).

The impacts of the hydrological changes in the Delta will be more water flushing through the system in the dry season and slightly reduced average flood heights in the wet. This may reduce saline encroachment up the channels in the dry season, which has become more severe as more and more freshwater is extracted from the channels for multicropping of rice by Vietnamese farmers. However, more availability of freshwater, unless there is an increase in resource management effectiveness, may simply lead to an increase in the area of irrigated agriculture, putting more pressure on biodiversity with little or no dry season water quality benefits.

Blockage to fish migration pathways will lead to severe impacts on many fish species. Many Mekong River fish species are known to migrate over long distances (Hortle 2009), and as previously noted a number of migratory species travel to, or through, the Delta. Species which have their migrations blocked will potentially undergo major declines, with the size of the decline for each species depending on the proportion of the population which is not longer able to reach its breeding site. There are numerous examples from around the world of the impacts of dams on migratory fish species (e.g. see Petts 1984), including some spectacular examples of crashes in migratory fish populations following dam construction, as happened, for example, for salmon in the Columbia River in the US (Robinson 1978).

The trapping of nutrient and sediment by dams can also cause dramatic changes to the downstream reaches of rivers. A reduction of sediment moving down the river will trigger two sets of changes. Within the channel there will be increased erosion, however that tends to occur for only distances from kilometres to tens of kilometres below the dam (Petts 1984), so is not likely to be an issue within the Mekong Delta. Of more concern is that a reduced sediment load passing down the river may lead to erosion at the seaward edge of the Delta, causing a loss of arable land. That phenomenon occurred in the Nile Delta following the closure of Aswan Dam (Kassas 1973).

Trapping of nutrients by dams has risks both for agriculture and for fisheries within the Delta. The risk to agriculture arises from the loss of the nutrient replenishment to the soil which normally arises from the nutrients dispersed across the floodplain by the annual flood. As a consequence farmers will increasingly have to purchase more fertilizers to maintain productivity. However the consequences may well be even greater for the coastal fishery. In the Nile the coastal sardine fishery crashed from 18,000 tonnes per year to virtually nothing following the closure of Aswan dam (George 1973). This amounted to a loss of nearly 60% of the total Egyptian Mediterranean fisheries catch. The fishery in the impoundment amounted to less than 10,000 tonnes per year during the post-filling fish "boom" – so was nowhere near sufficient to compensate for the loss.

In the case of the Mekong Delta, the dam impoundments which may support new fisheries will occur in Laos, Thailand and Cambodia, not in Viet Nam. Consequently even these fisheries, which will be far smaller than the coastal fishery that will be lost, will not be available to fishers from Viet Nam.

Overall the consequences of Mekong Dams, and especially dams in the lower reaches of the river in Cambodia, are likely to be a loss of some fish species in the delta, and changes in abundance in many other species, with many being reduced and probably a few species increasing in abundance as a result of reduced competition.

### *11.4.7 Climate Change*

The greatest potential impact of climate change on the Mekong Delta will occur through sea level rise. There have been a number of investigations attempting to predict the likely consequences of sea level rises for human populations within the delta (e.g. Wassman et al. 2004; Doyle et al. 2010). Less attention has been paid to the impacts on biodiversity.

Sea level rise will mean a reduction in land area within the Mekong Delta. The Ministry of Natural Resources and Environment (MONRE) (2009) predicted a range of loss of land between about 13% for a rise of sea level of 65 cm to about 38% for a rise of sea level of 100 cm, with their best guess for planning purposes being about 19% (or 7,580 km$^2$) loss of land. The major area which would be affected under such a scenario would be the Ca Mau peninsula.

For biodiversity that will mean a loss of some of the coastal habitats as sea water encroaches inland. Areas of aquatic habitat now brackish will become saline, and areas now fresh will become brackish or even saline. That will cause a redistribution of some species and a loss of others. Flooding regimes will change, which could alter the character of the parks and reserves which are critical to biodiversity conservation in the Delta at present. Loss of land elsewhere in the Delta may mean that encroachment on these reserves becomes an increasing problem as well.

While climate change is not likely to cause major issues in the next 5 years, over the longer term it has the potential to become the main cause of biodiversity within the Delta.

## 11.5 Conclusions

While biodiversity within the Mekong Delta has been imperfectly documented so far, it is clear that the Delta is an important biodiversity location for birds and fish, but much less so for mammals, reptiles and plants. The national parks and conservation reserves within the Delta, whilst comparatively small in area, play a disproportionately large role in maintaining biodiversity, particularly for iconic species such as Sarus Cranes.

The biodiversity of the Delta has been greatly reduced in the past by human encroachment converting the original ecosystems to rice paddies. The process has been going on for centuries, but has become more rapid in recent decades. However, in future, deteriorating water quality, the construction of dams on the Mekong upstream of the Delta, overfishing and sea level rise as a result of climate change constitute the greatest threats to biodiversity in the Mekong Delta.

**Acknowledgement** Thanks to Doan Trang Nguyen for assistance with translating some Vietnamese material.

## References

Adamson PT, Rutherfurd D, Peel MC, Conlan I (2009) The hydrology of the Mekong River. In: Campbell IC (ed) The Mekong. Biophysical environment of an international river basin. Academic, Amsterdam, pp 53–76

Buckton ST, Safford RJ (2004) The avifauna of the Vietnamese Mekong Delta. Bird Conserv Int 14:279–322

Campbell IC (2007) Perceptions, data and river management – lessons from the Mekong River. Water Resour Res 43. doi:10.1029/2006WR005130

Campbell IC (2009a) Introduction. In: Campbell IC (ed) The Mekong. Biophysical environment of and international river basin. Academic, Amsterdam, pp 1–11

Campbell IC (2009b) Development scenarios and Mekong River flows. In: Campbell IC (ed) The Mekong: biophysical Environment of an International River Basin. Academic/Elsevier, Amsterdam, pp 389–402

Campbell IC (2009c) The challenges for Mekong River management. In: Campbell IC (ed) The Mekong: biophysical environment of an international river basin. Academic/Elsevier, Amsterdam, pp 402–419

Campbell IC (2011) Managing international river basins: successes and failures of the Mekong River Commission. In: Grafton RQ, Hussey K (eds) Water resources planning and management. Cambridge University Press, Cambridge, pp 724–740
Campbell IC, Chessman BC, Resh VH (2009) The development and application of biomonitoring in the Lower Mekong River system. In: Campbell IC (ed) The Mekong: biophysical environment of an international river basin. Academic, Amsterdam, pp 321–334
Carling PA (2009a) The geology of the lower Mekong River. In: Campbell IC (ed) The Mekong; biophysical environment of an international river basin. Academic, Amsterdam, pp 13–28
Carling PA (2009b) Geomorphology and sedimentology of the Lower Mekong River. In: Campbell IC (ed) The Mekong: biophysical environment of and international river basin. Academic, Amsterdam, pp 77–111
Carvalho FP, Villeneuve JP, Cattini C, Tolosa I, Thuan DD, Nhan DD (2008) Agrochemical and polychlorobiphenyl (PCB) residues in the Mekong River delta, Vietnam. Mar Pollut Bull 56:1476–1485
Deap Loeung, Degen P, van Zalinge N (2003) Fishing gears of the Cambodian Mekong. Inland Fisheries Research and Development Institute of Cambodia, Phnom Penh
Delacour J (1970) The contribution of Gilbert Tirant to Indochinese ornithology. Nat Hist Bull Siam Soc 23:325–329
Doyle TW, Day RH, Michot TC (2010) Development of sea level rise scenarios for climate change assessments of the Mekong Delta, Vietnam. Open-file report 2010-1165. U.S. Geological Survey, Reston Virginia
George CJ (1973) The role of the Aswan High Dam in changing the fisheries of the southeastern Mediterranean. In: Taghi Farvar M, Milton JP (eds) The careless technology. Ecology and international development. Tom Stacey, London, pp 159–178
Gupta A (2009) Geology and landforms of the Mekong Basin. In: Campbell IC (ed) The Mekong. Biophysical environment of an international river basin. Academic, Amsterdam, pp 29–51
Hook J, Novak S, Johnston R (2003) Social atlas of the Lower Mekong Basin. Mekong River Commission, Phnom Penh
Hortle KG (2009) Fisheries of the Mekong River Basin. In: Campbell IC (ed) The Mekong. Biophysical environment of an international river basin. Academic, Amsterdam, pp 197–249
Hynes HBN (1970) The ecology of running waters. Liverpool University Press, Liverpool
Kassas M (1973) Impact of water control schemes on the shoreline of the Nile Delta. In: Taghi Farvar M, Milton JP (eds) The careless technology. Ecology and international development. Tom Stacey, London, pp 179–188
Krohn A (2009) The reptile and amphibian diversity of Tram Chim National Park, Dong Thap Province, Viet Nam. ISP collection paper 689. http://digitalcollections.sit.edu/isp_collection/689. Accessed Mar 2011
Lê Quang Khôi (2010) An introduction to the karst of Kien Giang. Nhà Xuât Bàn Nông Nghiêp, Hà Nôi
Mattson N, Kongpheng Buakhamvongsa, Naruepon Sukumasavin, Nguyen Tuan, Ouk Vibol (2002) Cambodia Mekong giant fish species: on their management and biology. MRC technical paper no. 3. Mekong River Commission, Phnom Penh
Minh NH, Minh TB, Kajiwara N, Kunisue N, Iwata H, Viet PH, Tu NPC, Tuyen BC, Tanabe S (2007) Pollution sources and occurrences of selected persistent organic pollutants (POPs) in sediments of the Mekong River delta, South Vietnam. Cheosphere 67:1794–1801
MONRE [Ministry of Natural Resources and Environment] (2009) Climate change, sea level scenarios for Vietnam. Ministry of Natural Resources and Environment, Hanoi
MRC [Mekong River Commission] (2003) The state of the basin report 2003. Mekong River Commission, Phnom Penh
MRC [Mekong River Commission] (2006) Biomonitoring of the Lower Mekong River and selected tributaries. MRC technical paper no 13. Mekong River Commission, Phnom Penh
MRC [Mekong River Commission] (2007) Diagnostic study of water quality in the Lower Mekong Basin. MRC technical paper no. 15. Mekong River Commission, Vientiane

MRC [Mekong River Commission] (2008a) Biomonitoring of the Lower Mekong River and selected tributaries. 2004–2008. MRC technical paper no 20. Mekong River Commission, Phnom Penh
MRC [Mekong River Commission] (2008b) An assessment of water quality in the Lower Mekong Basin. MRC technical paper no. 19. Mekong River Commission, Vientiane
MRC [Mekong River Commission] (2009) Report on the 2006 biomonitoring survey of the Lower Mekong River and selected tributaries. MRC technical paper no. 22. Mekong River Commission, Vientiane
MRC [Mekong River Commission] (2010) Report on the 2008 biomonitoring survey of the Lower Mekong River and selected tributaries. MRC technical paper no. 27. Mekong River Commission, Vientiane
Nguyen XV, Wyatt AB (2006) Situation analysis: plain of reeds, Vietnam. Mekong wetlands biodiversity conservation and sustainable use programme. Vientiane, Lao PDR
Petts GE (1984) Impounded rivers. Perspectives for ecological management. Wiley, Chichester
Pham Van Mien (2002) Draft report on review biological assessment of freshwater ecosystems in Viet Nam. Report to the Vietnam National Mekong Committee
Pham Van Toan, Sebesvari Z, Vo Phuong Hong Loan, Renaud F (2010) Monitoring and monitoring the fate of commonly used pesticides in surface water of the Lower Mekong Delta. Geophysical research abstracts. 12, EGU2010-743
Robinson WL (1978) The Columbia: a river system under siege. Or Wildl 33:3–7
Rundel PW (2009) Vegetation in the Mekong basin. In: Campbell IC (ed) The Mekong: biophysical environment of an international river basin. Academic, Amsterdam, pp 143–160
Skeat AJ, Oertzen I, Maddison J (1987) Control of *Mimosa pigra* in Kakadu National Park. In: Lemerle D, Leys AR (eds) Proceedings of the eighth Australian weeds conference. Sydney, New South Wales, Australia, pp 464–466
Sterling EJ, Hurley MM, Minh LD (2006) Vietnam. A natural history. Yale University Press, New Haven/London
Thang Nam Do, Bennett J (2008) Estimating wetland biodiversity values: a choice modelling application in Vietnam's Mekong River Delta. Environ Dev Econ 14:163–186
Tran Triet, Safford RJ, Phat TD, Van Ni D, Maltby E (2000) Wetland biodiversity overlooked and threatened in the Mekong Delta, Vietnam: grassland ecosystems of the Ha Tien Plain. Trop Biodivers 7:1–24
Tran Triet, Le Cong Kiet, Nguyen Thi Lan Thi, Pham Quoc Dan (undated) The invasion by Mimosa pigra or wetlands in the Mekong Delta, Vietnam. http://www.ento.csiro.au/weeds/pdf/mimosa_symposium/07Trietetal.pdf. Accessed Apr 2011
Valbo-Jørgensen J, Coates D, Hortle K (2009) Fish diversity in the Mekong basin. In: Campbell IC (ed) The Mekong: biophysical environment of and international river basin. Academic, Amsterdam, pp 161–196
Vidthayanon C (2008) Field guide to fishes of the Mekong Delta. Mekong River Commission, Vientiane
Wassman R, Hien NX, Hoanh CT, Tuong TP (2004) Sea level rise affecting the Vietnamese Mekong Delta: water elevation in the flood season and implications for rice production. Clim Chang 66:89–107
Wilson EO (1988) The current state of biological diversity. In: Wilson EO, Peter FM (eds) Biodiversity. National Academy Press, Washington, DC, pp 3–18

# Chapter 12
# Mangrove Ecosystems in the Mekong Delta – Overcoming Uncertainties in Inventory Mapping Using Satellite Remote Sensing Data

**Steffen Gebhardt, Lam Dao Nguyen, and Claudia Kuenzer**

**Abstract** In the Mekong Delta of Vietnam, losses of mangroves have been especially severe over the last decades mainly due to clearing for shrimp farming leading to a land use conflict between mangrove silviculture and shrimp aquaculture. Following the law on rules for land use and natural resource utilisation for reforestation of mangrove forests (1999), shrimp farming-forestry enterprises (SFFEs) have therefore been established where shrimp are cultured together with mangroves in integrated farming systems whereby 60% of the farm area must be protected mangrove. Monitoring tools are therefore required to map the extent of mangroves and aquaculture and to delineate the percentage of mangroves in mixed mangrove-aquaculture farms. Current methodologies in forest mapping and mangrove area estimation require a large amount of manual work. In addition, currently applied methodologies in base data acquisition and transformation into statistics or maps are varying. This is leading to different estimates of mangrove area of the same region in different statistical surveys. Similar problems arise for actual mangrove distribution maps which are generated based on different sources and methodologies. Precise and reliable mangrove maps are of increasing importance, since changes in these ecosystems

S. Gebhardt (✉)
Department of Remote Sensing, University of Würzburg, Muenchner Str. 20, Oberpfaffenhofen 82234 Wessling, Germany
e-mail: steffen.gebhardt@cimonline.de

L.D. Nguyen
GIS and Remote Sensing Research Center (GIRS), Vietnamese Academy of Science and Technology (VAST), 1 Mac Dinh Chi, Dist. 1, Ho Chi Minh City, Vietnam
e-mail: ldnguyen@vast-hcm.ac.vn

C. Kuenzer
German Aerospace Center (DLR), Earth Observation Center (EOC), German Remote Sensing Data Center, Oberpfaffenhofen, 82234 Wessling, Germany
e-mail: claudia.kuenzer@dlr.de

F.G. Renaud and C. Kuenzer (eds.), *The Mekong Delta System: Interdisciplinary Analyses of a River Delta*, Springer Environmental Science and Engineering,
DOI 10.1007/978-94-007-3962-8_12, 

often occur rapidly. There is a growing demand for detecting and assessing changes especially in the context of climate change, sea level rise and related threats to coastal ecosystems. International organizations and government agencies in several countries are, for this reason, urgently implementing programs of mapping and monitoring to measure the extent of decline of these important ecosystems. Remote sensing enables large-scale, rapid, long-term, and cost-effective mapping and monitoring of mangrove ecosystems and with that provides important information for resource inventory mapping, change detection, and monitoring aquaculture activities. It also supports various management requests, ecosystem evaluation, productivity measurement, water quality assessment, or disaster management.

## 12.1 Mangrove Development in the Mekong Delta

Mangrove ecosystems provide important products and services for the following ecosystem services (i) Regulation (e.g. shoreline protection); (ii) Provisioning (support fish population and therefore fisheries, aquaculture, construction material, fuel, tannins, honey, traditional medicine, paper, and textiles); (iii) Cultural (tourism and recreation, spiritual), and (iv) Support (e.g. nursery habitats, nutrient cycling) (Kathiresan and Bingham 2001; UNEP-WCMC 2006; Alongi 2008; Künzer et al. 2011).

Like other ecosystems, they are subject to various natural disturbances (e.g. geological, physical, chemical, biological), and are daily subject to changes in temperature, water and salt exposure, and varying degrees of anoxia brought about by tides. Mangrove forests ecosystems are fairly robust and highly adaptable to life in waterlogged saline soils within warm, subtropical and tropical seascapes and exhibit a high degree of ecological stability (Alongi 2008). Mangroves offer protection from waves and tidal bores, and can dampen shoreline erosion; they prevent the entry of seawater inland and thus protect the underground water systems, which constitute a source of drinking water supply to coastal populations (Kathiresan and Rajendran 2005; Alongi 2008).

Mangrove forests currently occupy areas between 167,000 $km^2$ – or 16,700 thousand ha (Valiela et al. 2001) and 181,000 $km^2$ – or 18,100 thousand ha (Spalding 1997) globally, with an economic value on the order of 200,000–900,000 USD per $km^2$ (UNEP-WCMC 2006). The Food and Agriculture Organization of the United Nations (FAO) estimated the global area of mangroves 15,237 thousand ha in 2005. This represents a decrease of 3,564 thousand ha when compared to 1980 statistics. Almost 50% of mangrove destruction in this period took place in the South East Asian region (1,690 thousand ha). According to FAO statistics, the mangrove area in Vietnam changed from 269 thousand ha in 1980 to 158 and 157 thousand ha in 2000 and 2005, respectively. From these about 109 thousand ha (70%) can be found in Southern Vietnam, dominantly in the coastal areas of the Ca Mau province (ca. 50%) and Ho-Chi-Minh City (ca. 25%) (Figs. 12.1 and 12.2).

Mangrove forests in Vietnam used to cover more than 400 thousand ha in the 1940s (Maurand 1943). War, forest fire, collection of fuel wood and other human

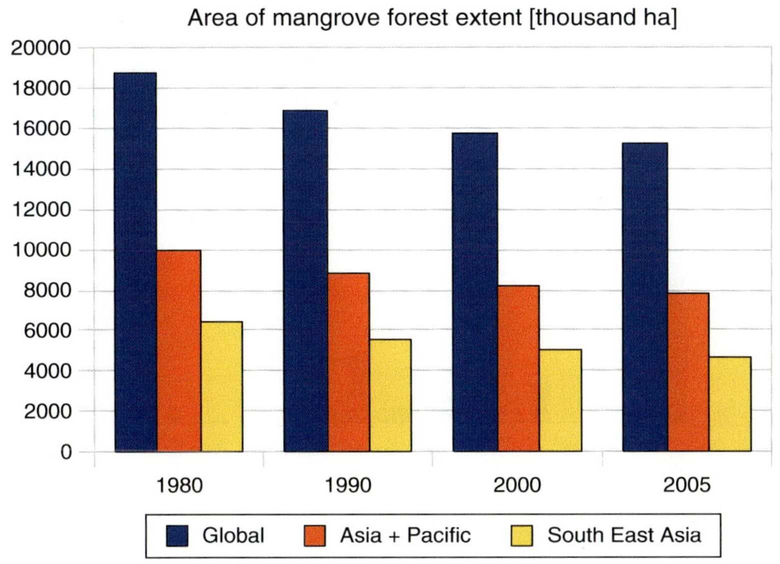


**Fig. 12.1** Area of mangrove forests from 1980 to 2005 globally and regionally (Source: FAO)

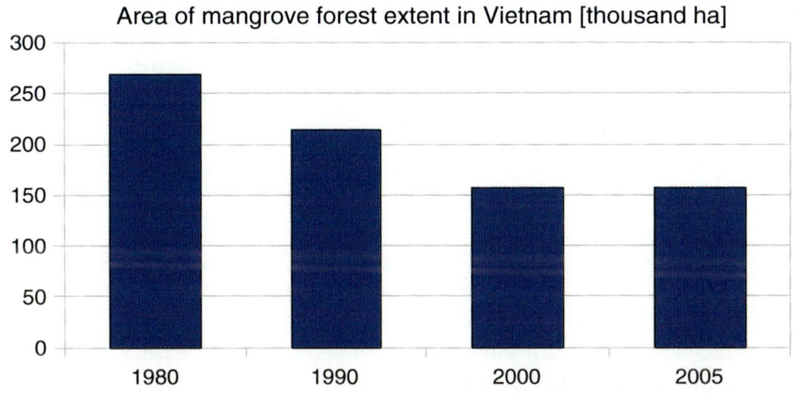


**Fig. 12.2** Area of mangrove forests from 1980 until 2005 in Vietnam (Source: FAO)

activities have resulted in the reduction of the mangrove forests in the Mekong Delta (Thu and Populus 2007). In 1968 about 600 thousand ha were sprayed with chemicals, and over 20% of the total forested regions of South Vietnam was defoliated (Pfeiffer 1970). After the end of the war in 1975 mangrove forests initially recovered as a result of both natural regeneration and manual planting. However, in the 1980s and early 1990s the mangrove forest was again destroyed due to the overexploitation of timber for construction and charcoal, and conversion of forest land into aquaculture (Clough et al. 2000).

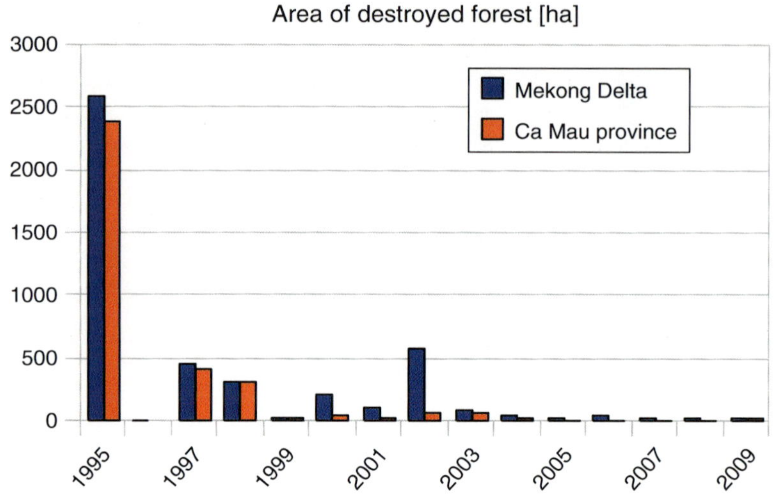


**Fig. 12.3** Area of destroyed forest in the Mekong Delta and in the Ca Mau province from 1995 until 2009 (Source: GSO n.d.)

State forest enterprises were established in the 1980s to ensure sustainable management and replanting but the pressure for shrimp aquaculture farming land was too great; by 1992 the forest area in Ca Mau was at its lowest historical level (51 thousand ha) its originally highly diverse mangrove forests having been turned into monoculture forests consisting mainly of planted *Rhizophora apiculata* (Christensen et al. 2008). By the mid-1990s forest felling bans were imposed and the forest enterprises were force to reforest and protect the mangrove (Christensen et al. 2008).

Especially, since the end of the 1990s, mangrove forests have been cleared for shrimp farming in many areas, however, although the rising importance of shrimp aquaculture in Vietnam has brought considerable financial benefits to the local communities, the rapid, and to a large extent, uncontrolled increase in pond area has contributed to considerable loss of mangrove forests and environmental degradation in the Mekong Delta (Johnston et al. 2000). Figure 12.3 shows statistics from the General Statistics Office of Vietnam (GSO n.d.) on the area of destroyed forest in the Mekong Delta and the Ca Mau province. Even though these measures are not representing mangrove forests exclusively this reflects the processes of land transformation described above. As the primary forest in the Ca Mau province is and has been mangrove forest, one can clearly identify peaks of forest destruction and conversion into shrimp farms in the mid and end 1990s. With the beginning of the new century, the forest destruction decreased from about 70 thousand ha in 2002 to only 7 thousand ha in 2008. Over the same time period the area of planted forests has been comparatively low in Ca Mau. They tended to decrease from about 11 thousand ha in 1995 to ca. 4 thousand ha in 2009 (Fig. 12.4). Since the year 2000, the plantation

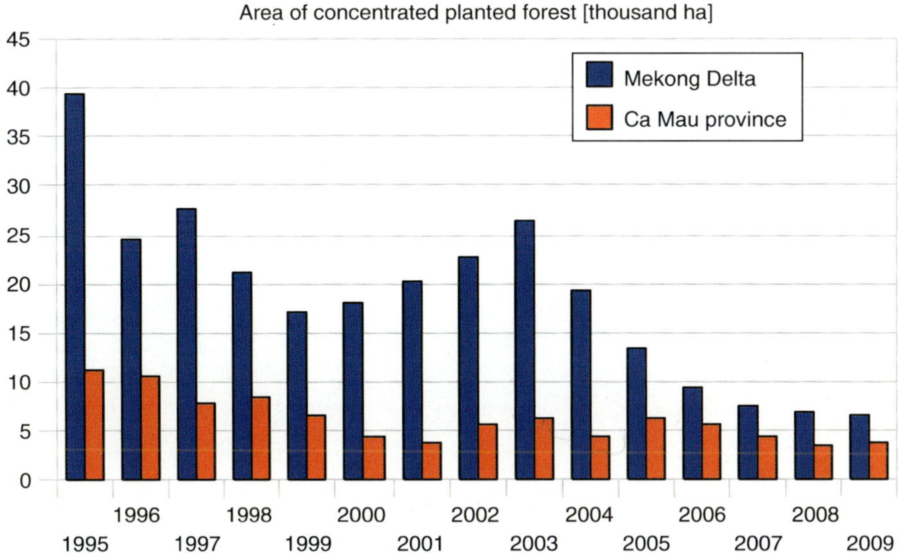


**Fig. 12.4** Area of concentrated planted forest in the Mekong Delta and in the Ca Mau province from 1995 until 2009 (Source: GSO n.d.)

rates are more or less constant at around 4–6 thousand ha. However, even after 2000 the area of forest destruction has been higher than of plantations, e.g. in the years 2002 and 2003 the area of destruction (110 and 572 thousand ha) has been 10 times as high as the plantation area (3.7 and 5.7 thousand ha). In 2009 the destruction area of 18 thousand ha was four times as high as the plantation area (4 thousand ha).

To ease the land use conflict between mangrove silviculture and shrimp aquaculture, a law on rules for land use and natural resource utilisation for reforestation of mangrove forests has been setup in 1999 (Government of Vietnam (GoV) 1999). As a consequence, a number of provincial authorities have established shrimp farming-forestry enterprises (SFFEs) where shrimp are cultured together with mangroves in integrated farming systems as shrimp and other cultured species depend mostly on natural food chains for their food supply (Clough et al. 2000; Johnston et al. 2000). For example, the Ca Mau province has 12 mangrove forest enterprises and depending on their location they are divided into two main landuse zones: (i) a conservation zone named the full protection zone (FPZ) on which all land must be forested and conserved with no human settlement allowed except for fishing communities at river mouths (Fig. 12.5), and (ii) a buffer zone (BZ) with a forest cover of 60%, and 40% for non-forest purposes, i.e. agriculture or aquaculture (Fig. 12.6) (GoV 1999; Christensen et al. 2008).

The official statistics from GSO underline the success of these activities. When comparing Fig. 12.3 with Fig. 12.7 showing the shrimp production in the Mekong Delta and the Ca Mau province. Before the law became effective the shrimp output was relatively low. Over the same period forest destruction was ongoing at a

**Fig. 12.5** Full protection mangrove zone in south-western Ca Mau (Photo: Gebhardt 2010)

**Fig. 12.6** Shrimp farming-forestry enterprises within buffer zone (Photo: Gebhardt 2010)

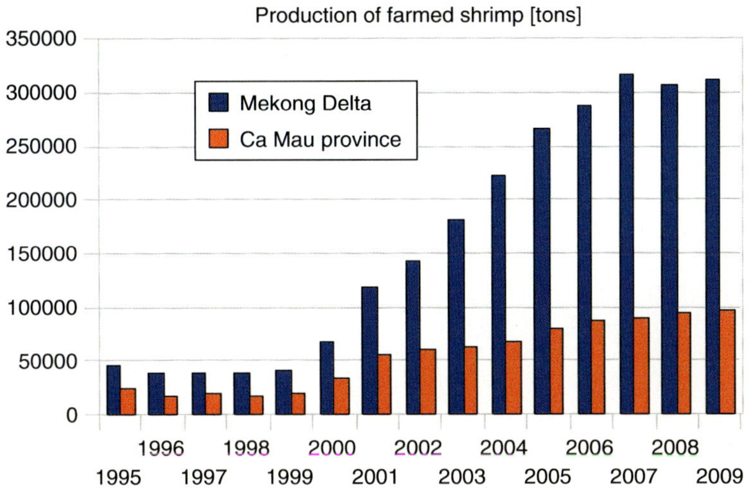


**Fig. 12.7** Production of farmed shrimp in the Mekong Delta and in the Ca Mau province from 1995 until 2009 (Source: GSO n.d.)

rapid pace. However, the conversion of mangrove to shrimp farms in Ca Mau did not directly lead to higher shrimp yields. With the law coming into force shrimp production increased. As low rates of deforestation over the same time period occurred, this indicates the successful implementation of the integrated shrimp – mangrove farming systems.

## 12.2 "Lies, Damned Lies, and Statistics": The Problem of Mangrove Inventory Mapping

The problem with statistics is that they are often hard to verify, especially when there is no metadata describing who collected the data, how they have been measured, who were the surveyors and how these measures have been aggregated and validated. Different surveys on the same target, with different observers' perspectives or different terminologies and understandings, or the use of different methodologies might lead to different findings for the same object.

Comparing, for example, the measures from FAO on mangrove forest area in Vietnam in 2005 (Fig. 12.2) to measures of GSO and Forest Inventory Planning and Investigation Institute (FIPI) for the same year in the Mekong Delta (Fig. 12.8), one can see a strong disagreement with 157 thousand ha for the whole country as given by the FAO study versus 216 thousand ha for the Mekong Delta only by the GSO/FIPI report. Which number is correct, or are they both correct? Do both assessments/

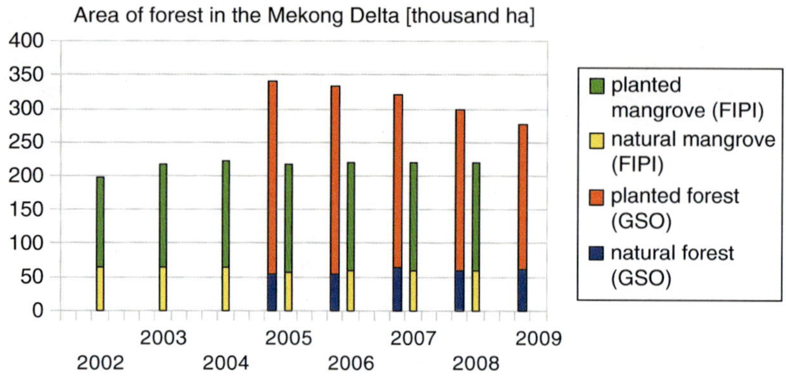


**Fig. 12.8** Area of (mangrove) forest in the Mekong Delta from 2002 until 2009 (Source: GSO and FIPI)

estimations really consider the same target? It is very difficult to answer these questions. One can only guess that different methodologies and scopes lead to different results with both measures telling their own truth.

Many approaches for mangrove extent estimation are used in Vietnam and are briefly reviewed here. The FAO estimate on mangrove extent builds on the FAO Global Forest Resources Assessment (FAO 2001, 2006).[1] "The world's mangroves 1980–2005" (FAO 2007) was prepared in collaboration with mangrove specialists throughout the world. It builds on a 1980 assessment by FAO and the United Nations Environment Programme (UNEP), the FAO Global Forest Resources Assessments, on an extensive literature search and on communication with mangrove and forest resources assessment specialists. Some 2900 national and sub-national data sets on the extent of mangrove ecosystems were collected during this process, permitting the compilation of an updated list of the most recent reliable estimates for each of the 124 countries and areas in which mangroves are known to exist. They show that changes in definitions and methodologies over time make it difficult to compare results from different assessments, and the extrapolation to 2005 was constrained by the lack of recent information for a number of countries.

The General Statistics Office (GSO) is an institution directly under the Ministry of Planning and Investment (MPI), performing an advisory function to help the MPI Minister in doing the task of state management in the field of statistics. The GSO is conducting statistical operations and provides socio-economic statistical information to all agencies, organizations and individuals in accordance with the law (GoV 2010). Article 3 of the "Decision On Promulgation of the National Statistical Indicator System" (GoV 2005) defines "The Director General of GSO is responsible for standardizing definitions, content, calculation methods, data sources of each

[1] The world's mangroves 1980–2005: http://www.fao.org/docrep/010/a1427e/a1427e00.HTM

indicator in the National Statistical Indicator System (NSIS) and defining provincial, district, commune statistical indicator system to implement consistently in the whole country".[2] Further, Article 4 regulates that "Ministries, ministerial level agencies, Government's agencies, People's supreme procuracy, People's Supreme Court, National Assembly Office, Organization Department of the Party Central Committee, central offices of all politic-social organizations are responsible for collecting and compiling assigned indicators in the NSIS and providing to GSO for compiling and publishing".

The Sub-National Institute for Agricultural Planning and Projection (Sub-NIAPP) in Ho-Chi-Minh city is organized under the National Institute of Agricultural Planning And Projection, Ministry of Agriculture And Rural Development. The Sub-NIAPP is amongst others responsible for agricultural, land use and infrastructural planning and also for maintaining respective databases and periodical mapping. Every 5 years landuse maps of the Mekong Delta have to be compiled at scales of 1:100,000 and 1:250,000. In order to create these maps, data are collected from the Departments of Agriculture Rural Development (DARD) in each individual province in the Mekong Delta. The DARDs themselves are collecting the data at district and commune levels. Sub-NIAPP is then responsible for the compilation of these data into a unique map representation whereby remote sensing data and on screen digitizing are utilized for data cross validation and final map composition. Besides several classes focusing on agricultural land use, aquaculture and forestry are determined. The Sub-NIAPP land-use map for the Ca Mau region from 2000 at a scale of 1:100,000 is shown in Fig. 12.10a.

The Vietnamese Forest Inventory Planning and Investigation Institute (FIPI) of the Ministry of Agriculture and Rural Development (MARD) is responsible for the implementation of the national program for forest resources assessment and monitoring and with that the updating of the nationwide forest map of Vietnam. They are also responsible for compiling statistical data for GSO. A national forest inventory has to be conducted every 5 years and transformed into maps at national, regional and provincial scales. The Southern Sub-Institute of Forest Inventory and Planning (Sub-FIPI) is responsible for field data collections to support the forest resources monitoring and mapping in the Mekong Delta. For 25 permanent sample plots distributed over the forest areas in the Mekong Delta, a field inventory is conducted on a regular basis. Each of these sample plots represents an area of 1,000 ha. Within each sample plot forest land-use, species composition, the number of trees, diameter at breast height, and height of trees are measured. These data are valuable input to update forest inventory maps. However, they are only single measures at distinct places and are of insufficient spatial density to support detailed area-wide mapping of the current situation. Based on these observations and on the reporting of state shrimp farming-forestry enterprises (SFFEs) at the communal and district levels, maps and statistics are compiled. Remote sensing data are an additional data

[2] http://www.gso.gov.vn/default_en.aspx?tabid=510&idmid=6&ItemID=3993

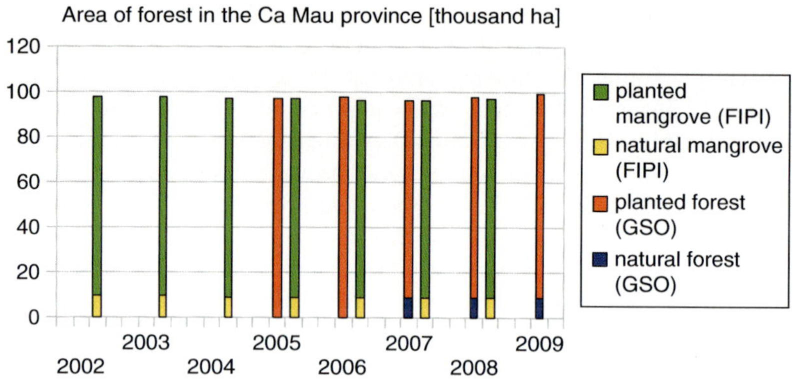


**Fig. 12.9** Area of (mangrove) forest in the Ca Mau province from 2002 until 2009 (Source: GSO and FIPI)

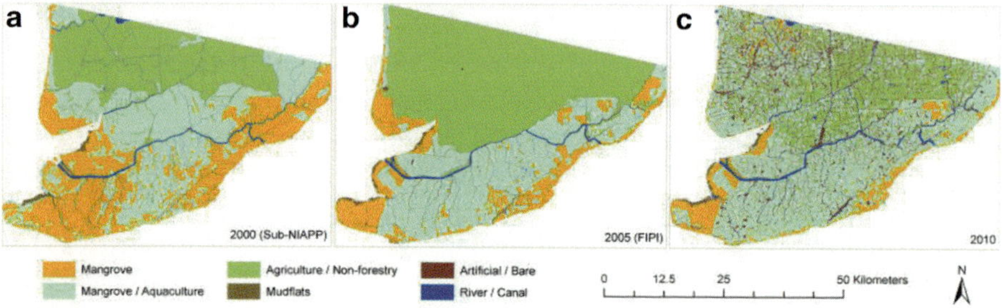


**Fig. 12.10** Land-use/forest maps of the Ca Mau region from (**a**) 2000 (Sub-NIAPP), (**b**) 2005 (FIPI), and (**c**) 2010 (DLR)

source for on-screen digitizing and cross-validation. The FIPI forest map for Ca Mau province is shown in Fig. 12.10b.

Figure 12.10 is completed by a recent mangrove map as calculated from remote sensing data (Fig. 12.10c). The visual comparison of the 2000, 2005 and 2010 maps (Fig. 12.10) indicate a strong decrease in forest (mangrove) area in the Ca Mau region. However, this trend is not obvious in the statistics of FIPI given in Fig. 12.9 where an almost constant mangrove cover is reported between 2002 and 2009.

Which of this information is correct? Which of the individual maps is really representing the current situation for the respective year? Are the statistics correct or do the maps provide the correct information? Each of these maps was compiled using a set of data from varying sources like field samples, local statistics or remote sensing. In addition the map production processes are actual chains of subprocesses of different individual editors, maybe having different background, capacities, or experiences. Different standards in field sampling, statistical reporting or also digitizing further influence and hinder the final map compilation processes.

Also the definitions of the class "Mangrove" might be different in these maps, with Sub-NIAPP focusing agricultural usage and FIPI looking at actual forest coverage. The problem is finally that too many factors are influencing the mapping. Effective mandatory standards and schemes are required for collecting, processing and transforming data. This is however, extremely difficult when covering a large area and can hardly be achieved based on field sampling and post-processing. Remote sensing can be an efficient tool for overcoming these gaps.

## 12.3 A Short-Review on Remote Sensing for Mangrove Mapping in Vietnam

Remote sensing enables large-scale, rapid, long-term, and cost-effective mapping and monitoring of mangrove ecosystems and with that provides important information for resource inventory mapping, change detection, and monitoring aquaculture activities (Green et al. 1998). It also supports various management requests, ecosystem evaluation, productivity measurement, water quality assessment, or disaster management (Künzer et al. 2011).

Over the last decades the number of applications utilizing remote sensing data for mangrove ecosystems mapping and analysis increased. They cover the application of various optical sensors, ranging from aerial photography, to high- and medium-resolution imagery, and hyperspectral data to active microwave (SAR) systems all over the world (Künzer et al. 2011). For a detailed review the reader may be referred to the original paper of Künzer et al. 2011. In the following we only provide a brief overview on remote sensing for mangrove mapping in Vietnam.

Some publications focus on mangrove mapping and change analysis in Vietnam. Seto and Fragkias (2007) utilized Landsat satellite images from 1975 to 2002 to analyse the rate of mangrove conversion and aquaculture development in Nature Reserve areas in the Red River Delta in Vietnam. Using pixel-based artificial neural network classification they derived the five land-cover classes of mangrove, aquaculture, sand, water, and agriculture. They found strong discrepancies between the image classification results and the official statistics for mangrove areas, with the image classification results indicating only half of the official area. Also they proved that a decrease of mangrove area correlated to an increase in aquaculture farming.

A similar study was performed by Thu and Populus (2007) for the Tra Vinh province in the Mekong Delta in Vietnam. They used SPOT satellite imagery from 1995 and 2001 and calculated mangrove distribution and density based on the Normalized Difference Vegetation Index (NDVI) derived from multi-spectral imagery and found a strong decrease (>50%) in mangrove extent over that period.

In a study by Tong et al. (2004), SPOT 4 satellite images from 2001 have been used to map the mangrove distribution in Ca Mau province based on visual interpretation (to identify and to delineate conspicuous homogeneous land-use units through tint, texture, morphology, environment and contrast effects) and Maximum

Likelihood classification into six classes of mangrove, non-mangrove, aquaculture, paddy, tidal flat and sediment, and water. They especially aimed at minimizing the confusion of the aquaculture class (mixed shrimp-mangrove forestry farms) with flooded paddy fields and therefore incorporated measures of NDVI, Normalized Difference Water Index (NDWI) and short-wave infrared (SWIR) into the classification procedure. They achieved an overall accuracy of 80% over all six classes.

A more detailed study for the Red River Delta area was carried out by Beland et al. (2006). They used Landsat images from 1986, 1992, and 2001 to investigate land-cover changes related to shrimp aquaculture. The applied automated unsupervised image classification resulted in 10 land-cover classes (mangrove, casuarina, sand, dry bare soil, humid bare soil, young or damaged mangrove, sea water, aquaculture, mixed mangrove with aquaculture, and jute). Reference data were used to assess classification accuracy and change detection for the two land-cover classes that were of most interest in this study: mangrove forest and aquaculture ponds. For the mangrove land-cover class, user's accuracy was 72% for 1986, 44% for 1992, and 80% for 2001. The producer's accuracies were respectively 88%, 95%, and 91%.[3] For the aquaculture land-cover class, the user's accuracy was 46% for 1992, and 77% for 2001. The producer's accuracies were 84% for 1992 and 92% for 2001.

A more recent study was carried out by the authors within the frame of the German-Vietnamese research projects WISDOM (A water-related information system for the sustainable development of the Mekong Delta) and RICEMAN (Rice and mangrove mapping in Vietnam).

In this application, SPOT5 multi-spectral and TerraSAR-X synthetic aperture radar (SAR) imagery from 2010 have been utilized for automatic mapping of mangrove species, aquaculture and other dominant land cover classes (Figs. 12.10c and 12.11). Our approach highlights the benefits of synergistic usage of multi-spectral and SAR data for robust mangrove mapping. We successfully detected aquaculture areas mixed with mangroves at high accuracies. Based on these results mangrove development, especially within shrimp farming-forestry systems, can be monitored.

An object-oriented classification approach was applied composed of image segmentation, feature extraction and decision-tree classification. Image segmentation of the SPOT5 image resulted in fairly generalized image objects which are not representing individual aquaculture farms but combine neighbouring farms having similar texture properties in the multi-spectral SPOT5 image.

---

[3] Overall Accuracy is the number of incorrect observations divided by the number of correct. This is a very crude measure of accuracy.

User's Accuracy is a measure of how well the classification performed in the field by category (rows). The user's accuracy details errors of commission. An error of commission results when a pixel is committed to an incorrect class.

Producer's Accuracy is a measure of how accurately the analyst classified the image data by category (columns). The producer's accuracy details the errors of omission. An error of omission results when a pixel is incorrectly classified into another category. The pixel is omitted from its correct class.

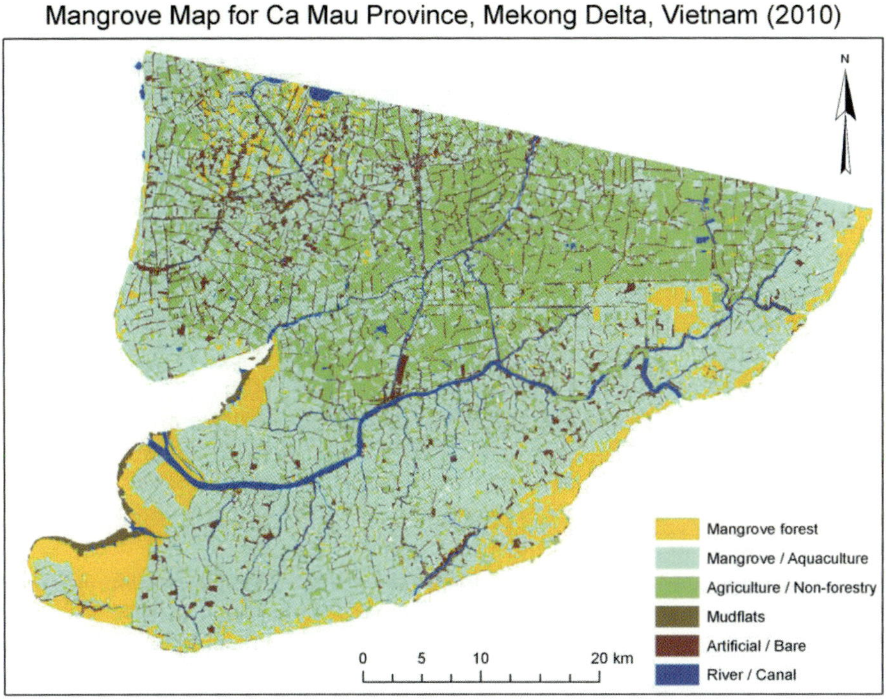


**Fig. 12.11** Remote sensing based mangrove classification result from 2010 for the Ca Mau province as a result of the German-Vietnamese WISDOM/RICEMAN projects

Training, feature selection, classification and validation were done automatically utilizing the Twinned Object- and Pixel-based Automated Classification Chain (Huth et al. 2012). The automatic feature extraction highlighted the benefit of using both sensors and derived vegetation indices and watermask products with the most dominant features origin from both, the optical and the radar data.

Validation of independent image objects yielded in overall accuracies of 73% when discriminating mangrove species classes, with only moderate accuracies of 63% and 71% for *Rhizophora* and *Avicennia* classes, respectively. Classification into one combined mangrove resulted in an overall accuracy of 89% with the mangrove class rating 84%. In both cases aquaculture farms were correctly classified with an accuracy of 91%. The comparison to field surveys taken during independent campaigns in 2004 and 2010 yielded in overall accuracies of 73% with the aquaculture class ranging from 64% to 73% and the mangrove class ranging from 74% to 77%.

The applied data and automatic classification method provided high accuracies over all classes and especially for the main classes of interest mangrove and mixed shrimp farming-forestry systems. With the utilized TWOPAC classification environment, a re-utilization of image samples and trained classifiers is possible, making a transfer of the methodology to neighbouring satellite imagery possible. The derived products are valuable for mangrove inventory mapping and form the basis for monitoring and analysing changes in the ecosystem.

## 12.4 Concluding Remarks on the Potential of Remote Sensing for Mangrove Inventory Mapping

Remote sensing appears to be the only suitable technology for monitoring large areas on the earth surface at one time and repetitively over a period of time. It readily competes with costly, time consuming, and non-area wide field data collection and subsequent analysis. In many studies remote sensing data have successfully been used for mapping mangroves at varying scales with high classification accuracies. It therefore provides a tremendous benefit for mangrove inventory mapping.

However, as can be seen from the above mentioned remote sensing studies, all of them utilized different methodologies and classification schemes. The application of standardized techniques and analysis as routine tools is a future challenge for a steady monitoring and mapping of mangroves (Malthus and Mumby 2003; Künzer et al. 2011). This will enhance the possibility to compare the suitability of methods for certain purposes on different data sets and locations. A higher transparency concerning the processing steps (and pre-processing steps) will also be beneficial for the transferability of a given method to another research site.

Scientists and users all over the world have different goals and requirements for their investigations. In many studies, accuracy assessment has not been performed or not considered necessary. Furthermore, accuracy assessments can be carried out by applying different methods (Malthus and Mumby 2003; Künzer et al. 2011). This is an additional factor that diminishes the comparability between studies. If management decisions depend on these results, an accuracy assessment is essential; otherwise, the findings could lead to inappropriate and cost-intensive actions for the user (Green et al. 1996).

The mapping results from the WISDOM/RICEMAN projects for the Ca Mau peninsula are a important data source for the updating of the FIPI forest inventory map. They allow for a more detailed discrimination, particularly between the mangrove forest types and aquaculture farming-forestry systems.

Over the last years various international development aid projects have been set up in the Mekong Delta targeting natural resources management in coastal zones, e.g. in Ca Mau province, focussing on mangrove management and reforestation. The mapping results provide the most recent, area-wide, and accurate mangrove inventory to date. The applied methodologies will enable monitoring mangrove development and the supervision of success of mangrove plantation activities.

## References

Alongi DM (2008) Mangrove forests: resilience, protection from tsunamis, and responses to global climate change. Estuar Coast Shelf Sci 76:1–13

Beland M, Goita K, Bonn F, Pham TTH (2006) Assessment of land-cover changes related to shrimp aquaculture using remote sensing data: a case study in the Giao Thuy District, Vietnam. Int J Remote Sens 27:1491–1510

Christensen SM, Tarp P, Hjortsø CN (2008) Mangrove forest management planning in coastal buffer and conservation zones, Vietnam: a multi methodological approach incorporating multiple stakeholders. Ocean Coast Manag 51:712–726

Clough B, Tan DT, Phuong DX, Buu DC (2000) Canopy leaf area index and litter fall in stands of the mangrove Rhizophora apiculata of different age in the Mekong Delta, Vietnam. Aquat Bot 66:311–320

FAO [Food and Agriculture Organization] (2001) Global forest resources assessment 2000: main report. FAO Forestry paper 140, Rome

FAO [Food and Agriculture Organization] (2006) Global forest resources assessment 2005–progress towards sustainable forest management. FAO Forestry paper 147, Rome

FAO [Food and Agriculture Organization] (2007) The world's mangroves 1980–2005. Food and Agriculture Organization of the United Nations, Rome. http://www.fao.org/docrep/010/a1427e/a1427e00.HTM. Accessed 2 Aug 2011

GoV [Government of Vietnam] (1999) Law on: the rule for land use and natural resource utilisation based on the decision no. 116/1999/OD-TTg on 3rd May 1999 for reforestation of mangrove forest in Camau, Baclieu, Soctrang and Travinh. Government of Vietnam

Green EP, Mumby PJ, Edwards AJ, Clark CD (1996) A review of remote sensing for the assessment and management of tropical coastal resources. Coast Manag 24:1–40

Green EP, Clark CD, Mumby PJ, Edwards AJ, Eellis AC (1998) Remote sensing techniques for mangrove mapping. Int J Rem Sens 19:935–956

GSO [General Statistics Office] (n.d.) Webpage of the General Statistics Office. www.gso.gov.vn. Accessed 2 Aug 2011

Huth J, Kuenzer C, Wehrmann T, Gebhardt S, Dech S (2012) Land cover and land use classification with TWOPAC: towards automated processing for pixel- and object-based image classification. Submission to Remote Sensing

Johnston D, Trong NV, Tien DV, Xuan TT (2000) Shrimp yields and harvest characteristics of mixed shrimp-mangrove forestry farms in southern Vietnam: factors affecting production. Aquaculture 188:263–284

Kathiresan K, Bingham BL (2001) Biology of mangroves and mangrove ecosystems. Adv Mar Biol 40(40):81–251

Kathiresan K, Rajendran N (2005) Coastal mangrove forests mitigated tsunami. Estuar Coast Shelf Sci 65:601–606

Künzer C, Bluemel A, Gebhardt S, Vo QT, Dech S (2011) Remote sensing of mangrove ecosystems: a review. Remote Sens 3:878–928

Malthus TJ, Mumby PJ (2003) Remote sensing of the coastal zone: an overview and priorities for future research. Int J Rem Sen 24:2805–2815

Maurand P (1943) L'Indochine Forestière. Imprimerie d'Extrême Orient, Hanoi

Pfeiffer EW (1970) Defoliation and bombing effects in Vietnam. Biol Conserv 2:149–151

Prime Minister, Government of Vietnam (2005) Decision no: 305/2005/Q§-TTg. On Promulgation of the National Statistical Indicator System

Prime Minister, Government of Vietnam (2010) Decision no: 54/2010/QĐ-TTg. Specify the function, duty, authority and organisational structure of The General Statistics Office directly under the Ministry of Planning and Investment

Seto KC, Fragkias M (2007) Mangrove conversion and aquaculture development in Vietnam: a remote sensing-based approach for evaluating the Ramsar convention on wetlands. Glob Environ Chang 17:486–500

Spalding M (1997) The global distribution and status of mangrove ecosystems. Int News Lett Coast Manag-Intercoast Netw 1:20–21

Thu PM, Populus J (2007) Status and changes of mangrove forest in Mekong Delta: case study in Tra Vinh, Vietnam. Estuar Coast Shelf Sci 71:98–109

Tong PHS, Auda Y, Populus J, Aizpuru M, Habishi AA, Blasco F (2004) Assessment from space of mangroves evolution in the Mekong Delta, in relation to extensive shrimp farming. Int J Remote Sens 25:4795–4812

UNEP-WCMC [United Nations Environment Programme World Conservation Monitoring Centre] (2006) In the front line: shoreline protection and other ecosystem services from mangroves and coral reefs UNEP-WCMC
Valiela I, Bowen JL, York JK (2001) Mangrove forests: one of the world's threatened major tropical environments. Bioscience 51:807–815

# Chapter 13
# Agriculture and Water Quality in the Vietnamese Mekong Delta

**Zita Sebesvari, Huong Thi Thu Le, Pham Van Toan, Ute Arnold, and Fabrice G. Renaud**

**Abstract** The economy of the Vietnamese Mekong Delta is based mainly on agriculture which is in turn heavily relying on the availability of fresh water of good quality. Under the *Doi Moi* (Renovation) process starting in 1986, agriculture and aquaculture production intensified rapidly partially through increased utilisation of agrichemicals such as fertilizer, pesticides, and veterinary products which contribute to non-point source pollution of aquatic ecosystems. Due to increased pollution by these agrichemicals, the capacity of freshwater systems to provide essential ecosystem services such as to maintain agriculture and aquaculture productivity and sustain human health is therefore under great pressure. This chapter reviews the main pathways of non-point source pollution to the water systems originating from agricultural production and addresses the trade-offs between agriculture and freshwater provisioning. Additionally, the chapter identifies possible measures that have been or could be implemented by the agricultural sector in order to enhance the sustainability of the water based agricultural activities in the Mekong Delta.

Z. Sebesvari (✉) • H.T.T. Le • F.G. Renaud
United Nations University - Institute for Environment and Human Security (UNU-EHS), UN Campus, Hermann-Ehlers-Str. 10, 53113 Bonn, Germany
e-mail: sebesvari@ehs.unu.edu; lehuong@ehs.unu.edu; renaud@ehs.unu.edu

P. Van Toan
United Nations University - Institute for Environment and Human Security (UNU-EHS), Bonn, Germany

College of Environment and Natural Resources, Can Tho University, Can Tho City, Vietnam
e-mail: pvtoan@ctu.edu.vn

U. Arnold
Section Plant Nutrition, Institute of Crop Science and Resource Conservation (INRES), University of Bonn, Karlrobert-Kreiten-Str. 13, 53115 Bonn, Germany
e-mail: uarnold@uni-bonn.de

F.G. Renaud and C. Kuenzer (eds.), *The Mekong Delta System: Interdisciplinary Analyses of a River Delta,* Springer Environmental Science and Engineering,
DOI 10.1007/978-94-007-3962-8_13, 

## 13.1 Introduction

Agriculture (i.e. crop and animal production, pastures, orchards, aquaculture, forestry) impacts water quality (e.g. Novotny 1999) as well as water-related ecosystem services worldwide (Gordon et al. 2010). The main processes at play include runoff and tile drainage which contribute suspended matter, phosphorus, nitrogen, plant protection products, metals, pathogens, salts, veterinary medicines, feed additives and hormones to the freshwater systems (Casali et al. 2008; Diaz 2001; Causape et al. 2004; Ongley 1996). Land management practices may also amplify natural processes such as oxidization of sulphides in acid sulphate soils leading to leaching of sulphuric acid into water bodies and consecutive immobilization of metal ions (Minh et al. 1997; Chap. 14). Agricultural management which would limit negative impacts on aquatic ecosystems and the services they provide as well as on other ecosystem services such as pollination, biodiversity, and water purification capacity, is a challenge and is especially relevant in areas where the landscape is shaped to a large extent by agricultural production, which is the case in the Mekong Delta (MD).

To date, 63% of the delta is used for agricultural production, compared with 28% for whole Vietnam and 38% for the Red River Delta (GSO 2008). Output value of agriculture in the Mekong Delta accounted for 33% of the total output value of the country in 2008. The majority of rural households in the Delta were involved in either annual crop production (48%) or aquaculture (40%) in 2009 (GSO 2009). Agriculture is practiced mainly in small family units (Wieneke 2005), with 50% of the farms being smaller than 0.5 ha and 70% smaller than 1 ha (GSO 2003; Cuc 2003). Agriculture is becoming increasingly market orientated as opposed to a means of direct subsistence as was the case in the past.

This chapter aims to review the main agricultural pollution sources and their impact on water quality, ecosystem services as well as on agricultural production itself in the Mekong Delta. In addition, the chapter discusses ongoing and possible pollution mitigation measures that may be implemented by the agricultural sector in the Mekong Delta.

## 13.2 Agricultural Production Systems in the Mekong Delta

More than 70% of the households are involved in annual crop production in the Mekong Delta provinces of Dong Thap, An Giang, Long An and Kien Giang and the same proportion stands for aquaculture in the coastal provinces of Tra Vinh, Bac Lieu and Ca Mau. Most of the livestock farms are located in Ben Tre, Long An and Vinh Long while Vinh Long, Ben Tre and Tien Giang are provinces where perennial crop production is important. A large proportion of households tend to diversify agricultural outputs by, for example, growing fruit trees next to the rice paddy or raising low numbers of livestock in backyards. The following sections aim to summarize some important characteristics of the respective production systems.

### 13.2.1 Crop Production

Rice is the main crop in the Mekong Delta cultivated on 1.9 million ha representing nearly 50% of the agricultural production land (GSO 2009). The region supplies 53% of national rice production (GSO 2009). Worldwide, Vietnam was the 5$^{th}$ largest producer for paddy rice and the 2$^{nd}$ largest exporter for milled rice in 2008 (FAOSTAT 2012).

Depending on the area and its water regime, one to tree rice crops per year are produced with double cropping being dominant (1.3 million ha) in the delta. The area with three rice crops per year covers about 0.4 million ha. Rainfed rice production account for *ca.* 30% of the planted area while the rest is irrigated (My Hoa et al. 2006). Typically, irrigated, direct seeded double-crop rice grown on alluvial soils is harvested in March (winter-spring crop) and July (summer-autumn crop). Alluvial soils, situated along the Tien and Hau rivers, accounting for one third of the delta's soils, are likely to achieve high yields. Rice grown on the other major soil type, i.e. acid sulphate soils (e.g. Dong Thap and Long Xuyen quadrangle) is often suffering from phosphorus deficiencies and aluminum toxicity (Chap. 14). In coastal regions where saline soils are dominant, large areas have been transformed for aquaculture production, as crop production is negatively affected by increased salinity (see Chap. 10).

Beside rice, the Mekong Delta is a major production area of fruits such as mango, longan, pineapple, bananas, and others. Annual crops, including sugar cane (*Saccharum officinarum*, 65,000 ha), maize (*Zea mays*, 41,000 ha), peanut (*Arachis hypogaea*, 14,000 ha), sweet potato (*Ipomoea batatas*, 13,000 ha) and cassava (*Manihot esculenta*, 7,000 ha) add up to less than 10% of the cultivated land area (GSO 2008).

### 13.2.2 Livestock Production

Livestock production is the second most important source of income after crop production in Vietnam. Pig production is dominating the livestock sector with 26.8 million pigs in 2008 in the country and around 3.6 million in the Mekong Delta, mainly in Tien Giang, Tra Vinh and Kien Giang provinces (GSO 2008). Pig production increased between 1995 and 2008 with average annual growth rates of 3.4% (GSO 2008). Pork accounts for more than 70% of total meat production in Vietnam (Fisher and Gordon 2008).

Similar to crop production, there are many small farming units owning 1–10 pigs (Thuy 2010). The average commercial farm is also small with 5–100 pigs and there are only few state owned enterprises with larger number of animals (Knips 2004). However, industrial pig production is subsidized by the Vietnamese government to increase exports and thus likely to gain larger share in terms of livestock production in the future (Drucker et al. 2006). In the Mekong Delta growth was mainly recorded in small-scale (5–20 sows) and medium-scale (20–100 sows) systems (Thuy 2010). Pigs raised in backyards are commonly fed with rice by-products and water plants (Kamakawa et al. 2006). The sewage is typically discharged into fish ponds or irrigation canals.

In 2008 there were 48.5 million poultry in the Delta, mainly in Long An, Tien Giang and Kien Giang provinces (GSO 2008). The outbreak of the avian influenza in 2003–2004 resulted in a rapid reduction in the number of poultry and recovery of poultry production was only achieved in 2009.

Cattle are of minor importance with approximately 700,000 cattle in 2008, mainly in Ben Tre and Tra Vinh provinces (GSO 2008). Ducks are often raised in cages above ponds from where excrements fall into the water. Chicken are free-roaming in small farms but are also raised in bigger units in cages. There, like in pig farms, excrements are usually flushed with water several times a day to clean the sty and to cool the animals. The effluent may be put into fishponds, or reach the canal system directly.

### 13.2.3 Aquaculture

Aquaculture is one of the most dynamically growing sectors in the last decade in Vietnam (see Chap. 14). Average annual percentage growth rate of fish and fishery products export was 20.9% for the period 1996–2006 (FAO 2008). This rapid growth enabled Vietnam to become one of the largest fishery exporters in the world (FAOSTAT 2012). In coastal areas intensive commercial and intensive family farms co-occur with more extensive brackish water polycultures and rice–shrimp farms (Joffre and Bosma 2009). Main areas of the shrimp production are Bac Lieu, Ca Mau and Soc Trang provinces while fish is mainly produced in An Giang, Dong Thap and Can Tho. Beside fish and shrimp cultures, mixed cultures (rice-fish, rice-shrimp) are popular. Catfish species such as *Pangasianodon hypophthalmus* (Sauvage) are among the most important species of freshwater aquaculture in the delta. Intensive *Pangasius* aquaculture consists typically in a small-size pond and water is mainly obtained from and discharged to the river (Phan et al. 2009).

### 13.2.4 Main Drivers of Growth

The rapid growth in rice production and aquaculture during the last two decades was a direct result of policy reforms starting in 1988. Politburo Resolution 10-NQ/TW allowed for land-use rights lasting for 15 years and for free trade of the produced goods instead of compulsory state procurement. The Land Law of 1993 permitted even longer term land use rights (20–50 years) and also allowed land users to transfer their land use right to another user within the duration of the lease (Tuyen 2010). These policy reforms led to more planning security and more responsibility for the land and thus to investments in soil quality as well as in enhanced productivity e.g. via the utilization of fertilizers, plant protection products or the introduction of additional rice seasons (Dang and Danh 2008). Free trade and access to the global markets greatly enhanced the role of agriculture as a promoter of growth in the Vietnamese economy by allowing for higher growth rates of agricultural production than domestic consumption and leading to increasing export volumes. These developments were accompanied by land conversion, changes in production patterns (e.g. shift to cash crops or aquaculture) and agricultural intensification (e.g. tree rice crops instead of two). The drawbacks of this development are widely recognized as environmental pollution (IRIN 2009), concerns for drinking water quality and food safety (Holland 2007; Neubacher 2007), failures in meeting international standards of exported goods (Yen 2006) and health concerns (Margni et al. 2002; Dasgupta et al. 2007).

## 13.3 Agricultural Related Sources of Water Pollution

The main agricultural activities linked to water pollution in the Mekong Delta are land preparation (e.g. puddling), fertilizer and pesticide use, waste water release from animal production (mainly pig and poultry) and aquaculture (mainly fish and shrimp).

### *13.3.1 Land Preparation*

At the beginning of the new rice season farmers routinely reconstruct small irrigation ditches and transfer the excavated soil to the canal leading to higher suspended sediment load in the water. Similarly, puddling of the topsoil in submerged rice fields leads to enhanced soil erosion and thus enhanced load of suspended matter.

In areas where acid sulphate soils occur (40% of the soils in the Delta – Tuong et al. 1998) puddling also increases the contact area between soil and water, thus increases the concentration of sulphuric acid and toxic metal ions in the ponded water and after drainage, in surface water. Flushing of acid sulphate soils with water to reduce soil acidity as management practice at the beginning of the rainy season leads to a transfer of acidity and toxic amounts of metal ions (mainly aluminium, iron, potassium and manganese) to the surface water by decreasing their pH to values lower than pH 3 (Kijne 2006; Chap. 14).

### *13.3.2 Fertilizer Use*

Intensive agriculture (high yield and short duration varieties, multiple crops per year) as practiced in large areas of the MD rapidly depletes soil nutrients and requires high volume inputs of fertilizers (Soong 2006). Organic fertilizers are rarely used in the MD. Instead, mineral fertilizers (mainly urea, ammonium sulphate, diammonium phosphate (DAP)) dominate the market (UNEP 2005). Urea is the most common nitrogenous fertilizer for rice. Compound fertilizers are also frequently used. Most of them contain three elements essential for growth, NPK (Nitrogen-Phosphorus-Potassium). The most popular mixture of NPK is 16-16-8 (Shaddick 2007). Vietnam is the largest market for DAP in South-East Asia importing about 760,000 tons in 2006 for direct application as well as for NPK production (Shaddick 2007). Total amounts of inorganic fertilizers input for rice in the MD are around 400,000 tons of nitrogen, 180,000 tons of phosphorus, and 120,000 tons of potassium annually while fertilizer use efficiency is low due to losses through volatilization, leaching, and runoff (Hach and Tan 2007). There was a sharp increase in imported fertilizers from 2.3 million tons in 1995 to around 4 million tons in 2004 for Vietnam (GSO 2005).

Recommended amounts of nitrogen fertilizer for alluvial soils range from 100 to 120 kg N/ha during the dry season and 80–100 kg N/ha during the rainy season. In acid sulphate soils nitrogen fertilizer recommendations are lower i.e. 80–100 kg N/ha in the dry season and 60–80 kg N/ha in the wet season (Hach and Tan 2007). However, according to a study conducted in 35 villages in 11 provinces in 2001/2002, 95 and 105 kg N/ha on average are typically applied to rice during the winter/spring and summer/autumn rice season, respectively while the amount of phosphorus fertilizer was around 55 kg P/ha for both seasons (Huan et al. 2005). Farmers tend to favour nitrogen fertilizer resulting in unbalanced nutrient content especially for wet season crop (Hach and Tan 2007). Nhan et al. (2002) showed that agricultural intensification (shift from two to three rice crops) substantially increases the amount of chemical fertilizers used which contributes to increased productivity on the one hand but reduces economic efficiency and increases the environmental impacts on the other hand.

### *13.3.3 Pesticide Use*

Pesticides include insecticides, herbicides, fungicides, and various other substances used to control pests. As a result of the transformation process in Vietnam, pesticide use has nearly doubled in the 1990s by reaching over 40,000 tons of pesticide active ingredients in 1998 (Meisner 2005). In addition, Berg (2001) and Van Mele and van Lenteren (2002) reported increasing pesticide use in the 1990s which was also reflected in the amount of imported pesticides (Fig. 13.1).

In 2005, 33% of pesticide active ingredients used were insecticides, 32% were fungicides, 29% were herbicides and 1% were other compounds such as molluscicide and acaricide (FAO 2007). Due to serious pest outbreaks in 2006, the Vietnamese Plant Protection Departments observed a noticeable increase in pesticide use since 2006 (Plant Protection Departments in Can Tho and Dong Thap, pers. comm.).

The average pesticide application dose in the Mekong delta is in the range of 3.1–7.0 kg a.i./ha (Phuong and Gopalakrishnan 2003) which is rather low when

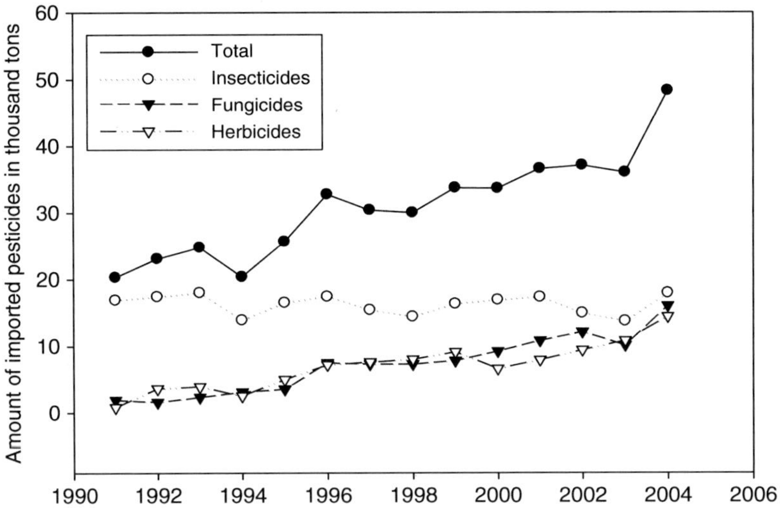


**Fig. 13.1** Amount of plant protection products imported in Vietnam (1991–2004), Huan (2005)

compared to developed countries such as Japan (14.30 kg a.i./ha) and South Korea (10.70 kg a.i./ha) but higher than in other developing countries like the Philippines (1.56 kg a.i./ha) or Bangladesh (1.50 kg a.i./ha) (UNEP 2005).

Research conducted by Pham Van Toan (UNU-EHS 2010) involving household interviews and participatory rural appraisals (PRA) between August 2008 and August 2009 at two study sites in Cai Rang (Can Tho, mixed suburban agriculture with double rice, vegetables and fruits) and Tam Nong (Dong Thap Province, double rice monoculture) surveyed current pesticide usage, application strategies, as well as crop and water management. For insect control, pyrethroids, carbamates, nicotinoides and abamectin were most frequently used. Most popular fungicides belonged to the group of conazoles. For weed control several herbicide types were used frequently such as phenoxy, amide, urea and aromatic acid herbicides. Compared to previous research on pesticide use patterns in the Mekong Delta (e.g. Huan et al. 1999; Berg 2001) organochlorines and organophosphates are today less frequently used, while pyrethroids, carbamates, and biopesticides have become popular.

### *13.3.4 Animal Production*

Livestock production impacts water quality heavily and has been reported to be one of the largest sectoral sources of water pollutants, mainly as animal wastes, antibiotics, hormones, pesticides, microbes and protozoa. It is estimated that in the USA, livestock and feed crop agriculture are responsible for 37% of pesticide use, 50% of antibiotic use, and a third of the nitrogen and phosphorus loads in freshwater resources (FAO 2006).

There is a considerable input of waste from animal husbandry into surface waters in the Mekong Delta. Until recently only a small amount of animal waste was used as organic fertilizer (e.g. as compost, biogas slurry), while the majority is discharged in water courses. Nutrients in animal manure could readily substitute some of the applied mineral fertilizer – enough to supply half the agricultural land with ~60 kg N/ha (Arnold 2010) and increase carbon contents in soils.

Many types of pharmaceutical products are used in animal production, antibiotics being the most common registered products (70% of all veterinary drugs). No comprehensive data exist on veterinary drug use but it is known to be widespread (GARP 2010).

Outbreaks of pig diseases are very common in the Delta. They are usually recognized in terms of their clinical symptoms and treated by antibiotics without precise diagnoses although the high mortality of pigs is caused in many cases by viral infections (Kamakawa et al. 2006). Beside their therapeutic use, antibiotics are also used for disease prophylaxis as well as to promote growth and increase feed efficiency. For example, pigs and poultry are fed with antibiotics such as tetracycline and tylosin. Beside antibiotics, other drugs such as Salbutamol and Clenbuterol have been frequently used as growth promoters. Some of them are banned to be used in animal production in Vietnam (e.g. Clenbuterol since 2002) due to their health risks to humans. Despite the ban, Clenbuterol was reported to occur in pork meat in 30% of the samples taken by Vietnamese authorities in January 2011.[1] Beside growth promotion, Clenbuterol is used with the intention to boost the ratio of muscle to fat tissues and thus to produce leaner meat.

### *13.3.5 Aquaculture*

Similar to livestock production, many inputs are required in aquaculture in the form of feed, pesticides, antibiotics and hormones. Excess fish feed and fish excrements are major contributors to surface water pollution since waste water of aquaculture farms is discharged directly into rivers (63%) or primary canals (19%) and in 90% of the cases without treatment (Phan et al. 2009). Only *ca.* 30% of the fish feed is transformed into a fish product (Jungersen 1996) while the rest is released via excrements back to the water. Bao et al. (2005) showed that nutrient concentration in rivers increased during the time of restoration of shrimp ponds in Ca Mau. Bosma et al. (2009) conducted an environmental impact assessment of the *Pangasius* sector in the Mekong Delta and concluded that the feed component (including feed production) had the largest environmental impact. Anh et al. (2010) showed that 1 ton of frozen *Pangasius* fillets accounted for the release of 2,050 kg total suspended solid (TSS), 106 kg nitrogen and 27 kg phosphorus to surface waters. The overall

[1] VietNamNet Jan 30, 2011 http://english.vietnamnet.vn/en/society/4486/city-pork-tainted-with-harmful-clenbuterol.htmlwith-harmful-clenbuterol.html

waste emissions from *Pangasius* production and processing accounted for only <1% of the total TSS, nitrogen and phosphorus loads in the Mekong Delta. However, the pollutant loadings from individual farms may exceed water quality standards locally, leading to local pollution problems. De Silva et al. (2010) calculated (based on the median nutrient discharge levels for commercial feeds) that *Pangasius* production discharged in the Mekong Delta 31,602 tons N and 9,893 tons P, and 50,364 tons N and 15,766 tons P in 2007 and 2008, respectively.

Additionally, veterinary drugs (e.g. antibiotics), chemical treatment (e.g. pesticides), hormones and feed additives (e.g. vitamins) were reported to be used frequently in *Pangasius* cultures. Kestemont (2007) surveyed chemical and drug use in shrimp farming in North and South Viet Nam reported on 155 different drugs and chemicals which are currently used, including 31 antibiotics, 49 pesticides and disinfectants, and 21 other compounds for water treatment. Antibiotics are used for disease prevention, treatment and also for growth promotion (Sapkota et al. 2008). The main application occurs before stocking (treatment of the seed), in case of disease outbreaks and as a reaction to high market prices of fish in order to promote fish growth. The occurrence of diseases was reported to be highest during the onset of the wet season in June and July (Phan et al. 2009). Hormones are used to control and induce spawning and thus ensure reproduction at the right time. In *Pangasius* breading farms, spawning is induced by hormonal treatment by injecting breeders with human chorionic gonadotrophin (hCG). Females receive doses of 200–6,500 IU/kg per treatment and are injected up to four or five times.[2] The combined dose rate for a female *Pangasius* is in average 5,400 IU/kg. Males receive in average 1,160 IU/kg, usually in a single injection at the time of the last injection given to females (Bui et al. 2010). Hormones are also used to create mono-sex cultures in case males or females have advantages in terms of food conversion rate or final product size. For example, sex conversion to males by feeding with methyl-testosterone (Chakraborty et al. 2011) is a common practice in many countries in *Tilapia* cultures. However, this practice was not frequently applied in the Mekong Delta in the recent past. In 2003 only 20% of the Tilapia seed producers around Ho Chi Minh City were aware of male-monoculture *Tilapia* production techniques (Huy et al. 2003). However, *Tilapia* production is on the increase (farmers moving *Pangasius* from cages to ponds while stocking the cages with *Tilapia*), thus more advanced cultivation techniques including hormonal treatment might follow.

In terms of chemical treatment the nursery stage is probably the most intensive stage. Due to high mortality rates, farmers are intensively using chemicals, such as copper sulphate, and antibiotics. Beside the high susceptibility of the fish seed to diseases, the high use of veterinary products in hatcheries and nurseries could be linked to the fact that they neither bear the responsibility for the final fish product nor they have to be prepared for residue controls at this stage of the production (Poisson 2005).

[2] IU means international unit which is a unit to measure the activity of vitamins and drugs. The unit is not based on volume or weight but on the biological effect expected with a dose of 1 I.U.

Aquaculture contributes also to the (micro)biological contamination of surface water via diseased and dead fishes. Farmers usually remove dead fish floating on the surface but do not dive to remove them from the sediment. Dead fishes collected from ponds are often eaten or released to the surface water (Poisson 2005).

Local state authorities (e.g. DONRE Can Tho, personal communication, 2009) consider aquaculture as one of the major pollutants in the Lower Mekong region. Fish processing factories are an additional and growing pollution source in the region.

## 13.4 Main Agriculture Related Water Pollutants in the Mekong Delta

### *13.4.1 Suspended Matter*

The Mekong has a rather high suspended matter load by nature. Mean value of total suspended solids (TSS) at the Mekong River Commission's (MRC) stations in the MD were around 75 mg/L (MRC 2008). The Delta stations had similar TSS concentrations as the mainstream stations but more than the stations in the tributaries. Recent measurements (2009–2010) conducted by the German Research Centre for Geosciences (GFZ) in flood plains and irrigation canals in Dong Thap province showed average TSS concentrations around 100 mg/L (Hung et al. 2011). These amounts of TSS are relatively high when comparing it to water quality guidelines (refer to QCVN 08:2008/BTNMT). However, Mekong aquatic species are adapted to high suspended sediment concentration. In addition, there is no evidence in the literature referring to the share of agricultural activities – as of interest in this chapter – in the amount of transported solids. An impact on water quality is rather to be expected indirectly via soil particles which reach the river mainly because of rainfall and runoff. Since considerable amount of the surrounding land is used for agriculture, soil particle-bound transport of non-polar agrichemicals applied in the fields will be significant as has been shown by Liu et al. 2008 for the Yangtze River. However, no data is available on the extent of particle-bound transport of agricultural pollutants in the MD.

Beside particle-bound pollutant transport, suspended solids reduce the aesthetic value of the water for personal use, enhance the need for water flocculation before use (e.g. alum treatment), cause a physical disruption of the hydraulic characteristics of the channel with impacts on navigation and potential destruction of habitats (Ongley 1996).

Despite of the above mentioned disadvantages farmers usually welcome suspended matter since it is beneficial in terms of soil fertility. It also limits eutrophication driven algal blooms since high suspended sediment concentration reduces light penetration in water bodies and sequesters phosphorus into particulate-associated phosphorus that is not bioavailable.

### 13.4.2 Nutrients

High amount of nutrients represent a threat to aquatic ecosystems via leaching, surface runoff, subsurface flow and soil erosion from agricultural land (more details are given in Chap. 14). The amount of N and P reaching water systems depends on the time and rate of fertilizer application together with land-use management and site characteristics (e.g. soil texture and profile, slope, vegetation cover) and climate (e.g. rainfall characteristics). While agricultural nutrient leaching is extensively studied in the temperate zone and in particular in Europe and North-America where eutrophication is a problem and where agriculture seems to be the major anthropogenic source of nitrogen and phosphorus pollution (Cartwright et al. 1991; Novotny 1999; EEA 2005), not much information exists on eutrophication in the Mekong Delta (Ongley 2009). A transfer of knowledge from the temperate zones could prove difficult since thresholds at which eutrophication might occur vary among ecosystems (Di and Cameron 2002). Similarly, the risk associated with a certain eutrophication level might be different in the tropics. For example the risk posed by toxic algal blooms and cyanobacteria to aquatic life and humans is higher in the tropics since in temperate regions cyanobacteria are usually not occurring in winter, whereas seasonal changes in temperature in tropical aquatic environments are often not large enough to significantly influence the population (WHO 1999). Also, the growth rates of cyanobacteria are often maximal above 25°C (Robarts and Zohary 1987) which favours growth in tropical regions compared to many temperate regions. Thus, potential threats posed by cyanobacteria to the aquatic ecosystem are larger in the tropics than in the subtropics (WHO 1999). On the other hand, in the Mekong Delta the high load of suspended matter immobilizes large amounts of phosphorus while the low light penetration in the water body hinders algal bloom. Also, although algal blooms result in a high amount of dead organic matter which starts to decay and deplete dissolved oxygen in the water, the effects on fauna may be less severe in the tropics, since there are many species which have evolved breathing organs that extract oxygen from the air. Obligate air breathers, such as the snakehead *(Channa striata)* must breathe air periodically to avoid suffocation (Vivekanandan 1977). Facultative air breathers, such as *Pangasius*, only breathe air if dissolved oxygen concentrations in the water are not sufficient and will otherwise rely on their gills for oxygen (Lefevre et al. 2011).[3]

The threshold values for total-P for the protection of aquatic life used by the MRC are 0.13 mg/L, and 0.7 mg/L for the sum nitrite and nitrate-N. Measurements at Delta stations of MRC often exceed this threshold. Stärz (2012) reported median concentrations of nitrate as 2.2 mg/L in the rainy season and 1.2 mg/L in the dry season in suburban areas of Can Tho while median ammonium concentrations were 0.3 and 0.2 mg/L, respectively. Nutrient concentrations were lower in streams and rivers than in smaller canals. Governmental water quality monitoring programmes

[3] Note that *Pangasius* is not considered anymore as an obligate air breather which was the case over a quarter of a century (Browman and Kramer 1985) with possible implications for the design of *Pangasius* ponds.

recorded 0.6–2.0 mg/L nitrate (1.2 mg/L in average), 0.12–1.01 mg/L ammonium (0.65 mg/L in average) and 0.08–0.22 mg/L nitrite (0.12 mg/L in average) in canals and rivers of Can Tho in 2009 (SOE Can Tho 2009). Nutrient concentrations in Vinh Long (a less populated and more agriculture dominated province) were lower for the same year (SOE Vinh Long 2009). Additional data are presented in Chap. 14. Despite the importance of nutrient pollution little is known about nutrient losses from agriculture in the delta (MRC 2008). Total annual outputs of nitrogen and phosphorus by the Tien and Hau Rivers were estimated as 0.24 million tons and 0.07 million tons, respectively (White 2002). The share of agriculture in total nutrient input to the rivers and canals is unknown. Major contributors to the nutrient load are livestock farming, aquaculture production and human waste. Around 3–5 kg N and 0.5–1 kg P per person per year are released to the environment (Wohlsager et al. 2010). So far no waste water treatment plant is in operation in the Mekong Delta – in Can Tho one is under construction for part of the city.

In summary, nitrogen and phosphorus levels are considered to be non-eutrophic in general (Liljeström 2007). Eutrophication occurs locally and at small scale (ponds, small canals) and is not reported in main canals or rivers. This is due to tidal exchange, high dilution rates and high turnover rates due to the tropical climate. While algal blooms are not a high concern in the MD, extensive growth of water hyacinth (*Eichhornia crassipes*) causes serious problems for irrigation and transport. Water hyacinth growth capacity achieve their maximum levels at nitrogen concentrations around 1 mg/l, which can be reached in smaller canals at the end of the dry season when temperatures are the highest.

### 13.4.3 Pesticides

Pesticides generally reach water courses as pulses generated at application time (due, for example to spray drift) or after major rainfall events (surface runoff, interlayer flow or drainflow). Once they are released into the aquatic environment, their fate depends largely on their physicochemical properties and on the prevailing environmental conditions influencing their degradation rates (Renaud et al. 2008). In the recent past, organophosphate and organochlorine insecticides were widely used in agricultural systems of the MD (Minh et al. 2007b; Berg 2001; Quyen et al. 1995). Some of these compounds are typically persistent in the environment (in particular the organochlorines) and can bio-accumulate in fat tissues, potentially threatening aquatic ecosystems for many years, and this despite the fact that many molecules from these classes have been banned in Vietnam. This is, for example, the case for DDT (dichlorodiphenyltrichloroethane) which has been banned in Vietnam since 1995 but can still be found in relatively high concentrations in various environmental compartments throughout Vietnam in general and in the MD in particular, an indication of the past widespread use of the molecule for agricultural purposes and subsequently for vector control (Minh et al. 2007a, b). In the MD, Minh et al. (2007a) reported DDT concentrations in sediments ranging from 0.01 to 110 ng/g dry weight along the Hau River with the higher concentrations found near urban areas such as Can Tho City. In some cases, p,p′-DDT was found in non-urban environments in greater concentrations than its degradation products such as p,p′-DDE suggesting recent inputs of DDT to the system (Minh et al. 2007a). Carvalho et al. (2008) monitored more than 70 pesticide residues and polychlorobyphenyl compounds (PCBs) and found sediment concentrations of ΣDDT ranging from 0.45 to 67.49 ng/g dry weight and concentration in the soft tissues of bivalve molluscs ranging from 5.46 to 123.03 ng/g dry weight. Other organochlorines detected in sediments but at much lower concentrations were hexacyclochlorohexane (HCHs), endosulfans and chlordane (Carvalho et al. 2008). The highest concentrations of DDT were found next to villages and towns and Carvalho et al. (2008) attributed this to vector control activities rather than use in agriculture. However, endosulfan was detected in highest concentrations in rural areas where they are or have been used for agricultural purposes (Carvalho et al. 2008). In a case study area near Can Tho City, Ikemoto et al. (2008) showed that concentrations of ΣDDT measured in crustaceans and fish were the highest when compared to other persistent organic pollutants (POPs) and were shown to bio-magnify across trophic levels.

Despite the increasing number of scientifically-proven pollution problems (e.g. Carvalho et al. 2008; Ikemoto et al. 2008; Minh et al. 2007a; UNU-EHS 2010), regular state-organized monitoring of pesticide pollution in the Mekong Delta is very limited in terms of sampling points, sampling frequency and elements considered (authors observations and discussions with DONREs of Can Tho and Dong Thap). Next to state authorities, the Mekong River Commission conducts water quality monitoring in the Delta, but MRC does not carry out systematic monitoring of pesticides at the MRC stations in Vietnam. However, the MRC conducted a diagnostic study of water quality in the Lower Mekong Basin in 2003–2004 where selected organochlorine, organophosphorous and triazine pesticides were monitored at three MRC stations (Tan Chau – Mainstream, My An – Plain of Reeds and Chau Doc – Bassac). The concentrations were all below the method detection limit but MRC stated in the study that the detection limits in the study were rather high and an assessment of the results was not possible.

In summary, currently used pesticides are not monitored at all or at very low frequencies, with few exceptions such as project-based monitoring. As an example, in ongoing studies by Pham Van Toan (UNU-EHS 2010), 233 water samples were collected in Ba Lang (Can Tho City) and 109 in An Long (Dong Thap Province) from August 2008 to August 2009 in irrigation canals and field outflows. Samples were analyzed for buprofezin, butachlor, cypermethrin, difenoconazole, endosulfan (alfa, beta and metabolite), fenobucarb, fipronil hexaconazole, isoprothiolane, profenofos, pretilachlor, propanil and propiconazole. Recently used pesticides were present at detectable levels throughout the year in the field outflows and in irrigation canals in two agricultural areas of the Mekong Delta. Median concentrations of the quantified compounds ranged from 0.01 to 2.72 μg/L in An Long and from 0.01 to 0.38 μg/L in Ba Lang. The frequent detection of pesticides in rural areas indicates that surface water quality may deserve special attention in these land-use settings particularly because surface water is often used in rural areas for drinking water.

Pesticide occurrence in surface water itself does not cause necessarily adverse effects on aquatic ecosystems or humans health. In order to determine the potential impact of the pesticide concentrations measured on the environment and aquatic life a comparison with published surface water and aquatic life guidelines or benchmarks namely with Vietnam's National technical regulation on surface water quality (QCVN 08: 2008/BTNMT) and EPA's Aquatic Life Benchmark Table (US EPA 2009) was undertaken. Vietnam's National technical regulation on surface water quality considers only endosulfan among the analyzed pesticides. Endosulfan or its metabolites were measured in 2.6% and 17.4% of the samples in Ba Lang and An Long respectively. In 1.3% and 9.2% of the samples the detected concentration exceeded the B1 quality guideline of this regulation (0.01 μg/L), meaning that the water would not be suitable for irrigation purposes. EPA's Aquatic Life Benchmark Table (US EPA 2009) contains estimates of concentrations below which pesticides are not expected to harm aquatic life. The Table considers propiconazole, profenofos, propanil and cypermethrin. Chronic invertebrate guideline values were exceeded for profenofos in 0.9% of the samples in Ba Lang and in 8% of the samples for Cypermethrin in An Long. Cypermethrin guideline values are regularly exceeded in samples taken after rainfall events. Here, all the detected concentrations exceeded also the acute and not only the chronic toxicity guideline values for invertebrates as well as for fish. In 90% of the samples taken in An Long, and in 45% of the samples from Ba Lang, more than three different pesticides were detected at the same time.

While for temperate climate many studies deal with the fate and toxicological endpoints of pesticides, such data and studies are rare for the tropics. Usually it was argued that pesticides degrade faster in the tropics mainly due to faster temperature-related biological processes. This view was recently challenged by Daam et al. (2008, 2009). Based on a systematic comparison of fate and ecological effects of chlorpyrifos and linuron in freshwater model ecosystems between tropical and temperate regions these authors reported more similarities between the fate and toxicological impacts of these pesticides in both climatic zones than expected previously and recommended the use of toxicity data generated in temperate countries also in the tropics.

### *13.4.4 Antibiotics*

Antibiotics are used in large amounts in aquaculture as well as in animal production (mainly pig and poultry). In aquaculture, the risk of bacterial infections among farmed fish is high mainly due to high fish and fish farm densities and sanitary shortcomings. Therefore, large amounts of antibiotics are used in fish feed for both prophylactic (disease prevention) and therapeutic (disease treatment) purposes in aquaculture worldwide (Sapkota et al. 2008). The use of large amounts of antibiotics leads to antibiotic residues in aquatic organisms (e.g. fish) and thus to unintentional ingestion by consumers. Unintended antibiotic consumption may affect human health directly or indirectly via development of antibiotic resistance in bacteria that are pathogenic to humans. Recognition of the risks associated with the consumption of antibiotics on human health has led to bans on the use of certain antibiotics in animal food production (particularly those antibiotics for which no safe residue levels can be determined, Decision No 07/2005/QD-BTS), and for example in the European Union to general bans of antibiotics for growth promotion (EC No 1831/2003) and to the establishment of maximum residue limits (MRLs) for those with known risks (FAO 2002).

Next to the effect on food safety and human health there is also an impact on the aquatic environment. The use of a wide variety of antibiotics in large amounts leads to detectable concentrations or even persistence in the environment and to the emergence of antibiotic-resistant bacteria in aquaculture environments, to the increase of antibiotic resistance in fish pathogens, and in alterations of the bacterial community both in sediments and in the water column (Cabello 2006; Huys et al. 2007; Martinez 2009).

Despite the widespread use of antibiotics in aquaculture, limited data are available on concentrations of antibiotics occurring in the aquatic environment in Vietnam. In shrimp ponds in mangrove areas, trimethoprim, sulfamethoxazole, norfloxacin and oxolinic acid were found in all sediment samples in both pond and surrounding canal samples at maximum concentrations of 73.6, 820.5, 2615.9, 426.3 μg/g wet weight, respectively, while in water samples taken from shrimp ponds and surrounding canals, these compounds were observed at concentrations of 1.0, 2.4, 6.1 and 2.5 mg/L, respectively (Tuan and Munekage 2004). In waste water of pig farms and in canals near chicken and pig farms sulfamethazine was reported at concentration from 18.5 to 19.2 μg/L (Managaki et al. 2007).

Currently, there is no regulation in place with regard to maximum allowed concentrations of veterinary drugs in surface water. The generation of sufficient scientific information on environmental concentration levels and the assessment of their relevance with regard to their impact on aquatic life is ongoing. As a point of reference we provide here the outcome of a survey undertaken by the U.S. Geological Survey (USGS) in 1999–2000 to assess the occurrence of human and veterinary drugs (including antibiotics), natural and synthetic hormones, detergent metabolites, plasticizers, insecticides, and fire retardants in 139 U.S. streams (Barnes et al. 2002). In this survey antibiotics were detected in every second sample with average concentrations of 0.1 μg/L in the stream water.

Several classes of antibiotics were detected (commonly at concentrations of >100 μg/l) in swine waste storage lagoons in the USA. In addition, antimicrobial compounds were detected in surface and groundwater samples collected near to swine and poultry farms (Campagnolo et al. 2002).

In terms of possible toxicological effects, plants, cyanobacteria and soil microorganisms have been shown to be sensitive to many of the antimicrobial groups (Boxall et al. 2003). Reported EC50 values[4] for cyanobacteria with amoxicillin, benzylpenicillin, sarafloxacin, spiramycin, tetracycline, and tiamulin are all below 100 μg/L (Holten-Lützhoft et al. 1999).

### *13.4.5 Microorganisms*

Coliforms are found throughout the environment. Although most of them are relatively harmless and are easily killed with chlorine or by boiling, total coliform

[4] EC50 refers to the molar concentration of an agonist (e.g. an antibiotic) that produces 50% of the maximal possible effect of that agonist.

counts are indicators of how much the water is polluted by human and animal wastes (FAO CDR 1997).

Agricultural activities, especially animal production are important sources of microbial contamination in surface water. Extensive studies on the quality of wastewater and water use in animal production in various parts of Vietnam revealed serious pollution problems (Tien et al. 2009). Wastewater contained very high number of *E. coli* and *Samonella* indicating a severe threat to human and animal health and to the environment. Only relatively low percentage of households or animal farms (15–36%) had waste water treatment. All analyzed parameters such as biological oxygen demand (BOD), coliform counts and *E. coli* exceeded the national technical standard. *Salmonella* was found in 34% of animal waste water samples (Tien et al. 2009).

Similarly, a study in Tan Phu Thanh, a village in the vicinity of Can Tho city, showed a high level of faecal contamination with the presence of aerobes, coliforms and *E. coli* in three types of drinking water of animals. Due to the lack of waste water treatment, animal faeces were directly discharged into canal or river resulting in the pollution of water bodies. River water samples had the highest microbial pollution. The concentration of aerobes, coliforms and *E. coli* in well water was lower than that of river water showing that many organisms were filtered by soil. The traditional approach to put pig manure and family waste directly to the river or to fish pond as feed for fish have made the river in this area more polluted. Bio-digester, used in less than 5% of households, reduced microbial contaminations in water by one log unit (from $10^8$ to $10^7$) MPN 100 $ml^{-1}$ (Arnold 2010; Phan et al. 2003). However, this degree of reduction is not sufficient to reach acceptable water quality without further treatment (Arnold 2010). Not only *E. coli* but also other enterobacteria were detected in water samples from canal, swine drinking water tanks, fish ponds and drilled wells in swine breeding environment in Can Tho (Kobayashi et al. 2003). These studies indicated potential health risk to humans and other organisms.

*Salmonella* spp. are found in intestinal tract of human, domestic and wild animals and known as pathogens to humans (FAO CDR 1997). The faecal excretion of *Salmonella* by domestic animals may lead to contamination of surface water in rivers and lakes and to spreading of the pathogen to other domestic animals. In Tan Phu Thanh village (Chau Thanh district, Hau Giang province) Phan et al. (2003) presented indirect evidence that pig excreta directly discharged into ponds or canals caused the pollution of the surface water and the water used for animal watering by *Salmonella* spp. which was isolated from 43%, 37%, and 38% in water from Ba Lang River, canal and pond, respectively.

## 13.5 Implications for Ecosystem Services

Water quality degradation by agriculture has the potential to negatively affect other water users at multiple scales as well as related ecosystem services. In general terms, agriculture increases the value of provisioning ecosystem services (fuel, food, timber etc.) at the expense of regulating (e.g. pest control, pollination), cultural (recreation etc.) and supporting (e.g. soil formation) services (Gordon et al. 2010).

Excess fertilizer and feed use, the discharge of animal wastes into surface waters and agricultural practices that degrade soil quality lead in many places of the world to eutrophication of water bodies with a depletion of dissolved oxygen in the water column, algal blooms, increased turbidity, reduction in aquatic animal populations and excess growth of undesired (and often invasive) aquatic vegetation. In terms of ecosystem services this means increasing costs for water purification (regulation), and decreasing fishery (food provision) and recreational values (cultural service) (Tilman et al. 2002). Large amounts of nutrients have the potential to impact on food provision by causing nutrient imbalances for the crops and enhanced susceptibility of the plants to some pests and diseases (Altieri and Nicholls 2003). Looking specifically at the Mekong Delta, eutrophication seems to be not a major problem to date and rather an issue in smaller water courses and canals then in the main distributaries. Additionally, beside the above mentioned negative influences of an enhanced nutrient concentration of the water, the addition of nutrients to the system also leads to increased productivity of e.g. microalgae although species diversity is likely to decrease (Hillebrand and Sommer 2000). Similarly, the spread of invasive species such as the water hyacinth has advantages by providing a new resource for animal feed, fertilizer and woven products such as baskets, bags and furniture. While enhanced nutrient content has also advantages, no positive effects can be attributed to pollution by pesticides, harmful microorganisms or antibiotics. Pesticides have the potential to reduce biodiversity and affect natural pest regulation by harming pest predators (Dale and Polasky 2007). Antibiotics influence microbial communities by changing their composition and allowing for the development of microbial resistances to antibiotics. This might have negative effects on nutrient cycling in soils as a supporting service. Metal ion leaching from acid sulphate soils used for crop production and low pH can potentially change aquatic organism communities by harming sensitive species after exceeding their threshold toxic concentrations (food provision and cultural services) (Minh et al. 1997). In general terms, any practice that changes species composition or reduces biodiversity in aquatic ecosystems may also negatively influence goods and services, since the provision of services depends both on the number and type of species in an ecosystem (Loreau et al. 2001; Tilman et al. 2002).

In the next section, some examples of intra-agricultural feedback loops are reviewed for the Mekong Delta with respect to low water quality as a threat to food provisioning.

### *13.5.1 Fish Production*

One of the major threats to aquaculture is poor water quality. Peaks of mortality are regularly observed in aquaculture during the flood season probably influenced by runoff from agriculture and urban waste water (Poisson 2005). Organisms in aquaculture are generally sensitive to water pollution. Degradation of water quality leads to an increase in the frequency and severity of diseases, mainly by reducing the resistance of the host organisms and the level of unspecific immunity to diseases

(Svobodová et al. 1993). Especially low oxygen levels, elevated ammonia concentrations, chronic sublethal exposure to heavy metals, pesticides or polychlorinated biphenyls were reported to enhance the predisposition of fish to viral diseases (Svobodová et al. 1993). Organic pollution of water, followed by a decreased content of dissolved oxygen, creates a favourable environment for the growth of bacteria. For catfish the most severe diseases are bacillary necrosis disease of Pangasius (BNP, white spot) caused by the bacteria *Edwardsiella ictaluri* (Crumlish et al. 2002), parasites (e.g. Trichodina spp. and Epistylis spp.) and red spot caused by different motile *Aeromonas* (bacterium) species (Phan et al. 2009; Khoi 2011). Reduced water quality is a predisposition for the infection with the bacteria *Edwardsiella ictaluri* in catfish which can cause 50–90% mortality (Crumlish et al. 2002). The same is the case for aeromonas (red spot) (caused by different motile aeromonas species), and parasites e.g. *Trichodina spp.* and *Epistylis spp.* (Phan et al. 2009; Khoi 2011).

In acid sulphate soil areas, acidity of water might cause direct fish kill, while aluminium damages the fish gills (Sammut et al. 1996). A combination of low dissolved oxygen content, high aluminium concentrations and slightly acidic water (pH around 5) was reported to be a lethal combination for fish (Sammut et al. 1996).

### *13.5.2 Crop Production*

Several of the impacts of decreased water quality are related to the impairment of agricultural production. Higher loads of suspended matter are advantageous for soil fertility while chemical and microbial pollution may negatively affect crops. Low quality water might also lead to the accumulation of contaminants in the soil, which may impact soil quality negatively in the long run. Beside direct effects, the presence of pesticides or pathogens in irrigation water affects the saleability and acceptance of the agricultural product for the market against the background of food safety as an emerging issue. Phytotoxic compounds such as trace metals and pesticides may influence the growth or yield of plants (Enderlein et al. 1997).

In acid sulphate soil areas low pH causes the mobilization of phytotoxic aluminium ions. Aluminum toxicity is the primary factor limiting crop productivity in acidic soils (Chap. 14). High concentrations of aluminum affect root system structure, whole plant growth and seed yield (R'bia et al. 2011).

## 13.6 Scale Considerations

Using MRC data, Campbell (2007) concluded that concerns about water quality in the delta are not based on empirical evidence although in this volume, Campbell notes that the problems of water pollution are one of the major threats to biodiversity

(Chap. 11). Similarly, Ongley (2009) analysed existing MRC data on water quality in the main streams of the river and concluded that the water quality of the Lower Mekong is good and suitable for aquatic life and for agricultural use. However, the assessment has been done with respect to the main stream. While the dataset with regard to dissolved oxygen, conductivity, phosphorus, nitrogen, ammonia and COD was large, other compounds like metals, polychlorinated biphenyls (PCBs), polycyclic aromatic hydrocarbons (PAHs), dioxins, pesticides were only analyzed in the frame of a diagnostic study with some limitations regarding detection limits and number of samples and sampling sites (MRC 2007). However, in smaller distributaries and canals the pollution is likely to be more severe and largely not monitored. UNU-EHS (2010) reported on the co-occurrence of several pesticides in irrigation canal water as well as in sediment. Similarly, the impacts of acid sulphate soils on water quality are negligible in the Tien and Hau Rivers but significant in the smaller canals (Chap. 14). The Vietnam-Netherlands Mekong Delta Masterplan project reported recently (2011) also that the water in smaller canals was more polluted.

With regard to aquaculture, De Silva et al. (2010) argue that the amount of nutrient discharge in the canals and rivers is small compared to the potential nutrient run-off from fertilizers used in rice farming in the Mekong Delta (7.48 million kilogram fertilizer used in total in 2000 (Truong 2003)) while Bosma et al. (2009) found that the nutrient load in the Mekong River did not change recently as compared to the period prior to the expansion of the catfish farming sector. However, during discharges of sludge from the ponds, nutrient output into the receiving canals reaches very high concentrations degrading water quality locally, especially when the ponds are located along small canals and during the dry season when the water level is low (Bosma et al. 2009). The contribution of *Pangasius* farming to eutrophication in small canals seems to be dependent mainly on the type of sludge management (Bosma et al. 2009). Anh (2010) also pointed to the fact that there is a considerable fluctuation with regard to water pollution among aquaculture units and that pollution levels depend very much on the applied technologies and management schemes.

It is therefore important to consider smaller canal systems when addressing the issue of water quality. People in the MD rely strongly on canal water for their livelihoods, domestic use and hygiene and are therefore constantly exposed to pollutants in these systems. Most people rely on the smaller canals for this which implies that monitoring programmes, when put in place, need to consider these smaller systems to inform local populations of potential pollution threats.

## 13.7 Mitigation Options

Looking at the agricultural sector in Vietnam, there is a (slow) move towards more sustainable agricultural production due to growing awareness of the government and the public for environmental and food quality issues and due to export limitations such as for fish and shrimps because of violation of guideline values (e.g. from the

European Union, Japan, USA). Mitigation measures include efforts to (i) reduce the use of agrichemicals which might end up as pollutants in water (e.g. via awareness raising, education, incentives, enforcement of regulations), (ii) reduce the likelihood that these substances reach aquatic ecosystems (e.g. mode and time of application, buffer zones), (iii) enhance the resilience of ecosystems to deal with pollution problems (e.g. support biodiversity, balance ecosystem services), and (iv) promote environmentally friendly products on the market. In the following sections some of the relevant mitigation measures for the MD are summarized (without claiming that this list is exhaustive).

The Vietnamese government has put considerable efforts into the promotion of different campaigns aiming to enhance farmers' income and simultaneously reducing pollution originating from agriculture. In the field of crop production, several programs were promoted in the last decade. The Centre for Agriculture and Fisheries Extension (CAFE) is one of the important organisations in agriculture and aquaculture in general and in plant protection in particular. It is the office which designs, implements and promotes important pest and disease control programs in Vietnam. In 2009, CAFE had a budget of 17.6 billion VND (10% of the total budget) for educational programs including cooperation with the national media (CAFE report 2009).

### *13.7.1 Ba Giam, Ba Tang (3 Reductions, 3 Gains)*

The Ministry of Agriculture and Rural Development (MARD) started to promote the programme "Three reductions, three gains" (3R3G) in 2003. CAFE promoted the programme with the participation of 4,140 households in a total area of 2,027 ha. 3R3G was started in few provinces in the Mekong Delta and later on expanded to the rest of the country. This rice production model has been successful in increasing rice yields, lowering costs of production thanks to the reduction in seed and 13–33% of pesticide and fertilizer use and thus achieved the overall net gain of US $35–58/ha in winter-spring and summer-autumn rice respectively (Huan and Chien 2010; Van 2010).

### *13.7.2 Mot Phai, Nam Giam (1 Must Do, 5 Reductions)*

This program builds on the 3R3G policy and encourages farmers to reduce seeding rate, fertilizer, pesticide, and water use as well as postharvest losses (5 Reductions). At the same time it advises farmers to use certified seeds (1 Must Do). This program was started in An Giang in 2006 and subsequently implemented in larger areas of the delta in 2009 with a strong support from MARD through CAFE. Many workshops have been conducted and posters have been displayed at public sites such as in schools, coffee shops, and farmer's clubs in different provinces. Furthermore, the Department of Agriculture and Rural Development (DARD) in each province also distributed brochures and booklets about the Mot Phai Nam Giam program (IRRI Annual Report 2009).

### 13.7.3 Bon Dung (4 Rights)

CAFE has also implemented the program "4 rights" to reduce the use of agrichemicals. CAFE recommended farmers to use the correct pesticide, at correct application doses, at the right time with adequate application methods to control brown planthopper.[5]

### 13.7.4 Cong nghe Sinh Thai (Ecological Engineering)

In 2009, the Plant Protection Department (PPD-MARD) launched in collaboration with the International Rice Research Institute (IRRI) and the Cuu Long Delta Rice Research Institute (CLRRI) the Ecological Engineering (Cong nghe Sinh Thai) program for sustainable rice production in the Mekong Delta with the financial support of the Asian Development Bank (ADB). In 2010 Ecological Engineering (EE) was first introduced by IRRI in two agriculture communities in Cai Be and Cai Lay districts in Tien Giang province. The aim of the program is to control rice pests by using insects' natural enemies to reduce insecticide use and strengthen ecosystem services by planting different flower plants which can attract the natural enemies of rice pests along rice field embankments (Van 2010).

The new EE program is now integrated with 3R3G or 1M5R models to enhance pest control and increase income for farmers while reducing chemical use. This program has been successful and thus expanded to other provinces in the delta (Huan and Chien 2010).

In addition, MARD and CAFE also encourage the use of biological control products against rice pests and diseases. Some biocontrol products introduced by the CLRRI have been shown to be effective to control brown planthopper and thus have helped farmers to increase their income by 7–11% (CAFE report 2009; CLRRI report 2009).

### 13.7.5 Organic Farming

Organic agriculture has emerged in Vietnam over the past decade, particularly in response to health concerns around inappropriate pesticide use. In 2007 MARD issued "National Standards for Organic Production and Product Processing" (Decision 07/2007/QD-BNN 2007). These standards are based on the Basic Standards of the International Federation of Organic Agriculture Movements (IFOAM). For some products (e.g. vegetables) the promotion of organic farming could be a viable option to reduce environmental pollution. Beside less pollution of the water sources, organic farming tends to increase soil fertility and water retention

---

[5] CAFE website http://www.khuyennongvn.gov.vn

capacity in the long run (Niggli et al. 2007). However, products from organic farming are generally more expensive to produce and thus the market capacity for these products will be most likely the limiting factor with respect to the future development of the sector.

### 13.7.6 Aquaculture

Similarly to crop production, the first choice is to avoid pollution by decreasing the inputs of possible pollutants in aquaculture by e.g. reducing the use of water, feed, pesticides and veterinary drugs. In addition, waste water and sludge should be treated before released to the aquatic environment. Sediment ponds can be used to hold the wastewater prior to discharge into the river. Although this method could be meaningful, too short retention times and too small pond sizes decrease its effectiveness. Lang et al. (2009) suggested a trickling-filter system as an effective and cost-friendly version for pre-treatment of aquaculture waste water while Anh (2010) suggested an integrative approach by describing the idea of an eco-agro-industrial cluster for pollution reduction by pooling processing industries, animal feed and composting plants and an anaerobic wastewater treatment plant in one cluster.

Pollution caused by the aquaculture sector in the delta received considerable public attention in Vietnam as well as in the receiving countries for the products. Bush and Duijf (2011) recently analyzed the role of political economy in the perceptions and regulations with regard to the quality of *Pangasius* and pointed to the fact that several recent studies (Anh et al. 2010; Bosma et al. 2009; De Silva et al. 2010) concluded that *Pangasius* and shrimp production would pollute water resources to a lesser extent then formerly anticipated although there was plenty of room for improvement.

### 13.7.7 Regulation

To date, 29 pesticides have been banned and 17 were restricted in use in Vietnam (latest version: Decision No 23/2007/QĐ-BNN). The existing law is already taking into account a similar list of hazardous compounds as, for example, the regulation in the European Union, by including many organochlorine compounds such as DDT, aldrin, lindane, chlordane, endosulfan and dieldrin. Vietnam is among the parties of the Stockholm (ratified in 2002) and the Rotterdam Conventions (accession in 2007). Although monitoring data suggests that some of the banned pesticides are still in use, compared to historical values, Minh et al. (2007a) concluded that DDT concentrations are overall decreasing in southern Vietnam but concentrations of some samples remained higher than environmental quality guidelines selected by the authors. It seems that measures other then improved regulation need to be applied to improve the water quality in the delta with respect to these compounds.

In aquaculture some antibiotics (e.g. Chloramphenicol, Metronidazole and Nitroimidazole) and pesticides (e.g. Trichlorfon) have been banned and others have been restricted (e.g. Difloxacin, Tetracycline and Cypermethrin, Deltametrin) for use per Decision 07/2005/QD-BTS. However, there is a need for more effective control as proven e.g. by the common use of fluoroquinolones 8 months after their ban in Vietnam (Poisson 2005). Additionally, next to the tests performed by the buyers (mostly for malachite, chloramphenicols and nitrofurans), regular monitoring for the use of permitted and non-permitted substances should be performed at the farm level and not only on the processing level.

### *13.7.8 Consumer's Trust in Safe and Environmental-Friendly Agricultural Products*

Consumers demand (in Vietnam as well as in importer countries of Vietnamese products) for safe products in terms of human health and environmental pollution is a possible driver of change. This can be supported by regulations like Decision No. 46/2007/QD-BYT on the "Maximum Level of Residue of Biological and Chemical Substances Allowed in Food". Beside regulations, voluntary measures like the implementation of best management practices accompanied by product certification or the setup of reliable supply chain management practices are of importance as described below for *Pangasius* in Vietnam.

At the international level the Pangasius Aquaculture Dialogue (PAD) which was initiated by the World Wildlife Fund (WWF) operates by including a large variety of actors and stakeholders aiming to develop standards of sustainable aquaculture. Products should be certified by the newly established Aquaculture Stewardship Council (ASC). Other approaches include the implementation of Best Aquaculture Practices (BAP) with ACC (Aquaculture Certification Council) certification, Code of Conduct for Responsible Aquaculture (CoC), Safe Quality Food 1000 (SQF 1000) and GlobalGAP. However, implementation of these practices is still in its infancy. GlobalGAP (formerly known as EUREPGAP), PAD and BAP are probably the largest programmes in Vietnam (Belton 2010). GlobalGAP is a recognized reference for Good Agricultural Practices (GAP) in the global market. In the last few years the Vietnamese government was also initiating activities to apply good agricultural practices and gain corresponding certificates such as GlobalGAP or VietGAP. In this context, the German Technical Cooperation (GTZ) supported the development of the GlobalGAP standard in Vietnam for *Pangasius* (GlobalGAP 2009). The test aquaculture sites were located in An Giang province, Vietnam. The new standard was announced in April 2009 (GlobalGAP 2009).

The WWF has set Vietnamese *Pangasius* on the "red list" in 2010 meaning that its consumption would not be recommended because of unsustainable farming practices. The decision hit the Vietnamese *Pangasius* industry strongly and led to a dialogue with WWF. As a result Vietnam and WWF have signed a Memorandum of Understanding (MoU) which committed both sides to an initial 5 years of co-operation to promote sustainable practices and gain the Aquaculture Stewardship Council (ASC) label. After the MoU was signed WWF moved *Pangasius* from the "red list" to a new category named "Moving towards certification". Since ASC is co-funded and promoted by WWF there were some criticisms on WWF's mode of operation in several internet fora.[6] Furthermore, Belton et al. (2011) analysed critically the Vietnamese *Pangasius* market with respect to the types of aquaculture units that would likely gain certificates and if this would be relevant for the market position of these actors as well as for the environment. They concluded that the certification will likely result in greater polarisation between larger and smaller farmers and that the advantages for the local environment could be of minor significance.

Beside certification processes, large international companies are establishing reliable supply chain management practices for fruits and vegetables for their own markets in Vietnam. For example, a large retailer and the German Technical Cooperation (GTZ) carried out a joint project to enhance farmer's capacity to produce safe fruits and vegetables and to develop a supply chain for the retailer's markets (GTZ 2008).

### *13.7.9 Payment for the Preservation of Ecosystem Services*

Internalization of the costs of environmental pollution is a repeatedly discussed option to sensitise farmers for the environmental aspects of their production. With regard to pesticide use, there would be several ways to place a tax (tax on rice output, tax on residue discharge or tax on pesticide input), the only practicable option being tax on pesticide input. Before 1996, there was no tax on pesticides in Vietnam and it was one of the high priority import items. Since 1996, tax rates are in the range of 1–5% on pesticide import and domestic production. When compared with other imported goods with high tax rates such as 100–150% for motorcycle, or 150–200% for car, the tax rate on pesticide appears negligible. A higher tax could be levied as an incentives to reduce the quantity of pesticides used. However, experiences of OECD countries with taxes on agrichemicals show that a direct tax (fix percentage of the pesticide price) does not reduce the amount of active ingredients used significantly. More promising approaches include tax systems which account for the different toxicity of pesticides (incentives for replacements of toxic chemicals with more environment friendly substitutes) or recycle the revenue, with revenues redirected to research and the promotion of more sustainable alternatives (Pretty et al. 2001).

Beside tax, incentives could be considered for facilitating mitigation options like the move to set aside some of the land for waste water and sediment treatment in aquaculture areas.

---

[6] e.g. www.fair-fish.ch; http://qualasaexpertise.wordpress.com

## 13.8 Outlook

Water quality in the Mekong Delta is an often articulated but not comprehensively monitored concern. While water quality appears to be surprisingly good in the main streams with regard to regularly monitored parameters, it is largely unknown or reported to be degraded in smaller water courses and canals. These smaller water courses deserve definitely larger attention since a large proportion of the rural population relies on them for their everyday life in terms of irrigation water, water for personal hygiene, washing and – in the dry season – partly also for drinking. The share of agriculture in the degradation of water quality is unknown at the Delta scale but is often obvious at smaller scales, especially in the dry season during low flows. In terms of ecosystem services, there is little scientific evidence published on the deterioration of specific services due to agricultural related water pollution in the delta. This might be due to the fact that specific studies have not been undertaken so far.

Severe droughts, as have repeatedly occurred in the recent past in the dry season have caused low water levels and thus exacerbated the water quality concerns. On the other hand, predicted effects of climate change will likely lead to enhanced agrichemical use in the Mekong Delta (Sebesvari et al. 2011) which increases again the threat to aquatic ecosystems. It is thus recommended not to take for granted that high dilution factors and fast degradation processes would continue to maintain the water quality within an acceptable range but to take good care of water quality by, for example, improving agricultural management approaches, implementing monitoring programs, accounting for important ecosystem services, and establishing incentives for the preservation or regeneration of key services.

## References

Altieri MA, Nicholls CI (2003) Soil fertility management and insect pests: harmonizing soil and plant health in agroecosystems. Soil Tillage Res 72:203–211

Anh PT (2010) Mitigating water pollution in Vietnamese aquaculture production and processing industry. The case of pangasius and shrimp. PhD-thesis, Wageningen University

Anh PT, Kroeze C, Mol A, Bush S (2010) Water pollution by *Pangasius* production in the Mekong Delta – Vietnam: cause and options for control. Aquacult Res 42:108–128

Arnold U (2010) Options for reuse of nutrients from waste water in the Mekong Delta, Vietnam. Bonner Agrikulturchemische Reihe, 42, Bonn. http://hss.ulb.uni-bonn.de/2010/2385/2385.htm. Accessed 2 Aug 2011

Bao TQ, Hao NV, Dehairs F (2005) Behavior of nutrients in natural river and shrimp farming area in Bac Lieu Province, Mekong Delta. In: Proceedings of the international symposium on sustainable development in the Mekong River Basin, 6–7 October 2005, Ho Chi Minh City. http://mekong.job.affrc.go.jp/0510067Proceedings.pdf#page=33. Accessed 15 Apr 2011

Barnes KK, Kolpin DW, Meyer MT, Thurman EM, Furlong ET, Zaugg SD, Barber LB (2002) Water quality data for pharmaceuticals, hormones, and other organic wastewater contaminants in U.S. streams, 1999–2000: U.S. Geological Survey Open-File Report 02–94

Belton B (2010) The Social Relations of Aquaculture Development in South and Southeast Asia. Ph.D. thesis, University of Stirling, Scotland, UK

Belton B, Haque MM, Little DC, Sinh LX (2011) Certifying catfish in Vietnam and Bangladesh: who will make the grade and will it matter? Food Policy 36:289–299

Berg H (2001) Pesticide use in rice and rice–fish farms in the Mekong Delta, Vietnam. Crop Prot 20:897–905

Bosma RH, Hanh CTT, Potting J (2009) Environmental impact assessment of the *Pangasius* sector in the Mekong Delta. Wageningen University, Wageningen, 50 pp

Boxall ABA, Kolpin DW, Halling-Sorensen B, Tolls J (2003) Are veterinary medicines causing environmental risks? Environ Sci Technol 37:286–294

Browman MW, Kramer DL (1985) Pangasius sutchi (Pangasiidae), an air-breathing catfish that uses the swimbladder as an accessory respiratory organ. Copeia 1985(4):994–998

Bui TM, Phan LT, Ingram BA, Thuy NTT, Gooley GJ, Hao NV, Phuong NT, De Silva SS (2010) Seed production practices of striped catfish, *Pangasianodon hypophthalmus* in the Mekong Delta region, Vietnam. Aquaculture 306:92–100

Bush SR, Duijf M (2011) Searching for (un)sustainabilty in pangasius aquaculture: a political economy of quality in European retail. Geoforum 42:185–196

Cabello FC (2006) Heavy use of prophylactic antibiotics in aquaculture: a growing problem for human and animal health and for the environment minireview. Environ Microbiol 8:1137–1144

CAFE [Centre for Agriculture and Fisheries Extension] Report (2009) http://www.khuyennongvn.gov.vn/b-vanban-tb/d-bao-cao-hang-nam. Accessed 5 Apr 2011

CAFE [Centre for Agriculture and Fisheries Extension] website (2011) http://www.khuyennongvn.gov.vn. Accessed 24 Mar 2011

Campagnolo ER, Johnson KR, Karpati A, Rubing CS, Kolpin DW, Meyer MT, Esteban JE, Currier RW, Smith K, Thu KM, McGeehin M (2002) Antimicrobial residues in animal waste and water resources proximal to large-scale swine and poultry feeding operations. Sci Total Environ 299:89–95

Campbell I (2007) Perceptions, data, and river management: lessons from the Mekong River. Water Resour Res 43:W02407. doi:10.1029/2006WR005130

Cartwright N, Clark L, Bird P (1991) The impact of agriculture on water quality. Outlook Agric 20:145–152

Carvalho FP, Villeneuve JP, Cattini C, Tolosa I, Thuan DD, Nhan DD (2008) Agrochemical and polychlorobyphenyl (PCB) residues in the Mekong River delta, Vietnam. Mar Pollut Bull 56:1476–1485

Casali J, Gastesi R, Alvarez-Mozos J, De Santisteban LM, Lersundi J, Valle D, de Lersundi R, Gimenez A, Larranaga M, Goni U, Agirre MA, Campo JJ, Lopez M, Donezar M (2008) Runoff, erosion, and water quality of agricultural watersheds in central Navarra (Spain). Agric Water Manag 95:1111–1128

Causape J, Quilez D, Aragues R (2004) Assessment of irrigation and environmental quality at the hydrological basin level II. Salt and nitrate loads in irrigation return flows. Agric Water Manag 70:211–228

Chakraborty SB, Mazumdar D, Chatterji U, Banerjee S (2011) Growth of mixed-sex and monosex Nile tilapia in different culture systems. Turk J Fish Aquat Sci 11:131–138

CLRRI [Cuu Long Delta Rice Research Institute]report (2009) Cuu Long Rice Research Institute, 2009. Annual report. http://clrri.org/lib/hangnam/TK09.pdf. Accessed 10 Mar 2011

Crumlish M, Dung TT, Turnbull JF, Ngoc NTN, Ferguson HW (2002) Identification of *Edwardsiella ictaluri* from diseased freshwater catfish *Pangasius hypophthalmus* (Sauvage) cultured in Mekong delta, Vietnam. J Fish Dis 25:733–736

Cuc NS (2003) Vietnam agriculture and rural area in the renovation period (1986–2002). Nha Xuat Ban Thong Ke, Hanoi, Vietnam

Daam MA, Crum SJH, Van den Brink PJ, Nogueira AJA (2008) Fate and effects of the insecticide chlorpyrifos in outdoor plankton-dominated microcosms in Thailand. Environ Toxicol Chem 27:2530–2538

Daam MA, Van den Brink PJ, Nogueira AJA (2009) Comparison of fate and ecological effects of the herbicide linuron in freshwater model ecosystems between tropical and temperate regions. Ecotoxicol Environ Saf 72:424–433

Dale VH, Polasky S (2007) Measures of the effects of agricultural practices on ecosystem services. Ecol Econ 64:286–296

Dang NH, Danh VT (2008) An overview of the development of the agricultural economy in the Mekong Delta. In: Lensink R, Mai VN (eds) Economic development of the Mekong Delta in Vietnam, CDS Research paper no. 27

Dasgupta S, Meisner C, Wheeler D, Xuyen K, Lam NT (2007) Pesticide poisoning of farm workers: implications of blood test results from Vietnam. Int J Hyg Environ Health 210:121–132

De Silva SS, Ingram BA, Phuong NT, Bui TM, Gooley GJ, Turchini GM (2010) Estimation of nitrogen and phosphorus in effluent from the striped catfish farming sector in the Mekong delta, Vietnam. Ambio 39:504–514

Decision No 07/2005/QD-BTS of February 24 (2005) Promulgation the list of chemicals and the list of antibiotics which are banned or restricted from use in fisheries production and business. The Ministry of Fisheries. http://www.imolahue.org/policy/deci07-2005-qdbts-en.pdf. Accessed 17 May 2011

Decision No 23/2007/QĐ-BNN of March 28 (2007) List of pesticides banned to use in Vietnam – 2007. Ministry of Agriculture and Rural Development. http://www.spsvietnam.gov.vn/EnglishSPS/Pages/List%20of%20Pesticides.aspx. Accessed 17 May 2011

Decision No. 46/2007/QD-BYT from December 19 (2007) Permitted maximum levels of biological/chemical residues in food. Vietnam Ministry of Health. http://www.fas.usda.gov/gainfiles/200808/146295469.pdf. Assessed 17 May 2011

Decision Number: 07/2007/QD-BNN. (2007) Regulations for the management and certification of safe vegetable production. Ministry of Agriculture and Rural Development (MARD), Hanoi

Di HJ, Cameron KC (2002) Nitrate leaching in temperate agroecosystems: sources, factors and mitigating strategies. Nutr Cycl Agroecosyst 64:237–256

Diaz RJ (2001) Overview of hypoxia around the world. J Environ Qual 30:275–281

Drucker AG, Bergeron E, Lemke U, Thuy LT, Valle Zárate A (2006) Identification and quantification of subsidies relevant to the production of local and imported pig breeds in Vietnam. Trop Anim Health Prod 38:305–322

EC No 1831/2003 of the European Parliament and of the Council of 22 September 2003 on additives for use in animal nutrition. http://eur-lex.europa.eu/LexUriServ/site/en/oj/2003/l_268/l_26820031018en00290043.pdf

EEA [European Environment Agency] (2005) Source apportionment of nitrogen and phosphorus inputs into the aquatic environment. EEA report no. 7/2005. European Environment Agency, Copenhagen, Denmark

Enderlein US, Enderlein RE, Peter W (1997) Water quality requirements. In: Helmer R, Hespanhol I (eds) Water pollution control: a guide to the use of water quality management principles. UNESCO, WHO and UNEP, London

FAO [Food and Agriculture Organization of the United Nations] (2002) The state of world fisheries and aquaculture. ftp://ftp.fao.org/docrep/fao/005/y7300e/y7300e00.pdf. Accessed 12 Mar 2011

FAO [Food and Agriculture Organization of the United Nations] (2006) Livestock's long shadow. Environmental issues and options. http://www.fao.org/ag/magazine/0612sp1.htm. Accessed 5 Apr 2011

FAO [Food and Organization of the United Nations] (2007) Plant protection profiles from Asia-Pacific countries. 3.14 Vietnam. http://www.fao.org/docrep/010/ag123e/AG123E18.htm. Accessed 9 Apr 2011

FAO [Food and Agriculture Organization of the United Nations] (2008) The state of world fisheries and aquaculture. ftp://ftp.fao.org/docrep/fao/011/i0250e/i0250e.pdf. Accessed 28 Mar 2011

FAO CDR [Food and Agriculture Organization of the United Nations Corporate Document Repository] (1997) Quality control of wastewater for irrigated crop production. http://www.fao.org/docrep/W5367E/W5367E00.htm. Accessed 15 Mar 2011

FAOSTAT (2012) http://faostat.fao.org/site/342/default.aspx. Accessed 16 March 2012

Fisher H, Gordon J (2008) Breeding and feeding pigs in Vietnam: assessment of capacity building and an update on impacts. ACIAR impact assessment series report 52. http://aciar.gov.au/files/node/8921/IAS52%20full%20text.pdf. Accessed 17 May 2011

GARP National Working Group-Vietnam (2010) Situation analysis: antibiotic use and resistance in Vietnam. GARP report

GlobalGAP (2009) http://www.globalgap.org/cms/front_content.php?idart=821. Accessed Aug 2009

Gordon LJ, Finlayson CM, Falkenmark M (2010) Managing water in agriculture for food production and other ecosystem services. Agric Water Manag 97:512–519

GSO [General Statistical Office of Vietnam] (2003, 2005, 2004, 2008 and 2009) Statistical handbook. Statistical Publishing House, Hanoi, Vietnam. http://www.gso.gov.vn/default_en.aspx?tabid=466&idmid=3&ItemID=8597. Accessed 3 Apr 2011

GTZ [Deutsche Gesellschaft für Technische Zusammenarbeit] (2008) Progress report for PPP measures, fruit and vegetable value chain development and improvement of domestic market policy for the distribution network. http://www.csr-weltweit.de/uploads/tx_jpdownloads/progress_report_PPP_Metro-MoT-GTZ-4._FINAL.pdf. Accessed Aug 2009

Hach CV, Tan PS (2007) Study on site-specific nutrient management (SSNM) for high-yielding rice in the Mekong Delta. Omonrice 15:144–152

Hillebrand H, Sommer U (2000) Diversity of benthic microalgae in response to colonization time and eutrophication. Aquat Bot 67:221–236

Holland J (2007) Walking a tightrope. Seafood International, July 14

Holten-Lützhoft HC, Halling-Sorensen B, Jorgensen SE (1999) Algal toxicity of antibacterial agents applied in Danish fish farming. Arch Environ Contam Toxicol 36:1–6

Huan NH (2005) Nhin lai bien phap su dung thuoc bao ve thuc vat trong cong tac quan ly dich hai. Plant Protection Department, Ministry of Agriculture and Rural Development, p 10

Huan NH, Chien HV (2010) Ecological engineering in an Giang province, Vietnam. http://ricehoppers.net/2010/07/ecological-engineering-in-an-giang-province-vietnam/. Accessed 20 Mar 2011

Huan NH, Mai V, Escalada MM, Heong KL (1999) Changes in rice farmer's pest management in the Mekong Delta, Vietnam. Crop Prot 18:557–563

Huan NH, Thiet LV, Chien HV, Heong KL (2005) Farmers' participatory evaluation of reducing pesticides, fertilizers and seed rates in rice farming in the Mekong Delta, Vietnam. Crop Prot 24:457–464

Hung NG, Güntner A, Merz B, Apel H (2011) Sediment dynamics in the floodplain of the Mekong Delta. European Geoscience Union Annual Assembly 2011, Vienna, Austria

Huy HP, MacNiven VA, Tu NV, Bhujel RC, Little DC (2003) Tilapia seed production in Ho Chi Minh City, Southern Vietnam. http://library.enaca.org/AquacultureAsia/Articles/Oct-Dec-2003/6tilapiavietnam.pdf. Accessed 6 June 2011

Huys G, Bartie K, Cnockaert M, Hoang ODT, Phuong NT, Somsiri T, Chinabut S, Yusoff FM, Shariff M, Giacomini M, Teale A, Swings J (2007) Biodiversity of chloramphenicol-resistant mesophilic heterotrophs from Southeast Asian aquaculture environments. Res Microbiol 158:228–235

Ikemoto T, Tu NPC, Watanabe MX, Okuda N, Omori K, Tanabe S, Tuyen BC, Takeuchi I (2008) Analysis of biomagnifications of persistent organic pollutants in the aquatic food web of the Mekong Delta, South Vietnam using stable carbon and nitrogen isotopes. Chemosphere 72:104–114

IRIN [Integrated Regional Information Networks] (2009) VIETNAM: even bottled water unsafe. http://www.irinnews.org/Report.aspx?ReportId=83965. Accessed 18 Feb 2010

IRRI Annual Report (2009) GAPs fill the gap. http://irri.org/about-irri/annual-reports/annual-report-2009/gaps-fill-the-gap?print=1&tmpl=component. Accessed 10 Apr 2011

Joffre OM, Bosma RH (2009) Typology of shrimp farming in Bac Lieu Province, Mekong Delta, using multivariate statistics. Agric Ecosyst Environ 132:153–159

Jungersen G (1996) Resource-saving and ecological aspects of using wastewater from eel breeding as a source of nutrients, water, and carbon dioxide for plant production. In: Etnier C, Etnier E (eds) Ecological engineering for wastewater treatment, 2nd edn. CRC Press, Inc., Boca Raton

Kamakawa A, Thu HTV, Yamada S (2006) Epidemiological survey of viral diseases of pigs in the Mekong delta of Vietnam between 1999 and 2003. Vet Microbiol 118:47–56

Kestemont P (2007) Improvement of shrimp production sustainability and safety in Vietnam. (BL/13/V06 – BL/12/V07)007). Bilateral collaboration between Belgian Science Policy Office

(BSPO) and Ministry of Science and Technology of Vietnam (MOST). http://www.belspo.be/belspo/fedra/BL/Vietnam/infoficheKESTEMONT_en.pdf. Accessed 15 Aug 2009

Khoi LND (2011) Quality management in the Pangasius export supply chain in Vietnam. The case of small-scale Pangasius farming in the Mekong River Delta. University of Groningen, Groningen

Kijne WJ (2006) Abiotic stress and water scarcity: identifying and resolving conflicts from plant level to global level. Field Crop Res 97:3–18

Knips V (2004) Review of the livestock sector in Mekong Countries, FAO Rome

Kobayashi H, Khai LTL, Phan TT, Yamasaki S, Taniguchi T (2003) Prevalence of pathogenic Escherichia coli in a swine breeding environment in Can Tho province, Vietnam. Jpn Agric Res Q 37:59–63

Lang VT, Vinh KQ, Truc NTT (2009) Environmental consequences of and pollution control options for Pond "Tra" fish production in Thotnot district, Cantho City, Vietnam, EEPSEA, research report no. 20009-RR3

Lefevre S, Huong TT, Wang T, Phuong NT, Bayley M (2011) Hypoxia tolerance and partitioning of bimodal respiration in the striped catfish (Pangasianodon hypophthalmus). Comp Biochem Physiol A Mol Integr Physiol 158:207–214

Liljeström I (2007) Nitrogen and phosphorus dynamics in the Mekong Basin: nutrient balance assessment in a catchment scale. Helsinki University of Technology, Finland. http://civil.tkk.fi/fi/tutkimus/vesitalous/opinnaytteet/liljestrom2007.pdf. Accessed 17 May 2011

Liu M, Cheng S, Ou D, Yang Y, Liu H, Hou L, Gao L, Xu S (2008) Organochlorine pesticides in surface sediments and suspended particulate matters from the Yangtze estuary, China. Environ Pollut 156:168–173

Loreau M, Naeem S, Inchausti P, Bengtsson J, Grime JP, Hector A, Hooper DU, Huston MA, Raffaelli D, Schmid B, Tilman D, Wardle DA (2001) Biodiversity and ecosystem functioning: current knowledge and future challenges. Science 294:804–808

Managaki S, Murata A, Takada H, Tuyen BC, Chiem NH (2007) Distribution of macrolides, sulfonamides and trimethoprim in tropical waters: ubiquitous occurrence of veterinary antibiotics in the Mekong Delta. Environ Sci Technol 41:8004–8010

Margni M, Rossier D, Crettaz P, Jolliet O (2002) Life cycle impact assessment of pesticides on human health and ecosystems. Agric Ecosyst Environ 93:379–392

Martinez JL (2009) Environmental pollution by antibiotics and by antibiotic resistance determinants. Environ Pollut 157:2893–2902

Meisner C (2005) Poverty environment report: pesticide use and poverty in Vietnam. World Bank, Washington, DC. http://siteresources.worldbank.org/NIPRINT/Resources/PEN_Report_PesticideUse_in_Vietnam_Mar2005.doc. Accessed 20 Aug 2010

Minh LQ, Tuong TP, van Mensvoort MEF, Bouma J (1997) Contamination of surface water as affected by land use in acid sulphate soils in the Mekong Delta, Vietnam. Agric Ecosyst Environ 61:19–27

Minh NH, Minh TB, Kajiwara N, Kunisue T, Iwata H, Viet PH, Tu MPC, Tuyen BC, Tanabe S (2007a) Pollution sources and occurrences of selected persistent organic pollutants (POPs) in sediments of the Mekong River delta, South Vietnam. Chemosphere 67:1794–1801

Minh TB, Minh NH, Iwata H, Takahashi S, Viet PH, Tuyen BC, Tanabe S (2007b) Persistent organic pollutants in Vietnam: levels, patterns, trends, and human health implications. Dev Environ Sci 7:515–555

MRC [Mekong River Commission] (2007) Diagnostic study of water quality in the Lower Mekong Basin. MRC technical paper no. 15, Mekong River Commission, Vientiane, 57 pp

MRC [Mekong River Commission] (2008) An assessment of water quality in the Lower Mekong Basin. MRC technical paper no. 19. Mekong River Commission, Vientiane

My Hoa N, Janssen BH, Oenema O, Dobermann A (2006) Potassium budgets in rice cropping systems with annual flooding in the Mekong River Delta. Better Crop 90:25–29

Neubacher H (2007) Pangasius 2007: growth on shaky grounds? Seafood International, 14–15 Aug 2007, pp 27–30

Nhan DK, De NN, Thanh DN (2002) Socio-economic and environmental impacts of intensive rice culture and opportunities for sustainable rice production in the Mekong Delta, Vietnam.

Presented in workshops on Trade liberation and Socio-economic and environmental impacts of rice production, HUA and UNDP, 8–9 Nov 2002 in Hue city, Vietnam
Niggli U, Schmid H, Fliessbach A (2007) Organic farming and climate change. Retrieved from http://www.blauen-institut.ch/tx_blu/tp/tpg/g1695_itc-climate.pdf. Accessed 2 Aug 2011
Novotny V (1999) Diffuse pollution from agriculture – a worldwide outlook. Water Sci Technol 39:1–13
Ongley ED (1996) Control of water pollution from agriculture – FAO irrigation and drainage paper 55. Food and Agriculture Organization of the United Nations, Rome
Ongley ED (2009) Water quality of the lower Mekong River. In: Campbell IC (ed) The Mekong: biophysical environment of an international river basin. Academic, New York
Phan TT, Khai LTL, Loc BC, Hayashidani H, Sameshima T, Watnanabe T, Tanigumchi T, Kobayashi H, Ito H, Akiba M (2003) Isolation of *Salmonella* strains from the aquatic environment and comparison with those of animal origin in Tan Phu Thanh village, Mekong delta, Vietnam. Jpn Agric Res Q 37:237–241
Phan LT, Bui TM, Thuy NTT, Gooley GF, Ingram BA, Hao NVN, Phuong NT, De Silva SS (2009) Current status of farming practices of striped catfish, *Pangasianodon hypophthalmus* in the Mekong Delta, Vietnam. Aquaculture 296:227–236
Phuong DM, Gopalakrishnan C (2003) An application of the contingent valuation method to estimate the loss of value of water resources due to pesticide contamination: the case of the Mekong Delta, Vietnam. Int J Water Resour Dev 19(4):617–633
Poisson G (2005) Pangasius farming practices in Vietnam. A study in view of sustainability issues. Deutsche Geselschaft für Technishe Zussammenarbeit GTZ/Binca PPP Programme Report, 38 pp
Pretty J, Brett C, Gee D, Hine R, Mason C, Morison J, Rayment M, Van Der Bijl G, Dobbs T (2001) Policy challenges and priorities for internalizing the externalities of modern agriculture. J Environ Plan Manag 44:263–283
QCVN 08:2008/BTNMT, QUY CHUẨN KỸ THUẬT QUỐC GIA VỀ CHẤT LƯỢNG NƯỚC MẶT (National technical regulation on surface water quality) (2008) Hanoi 2008
Quyen PB, Nhan DD, Sanh NV (1995) Environmental pollution in Vietnam: analytical estimation and environmental priorities. Trends Anal Chem 14:383–387
R`bia O, Horchani F, Smida I, Mejri M, Aschi-Smiti S (2011) Aluminium phytotoxicity and plant acclimation to acidic soils. Int J Agric Res 6:194–208. doi:10.3923/ijar.2011.194.208
Renaud FG, Bellamy PH, Brown CD (2008) Simulating pesticides in ditches to assess ecological risk (SPIDER): I Model description. Sci Total Environ 394:112–123
Robarts RD, Zohary T (1987) Temperature effects on photosynthetic capacity, respiration, and growth rates of bloom-forming cyanobacteria. N Z J Mar Freshw Res 21:391–399
Sammut J, White I, Melville MD (1996) Acidification of an estuarine tributary in eastern Australia due to drainage of acid sulphate soils. Mar Freshw Res 47:669–684
Sapkota A, Sapkota AR, Kucharski M, Burke J, McKenzie S, Walker P, Lawrence R (2008) Aquaculture practices and potential human health risks: current knowledge and future priorities. Environ Int 34:1215–1226
Sebesvari Z, Le TTH, Renaud FG (2011) Climate change adaptation and agrichemicals in the Mekong Delta, Vietnam. In: Stewart M, Coclanis P (eds) Environmental change and agricultural sustainability in the Mekong Delta. Springer, Dordrecht/New York
Shaddick P (2007) Compound fertilizer markets of South East Asia. IFA Crossroads Asia-Pacific, 17–19 Dec 2007, Bali, Indonesia
SOE Can Tho (2009) Status of the environment. Can Tho, Vietnam
SOE Vinh Long (2009) Status of the environment. Vinh Long, Vietnam
Soong J (2006) Soil fertility and changes in fertilizer use for intensive rice cultivation in the Red River Delta and Mekong Delta of Vietnam (2006). ISP collection. Paper 340
Stärz C (2012) Raumzeitliche Variabilität der Oberflächenwasserqualität in der Provinz Can Tho, Vietnam – Ein Surface Water Quality Tool als Beitrag zum integrierten Wasserressourcenmanagement. PhD-Thesis University of Bonn
Svobodová Z, Lloyd R, Máchová J, Vykusová B (1993) Water quality and fish health. EIFAC Technical paper, EIFAC/T54, FAO, Rome

The Vietnam-Netherlands Mekong Delta Masterplan project (2011) Mekong Delta water resources assessment studies

Thuy NT (2010) Evaluation of catfish by-products as protein sources for pigs in the Mekong Delta of Viet Nam. Doctoral dissertation. Department of Animal Nutrition and Management, SLU. Acta Universitatis agriculturae Sueciae, vol 2010, 69

Tien PD, Dieu ND, Loc HV, Dan BTT (2009) Đánh giá thực trạng ô nhiễm môi trường trong chăn nuôi. Tạp chí chăn nuôi số 4:10–16

Tilman D, Cassman KG, Matson PA, Naylor R, Polasky S (2002) Agricultural sustainability and intensive production practices. Nature 418:671–677

Truong VT (2003) A draft inception report on integrated assessment of trade liberalization in rice sector, Vietnam. UNEP, Hanoi

Tuan XL, Munekage Y (2004) Residues of selected antibiotics in water and mud from shrimp ponds in mangrove areas in Vietnam. Mar Pollut Bull 49:922–929

Tuong TP, Minh LQ, Ni DV, van Mensvoort MEF (1998) Reducing acid pollution from reclaimed acid sulfate previous termsoils: experiences from the Mekong delta, Vietnam. In: Pereira LS, Gowing JW (eds) Water and the environment: innovative issues in irrigation and drainage. E&FN Spon/Routledge, London, pp 75–83

Tuyen NQ (2010) Land law reforms in Vietnam. Past and present. Asian Law Institute, ASLI, working paper no. 015. http://law.nus.edu.sg/asli/pdf/WPS015.pdf. Accessed 18 Mar 2011

UNEP [United Nations Environment Programme] (2005) Integrated assessment of the impact of trade liberalization: a country study on the Viet Nam Rice Sector, UNEP, Geneva, 80 p

UNU-EHS [United Nations University, Institute for Environment and Human Security] (2010) Final report of the contribution of UNU-EHS to the first phase of the BMBF-funded WISDOM project. Report to the German Federal Ministry of Education and Research

US EPA [U.S. Environmental Protection Agency] (2009) Aquatic life benchmark table. http://www.epa.gov/oppefed1/ecorisk_ders/aquatic_life_benchmark.htm. Accessed 17 May 2011

Van N (2010) Phát động "Ứng dụng công nghệ sinh thái trong sản xuất lúa bền vững". http://www.tiengiang.gov.vn/xemtin.asp?cap=3&id=13955&idcha=9662. Accessed on 12 Apr 2011

Van Mele P, van Lenteren JC (2002) Survey of current crop management practices in a mixed-ricefield landscape, Mekong Delta, Vietnam – potential of habitat manipulation for improved control of citrus leaf miner and citrus red mite. Agric Ecosyst Environ 88:35–48

VietNamNet (30-01-2011) City pork tainted with harmful clenbuterol. http://english.vietnamnet.vn/en/society/4486/city-pork-tainted-with-harmful-clenbuterol.htmlwith-harmful-clenbuterol.html. Accessed 2 Aug 2011

Vivekanandan E (1977) Ontogenetic development of surfacing behaviour in the obligatory air-breathing fish Channa (=Ophiocephalus) striatus. Physiol Behav 18:559–562

White I (2002) Water management in the Mekong Delta: changes, conflicts and opportunities, vol 61, IHP-VI technical documents in hydrology. UNESCO, Paris

WHO [World Health Organization] (1999) Toxic cyanobacteria in water: a guide to their public health consequences, monitoring and management. E & FN SPON on behalf of WHO, London/New York

Wieneke F (2005) Acceptance analysis of new technology for sustainable water management and sanitation. A case study of operating farm households in the Mekong Delta, Viet Nam. Dissertation, University of Bonn, Germany

Wohlsager S, Clemens J, Nguyet PT, Rechenburg A, Arnold U (2010) Urine – a valuable fertilizer with low risk after storage in the tropics. Water Environ Res 82:840–847

Yen H (2006) Japan exasperated by substandard VN seafood shipment. Vietnam Net 25 Aug 2006

# Chapter 14
# Aquaculture and Agricultural Production in the Mekong Delta and its Effects on Nutrient Pollution of Soil and Water

**Vo Thi Guong and Nguyen My Hoa**

**Abstract** The Mekong Delta (MD) is the most important area for agriculture and aquaculture production in Vietnam, especially in terms of the production of rice, fruits, shrimp and catfish. However, intensification of rice and shrimp production on both alluvial soils and acid sulphate soils (ASS) has resulted in degradation of soil and water environments. Results of the studies on water quality showed that surface water in two branches of the Mekong River had high biological oxygen demand, chemical oxygen demand and nitrate concentrations, which exceeded the limits of the Vietnamese standard for surface water. In ASS areas, water in the canals showed high levels of aluminium, iron and manganese, and toxic metals such as Arsenic, Cadmium, Copper, Nickel, Lead and Zinc. The concentrations of these metals were higher than that in non ASS areas. For shrimp cultivation, rice-shrimp and extensive shrimp systems were more sustainable in-terms of soil and water quality when compared to the intensive shrimp system. In ASS areas, shrimp cultivation had high risk of failure due to low pH and low alkalinity. Based on recent studies, rice rotation with upland crops and compost amendment were the best practices for maintaining soil fertility and improving rice yield in the intensive rice cultivation areas. Organic amendment also improved soil properties and fruit yields in fruit orchards. In intensive vegetable growing areas, soil available phosphate was very high due to high phosphate fertilization, therefore reducing P fertilizer application is a strategy to both save P resources and reduce production costs for farmers. Further studies on the approaches for sustainability in agriculture and aquaculture production in the MD are needed to adapt to climate change.

V.T. Guong (✉) • N.M. Hoa
Soil Science Department, Can Tho University (CTU), Campus II, 3/2 Street, Ninh Kieu District, Can Tho City, Vietnam
e-mail: vtguong@ctu.edu.vn; nmhoa@ctu.edu.vn

F.G. Renaud and C. Kuenzer (eds.), *The Mekong Delta System: Interdisciplinary Analyses of a River Delta*, Springer Environmental Science and Engineering, DOI 10.1007/978-94-007-3962-8_14, © Springer Science+Business Media Dordrecht 2012

## 14.1 Aquaculture and Agricultural Production in the Mekong Delta

### *14.1.1 Role of Aquaculture and Agricultural Production in the Mekong Delta*

The Mekong Delta of Vietnam is a large floodplain with an elevation of about 0–4 m above mean sea level, and covers about 74% of the total area of the Mekong River Delta. The land was formed relatively recently from alluvial deposits within the past 10,000 years. In Vietnam, it covers an area of about 4 million ha, between longitude 104°30′ to 107°E and latitude 8°30′ to 11°N. The Mekong River system reaches the delta with nine estuaries and is characterized by a dense network of canals. The Mekong meets the Tonle Sap River west of Phnom Penh, and then splits into the Tien and Hau Rivers which flow into the East Sea of Vietnam. The monsoon climate is characterized by a short but distinct dry season and a long wet season. Flooding occurs annually from July to December in large parts of the delta due to the overflow of the Mekong River, with maximal flooding depths occurring in the upstream areas of the delta. On the basis of the study of Vo Quang Minh (2006), soils in the Mekong Delta can be classified into five major groups (Fig. 14.1). Alluvial soils dominate along the Mekong River. These are the most productive soils for rice, upland crops and fruit orchards in the Mekong Delta. Acid sulphate soils (ASS) are found in depressions (backswamps), where there is no or little cover by river sediment deposits. This soil group can be further divided into different subgroups, such as actual and potential acid-sulphate soils (ASS) with and without saline intrusion, and peaty ASS. Soils with saline intrusion occur along the coast from the eastern to the western part. This group of soils can be divided into two subgroups: dry-season saline soils and permanent saline soils. These soils are rich in nutrients but the salinity limits plant growth. Grey soils (degraded soils) occupy only a small area in the delta, but play an important role as land resources of some provinces. They can be divided into two groups: soils developed on acidic magmatic rocks and soils developed on old alluvial deposits. Nutrients are depleted in these soils and soil microbial activity is generally low. Sand ridge soils are also distributed along the eastern coast. Rice and upland crops are cultivated on these soils. The position on a sandbar or between sandbars dictates the texture of the soil (Nguyen My Hoa 2003).

The total population of the Mekong Delta (MD) was 17.2 million in 2009, compared with over 86 million in the whole of Vietnam. The Mekong Delta is considered the main area of agricultural and aquaculture production in the country. On the basis of data from the General Statistics Office of Vietnam (2009), agricultural land constitutes about 65% of the total area of the Mekong Delta. Most prominent is the production of lowland rice with 20.5 million tons, which is 53% of the paddy production in Vietnam. Rice production has increased from 12.8 million tons in

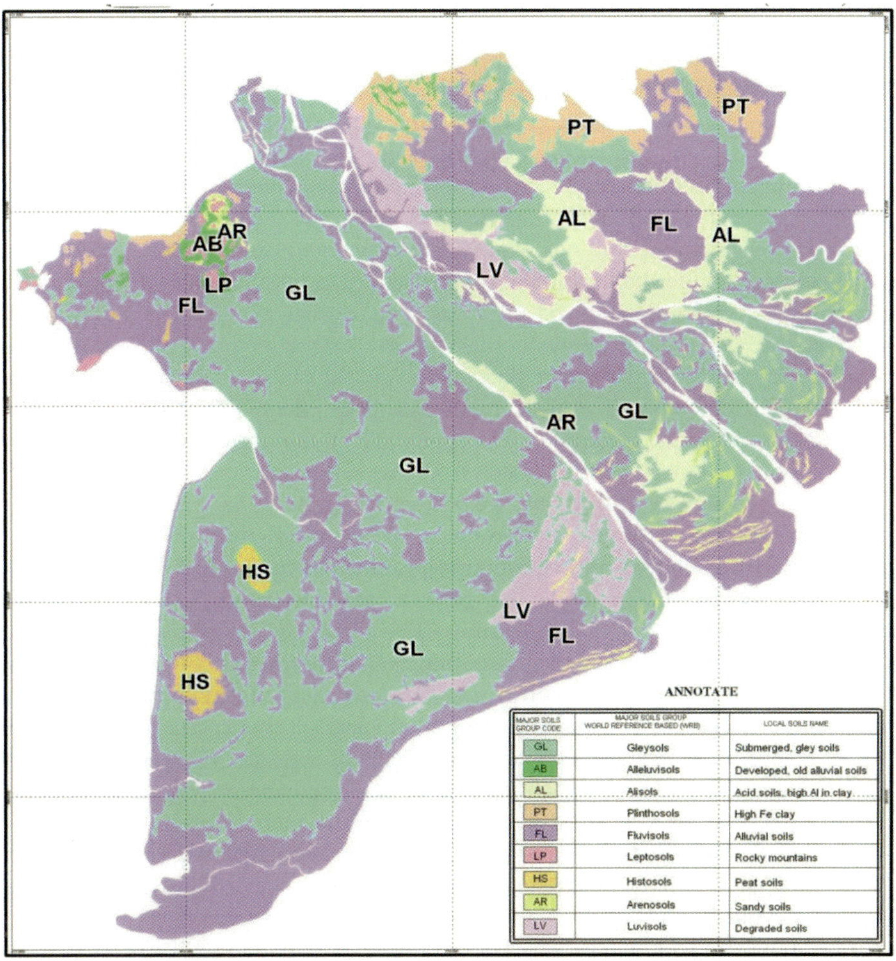


**Fig. 14.1** The major soil groups in the Mekong Delta (Vo Quang Minh 2006)

2003 to 19.2 million tons in 2007 (General Statistics Office 2007). During and after the Vietnam (or American) War and until 1985, Vietnam was a rice-importing country, but in 1989, 1.4 million tons were exported (Le Anh Tuan et al. 2007). From 1989 onwards, Vietnam became the second most important rice-exporting country in the world. The MD played an important role in this change, contributing about 90% of the rice exported from Vietnam.

Beside rice cultivation, aquaculture production in the delta provides 2.8 million tons of fish and shrimps, representing 74% and 75% of the totals of fish and shrimp production nationally, respectively. Fish-shrimp products have contributed 60% of the total export value of the whole country (General Statistics Office 2009).

### 14.1.2 *Opportunities and Challenges for the Region*

The MD has played an important role in the economic development of Vietnam, in terms of rice, fruit and aquatic products and of natural resource-based livelihoods (Tran Thanh Be et al. 2007). The export of agricultural and aquatic products brings significant benefits to farmers. In 2009, the export value of agriculture was about 9.0 billion USD, and 4.2 billion USD for aquatic products of the whole country (General Statistics Office 2009), with rice, black tiger shrimp and Pangasius catfish being the major globally traded products from the delta. Catfish culture has recently developed and contributed about 65% of the total catfish exported in 2003, which increased to 89% in 2006 (Ministry of Fisheries 2003, 2006).

Since the 1990s, government policies in Vietnam have changed to reform the agricultural sector, which has accelerated the development of this sector in the MD. Intensification of rice production was promoted by the construction of dikes and canal systems. Floods were prevented in the upstream part of the delta, and triple rice cropping was expanded, particularly on alluvial soils. However, soil and environmental degradation has become a serious problem, threatening the sustainability of livelihoods and agricultural production, because of intensive use of inorganic fertilizers and pesticides and the fact that sediment was no longer deposited as a result of flood-prevention measures. This is particularly associated with the aquaculture sector, which is the main culprit for mangrove deforestation and the release of antibiotics into the environment (Chap. 13). The intensification of shrimp production has accelerated since the year 2000, particularly in brackish water areas, where mangrove forests fulfil important ecological functions. The development has been driven by the high value and profit of shrimp production compared with those of rice (Nguyen Duy Can et al. 2007), which leads farmers to ignore the risks both of failure and of environmental hazards associated with this type of production system (Hossain et al. 2006). In the MD, soil and water are the key resources determining agricultural land use. Both intensive rice farming and aquaculture practices are water-intensive (To Phuc Tuong 2005; Dang Kieu Nhan et al. 2007); during the dry season in particular, water scarcity can have negative effects on crop yields, especially in rainfed areas.

## 14.2 Water Quality: A Case Study of the Mekong River and Integrated Rice-Shrimp Cultivation on ASS

### 14.2.1 *Surface Water Quality in the Dry and Rainy Seasons in the Mekong River in Vietnam*

In Vietnam, the quality of surface water has been frequently monitored by Provincial Departments of Resources and Environment in the Mekong Delta. Nguyen My Hoa and Dang Duy Minh (2009) investigated the existing data on water quality in 2007

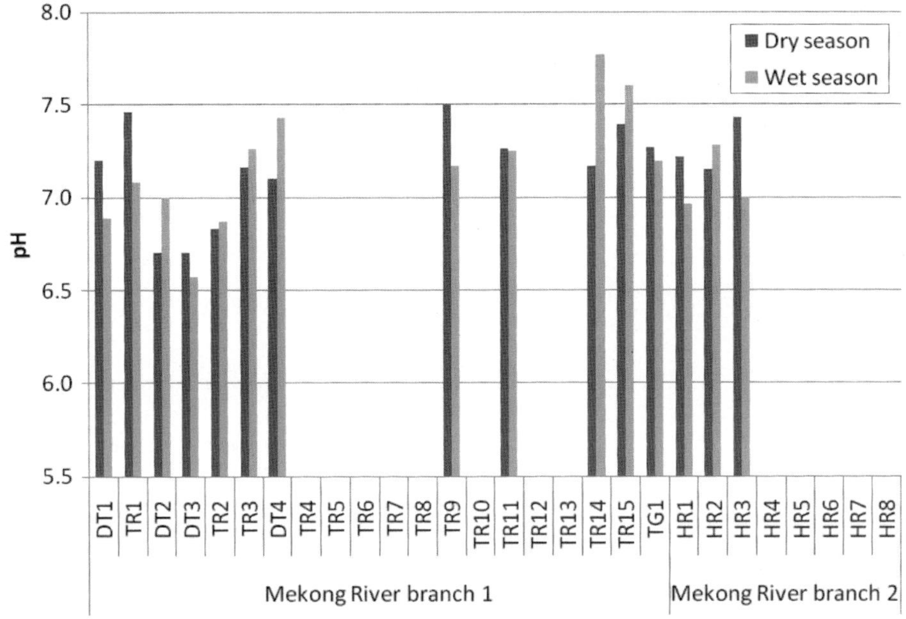


**Fig. 14.2** pH of surface water at sampling sites in Mekong River branch 1 (Tien River: TR) and Mekong River branch 2 (Hau River: HR) in 2007 (TR1 to TR15 were sampling sites from upstream to the middle section of Tien River; HR1 to HR8 were sampling sites from the middle section to downstream of Hau River). Data were missing for some sites (Source: Nguyen My Hoa and Dang Duy Minh 2009)

in four provinces along the Mekong River in Vietnam to study surface water quality and seasonal effects on water quality, namely, in Dong Thap Province upstream of the Mekong River branch 1 (Tien River), in Tien Giang and Vinh Long Provinces in the middle section of the Mekong River branch 1 and in Soc Trang Province at the downstream section of the Mekong River branch 2 (Hau River). Results showed that the pH of surface water was above 7 in both branches of the Mekong River (Fig. 14.2). Nguyen My Hoa and Dang Duy Minh (2009) reported that the change in pH was not associated with upstream effects but rather with soil types and land use in areas along the Mekong River. Effects of dry and wet seasons differed between sites. Again, other factors such as soil type, land use and hydrology at specific sites may be the reasons for the seasonal change in water pH.

The five day biological oxygen demand ($BOD_5$) exhibited high values upstream and lower values downstream in both the first and the second branches of the Mekong River (Fig. 14.3). Upstream, $BOD_5$ was above 20 mg $O_2$/l, an indication of high organic contamination in upstream areas. Aquaculture and industrial activities for processing aquatic products along the river may be the reasons for this contamination.

Chemical Oxygen demand (COD) in surface water in the two branches of the Mekong River was also very high, exceeding the limits of the Vietnamese standard

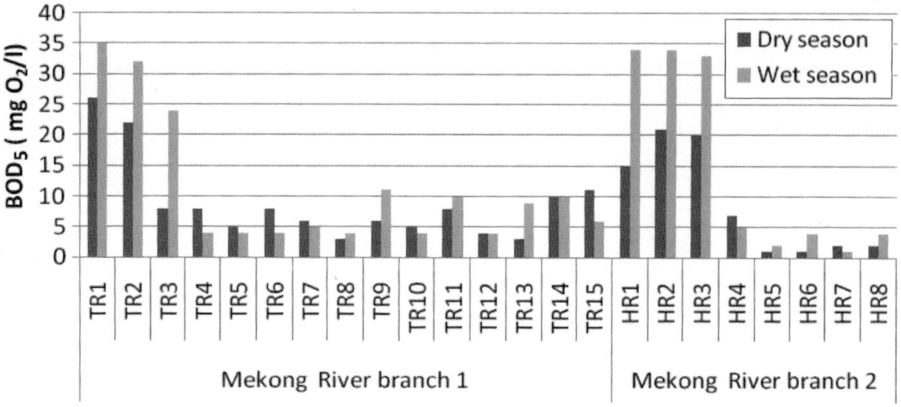


**Fig. 14.3** Five day biological oxygen demand ($BOD_5$) of surface water at sampling sites in Mekong River branch 1 (Tien River) and Mekong River branch 2 (Hau River) in 2007 (TR1 to TR15 were sampling sites from upstream to the middle section of Tien River; HR1 to HR8 were sampling sites from the middle section to downstream of Hau River) (Source: Nguyen My Hoa and Dang Duy Minh 2009)

for surface water (<10 mg/l). A trend similar to that for $BOD_5$ in surface water was also observed, with the exception of high values in the middle of the river at TR9, TR11, TR14 and TR15 (Fig. 14.4).

The concentration of nitrate ($NO_3$) exceeded the limit for surface water (10 mg/l) in the upstream regions of the two branches and was lower at the middle sections. Ammonium ($NH_4^+$) concentration was almost over the limit for surface water according to the Vietnamese standard (0.05 mg/l), except for some sites in the middle of the river in Mekong River branch 1 (Fig. 14.5).

In summary, changes in the season affected the quality of surface water in both branches of the Mekong River. The extent of this effect, however, varies between sites along the Mekong River. There is a clear trend in the water quality parameters. While changes in $BOD_5$, COD and $NO_3$ concentrations appear to be related to location along the river course, with the highest values upstream and lower values downstream, soil types and land use appear to be the primary factors that affected the water pH.

### *14.2.2 Surface Water in ASS Area*

In the Mekong Delta acid sulphate soils account for about 1.6 million ha, occupying about 40% of the total agricultural area, mainly located in Long Xuyen Quadrangle in the North-West, the Plains of Reeds in the North-East and the central of Ca Mau Peninsula in the South of the Mekong Delta. Upon oxygenation, the oxidation of reduced sulphur compounds leads to a rapid acidification of these soils with resulting

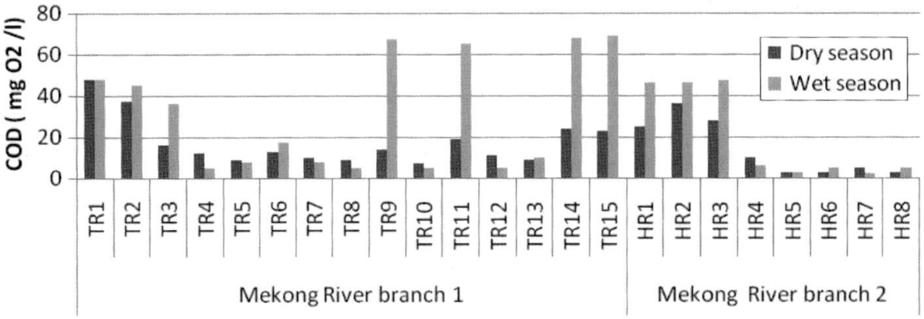


**Fig. 14.4** Chemical oxygen demand (COD) of surface water at sampling sites in Mekong River branch 1 (Tien River: TR) and Mekong River branch 2 (Hau River: HR) in 2007 (TR1 to TR15 represent sampling sites from upstream to the middle of Tien River; HR1 to HR8 represent sampling sites from the middle section to downstream of Hau River) (Source: Nguyen My Hoa and Dang Duy Minh 2009)

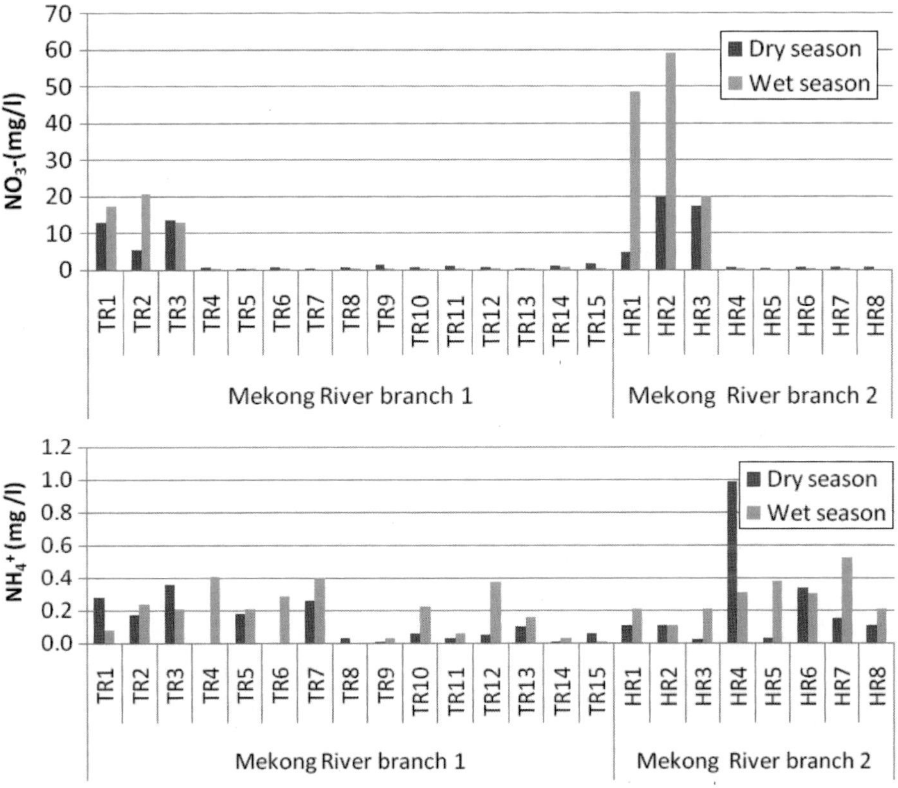


**Fig. 14.5** Concentration of nitrate ($NO_3^-$) and ammonium ($NH_4^+$) in surface water at sampling sites in Mekong River branch 1 (Tien River) and Mekong River branch 2 (Hau River) in 2007 (TR1 to TR15 were sampling sites from upstream to downstream of Tien River; HR1 to HR8 were sampling sites from the middle section to downstream parts of Hau river) (Source: Nguyen My Hoa and Dang Duy Minh 2009)

changes in the solubility of metal cations. Artificial drainage of areas for crop production either by digging canals or by leaching surface soil, as well as the construction of dikes to prevent yearly floods in the rainy season, have resulted in the pollution of water by toxic substances. Seasonal hydrological changes, especially annual floods, affect the concentration of toxic metals in canals of the ASS areas. Sunström et al. (2002) reported that the median concentrations of aluminium (Al), cadmium (Cd), cobalt (Co), manganese (Mn), nickel (Ni) and zink (Zn) in the drainage ditches around ASS in Finland were 135–1,044 times higher than the median concentrations in Finnish headwater stream. This is an example of environmental contamination, which is also very likely to occur in other ASS areas.

Nguyen My Hoa et al. (2007) investigated the concentrations of nine metal elements (Al, As, Cd, copper (Cu), iron (Fe), Mn, Ni, lead (Pb) and Zn) in the surface water of first-, second- and third-order canals and in the water table in the ASS Long Xuyen Quadrangle in the dry and rainy seasons and their relationship to concentration in soils. Their results showed that water pH was high in the first-order canals, low in the second-order canals and lowest in the third-order canals, where the first-order canals are the largest canals and third-order canals are the smallest and bring water to the fields. This reflects the highest acidity in canals directly drained from ASS and the dilution of this acidic water in the first-order canals by non-acidic water from the river. The pH in canal water was high in slightly ASS, lower in moderately ASS and lowest in the severely ASS (Fig. 14.6).

Nguyen My Hoa et al. (2007) found that the hydrology and management of ASS in the Long Xuyen Quadrangle also affect the acidity of canal water. At the beginning of the rainy season during May and July, pH in the first-order canals was below 3.5, whereas it was often in the range of 5–7 in other months of the year. During this period, rainwater dissolves and leaches hydrogen ions ($H^+$), which form during the dry season into the canal, resulting in severe acidity. pH in groundwater was as low as that recorded in third-order canals. Electrical Conductivity (EC) was highest in groundwater and in the third-order canals, especially in May and July, and was inversely related to pH. Very low pH of canal water was mainly due to acidification of the soil, which is caused by oxidation of materials containing sulphides during the formation of ASS, and the reclamation of ASS for agricultural production, such as by the excavation of artificial drainage canals and the construction of closed dikes to prevent flooding. The low pH of water in areas of ASS closed off by dikes compared with that in ASS in open areas indicates that management of ASS was important in controlling the acidification of canal water.

Nguyen My Hoa et al. (2007) also found that the concentrations of heavy metal ions in the first-, second- and third-order canals, and in groundwater in ASS of Long Xuyen Quadrangle, were 17–296 (Al), 4.5–37.0 (Cd), 3.0–15.3 (Cu), 5.3–58.5 (Fe), 1.9–35.8 (Mn), 12–233 (Ni) and 6.8–36.7 (Zn) times higher than those in non-ASS, ranging from 230 to 103,000 μg Al/l, 0.09–1.47 μg Cd/l, 4.55–42.75 μg Cu/l, 130–87,100 μg Fe/l, 1.7–224 μg Ni/l and 10–590 μg Zn/l. Because of water contamination in canals, farmers in these areas have to use water in water spouts, but poor farmers in remote areas also use contaminated water for their everyday practices.

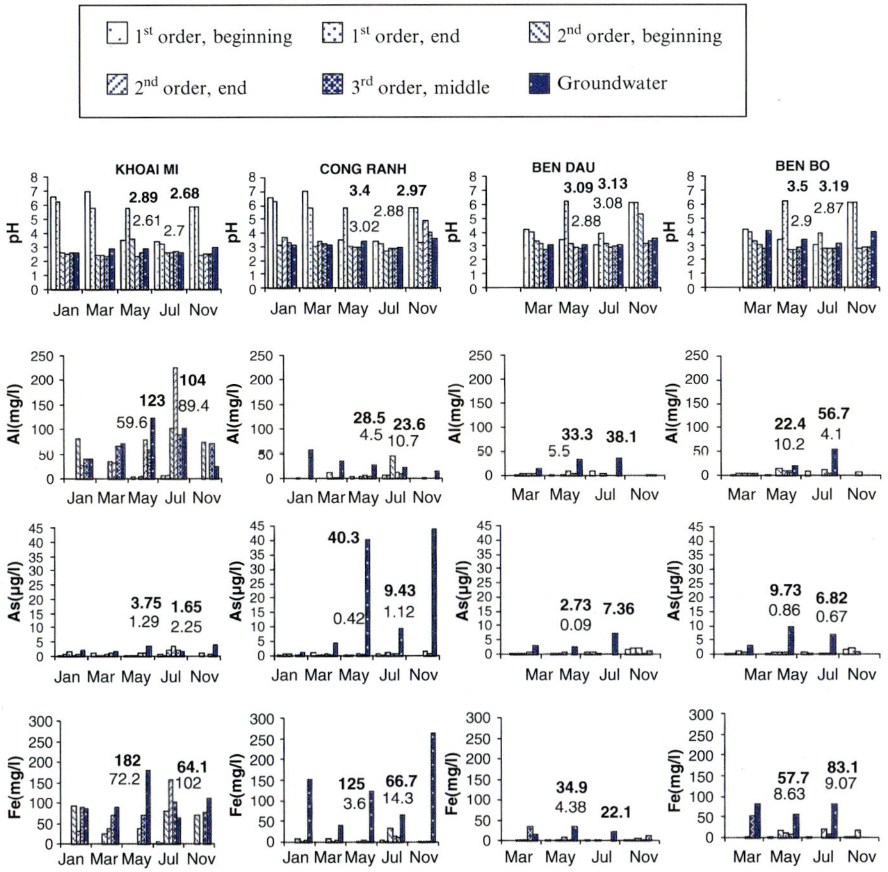


**Fig. 14.6** pH and concentrations of Al, As and Fe in first-, second- and third-order canals in severely ASS in Khoai Mi and Cong Ranh, and in moderately ASS in Ben Bo and Ben Dau. Numerical data shown in the graphs are for groundwater (*bold*) and third-order canals (*not bold*) in May and July only. Some data in July and November are missing (Source: Nguyen My Hoa et al. 2007)

Nguyen My Hoa et al. (2007) also reported that the concentrations of toxic metals (Al, Cu, Fe, Mn, Ni, Zn) were often highest in severely ASS in the areas closed off by dikes in comparison with those in open areas where toxic substances can be leached by annual floods. The results again showed the high contamination of toxic metals in severely ASS in comparison with that in moderately ASS and the effect of ASS management on contamination by toxic metals in canal water draining ASS. In general, the highest concentrations of metal ions were observed in May and July and this tendency was mostly observed in open areas.

It is well known that many factors control the release of metals from soil to water environments: pH, mobility of metals in soils, oxidation-reduction reactions, metal

**Table 14.1** Average Fe concentrations in A, B and C horizons in different soil groups in studied areas

| | Fe extracted by 0.05 M $NH_4$-EDTA (mg/kg) | | |
|---|---|---|---|
| Soil group | A horizon | B horizon | C horizon |
| Actual ASS | 742 | 1,600 | 2,703 |
| Potential ASS | 495 | 515 | 1,616 |
| Non-ASS | 549 | 105 | 414 |
| | Fe extracted by 0.43 M $HNO_3$ (mg/kg) | | |
| Soil group | A layer | B layer | C layer |
| Actual ASS | 2,418 | 1,788 | 2,892 |
| Potential ASS | 1,486 | 492 | 3,162 |
| Non-ASS | 2,238 | 193 | 428 |

Source: Nguyen My Hoa and Huynh Tri Cuong (2009)

content in soil, season, and hydrology. The pH value is considered the main factor affecting the mobility and the forms of metals in soils (McBride 1994). In general, metals are dissolved in acidic conditions, hence their release increases when pH decreases. When pH increases, the availability of metal decreases owing to precipitation as oxide and hydroxide forms. Åström (1998) reported that pH was the main factor controlling metal concentration in canal water leached out from ASS, except for As. Sunström et al. (2002) reported that Cd, Co, Cu, Mn, Ni and Zn were present at high levels in water in ASS, but their concentrations were weakly correlated with pH. Åström (1998) reported that Co, Mn, Ni and Zn were highly mobile, while Al and Fe were less mobile, but because of high levels of Al and Fe in the soil, their levels of release from soils were high. The availability of some metals in soil was found not to be important in controlling metal release (Åström 1998). Water flow and season affect metal concentration in water (Åström 2001; Nguyen My Hoa et al. 2007).

Determining concentrations of metals (As, Cd, Cu, Fe, Mn, Ni and Zn) in different soil groups of ASS and of the factors controlling the leaching of toxic metals from soil to water is essential to diagnose and assess the problem of water resource degradation. Fe concentration extracted by $NH_4$- Ethylenediaminetetraacetic acid (NH4-EDTA) and $HNO_3$ at 0.43 N following the method described by Houba et al. (1995) was found to be higher in the sulphidic C horizon in ASS areas than in non-ASS areas (Table 14.1).

Table 14.2 shows As concentration extracted by NH4-EDTA and $HNO_3$ at 0.43 M following the method described by Houba et al. (1995). Arsenic concentrations extracted by NH4-EDTA (588 μg/kg) and HNO3 at 0.43 M (1,023 μg/kg) were high in the sulphidic C horizon of actual ASS areas compared with those in non-ASS areas. Oxidation of As-containing sulphide compounds due to drainage and excavation of canals for irrigation could be the reason for high As in soil. Clemente et al. (2006) and Wang and Mulligan (2006) reported that As released from soil can combine with oxides of Fe, Al and Mn, and then be released into water when soils are submerged by flooding or during rice cultivation. Gustafsson and Nguyen Thanh Tin (1994) also found that the concentration of soluble As at depths of 80–120 cm was about 1,000 μg/kg. Gustafsson and Nguyen Thanh Tin (1994) reported that As

**Table 14.2** Average As concentration in A, B and C horizons in different soil groups in studied areas

| | As extracted by 0.05 M $NH_4$-EDTA (mg/kg) | | |
|---|---|---|---|
| Soil group | A horizon | B horizon | C horizon |
| Actual ASS | 44.3 | 75.3 | 588.0 |
| Potential ASS | 31.7 | 18.0 | 131.8 |
| Non-ASS | 17.7 | 21.5 | 80.0 |
| | As extracted by 0.43 M $HNO_3$ (mg/kg) | | |
| Soil group | A horizon | B horizon | C horizon |
| Actual ASS | 253 | 279 | 1,023 |
| Potential ASS | 279 | 79 | 634 |
| Non-ASS | 171 | 261 | 620 |

Source: Nguyen My Hoa and Huynh Tri Cuong (2009)

concentration in ASS in the Mekong Delta was 6–41 mg/kg, which was higher than the As concentration in non-contaminated areas (5 mg/kg). Dudas (1984) also reported a high As concentration in ASS in Canada of 8–45 mg/kg, whereas its concentration was about 4.8–13.6 mg/kg in non-contaminated areas. Therefore, the sulphidic C horizon in ASS in the Mekong Delta can be the source of As in soil and oxidation of this horizon by water drainage can have resulted in high As in soil. Mobilization of As depends on the concentrations of Fe, Al and Mn oxides/hydroxides and the reduction conditions; release of As into surface water was low due to the oxidation status and the presence of Fe and Al hydroxide in surface water. This explains the low As levels in surface water in canals in ASS areas in the Mekong Delta (0.41–1.31 μg/l), as found by Nguyen My Hoa and Huynh Tri Cuong (2009).

Concentrations of Cd, Cu, Fe, Mn, Ni and Zn extracted by 0.4 M $HNO_3$ were also found to be high in the sulphidic C horizon compared with those in non-ASS areas (Nguyen My Hoa et al. 2006). Nguyen My Hoa and Huynh Tri Cuong (2009) found that acidification of soil and water was the main factor controlling mobility of Cd, Fe and Ni in soil. Available and gradually available Cd, Fe and Ni played a minor role in controlling the release of these metals to water environments.

In summary, high concentrations of toxic elements in canal water in ASS areas were related to low pH. Therefore, controlling the removal of these elements from the top soil by liming may be important in limiting the contamination of the environment by these elements. Another important activity that should be considered in order to prevent deterioration of water quality is to stop the drainage of actual/potential ASS fields (Sunström et al. 2002).

### *14.2.3 Water and Soil Quality of Shrimp Ponds in ASS Area*

In the MD, transformation of cropping systems and animal production was carried out by local governments in order to diversify agricultural products and increase profit for farmers. Shrimp cultivation alternated with other crops was conducted spontaneously by farmers in areas with saline ASS. However, cultivation of shrimp in areas where the salinity of water was low and changed over time tended to affect shrimp growth, causing diseases that resulted in low productivity. Oxidation of pyrite due to digging of the soil to prepare shrimp ponds in ASS resulted in acidification of the surface water in the area (Vo Thi Guong et al. 2004). Nguyen My Hoa et al. (2010a) studied soil and water quality in areas that had cropping systems consisting of shrimp-upland crop, shrimp-rice and two rice crops in order to evaluate the suitability of the systems and impacts of shrimp cultivation in ASS. The results of the study showed that, in potential ASS (where sulphidic material occurs at a depth of 50–55 cm in Hau Giang Province) and with the farmers' current practices, water in shrimp ponds had low pH ($pH < 7$), even though lime was applied, low salinity (<2.5 g/kg), low alkalinity (<80 mg/l), high COD (10–20 mg/l), but low $BOD_5$ (<10 mg/l). The biggest problems for shrimp cultivation in this area were the low pH, low salinity and low alkalinity, which were not easy to control. For shrimp ponds in ASS, diffusion of $H^+$, $Fe^{2+}$ and $Al^{3+}$ from the bottom horizon and the removal of these toxic substances from the ponds formed by dikes during rainfall occurred continuously; therefore, the pH in pond water was often low and changed frequently, affecting the alkalinity of the system and hence shrimp growth. Boyd (1998) reported that optimum salinity for shrimp growth was 20–25 g/kg; therefore, low salinity in pond water in the studied site was not suitable. According to Chanratchakool et al. (1995), optimum alkalinity was 80–120 mg $CaCO_3$/l. However, pond water in ASS areas was often below 80 mg/l; therefore, it was not suitable for shrimp growth and lime should have been used frequently, especially after heavy rain.

A study of water and soil quality of shrimp and shrimp-rice systems on alluvial, potentially acid sulphate and ASS in Ca Mau Province was undertaken by Vo Thi Guong et al. (2004). Adverse conditions of soil and water in ASS were considered as one of the factors leading to failure of shrimp cultivation. The soil pH of potential and actual ASS in saline areas ranged from 7.0 to 7.4, but dropped rapidly to 4.5 when the soil dried during land preparation (Fig. 14.7). By contrast, the pH of saline alluvial soil remained at about 7.5 and there was no significant drop in pH below this value when the soil dried. Exchange sodium percentage (ESP) was below 15 in shrimp-rice systems of ASS. However, in the shrimp systems on potentially ASS and alluvial soils, the ESP ranged from 17 to 25, indicating soil sodification. Water alkalinity ranged from 58 to 70 mg of $CaCO_3$/l. This value was below the optimum level for shrimp growth during the first half of the shrimp culture period. Hydrogen sulphide ($H_2S$) had a range of 0.03–0.71 mg/l, which was harmful for shrimp growth. The highest level of $H_2S$ was found in ASS (Fig. 14.8). Water salinity varied widely (from 6 to 37 g/l). Dissolved nitrogen in water ranged from 0.2 to 0.4 mg/l. The level of phosphorus available in water was low, ranging from 0.01 to 0.04 mg/l. This low nutrient status was a factor related to low chlorophyll-a content and low natural food in the water column. Other factors such as BOD, COD and dissolved oxygen content were at optimum levels for shrimp.

Chemical and nutritional analysis of bottom soil of the shrimp ponds in ASS showed that fresh soil pH, (the soil pH measured just after soil sampling from the field, soil samples were not yet dried) of shrimp ponds was high ($pH = 7–8$), but

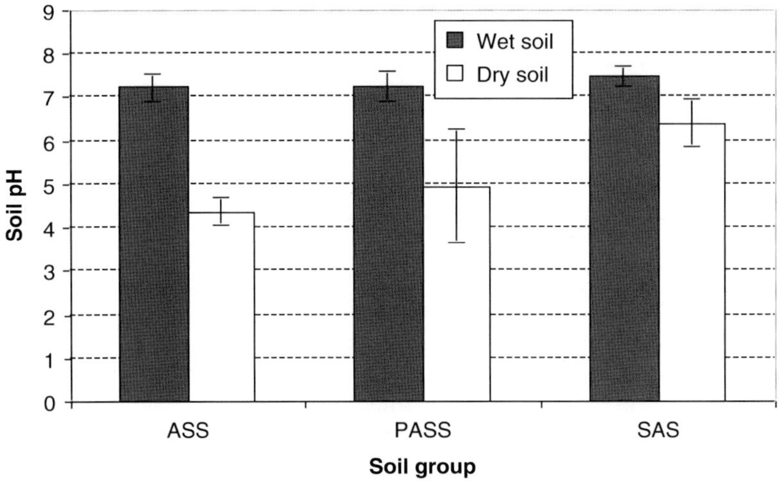


**Fig. 14.7** Wet and dry soil pH of shrimp ponds in: ASS, potential acid sulphate soils (*PASS*) and saline alluvial soil (*SAS*). Mean (± Standard Deviation)

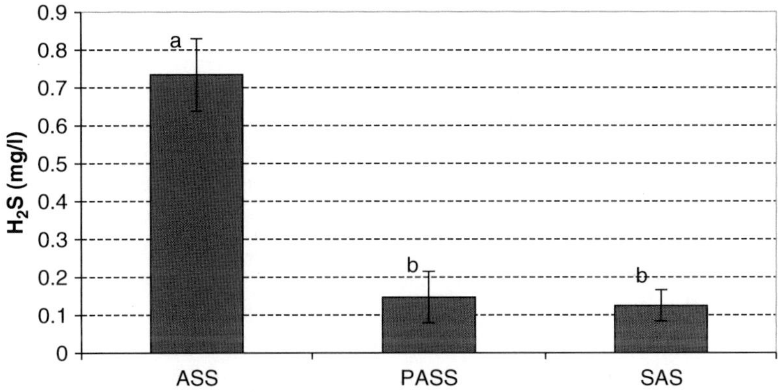


**Fig. 14.8** Hydrogen sulphide in deep water of shrimp pond in: Acid sulphate soils (*ASS*), potential acid sulphate soils (*PASS*) and saline alluvial soils (*SAS*) at the beginning of crop cycle. Mean (±Standard Deviation). Data with different letters a and b are significantly different ($P \leq 0.05$)

surface mud and bottom soil had high concentrations of mobile Fe and Al. This resulted in the release of high levels of soluble Fe and Al into pond water, and is the reason for low pH and low alkalinity in pond water. High concentrations of organic C, labile N and available P in surface mud samples were also the reasons for high organic C, and high N and P in pond water, which caused algal blooms and affected shrimp growth in the studied ponds. Therefore, it is suggested that shrimp cultivation in potential ASS where input water has low salinity is risky and not recommended. Suitable fish cultivation may be an option to be considered in such areas to improve the income of farmers.

## 14.3 Water Quality: A Case Study of Intensive Shrimp and *Artemia* Cultivation in Areas with Salinity Intrusion

### *14.3.1 Intensive Shrimp Cultivation*

The MD is considered a region rich in aquatic resources and with high potential for aquaculture production. Shrimp culture has developed rapidly since the 1990s. Very large benefits can be obtained from shrimp cultivation compared with those from other agricultural production systems; therefore, the conversion of rice fields to shrimp culture in areas with saline intrusion has been widespread in coastal provinces. These areas include alluvial soils as well as potential and actual ASS, which have been used for shrimp in rotation with rice as well as for intensive shrimp cultivation. However, these systems remain a high-risk option for farmers. There is little documentation on the characteristics of bottom soil sediments and the water column for these systems (Vo Thi Guong et al. 2004). The analysis of technical and economic aspects of intensive culture of black tiger shrimps in the coastal areas of the delta showed that shrimp yield was roughly 7 t/ha for each harvest, resulting in very high economic returns compared with those of semi-intensive culture; however, this approach has a high risk of failure. In addition, nutrients such as nitrogen and phosphorus accumulate in the muddy bottom soil and are subsequently released into the water column. This can result in lower water quality due to algal blooms and lack of oxygen, which have adverse effects on the food chain in shrimp ponds. In addition, social and economic constraints on this approach were identified as follows: insufficient capital investment, inadequate knowledge and technical skills for shrimp cultivation after transferring from rice to aquaculture, low level of development of infrastructure and water management systems, and low post-larva quality. All these factors are strongly related to the success of conversion from rice production to aquaculture in areas with salinity intrusion (Le Quang Tri and Vo Thi Guong 2005).

A recent study on the changes of water quality in 12 shrimp ponds with intensive, semi-intensive and extensive shrimp culture systems in Soc Trang Province was undertaken during the first 2 months of the shrimp cycle (Tat Anh Thu and Vo Thi Guong 2010). Results showed that the salinity after 1 month of shrimp stocking reached 8–16 g/kg and decreased to about 12 g/kg at the end of the shrimp crop. A shrimp cycle lasts about 5 months, and at the end of the cycle, rainfall reduced water salinity. pH of water was in the range of 7.8–8.3, which was related to the water alkalinity. The concentration of $CaCO_3$ was in the range of 80–110 mg/l. These ranges of salinity, pH and alkalinity were the optimal conditions for shrimp growth (Chanratchakool et al. 2002). Nitrogen and phosphorus are important nutrients in the food chain of shrimp ponds. Total nitrogen dissolved in water was high in intensive shrimp systems at 14–30 mg/l, while it ranged between 2 and 22 mg/l in extensive shrimp systems. Dissolved $NH_4^+$ was 3.2 mg/l in intensive systems, which is greater than the optimal level of 0.2–2 mg/l for shrimp (Whetstone et al. 2002; Chanratchakool et al. 2002). A high level of $NH_4^+$ in water can lead to algal blooms with subsequent negative effects on shrimp growth. In the extensive shrimp ponds, $NH_4^+$ concentration was 0.3 mg/l, which is an acceptable level for shrimp growth. In surface water, dissolved phosphorus (P) was in the range of 0.005–0.02 mg/l (Chapman 1997). In shrimp pond, P concentration in the range 0.05–0.25 mg/l is suitable for shrimp. We found a higher level of P in water of intensive shrimp culture, with 0.4 mg/l in the first month of growth, subsequently being reduced to 0.1 mg/l. High concentration of P also leads to algal bloom in shrimp ponds. In extensive shrimp ponds, P concentration was much less at 0.02 mg/l. Concerning $H_2S$ concentration in shrimp ponds, under reducing conditions, with high organic matter due to the waste from shrimp feeding, transformation of $SO_4^{2-}$ in the bottom soil into $H_2S$ is enhanced. The presence of this compound negatively affects shrimp production when the concentration reaches 0.05 mg/l (Chanratchakool et al. 2002; Boyd 1998). In the 12 studied ponds, we found that $H_2S$ concentration in water at the bottom of each pond was 0.62–0.68 mg/l in intensive, semi-intensive and extensive shrimp ponds. One month later, the concentration was 0.5 mg/l, and 3 months after the shrimp fry were released, the concentration was 0.2 mg/l. For optimal shrimp growth, $H_2S$ concentration must be lower than 0.03 mg/l. A high concentration of $H_2S$ is detrimental to shrimp health through oxygen deficiency and reduced tolerance to diseases (Andrew 2007). We determined the COD in different shrimp systems, and results showed that COD was in the range of 25–35 mg/l in intensive ponds, while it was less in the two other systems, with concentrations of 16–30 mg/l for semi-intensive systems and 10–21 mg/l in extensive systems. This high value indicates that eutrophication has occurred. A suitable level of COD is about 20 mg/l (Boyd 1998; Smith et al. 2002). On the basis of this study, we can conclude that the environment for shrimp growth in intensive shrimp ponds was not optimal because of low water quality, inappropriate concentrations of nutrients that led to algal bloom, lack of oxygen and high $H_2S$ concentration.

Sludge sediment in shrimp ponds was removed after every shrimp harvest. It is considered to cause pollution in ponds. Knowing the characteristics of this large amount of waste and whether or not it is possible to use it for crop production is an important factor to determine. Twelve large sediment waste samples were collected to evaluate selected chemical properties and their changes after a few months of leaching the salinity during the rainy season (Tat Anh Thu and Vo Thi Guong 2010). The results showed that waste sediment from the intensive and semi-intensive shrimp cultivation systems was mainly a loam clay sediment, had low total carbon (C) of 0.8–1.4% in soil, total nitrogen was about 0.14%, and total phosphorus was 0.10%, but there were high concentrations of available nitrogen (35.2 mg/kg) and available phosphorus (25.3 mg/kg). The sediments were also characterized by high salinity with an electric conductivity in the range of 4.0–20.7 mS/cm. ESP, indicating the degree of sediment sodicification, was extremely high at 78% in the waste sediment of intensive shrimp systems and 45% in semi-extensive shrimp systems, while the value was less in extensive shrimp systems (16%). The contamination by Cd was lower than the critical level of soil pollution. Total Cd concentration in the three systems ranged from 25.2 to 33.5 μg/kg, that is, much less than 5 mg/kg, the level of soil pollution. Similarly, Pb concentration was found at 1.0–1.75 mg/l,

**Table 14.3** Dissolved oxygen in water column of two groups of *Artemia* ponds from 1 to 4 weeks after inoculation

| No. | Sampling times | Dissolved $O_2$ (mg/l) | | $T_{value}$ | $P_{Prob}$ |
|---|---|---|---|---|---|
| | | Group 1 | Group 2 | | |
| 1 | 1 week | 6.93 | 8.54 | −8.03 | $0.0002^{**}$ |
| 2 | 2 weeks | 8.25 | 8.42 | −0.94 | $0.39^{ns}$ |
| 3 | 3 weeks | 5.71 | 5.52 | 0.40 | $0.704^{ns}$ |
| 4 | 4 weeks | 5.02 | 6.10 | −2.05 | $0.086^{*}$ |

*Group 1* Problem ponds; *Group 2* Normal ponds
*Significantly different ($P \leq 0.05$)
**Significantly different ($P \leq 0.01$)

much lower than the value for contaminated soil. In intensive shrimp ponds, salinity was reduced to 2 mS/cm after 3 months of exposure to rain. ESP remained at about 10%, implying no more sodicification. In terms of nutrients, levels of available phosphate, available nitrogen and labile nitrogen were decreased by about 50% after 3 months of leaching. On the basis of these results, the waste sediment in the shrimp cultivation can be used for agricultural production, after leaching away the salinity in about 3 months during the rainy season. Sediments were found to be relatively rich in nutrients and were not contaminated with heavy metals.

### 14.3.2 Low Water Quality, a Constraint on Artemia Cultivation

*Artemia* is a small branchiopod crustacean, living in hyper-saline waters in coastal lagoons and in the field where farmers produce salt at solar saltworks (Nguyen Van Hoa 2002). The cysts of *Artemia*, namely, dormant eggs, have been considered as excellent food for the aquatic larvae of shrimp and fish. With their digestibility and high nutritional quality, dry cysts contain 28% crude protein, 10% crude fibre and 10% crude fat (Treece 2000). The consumption of *Artemia* cyst has been increased largely to meet the requirements of shrimp hatcheries and fish larva culture (Nguyen Van Hoa 2002; Nguyen Thi Ngoc Anh et al. 2010). In the Mekong Delta, *Artemia* culture brings greater returns on investments for farmers than salt-making. However, many farmers are confronted with the problem of high mortality and low cyst yield. Management of nutrients in the field was thought to be a key for success. To test this hypothesis, six water samples of extremely low *Artemia* cyst yield and six samples of high *Artemia* cyst yield were collected in the same area in Vinh Chau district, Soc Trang Province to determine water quality and algal components (Tat Anh Thu and Vo Thi Guong 2010). In problem ponds with low yields (group 1), dissolved oxygen in water was low compared with that in areas with good productivity (group 2) (Table 14.3); alkalinity was also significantly lower than in group 2 (Fig. 14.9). Low water alkalinity is known to be related to water pH instability (Wurts and Durborow 1992). The changes in water pH can lead to change in the availability of nutrients, which is related to algal composition and therefore may inhibit *Artemia* growth.

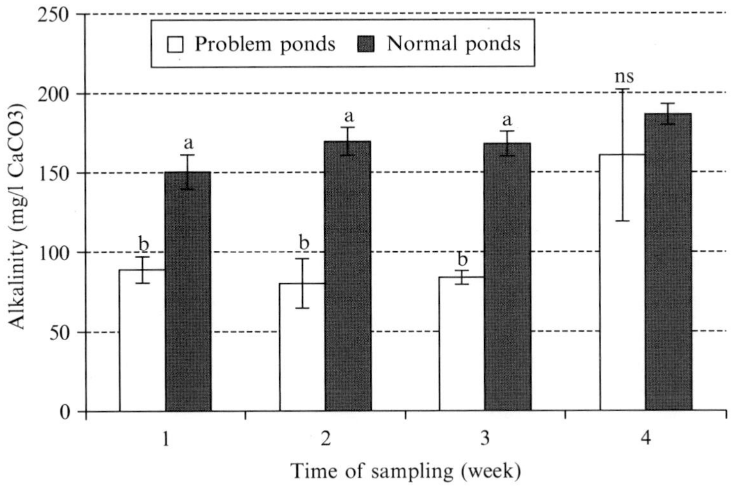


**Fig. 14.9** The alkalinity in ponds of two groups of Artemia (Tat Anh Thu and Vo Thi Guong 2010). Mean (±SD). Data with different letters are significantly different ($P \leq 0.05$)

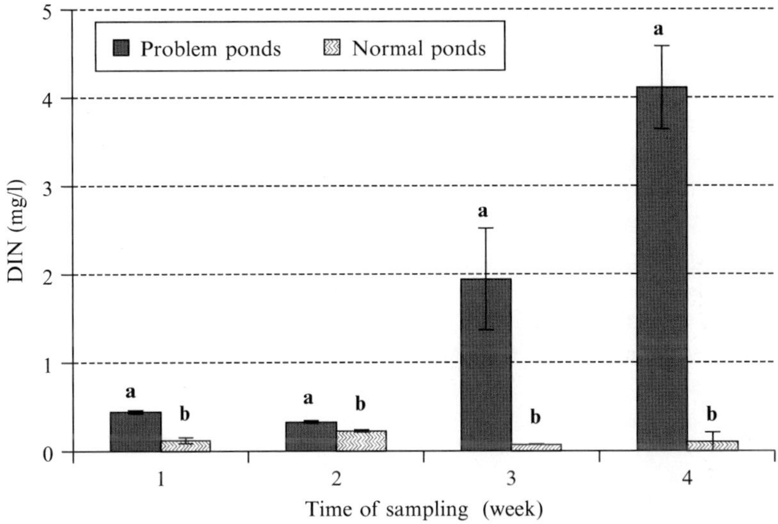


**Fig. 14.10** Concentration of dissolved inorganic nitrogen (*DIN*) in water column in ponds of two groups of Artemia (Tat Anh Thu and Vo Thi Guong 2010). Data with different letters are significantly different ($P \leq 0.05$)

Dissolved inorganic nitrogen (DIN) was rather high in group 1 compared with that in group 2, especially after 4 weeks of inoculation (Fig. 14.10). The level of DIN was greater than 2 mg/kg, indicating the presence of high nitrogen concentrations in the environment (Boyd 1998; Chanratchakool 2003). In the extremely high-salinity environment of *Artemia* ponds (70–120 g/kg), dissolved reactive phosphorus (DRP)

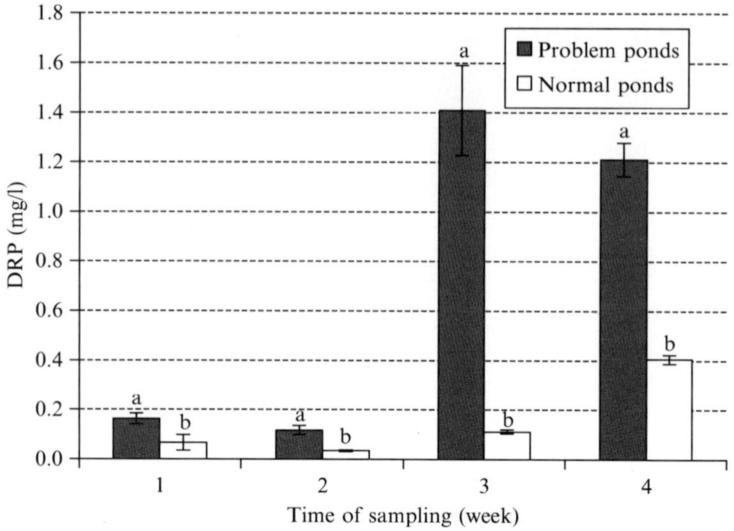


**Fig. 14.11** Concentration of dissolved reactive phosphorus (*DRP*) in water column in ponds of two groups of *Artemia* (Tat Anh Thu and Vo Thi Guong 2010). Data with different letters are significantly different ($P \leq 0.05$)

fraction was released from sediments in larger amounts than in less saline conditions (Chau Minh Khoi 2006; Chau Minh Khoi et al. 2008). DRP was compared between the two *Artemia* pond groups. The result indicated that DRP reached 1.2–1.4 mg/kg in group 1 (Fig. 14.10). The optimum for algal development is typically in the range of 0.03–0.1 mg/kg (Newton and Jarrell 1999; Shock and Pratt 2003). Above this concentration, algal bloom will occur (Seroka 2004). Algal density was determined in these two pond groups. In group 1, it reached $13 \times 10^4$ cells/ml, and was $3 \times 10^4$ cells/ml in group 2.

Our study highlighted that the low quality of water due to high levels of dissolved phosphorus and nitrogen contributed to limiting the growth of *Artemia* and led to reduced cyst production of *Artemia* in the Mekong Delta (Fig. 14.11).

## 14.4 Soil Quality in Agriculture and Aquaculture in the Mekong Delta

### *14.4.1 Soil Characteristics of Mono-Rice, Shrimp-Rice and Shrimp-Watermelon Cropping Systems in Potential ASS*

In mono-shrimp cultivation, soils can become sodic and cannot be used for crop cultivation, whereas in shrimp-rice or shrimp-upland systems, soils may be less saline and the growth of rice and upland crops may be less affected.

**Table 14.4** (a) Chemical characteristics of topsoil in different cropping systems; (b) nutritional characteristics of topsoil in different cropping systems

(a)

| | pH | EC (mS/cm) | CEC | Na | K | Mg | Ca |
|---|---|---|---|---|---|---|---|
| Studied site | 1:2.5 | 1:2.5 | cmol/kg | | | | |
| Rice field 1 (in mono-rice cropping system) | 2.76 | 6.32 | 4.25 | 1.93 | 0.06 | 3.94 | 3.41 |
| Rice field 2 (in shrimp-rice system) | 4.86 | 2.24 | 26.91 | 1.80 | 0.38 | 17.52 | 8.27 |
| Watermelon field 1 (in watermelon-shrimp system) | 3.04 | 3.60 | 19.53 | 0.67 | 0.15 | 3.34 | 8.51 |
| Watermelon field 2 (in watermelon-shrimp system) | 2.99 | 4.29 | 11.08 | 0.92 | 0.25 | 11.08 | 9.35 |

(b)

| | Organic C | N total | Labile N[a] | Available P (Bray 1) | P total |
|---|---|---|---|---|---|
| Studied soil | mg/kg | | mg N/kg | mg P/kg | g/kg |
| Rice field 1 (in mono-rice cropping system) | 128 | 4.9 | 17.5 | 6.01 | 0.4 |
| Rice field 2 (in shrimp-rice system) | 161 | 7.5 | 43.87 | 6.28 | 0.5 |
| Watermelon field 1 (in watermelon-shrimp system) | 179 | 6.1 | 37.41 | 32.41 | 5.4 |
| Watermelon field 2 (in watermelon-shrimp system) | 145 | 27.7 | 22.93 | 12.64 | 1.5 |

Source: Nguyen My Hoa et al. (2010a)

*EC* electric conductivity, *CEC* cation-exchange capacity

[a]Labile N was extracted by hot KCl at 2 M

Nguyen My Hoa et al. (2010a) reported that soil pH was low in mono-rice culture systems and in watermelon-shrimp culture systems in potential ASS in Hau Giang Province, ranging from 2.74 to 3.04. In shrimp-rice systems, pH was higher (4.86) due to the repeated leaching of the soil before rice cultivation. This is the result of soil acidification of the potential acid sulphate topsoil material (sulphidic material occurred at a depth of 40–50 cm at this research site) when soil was dried out. pH in soil in which watermelons were grown was improved to pH 4.2 after 2 months of cultivation thanks to liming and leaching by rain. EC of the soils was high (2.24–6.32 mS/cm), but the level of soluble and exchangeable Na was relatively low (<2 meq/100 g). High levels of soluble Fe and Al were the likely reasons for the high level of EC in soil. The levels of soluble and exchangeable Mg and Ca were higher in shrimp-rice and shrimp-watermelon cultivation systems owing to soil liming during shrimp cultivation. Soil organic carbon was high in mono-rice cultivation systems because the topsoil consisted of mostly half-decomposed organic material. However, in shrimp-rice and shrimp-watermelon systems, the level of organic C was high in topsoil because of the accumulation of organic residues during shrimp cultivation. Cation exchangeable capacity (CEC) of organic topsoil was low in mono-rice cultivation compared with that in mineral topsoil in shrimp-rice and shrimp-watermelon systems. The level of nitrogen was high in all soils, whereas total P was low (Table 14.4a, b; Fig. 14.12).

**Fig. 14.12** (**a**) Shrimp pond where pond dike was covered by plastic sheeting to prevent the leaching of acid substances from dike to pond water during rain, (**b**) shrimp-rice system and (**c, d**) shrimp-watermelon system (Source: Nguyen My Hoa et al. 2010a)

In summary, in potential ASS areas where farmers grew shrimp in shrimp-rice or shrimp-watermelon systems, soil pH dropped to low levels when the soil dried out. Liming and soil leaching are needed to cultivate shrimp or upland crops. Soils did not become sodic because the input water was low in salinity and the leaching of salinity was relatively easy in these soils. Cultivation of rice or watermelon in combination with aquaculture in these soils can be an option to increase income for farmers; however, suitable fish cultivation is a better choice than shrimp cultivation.

### *14.4.2 Soil Characteristics of Vegetable-Growing Areas in the Mekong Delta*

Intensive vegetable cropping may result in adverse physical, chemical and biological properties of soil if it is not well managed, especially when large amounts of chemical fertilizers and pesticides are applied for a long time without the addition of organic matter. Nguyen My Hoa et al. (2006) found that, in vegetable-growing areas in non-acidic alluvial soils in Tien Giang Province, total N, labile N and organic matter levels were low, and soil respiration was low at all sites with and without cow manure application (in these areas, farmers often apply about 0.5 t/ha

**Table 14.5** (a, b) Some soil chemical characteristics of the vegetable-growing areas in non-acidic alluvial soils in Tien Giang Province in Mekong Delta, Vietnam

(a)

| Studied sites | pH in water (1:2, 5) | CEC (unbuffered) (Cmol/kg) | CEC (buffered at pH = 8.1) (Cmol/kg) | N total (g/kg) | Labile N[a] (mgN/kg) |
|---|---|---|---|---|---|
| Basil field (with cow manure) | 4.13 | 17.17 | 25.13 | 1.4 | 9.53 |
| Chili field (no organic fertilizer) | 5.60 | 19.50 | 23.66 | 0.9 | 6.38 |
| Cabbage field (no organic fertilizer) | 5.08 | 18.54 | 24.73 | 1.0 | 7.51 |

(b)

| Studied sites | P Bray 1 (mg P/kg) | Organic matter (g/kg) | Ca (Cmol/kg) | Mg (Cmol/kg) |
|---|---|---|---|---|
| Basil field (with organic fertilizer) | 186.0 | 18.4 | – | 5.50 |
| Chili field (no organic fertilizer) | – | 14.4 | 8.38 | 9.24 |
| Cabbage field (no organic fertilizer) | 129.0 | 15.8 | 7.06 | 8.47 |

Source: Nguyen My Hoa et al. (2010b)
*CEC* cation-exchange capacity
[a]Labile N was extracted by hot KCl at 2 M

of cow manure (Table 14.5a, b). High levels of available P were recorded to the application of a high level of P by farmers in the area. The findings included poor soil quality in terms of organic C and N levels in soil, and low activity of soil micro-organisms due to a high level of chemical fertilizers and pesticides use, but a low level of organic matter had been applied or returned to the soil. Therefore, the use of organic fertilizer should be recommended in the areas in which intensive vegetable growing is undertaken.

Further investigations by Vo Thi Thu Tran et al. (2010) in the four major vegetable-growing areas in the Mekong Delta showed that P fertilizer was used at high rates. It was found that 40% of farmers interviewed in Binh Tan District, Vinh Long Province, applied 205 kg $P_2O_5$/ha/crop for corn and 3% of farmers applied an average of 98 kg $P_2O_5$/ha/crop for cucumber; in Thot Not Cantho, 53% of farmers applied 500–1,500 kg $P_2O_5$/ha/year for shallot; in Chau Thanh, Tra Vinh, all the farmers applied an average of 148 kg $P_2O_5$/ha for cucumber. These practices led to increases in soil P levels in these areas. Results of the analysis of 123 soil samples from An Giang, Vinh Long, Tra Vinh and Can Tho in the Mekong Delta, Vietnam, showed that 93.6% of studied soil samples had high total P content (Table 14.6). Available P as determined by the Bray 1 method was found to be high in more than 50% of studied samples (Table 14.7). The study showed that P was enriched in the vegetable-growing areas in the Mekong Delta; therefore, further studies are needed on the recommended rates of use of

**Table 14.6** Total P and rating scales of P in the studied vegetable-growing areas of the Mekong Delta (± standard deviation)

| Studied area | Total P (g/kg P) | Rating | Distribution (% of studied samples) |
|---|---|---|---|
| Thot Not District– Can Tho City | 0.74±0.13 | Very high | 65.63 |
| | 0.48±0.04 | High | 34.38 |
| | – | Medium | – |
| Cho Moi District- An Giang Province | 0.74±0.09 | Very high | 35.48 |
| | 0.52±0.04 | High | 54.84 |
| | 0.30±0.04 | Medium | 9.68 |
| Binh Tan District- Vinh Long Province | 0.65±0.04 | Very high | 26.67 |
| | 0.48±0.04 | High | 63.33 |
| | 0.30±0.04 | Medium | 10.00 |
| Mean value | 0.58±0.04 | Very high | |

Source: Nguyen My Hoa et al. (2010b)

**Table 14.7** Available P (Bray) with standard deviation and rating scales of P in the studied areas in vegetable-growing areas in the Mekong Delta (± standard deviation)

| Studied area | Available P (Bray 1) mg P/kg | Rating | Distribution (% of studied samples) |
|---|---|---|---|
| Thot Not District– Can Tho City | 75.34±43.34 | High | 90.63 |
| | 14.03±0.96 | Medium | 9.38 |
| | | Low | |
| Cho Moi District- An Giang Province | 41.33±15.48 | High | 70.97 |
| | 14.70±4.39 | Medium | 25.81 |
| | 6.82±0.00 | Low | 3.23 |
| Binh Tan District- Vinh Long Province | 40.12±16.74 | High | 53.33 |
| | 13.84±3.12 | Medium | 33.33 |
| | 5.36±1.06 | Low | 13.33 |
| Chau Thanh District- Tra Vinh Province | 84.2±57.6 | High | 80.0 |
| | 14.37±1.96 | Medium | 13.33 |
| | 5.05±2.06 | Low | 6.67 |

Source: Nguyen My Hoa et al. (2010b)

phosphorus fertilizer for crops, on the increase of P fertilizer use coefficient and the decrease of environmental risk due to the possibility of P run-off from agricultural soil for better P management in the area.

### *14.4.3 Soil Quality Improvement of Rice Fields*

Sustainable rice production in the Mekong Delta is under threat owing to intensified continuous rice cropping systems with two or three rice crops per year. Farmers apply high doses of inorganic fertilizers and pesticides to maintain rice yields. Therefore,

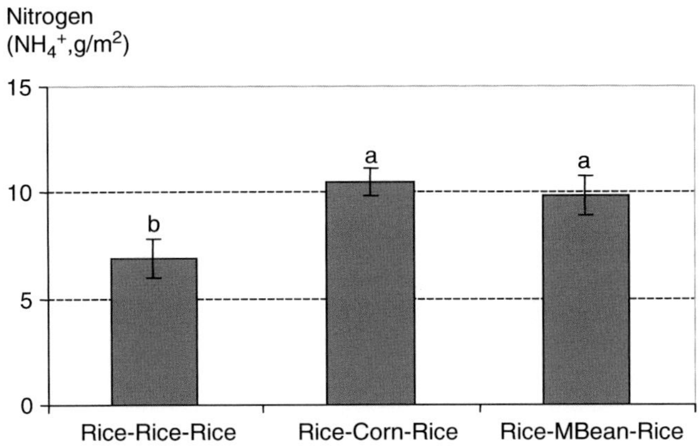


**Fig. 14.13** Mineralized nitrogen supplied from soil ($NH_4^+$,g/m$^2$) in different rice-based systems (Nguyen Minh Dong 2006). Data with different letters are significantly different ($P \leq 0.05$)

when long-term mono-rice cropping with intensive use of inorganic fertilizers and pesticides is applied, the land-use system has negative effects on soil fertility and rice production. Continuous harvesting of rice three times a year behind dikes that limit or stop sedimentation processes led to reduced soil organic matter quality and nitrogen supplying capacity, to loss of micro-organism biodiversity and to a reduction in rice yield. Physical soil degradation, such as soil compaction and low aggregate stability, was found in areas with intensive rice cultivation. Rice-upland crop systems, organic amendments and soil drying before sowing resulted in improvements of soil N, bulk density, soil water retention and soil aggregate stability, and therefore contributed to significantly increased rice yield (Vo Thi Guong and Roel Merckx 2008; Vo Thi Guong et al. 2010a). A study from Nguyen Minh Dong (2006), based on labelled nitrogen ($^{15}N$) field test, showed that nitrogen supplying capacity in soils of triple rice crops was lower than that of double rice crops rotated with upland crops (Fig. 14.13). The application of cultural practices such as rice crop rotation and organic amendments (Figs. 14.14 and 14.15) showed positive effects to improve the chemical and physical properties of soil. These effects played an important role in terms of significantly improving rice yield (Vo Thi Guong et al. 2010a) (Fig. 14.16).

### *14.4.4 Soil Properties in Orange Orchard Raised Beds*

To avoid damage from submergence during floods, raised beds are typically constructed for the planting of fruit trees in the Mekong Delta. Soil degradation can be a factor leading to low fruit yield and fruit quality. Compacted soil layers, which are highly resistant to penetration, are one of the most common problems affecting root growth. Buttery et al. (1998) found that limited root penetration on compacted

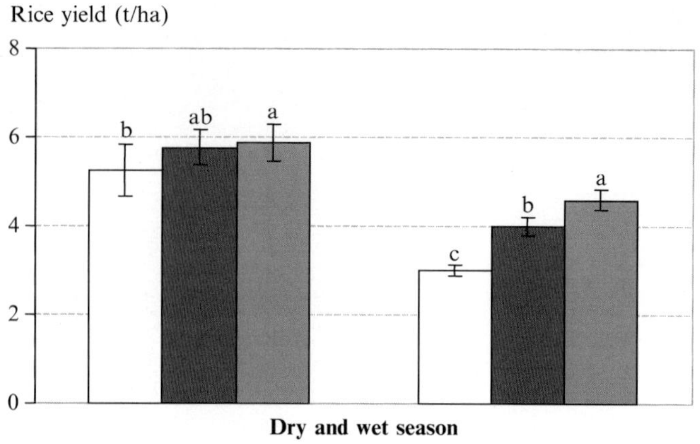


**Fig. 14.14** Effect of crop rotation on improvement of rice yield in system with three continuous annual harvests of rice. *First column*: three harvests of rice; *second*: rice-baby corn-rice; *third column*: rice-mungbean-rice (Vo Thi Guong et al. 2010b). Data with different letters are significantly different ($P \leq 0.05$)

**Fig. 14.15** Effect of organic amendment on rice yield in triple rice system (Vo Thi Guong and Roel Merckx 2008). Data with different letters are significantly different ($P \leq 0.05$)

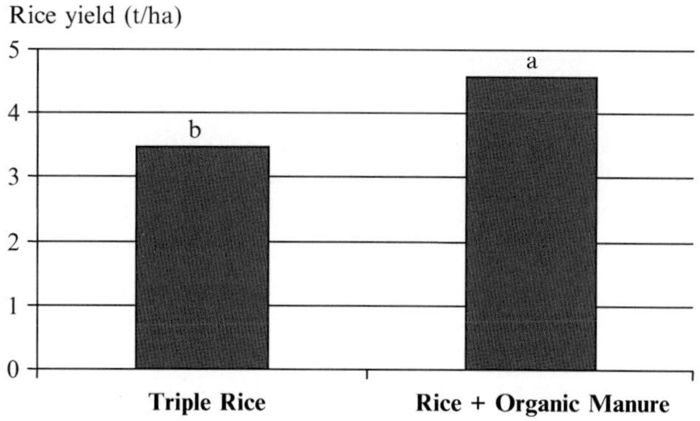


soils led to reduced soybean yield. Deep-rooted cover crops are one possible solution to problems of compaction, especially in farming systems without tilling (Unger and Kaspar 1994). Some selected soil chemical properties from differently aged raised beds were studied in Can Tho Province to clarify the constraints of orchards growing on old raised beds (Vo Thi Guong et al. 2006).

We found that the soil pH of the youngest raised beds was 5.3, which was significantly higher than that of raised beds of 16–33 years of age, which ranged from pH 3.5–4.6. This range of soil pH leads to limited macronutrient availability and soil microbial activity. Soil organic matter, labile organic nitrogen (easily turned-over fraction of soil organic nitrogen), available nitrogen and percentage of base saturation (the percentage of total exchangeable cation base on the soil exchangeable complex) were all significantly lower in the oldest 33-year-old raised beds (Table 14.8). Available phosphorus in the soil was found to be in the range of 0.9–1.2 mg/l and there was no statistical difference between the values for raised

**Fig. 14.16** Picture of rice field rotated with upland crops (Vo Thi Guong and Roel Merckx 2008)

**Table 14.8** Selected soil chemical properties of orange plantation of different raised bed ages. Data with different letters a and b are significantly different ($P \leq 0.05$)

| Raised bed | $pH_{H2O}$ | Organic matter (%) | Labile org. N (mg/kg) | $N_{available}$ | Base saturation (%) |
|---|---|---|---|---|---|
| 7 years old | 5.3 a | 5.4 a | 226.1 c | 25.4 a | 89.0 a |
| 16 years old | 4.6 b | 4.0 ab | 281.5 a | 19.5 b | 82.8 a |
| 26 years old | 4.6 b | 4.9 ab | 258.2 b | 19.1 b | 81.5 a |
| 33 years old | 3.5 c | 3.3 b | 210.7 c | 16.0 c | 54.6 b |

beds of different ages. Fungal and bacterial density in soil from raised beds was significantly lower in the oldest raised beds (33 years) than in the other raised beds. The youngest (7 years old) raised beds exhibited the highest bacterial and fungal density, while there was no significant difference in the density of the 16- and 26-year-old raised beds. Soil bulk density in raised beds of 7 and 16 years of age varied from 0.9 to 1.1 g/cm$^3$, while raised beds of 26 and 33 years of age had bulk densities of approximately 1.3 g/cm$^3$. Soil penetration resistance of the raised beds of 16, 26 and 33 years of age was higher than 3 MPa in the 15 cm closest to the surface. As a penetration resistance of greater than 2 MPa is considered to restrict root growth (Bengnough and Mullines 1990), it can be seen that soils in raised beds constructed 26–33 years ago for orange production have undergone physical degradation leading to an increase in compaction, which is likely to limit root growth. Therefore, the degradation of soil physical, chemical and biological properties was shown to have occurred on old raised beds of fruit orchards.

To confirm the above results, the study was complemented with a soil survey and soil sampling on 60 raised beds in citrus orchards that had been constructed less than 10 years ago, between 12 and 18 years ago, between 22 and 28 years ago and

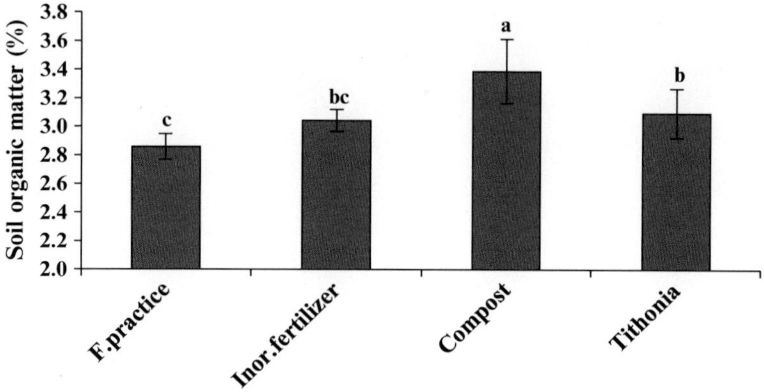


**Fig. 14.17** Effect of fresh *Tithonia* and sugarcane filter cake *compost* on soil organic matter (Vo Thi Guong et al. 2010c). F. practice: conventional farmers' practices. Data with different letters are significantly different ($P \leq 0.05$)

more than 30 years ago in Chau Thanh, Hau Giang Province (Vo Thi Guong et al. 2010c). Soil analyses indicated that soil degradation occurred in raised beds constructed more than 30 years ago. Soil aggregate stability was low and soil strength resistance was in the range that indicates soil compaction. Soil organic matter, CEC and base saturation percentage were very low compared with those of the younger raised beds ($P < 0.05$). These results emphasized that soil degradation in fruit orchards on long raised beds should be a focus for improvement of soil quality to increase the fruit quality and fruit yields.

## 14.4.5 Improvement of Soil Properties in Fruit Orchards

The addition of relatively high quantities of organic matter over a short time period has been shown to promote biological loosening of the soil and also to prevent soil and water loss, thus contributing to fertility and the maintenance or recovery of soil organic matter content (Rosolem et al. 2002). Adding fresh and composted organic substrates usually has beneficial effects on soil aggregate stability, humidification and microbial activity (Bipfubusa et al. 2008). A study was carried out to clarify whether soil physical and chemical properties and fruit yield of citrus on old raised beds can be improved by adding fresh and composted organic substrate. The effects of 10 t/ha sugarcane filter cake compost plus Trichoderma spp. and 20 t/ha of fresh *Tithonia diversifolia* were studied in orange orchards where the raised beds had been constructed more than 26 years ago. Both these organic treatments were combined with recommended inorganic fertilizer and compared with conventional farmers' practices (Vo Thi Guong et al. 2010c).

The results highlighted that addition of sugarcane filter cake compost and fresh *Tithonia diversifolia* led to significant changes in soil properties, such as increases of soil organic matter content (Fig. 14.17), available nitrogen and phosphorus,

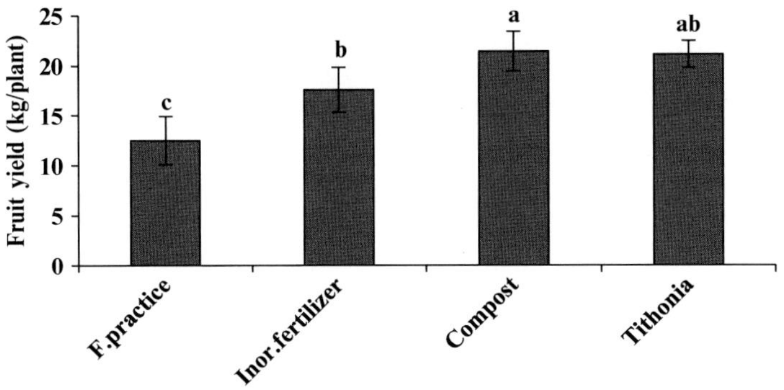


**Fig. 14.18** Effect of fresh Tithonia and sugarcane filter cake compost on orange fruit weight per plant (Vo Thi Guong et al. 2010c). Data with different letters are significantly different ($P \leq 0.05$)

percentage of base saturation, soil respiration and soil aggregate stability, as well as reduced soil compaction ($P<0.05$). Fruit yield was also improved with both sugarcane filter cake compost and fresh *Tithonia diversifolia* compared with that for conventional farmers' practices (Fig. 14.18), which involved application of a high level of inorganic fertilizer but gave low fruit yield ($P<0.05$) and resulted in poor soil quality in terms of its physical and chemical properties.

## 14.5 Prospects for the Future

In the future, the Mekong Delta will remain the most important area contributing to the economic development of Vietnam through agricultural and aquaculture production. Intensive rice production is the major farming activity. With its relatively high profits, aquaculture continues to be promoted as a means for economic growth for the region (Dang Kieu Nhan et al. 2007).

Continuous intensive harvesting of rice three times a year and the construction of dikes to prevent flooding have been major causes of reductions in soil quality and rice yield. To alleviate these problems, our recommendation to farmers is to apply rice-upland crop rotation systems as well as organic amendments. These practices considerably improve rice yields and soil fertility. Concerning fruit orchards, soil constraints were mainly linked to soil fertility degradation in terms of physico-chemical and biological properties. Organic amendments and balanced nutrients are the best ways to overcome the problem of declining soil quality and fruit yield.

In aquaculture systems, eutrophication is the main problem causing high mortality and failure in shrimp and *Artemia* cultures. In ASS, aquatic production face problems of high H2S concentrations in water and the soil pH dropped to low levels when soil dried out. Liming and ventilation to increase oxygen concentration in such systems are good approaches to counter this.

Concerns in this region are mainly focused on the production of high-quality and safe food and on ensuring safe working conditions for farmers. Therefore, several agricultural and aquaculture products have obtained VietGAP and GlobalGAP certifications to meet the requirements of global and domestic markets. Limited water resources in the dry season are a problem in relation to sustainable agriculture production and livelihoods. Several storage dams have been constructed upstream of the Mekong River in China, Thailand and Laos. Climate change also contributes to changes in the flood regime, reduction in dry season flows, changes in water discharge and sediment fluxes. All these factors may lead to extreme water scarcity in the future. Therefore, water-saving practices in agricultural and aquaculture production will be major concerns for high efficiency of water use and high economic returns. Floods and saline intrusion cause damage to rice and upland crops but also bring benefits (Dang Kieu Nhan et al. 2007). Therefore, agro-aquatic production systems need to be developed to adapt to the saline environment for food security and poverty alleviation. In addition, poor water quality and problem soils in areas with ASS and degraded soils, as well as intensive agro-aquatic cultivation, are still the major constraints to achieving sustainable production, and require further study.

## References


Andrew L (2007) Grow out pond and water quality management. JIFSAN good aquaculture practices program. University of Maryland, USA

Åström M (1998) Mobility of Al, Co, Cr, Cu, Fe, Mn, Ni, and V in sulfide-bearing fine-grained sediments exposed to O2: an experimental study. Environ Geol 36(3–4):219–226

Åström M (2001) Effect of widespread severely acidic soils on spatial features and abundance of trace elements in streams. J Geochem Explor 73:181–191

Bengnough AG, Mullines CE (1990) Mechanical impedance to root growth: a review of experimental techniques and root growth responses. J Soil Sci 41:341–358

Bipfubusa M, Angers DA, N'Dayegamiye A, Antoun H (2008) Soil aggregation and biochemical properties following the application of fresh and composted organic amendments. Soil Sci Soc Am J 72(1):160–166

Boyd CE (1998) Water quality for pond aquaculture. Research and development series no. 43. International Center for Aquaculture and Aquatic Environments, Alabama Agricultural Experiment Station, Auburn University, Alabama

Buttery BR, Tan CS, Drury CF, Park SJ, Armstrong RJ, Park KY (1998) The effects of soil compaction, soil moisture and soil type on growth and nodulation of soybean and common bean. Can J Plant Sci 78:571–576

Chanratchakool P (2003) Problem in *penaeus monodon* culture in low salinity areas. Advice on aquatic animal health care. Aquac Asia VIII(1):54–55

Chanratchakool P, Turnbull JF, Limsuwan C (1995) Health management in shrimp ponds, 2nd edn. Aquatic Animal Health Research Institute. Department of Fisheries, Kasetsart University, Bangkok

Chanratchakool P, Turnbull JF, Funge-Smith SJ, MacRae IH, Limsuwan C (2002) Health management in shrimp ponds, 3rd edn. Agricultural Publisher, Can Tho, Vietnam (Translated to Vietnamese by Aquaculture and Fisheries Faculty), Can Tho University

Chapman D (1997) Water quality assessment. A guide to the Use of biota, sediments and water in environmental monitoring, 2nd edn. E & FN Spon, London

Chau Minh Khoi (2006) Management of *Chaetoceros calcitrans* growth in hypersaline *Artemia franciscana* ponds by optimizing nitrogen and phosphorus availability. PhD thesis. Catholic University Leuven

Chau Minh Khoi, Vo Thi Guong, Drouillon M, Pypers P, Merckx R (2008) Chemical estimation of phosphorus released from hypersaline pond sediments used for brine shrimp Artemia franciscana production in the Mekong Delta. Aquaculture 274(2–4):275–280

Clemente R, Almela C, Bernal MP (2006) A remediation strategy on active phytoremediation followed by natural attenuation in a soil contaminated by pyrite waste. Environ Pollut 143:397–406

Dang Kieu Nhan, Nguyen Van Be, Nguyen Hieu Trung (2007) Water use and competition in the Mekong delta, Vietnam. In: Tran Thanh Be, Bach Tan Sinh, Miller F (eds) Literature analysis: challenges to sustainable development in the Mekong delta: regional and national policy issues and research needs. The sustainable Mekong research network Publisher, Bangkok

Dudas MJ (1984) Accumulation of native arsenic in acid sulfate soils in Alberta. Can J Soil Sci 67:317–331

General Statistics Office (2007) Statistical yearbook of Vietnam. Statistical Publishing House, Ha Noi

General Statistics Office (2009) Statistical yearbook of Vietnam. Statistical Publishing House, Ha Noi

Gustafsson JP, Tin NT (1994) Arsenic and selenium in some Vietnamese acid sulfate soils. Sci Total Environ 151:153–158

Hossain M, Ut TT, Bose ML (2006) Livelihood systems and dynamics of poverty in coastal province of Vietnam. In: Hoanh CT, Tuong TP, Gowing JW, Hardy H (eds) Environment and livelihoods in Tropical coastal zones: managing agriculture –fisheries-aquaculture conflicts. Cabi Publishing, Thailand, pp 30–47

Houba VJG, van der Lee JJ, Novozamsky I (1995) Part 5b: soil analysis procedures. In: Soil and plant analysis. A series of syllabi. Department of Soil science and Plant nutrition. Wageningen Agricultural University, Wageningen

Le Anh Tuan, Chu Thai H, Miller F, Bach Tan Sinh (2007) Flood and salinity management in the Mekong Delta, Vietnam. In: Tran Thanh Be, Bach Tan Sinh, Miller F (eds) Literature analysis: challenges to sustainable development in the Mekong delta: regional and national policy issues and research needs. The sustainable Mekong research network Publisher, Thailand, pp 15–68

Le Quang Tri, Vo Thi Guong (2005) Present situation of shrimp culture in the Mekong delta: social economic. Published in (Vietnamese). Vietnam Soil Sci J 21:162–165

McBride MB (1994) Environmental chemistry of soils. Oxford University Press, Oxford

Ministry of Fisheries (2003) Fisheries extension in Vietnam: ten years of operation and development. Ministry of Fisheries, Ha Noi

Ministry of Fisheries (2006) Guidance decree on implementation of development policies to fisheries. Ministry of Fisheries, Ha Noi

Newton BJ, Jarrell WM (1999) A procedure to estimate the response of aquatic systems to changes in phosphorus and nitrogen inputs. USDA natural resource conservation service

Nguyen Duy Can, Le Le Thanh Duong, Nguyen Van Sanh, Miller F (2007) Livelihoods and resource use strategies in the Mekong delta. In: Tran Thanh Be, Bach Tan Sinh, Miller F (eds) Literature analysis: challenges to sustainable development in the Mekong delta: regional and national policy issues and research needs. The sustainable Mekong research network Publisher, Bangkok, pp 69–98

Nguyen Minh Dong (2006) Effect of upland crop rotation with rice on rice yield in relation to the improvement of soil N available. Master thesis. Can Tho University, Can Tho

Nguyen My Hoa (2003) Soil potassium dynamics under intensive rice cropping. A case study in the Mekong delta, Vietnam. PhD thesis, Wageningen University, Wageningen

Nguyen My Hoa, Dang Duy Minh (2009) Surface water quality in dry and rainy season in Mekong river in Vietnam. In: The second international workshop on "Climate change impacts on surface water quality in East Asian watershed", The Gallery of University Malaysia, Kota Kinabulu, Malaysia, pp 84–89, 18–20 Feb 2009

Nguyen My Hoa, Huynh Tri Cuong (2009) Pollution of Cd, Fe and Ni in canal water in Acid Sulphate Soils as affected by soil acidification and available metal concentration in soils. In: Takizawa S, Kurisu F, Satoh H (eds) Southeast Asian water environment, vol 3. IWA Publishing, London, pp 25–30

Nguyen My Hoa, Tran Kim Tinh, Huynh Tri Cuong (2006) Pollution of toxic metals in canal water leached out from acid sulphate soils in Long Xuyen Quadrangle in the Mekong Delta-Vietnam and remediation. Report of the research program of the Vietnamese Ministry of Education and Training in 2005

Nguyen My Hoa, Tran Kim Tinh, Astrom M, Huynh Tri Cuong (2007) Pollution of some toxic metals in canal water leached out from acid sulphate soils in the Mekong Delta, Vietnam. In: Furumai H, Kurisu F, Katayama H, Satoh H, Ohgaki S, Thanh NC (eds) Southeast Asian water environment, vol 2. IWA Publishing, London, pp 99–106

Nguyen My Hoa, Ta Van Phuong, Phan Thanh Bang (2010a) Physical and chemical characteristics of soil and water and metal release from soil in shrimp pond in Acid Sulphate Soil. In: Vo Thi Guong, Nguyen My Hoa (eds) Overview of results on management of Acid Sulphate Soils in The Mekong Delta. Agriculture Publisher, Ho Chi Minh City, pp 65–102

Nguyen My Hoa, Pham Thi Phuong Thuy, Vo Thi Thu Tran (2010b) Available P in vegetable growing area in the Mekong Delta by Bray1, Mehlich 2 and Olsen method. In: Proceedings of agricultural conference in the Mekong Delta on sustainable agriculture development in adaptation to climate change- Part 2. Agriculture Publisher, Can Tho University, Can Tho City, Vietnam, pp 337–344

Nguyen Thi Ngoc Anh, Nguyen Van Hoa, Stappen GV, Sorgeloos P (2010) Effect of partial harvesting strategies on *Artemia* biomass production in Vietnamese salt works. Aquac Res 41:289–298

Nguyen Van Hoa (2002) Seasonal farming of the brine shrimp Artemia franciscana in artisanal salt ponds in Vietnam: effects of temperature and salinity. PhD thesis, Ghent University, Ghent, Belgium

Rosolem CA, Foloni JSS, Tiritan CS (2002) Root growth and nutrient accumulation in cover crops as affected by soil compaction. Soil Tillage Res 65:109–115

Seroka G (2004) The relationship between dissolved oxygen, nitrate, and phosphate concentrations and chlorophyll a concentrations in the Rhode River, a Sub-Estuary of the Chesapeake Bay. Thomas Jefferson High School for Science and Technology. Alexandria, VA STAR program

Shock C, Pratt K (2003) Phosphorus effects on surface water quality and phosphorus TMDL development. In: Western nutrient management conference, vol 5, Salt Lake City, pp 211–219

Smith DM, Burford MA, Tabrettirvin SJ, Ward L (2002) The effect of feeding frequency on water quality and growth of the black tiger shrimp (Penaeus monodon). Aquaculture 207(1/2):125–136

Sunström R, Åström M, Österholm P (2002) Comparison of the metal content in acid sulfate soil runoff and industrial effluents in Finland. Environ Sci Technol 36(20):4269–4272

Tat Anh Thu, Vo Thi Guong (2010) Soil, water quality and nutrients accumulation in aquatic ponds in Vinh Chau My Xuyen district, Soc Trang province (in Vietnamese). Can Tho University Publisher, Can Tho

To Phuc Tuong, Bouman BAM, Mortimer M (2005) More rice, less water- integrated approaches for increasing water productivity in irrigated rice-based system in Asia. Plant Prod Sci 8:229–239

Tran Thanh Be, Bach Tan Sinh, Miller F (2007) Challenges to sustainable development in the Mekong delta: regional and national policy issues and research needs. The sustainable Mekong research network Publisher, Thailand

Treece GD (2000) Artemia production for marine larval fish culture. SRAC publication no. 702. USDA, Washington, DC

Unger PW, Kaspar TC (1994) Soil compaction and root growth: a review. Agronomy J 86:759–766

Vo Quang Minh (2006) Setting up the fertility capability classification system for intensive rice areas in the Mekong delta (in Vietnamese). PhD thesis, Can Tho University, Can Tho, Vietnam

Vo Thi Guong, Roel Merckx (2008) Soil dynamics in Terrestrial/Aquatic environments. Final report of VLIR- IUC- R3 project, presented at workshop on 12/2008 at Can Tho University

Vo Thi Guong, Thai Truong Giang, Le Quang Tri (2004) Soil and water properties of rice-shrimp and intensive shrimp system at Ca Mau province (in Vietnamese). Vietnam Soil Sci J 20:36–41

Vo Thi Guong, Nguyen Khoi Nghia, Tran Kim Tinh, Duong Minh (2006) Improvement of soil physical and chemical degradation in raised beds of orchards by using organic amendments and cover crops. Vietnam Soil Sci J Special:25–27

Vo Thi Guong, Duong Minh Vien, Huynh Dao Nguyen, Nguyen Minh Dong (2010a) Improvement of soil properties and rice yield in triple rice in the Mekong delta (in Vietnamese). Agriculture Publisher, Ho Chi Minh City

Vo Thi Guong, Nguyen Minh Dong, Chau Minh Khoi (2010b) Soil organic matter quality and N supplying capacity in intensive rice and rice rotation system (in Vietnamese). Vietnam Soil Sci J 16b:147–154

Vo Thi Guong, Ngo Xuan Hien, Duong Minh (2010c) Effect of fresh and composted organic amendment on soil compaction and soil biochemical properties of citrus orchards in the Mekong delta, Vietnam. Paper presented at the 19th World soil congress, Brisbane, Australia

Vo Thi Thu Tran, Ph m Th Phuong Thuy, Dang Duy Minh, Nguyen My Hoa (2010) Present cultivation practice and available phosphorus in vegetable growing area in The Mekong Delta in Vietnam (in Vietnamese). Report of the research program of Can Tho University 2009

Wang S, Mulligan CN (2006) Occurrence of arsenic contamination in Canada: sources, behavior and distribution. Sci Total Environ 366:701–721

Whetstone JM, Treece GD, Browdy CL, Stokes AD (2002) Opportunities and constraint in marine shrimp farming. Southern Regional Aquaculture Centre (SRAC) publication no. 2600. USDA, Washington, DC

Wurts WA, Durborow RM (1992) Interactions of pH, carbon dioxide, alkalinity and hardness in fish ponds. Southern Regional Aquaculture Centre (SRAC) publication no. 464. USDA, Washington, DC

# Part V
# Water Knowledge and Information

# Chapter 15
# Managing Knowledge for the Development of the Mekong Delta

**Solvay Gerke, Hans-Dieter Evers, Bui The Cuong, Tatjana Bauer, and Judith Ehlert**

**Abstract** The Mekong Delta in Vietnam is one of the most productive areas in terms of rice cultivation and aquaculture in the world. Ambitious water resources management and flood control have largely contributed to this success. Nevertheless, agricultural intensification and diversification have generated new ecological and socio-economic challenges in the region. Vietnam's science and technology policy, described in this chapter, emphasises research on the water sector, focusing on the delta's sustainable and integrated water resources management. Several research institutes have been established and numerous research programmes, partly in cooperation with foreign organisations, have been initiated. The emergence of knowledge clusters of water-related research in these regions is the result of Vietnam's science policy, in which Ho Chi Minh City and the Mekong Delta are playing decisive roles. Clustering has a positive effect not only on the increase of knowledge output, but also on the economic growth of these regions. Ho Chi Minh City and Can Tho City are knowledge hubs with favourable conditions and a large pool of skilled people and advanced infrastructures. This chapter will analyse to what extent proximity and clustering have led to inter-organisational networking and knowledge sharing in academia. It will be shown that limited knowledge sharing has greatly reduced the effectiveness of water-related knowledge production and knowledge output in the south of Vietnam. This finding is mirrored at the local level, where the

S. Gerke (✉) • H.-D. Evers • T. Bauer • J. Ehlert
Center for Development Research (ZEF), University of Bonn,
Walter-Flex-Str. 3, 53113 Bonn, Germany
e-mail: solvay.gerke@uni-bonn.de; hdevers@uni-bonn.de; tatjana.bauer@uni-bonn.de; judith.ehlert@uni-bonn.de

B.T. Cuong
Southern Institute of Sustainable Development (SISD),
270 Nguyen Trong Tuyen Street, Phu Nhuan District, HCM City, Vietnam
e-mail: cuongbuithe@yahoo.com

F.G. Renaud and C. Kuenzer (eds.), *The Mekong Delta System: Interdisciplinary Analyses of a River Delta,* Springer Environmental Science and Engineering,
DOI 10.1007/978-94-007-3962-8_15, 

access of rural communities to water-related innovations and technologies will be examined and alternative local knowledge strategies of managing livelihood insecurities investigated. The current transformation process of agricultural modernisation – as heavily enforced in the Mekong Delta – will aggravate livelihood insecurities in the future. Water-related knowledge needs thus to be produced and disseminated more effectively to the targeted communities. Further, the importance of local knowledge has to be taken into account for a coherent strategy to develop a knowledge-based economy.

## 15.1 Knowledge for Development

Research in the WISDOM project (water related information system for the sustainable development of the Mekong Delta – Vietnam) shows that water plays a central role for the population and its diverse livelihoods in the Mekong Delta. Water is economically and socio-culturally omnipresent. Recent developments in the delta region are crucially linked to the management of the water resources. Flood control measures ensured successful rice intensification, fostered the aquaculture boom and enabled transportation networks – all accompanied by urbanisation and industrialisation. The success story of the Mekong Delta, however, also entails negative side effects such as growing water pollution due to agricultural run-offs and industrial wastewater release, soil degradation and the decrease in wild fish resources (Chaps. 13 and 14). The achievements of the Green Revolution and its water control paradigm have created new environmental uncertainties and risks. The dependency on nature in agricultural production has gradually been displaced by new dependencies on science and technology in agricultural intensification and diversification (Ehlert 2011). This indicates the role of scientific knowledge and technology for development. In general, numerous studies have argued that knowledge has replaced capital and labour as the main driver of development (World Bank 1999; Menkhoff et al. 2011). Building an ICT (information and communication technology) infrastructure has usually been one of the leading policy measures, in addition to developing universities and research institutes.

There is hardly any country in Asia that has not come forward with a plan, or at least a vision of developing a knowledge-based economy (KBE), leapfrogging various stages of rural development and industrialisation. Knowledge as a factor of production appears to be available in abundance and ready to be used to raise the level of productivity through knowledge-driven innovations. Vietnam is no exception to this trend and has emphasised its own development, or at least its support of science and technology research.

The Mekong Delta water sector is extremely complex, with many facets that the WISDOM information system tries to capture. Several large research institutes produce data that are necessary for planning the waterways, for monitoring and improving water quality, regulating the distribution of water for irrigation and guiding and regulating the hydraulic society (Evers and Benedikter 2009). Although global, formalised knowledge is absorbed and produced in universities, research institutes and research and development divisions in private sector and state enterprises, the local knowledge and work experience of the rural and urban population should be an important component in any strategy used to develop and govern a knowledge-based economy. The importance of local knowledge, especially for the development of local communities, has become increasingly recognised by research worldwide (Antweiler 1998; Evers 2000; Neubert and Macamo 2002). The broader term "local knowledge" includes any "locally and culturally situated knowledge that was and still is produced in local communities" (Antweiler 1998: 490), but local knowledge does not necessarily result in sustainable or socially just activities, nor is it always shared by all members of the community. Tacitness thereby designates the spheres of knowledge that are difficult to articulate; hence, they are to a great extent internalised routine, the 'know-how' of doing something based on experience. It is thus a highly context-bound and situated form of knowledge (Gerke and Ehlert 2011: 5).

One of the most difficult tasks of knowledge management is the transfer of this "tacit knowledge" into explicit knowledge, or transferring personal into organisational knowledge, which provides an essential challenge to the practice of knowledge management. The best way to transmit tacit knowledge or experience is still by observation, by face-to-face contacts and learning from doing. Routine work can easily be outsourced, but innovative, knowledge-based work needs team work and the existence of communities of practice, frequent social interaction and capacity building by direct face-to-face learning.

In Vietnamese culture, proverbs play an important role in passing on common knowledge, e.g. in the form of rules or advice. Proverbs thus can be a reflection of cultural common sense. "*Cho vàng cho bạc chứ ai đi buôn*" is a Vietnamese proverb, which when translated means "*You can give people gold and silver but you should not show them how to trade*". This proverb relates vividly to the canon throughout this whole chapter in that it advises strategically passing on knowledge by giving only gold and silver – knowledge artefacts – while holding back knowledge on the procedure, on the 'know- how,' in order to obtain gold and silver in the first place – the contextual process-knowledge, or 'business secrets'. Furthermore, it once again illustrates the technical limit of knowledge sharing inasmuch that you can give people a helping hand to start with, but then they have to go through the process of trial and error and making mistakes themselves in order to become experienced business people. The strategic and technical notions of limits to knowledge sharing are therefore précised perfectly in that saying, and since this advice made it into a proverb the content certainly is of cultural relevance in that knowledge transfer is strategic.

In this chapter we shall attempt to show the inner workings of this "machinery of water-related knowledge production and dissemination". Knowledge sharing within the epistemic community will be looked at as one important factor constraining or enabling the international profile of Vietnamese research institutes and universities. In a second step, the focus will be on the shortcomings of water-related knowledge dissemination and projects targeting local farming and fishermen communities. Along the case study of landless people without access to agri- and aquaculture technology, but totally relying on their tacit fishery knowledge, this knowledge turns out to be as much a strategic resource in mundane rural life as in the Vietnamese scientific community.

## 15.2 Knowledge Clusters as Centres of Development

Building a knowledge infrastructure means initially creating knowledge-producing and disseminating organisations such as research institutes, universities and colleges. To be effective, these have to be located closely in order to make use of common types of infrastructure such as laboratories, libraries and computing facilities. The geographical clustering theory assumes that proximity increases an organisation's innovation capacity when employees – especially researchers – can share ideas, products and services (Evers et al. 2011).

Developing industrial regions, clusters or knowledge hubs is, indeed, standard practice in many regional planning departments around the world. The allocation of human and financial resources creates knowledge-producing and disseminating organisations that can be measured, mapped and made to depict the contours of an epistemic landscape.

The assumption underlying these policies is that the clustering of knowledge-producing organisations increases knowledge output. In other words, isolated knowledge-producing institutes in knowledge-intensive industries are detrimental to innovation and economic growth. Clustering knowledge organisations is therefore the most effective policy for the way towards a knowledge-based economy and society. Ho Chi Minh City and the adjacent Mekong Delta are both destined to play a major role in Vietnam's effort to build a knowledge economy.

### *15.2.1 The Development of the Epistemic Landscape in the South of Vietnam*

Ho Chi Minh City is not only considered to be the financial centre of Vietnam, but also as an important cultural and industrial centre for the entire Southeast Asian region (Truong 2007: 24). The Mekong Delta, directly situated to the south-west of Ho Chi Minh City, is one of the world's most productive areas in terms of agriculture and aquaculture and ensures food security for the whole country. Given this background, we will show the importance of knowledge production for the region's socio-economic development.[1]

[1]Even though the largest concentration of knowledge-producing organisations is located in Vietnam's capital Hanoi, this chapter will focus mainly on the south of Vietnam, as this field research is an innovative investigation in this area. Certainly, knowledge production as such has led to the overall development of Vietnam contributing to different regions.

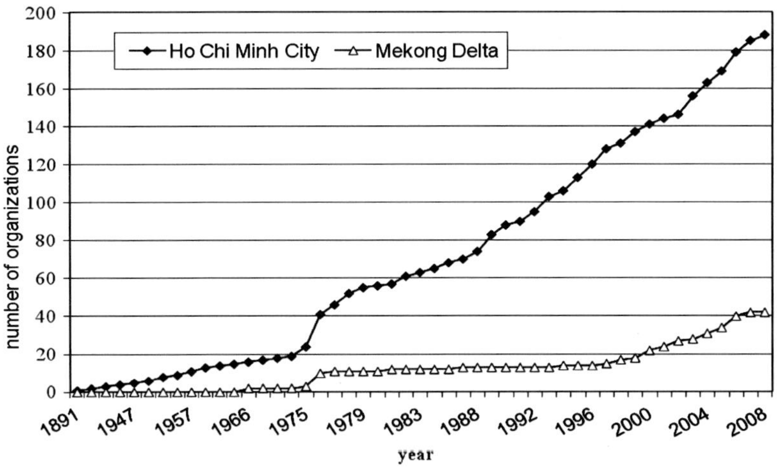


**Fig. 15.1** Number of knowledge-producing organisations in Ho Chi Minh City and the Mekong Delta (1891–2008) (Source: Own presentation based on data collected in the field from April 2008 to March 2009)

In the literature, we could find no comprehensive elaboration on the composition and allocation of knowledge-producing organisations, particularly from the points of view of Ho Chi Minh City and the Mekong Delta. Thus, an original dataset had to be compiled to acquire an overview of the status quo of Ho Chi Minh City's research environment. All datasets are based on an intensive listing of knowledge-producing organisations in Ho Chi Minh City, and were completed during a 1-year field research study in Vietnam from April 2008 to March 2009. This listing was finalised by means of various sources such as Ho Chi Minh City's Yellow Pages (Yellow Pages 2007), directories of scientific organisations published by the city's Department of Science, Technology and Environment (DOSTE 1998) and the Ministry of Science and Technology (MoST 2004, 2008). Through an extensive internet search and telephone campaign, every knowledge-producing organisation was verified and crosschecked.

The epistemic landscape of Ho Chi Minh City is in its infancy (Pham 2006: 238), highlighted by the fact that 78% of all currently operating knowledge-producing organisations were founded after 1975, at a time when the north and the south of Vietnam were united (see Fig. 15.1). In contrast, only 8% existed before 1975. For the remaining 14%, no data were available.[2] These figures indicate that knowledge-producing organisations in Ho Chi Minh City and the Mekong Delta have developed merely within the past 35 years.

[2]It can be assumed that a small number of organisations will have been dissolved, merged and renamed, indicating that these organisations are not all newly established.

A second peak was reached after 1986 with the introduction of the 'Renovation policy', otherwise known as *Doi Moi*. During Vietnam's subsequent transition to a market economy, local or provincial authorities, ministries and universities were allowed to create research and development (R&D) centres, without compulsory registration with the government, as had been the case before *Doi Moi* (Annerstedt and Nguyen 1996: 246). Although, as a consequence, the establishment of research and technology service centres has accelerated immensely, no reliable data relating to existing Vietnamese knowledge-producing organisations in Vietnam have been available until today.

We identified 218 knowledge-producing organisations, comprising 93 educational and 125 research organisations in Saigon (up to 1975) or Ho Chi Minh City (since 1975), broken down further into 49 universities (trường đại học), 48 centres (trung tâm), 44 institutes (viện), 34 colleges (trường cao đẳng), 29 sub-institutes (phân viện), ten academies (học viện) and four companies (công ty). In contrast, the Mekong Delta accounts for merely 42 knowledge-producing organisations, of which there are 20 colleges, 11 universities, seven research centres and four research institutes.[3]

According to our data, 7,736 staff members work for research organisations and 26,970 staff members for educational organisations.[4] The number of staff working for knowledge-producing organisations in Ho Chi Minh City and the Mekong Delta is rather high compared to other Southeast Asian countries such as Thailand, which, for instance, has only one-third of the capacity but has developed much faster than Vietnam. Despite large numbers of research personnel, scientific research in Vietnam results in limited scientific outcomes (Dang Mong Lan 2006; Gerke and Evers 2006: 17).

There is a large gap between the staff numbers of organisations in Ho Chi Minh City and the Mekong Delta, which would be expected due to the much lower amount

[3]For practical reasons, knowledge-producing organisations will be distinguished according to their main function, classified as either education (academies, colleges, universities) or research (companies, centres, sub-institutes, research institutes). This formal distinction, according to the key task of these organisations, does not have to exclude the other. As our field research has shown, in practice researchers from research organisations normally undertake teaching assignments at universities and university staff can also be involved in research projects. The term 'research organisations' is designated to all *research institutes* under ministries or directly under the government, *sub-institutes* belonging to parent organisations in Hanoi or to a functional ministry, *research centres* under the administration of research institutes or universities, and *companies* involved in privately run research projects. In contrast, 'educational organisations' comprise musical, military or political *academies*, *colleges* where students can gain a degree after 3 years' study, and *universities* where students have to study for 4–6 years to obtain a degree. Every one of these educational organisations is under the control of the Ministry of Education and Training (MoET) or a functional ministry such as the Ministry of Agriculture and Rural Development (MARD).

[4]The dataset includes only 80% of all identified knowledge-producing organisations. For the remaining 20% of organisations, staff numbers could not be ascertained and therefore will be ignored in the following analysis. Our data show that approximately one-sixth of staff numbers account for support staff, e.g. drivers, security guards and other service personnel.

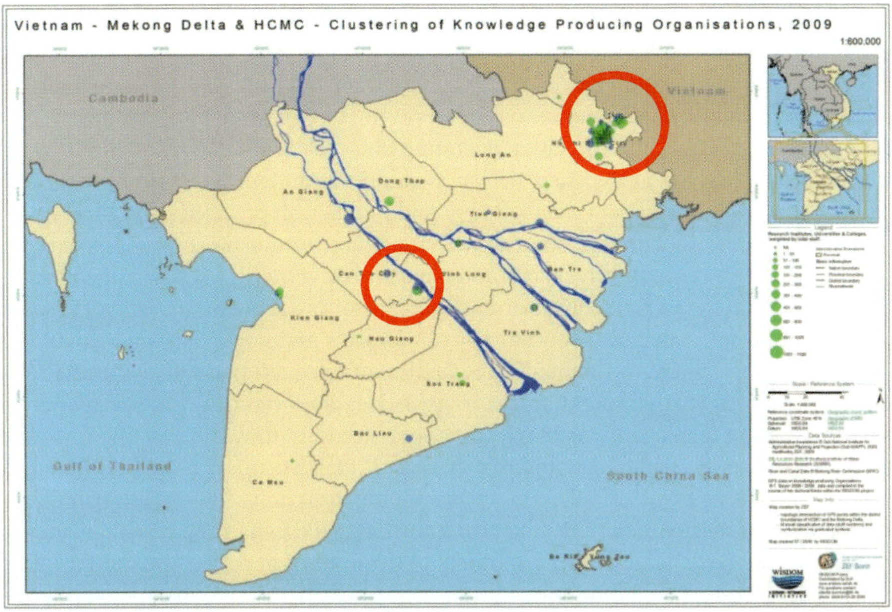


**Map 15.1** Knowledge-producing organisations in Ho Chi Minh City and the Mekong Delta (Map by Sven Genschick based on data of the WISDOM project. International boundaries for illustration only)

of knowledge-producing organisations in the latter. Nevertheless, this difference turns out to be even bigger when considering the populations of each area – Ho Chi Minh City with 6.4 million inhabitants and the Mekong Delta with 17.5 million (GSO 2007) – revealing an immense concentration of highly qualified staff in Ho Chi Minh City.

### *15.2.2 Proximity Matters – The Exchange of 'Tacit Knowledge'*

Using GIS-based mapping (Evers et al. 2009), we can identify two major knowledge clusters – Ho Chi Minh City and Can Tho City – and consider the proximity of these organisations as the determining factor in cluster formation (Map 15.1).

Looking at Vietnam's southern provinces within the Mekong Delta and adjacent areas, our data show that Ho Chi Minh City harbours one of the country's major knowledge clusters, followed by the much smaller cluster of Can Tho City. The distribution of knowledge-producing organisations within Ho Chi Minh City also shows clustering, in the sense that universities and research institutes are concentrated in adjacent urban districts. Clustering in these two urban areas is, to a large degree, a function of population density, or 'urbanism', i.e. the availability of urban institutions and of government policy.

Companies and organisations are attracted by cluster advantages for practical and economic reasons. State investments are made in order to create an efficient infrastructure, including excellent road networks and airports such as Vietnam's largest airport, the Tan Son Nhat International Airport in Ho Chi Minh City, or the Tra Noc Airport of Can Tho City, the latter operating since December 2008. Moreover, IT infrastructures are enhanced and the application of modern technology facilitates exchange as well as efficient management and production processes.

Further to the fact that economic activities tend to cluster, our study exposes similar trends in respect to research and educational activities. Knowledge clusters are agglomerations of production-oriented organisations. Having the ability to share knowledge assets such as laboratories or libraries reduces costs and enables a knowledge-sharing environment. The reduction of transaction costs, emphasised by classical industrial agglomeration theorists, is less important for knowledge-intensive production because transaction costs are extremely low. Conversely, however, a concentration of researchers and the sharing of tacit knowledge are facilitated by proximity (Evers et al. 2010). By virtue of the proximity of organisations, the recruitment of highly qualified staff[5] and knowledge exchange can be enhanced and higher productivity achieved which points to the important role of cluster building.

In fact, face-to-face communication is highly likely to take place, ensuring the transfer of valuable 'tacit knowledge'.[6] Highly skilled staff is available on the spot and therein approachable for organisations in terms of consulting services, sharing experiences and elaborating new ideas with these experts. Our survey,[7] carried out among Vietnamese researchers, shows that informal personal meetings and the telephone are by far the most important means of communication in Vietnam. It should be noted at this point that the effective usage of the telephone as a communication tool is only viable when the contact person is known beforehand, which also applies to email communication. The reasons for this are deeply embedded in the Vietnamese cultural suspicion of impersonal interactions, which are regarded as wholly untrustworthy. As personal relationships are inherent in professional life and the key to a project's success, they can consequently involve high transaction costs, as explained previously. The crucial advantage of clusters, therefore, is that people can meet over short distances, which saves a lot of wasted time and related

[5]Qualified people move to organisational agglomerations as the result of a better job market and opportunities for multiple jobs, e.g. in the consultancy business, because, as stressed by the interviewees, salaries for academics and scientists are exceptionally low, which means that they very often rely on an auxiliary income.

[6]'Tacit' knowledge lies in the experiences gained through action *('know-how')* embedded in specific cultural and social settings which shape and impact the innovation process (Boland and Tenkasi 1995: 350). For instance, tacit knowledge about organisational or managerial processes is difficult to transmit given that 'tacit' knowledge is deeply rooted in the individual's mind, which is applied daily in processes and routines. Therefore, it provides the competitive advantage of organisations in comparison with organisations outside the knowledge cluster.

[7]In total, 282 questionnaires were answered by Vietnamese staff members from seven different universities, colleges, research institutes and local authorities in which water-related activities take place.

travel expenses. Another advantage is the ease and comfort of attending seminars, workshops or conferences taking place in the same city, rather than travelling many hours to the countryside to visit perhaps only one organisation or workshop. Time and costs are too high to be beneficial, but by reducing these costs and time constraints through proximity, it is possible to build networks with many organisations working in the same field. Collaborations, meetings and face-to-face interactions take place actively as a result of advantageous facilities nearby such as coffee shops and recreational after-work establishments, where people can invest in valuable personal relationships that inevitably spill over into business. Evidence is mounting that, through the logistical proximity of clusters, opportunities accumulate and advantageous conditions are provided.

### *15.2.3 Achievements of Cluster Building – Rise of Scientific Output and Economic Growth*

Clusters provide a platform not only for Vietnamese industrial companies, research organisations and skilled people, but also as a main access point for international companies and organisations. It is understandable that international organisations prefer to find as many convenient conditions as possible if they are to pursue business connections in other countries. Facing inconvenient or impassable roads, long distances, unclear procedures, as well as language and cultural barriers will make transaction costs too high and render projects economically unviable. Furthermore, organisations located outside cluster areas are less ambitious in terms of innovation; they lack updated information about new technologies and management systems, while innovation seems to be non-profitable in a static environment that itself discourages change. Located away from clusters in such a way means that companies are isolated from a network of organisations that interact almost exclusively with the main customer base and potential collaboration partners. In this sense, regions without clusters risk being cut off from development and innovation processes and tend to lack capacity.

The survey confirms a high international influence on organisations in Ho Chi Minh City and Can Tho City.[8] Both areas create hubs in the south of Vietnam, with favourable conditions and a large pool of skilled people and advanced infrastructures. A positive effect of international cooperation is that large-scale international projects bring together different Vietnamese partners, which otherwise would never have collaborated. This intervention can be traced in the analysis of international publication output, as provided by the online academic database ISI Web of Knowledge,[9] which indicates that the majority of scientific articles with at least one

[8] During the field research, almost all interviewed organisations were directly or indirectly involved in international science and technology cooperation.

[9] "ISI Web of Knowledge" is an online academic database that only considers scientific articles published in English.

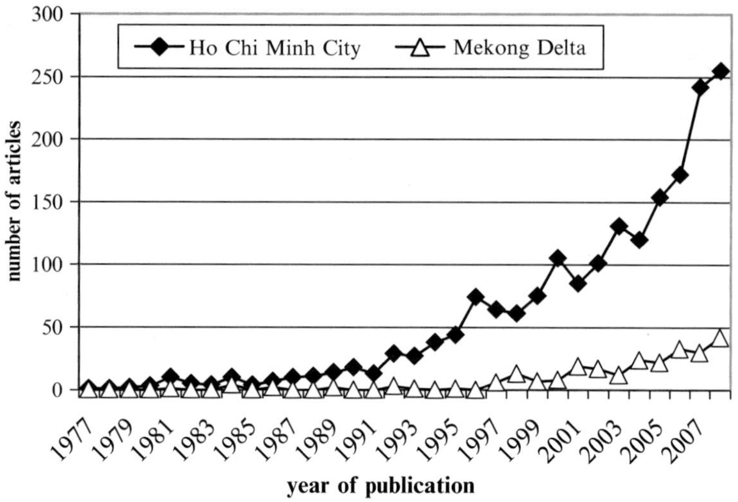


**Fig. 15.2** ISI journal articles published by Vietnamese authors (1977–2008) (Source: Own presentation based on information from www.isiknowledge.com)

Vietnamese author are joint productions. Many contributions involve Vietnamese authors from different institutes, indicating an exchange of ideas and information in terms of joint publication, as well as a learning effect among the authors. In addition to the enhancement of collaboration activities among Vietnamese scientists, knowledge exchange within the international research and science arena is also encouraged, because interacting with international experts gives Vietnamese academic and research staff the opportunity to improve their skills and methods and to work according to international standards in order to be recognised internationally. Business trips and study programmes abroad strengthen the capacity of Vietnamese organisations and more than likely create and maintain contacts for further projects.

The quantity and quality of scientific output are means by which the productivity and innovation of a region can be measured. In terms of knowledge clusters, it is a common standard to use international publications by Vietnamese authors. However, most knowledge output is produced in the Vietnamese language, which is often very roughly translated at best; with international publications, very few Vietnamese authors write directly in English. In general, there is still a lack of recognition of the vast Vietnamese knowledge output,[10] but the list of journal articles in the ISI is a measurement indicator (Fig. 15.2).

[10] It is a more complicated method to measure Vietnam's national output, which until now has been difficult to access. ISI includes a bias by virtue of the fact that English language journals dominate this database. Unfortunately, Vietnam's national scientific output is often neglected by international researchers due to language and administrative barriers. However, the results of our survey show that the national output is much higher than on an international scale. Since 2008, 23 national scientific journals have been uploaded on to the internet. This online databank *VJOL – Vietnamese Journal Online* – is a British initiative based at NACESTI, the National Centre for Scientific and Technological Information, under the administration of MoST, and is a first step for national as well as international readers to gain an insight into updated Vietnamese research findings.

The amount of Vietnamese knowledge output acknowledged internationally has increased significantly over the last 20 years. Compared to the Mekong Delta, Ho Chi Minh City's output production is almost ten times higher, which shows the superior role of Ho Chi Minh City in the development of southern Vietnam. Considering the Mekong Delta, with only 13 provinces, Can Tho City plays a crucial role for the region because its output accounts for 72% of the whole delta.[11] Obviously, there seems to be a correlation between the number of knowledge-producing organisations and international publications – the more organisations located in an area, the higher the output.

Clustering has an effect not only on the increase of knowledge output, but also on the economic growth of these regions.[12] Bearing in mind that the strategic locations of Ho Chi Minh City[13] and Can Tho City[14] originate from the flourishing development of southern Vietnam, the importance of the emergence of knowledge clusters in these regions is evident.

Statistical data show that the GDP of Ho Chi Minh City contributes one quarter of the country's GDP, even though the city holds just 7.8% of the total national population. More significant is the comparison of the GDP per capita of Ho Chi Minh City with that of Vietnam as a whole, the former being much higher for a long time with $2,100 against $835 in 2007. This proves that Ho Chi Minh City has enjoyed a far higher level of economic development than any other region in Vietnam. One of the contributing factors of this rapid development may be the positive effects of clustering. It is, however, difficult to assess whether this clustering

[11] Given that this development would not be possible without qualified people, Can Tho City took the initiative of opening the first university in the Mekong Delta. Consequently, science and research are now able to develop with the assistance of international cooperation, which is vital if the region is to interact on a global level.

[12] An attempt to measure the economic growth of Vietnam's provinces is the Provincial Competitiveness Index (PCI), which gives the opportunity to compare the economic development of each province in Vietnam and promotes local competition in order to improve the area's economy (www.pcivietnam.org). This index is based on surveys conducted in privately run companies and allows only educated guesses.

[13] We have demonstrated the importance of location for knowledge production and economic growth; nonetheless, this development would not have been possible without the strategic location of these regions. The Mekong River not only enables ships to enter the inner city of Ho Chi Minh City, but also provides cheap and effective transportation routes into the Mekong Delta. The subsequent strategic value of the region lies in economic advantages and opportunities for traders and manufacturers alike (Chong 2002: 21), and has transformed the region into a modern hydraulic society (Evers and Benedikter 2009). Today, Ho Chi Minh City is the focus of foreign investment in Vietnam, with half of all foreign investments flooding into the metropolis (Chong 2002: 22).

[14] Can Tho City is following closely on the heels of Ho Chi Minh City and fast becoming the economic, political, cultural and technological centre of the Mekong Delta (Le 2006: 118). In 2002, a World Bank loan was approved to restore waterway routes and ports in the Mekong Delta. Interestingly, the 600 km waterway network links Ho Chi Minh City with Can Tho City and Ca Mau, the most southerly part of Vietnam, on the one hand, and Ho Chi Minh City with Can Tho City and Ha Tien, which is close to the Cambodian border, on the other (Chong 2002: 93). In addition, Can Tho City is located at the crossroads of the highway network that interlinks the Mekong Delta, and another advantage is the recently opened Tra Noc Airport in Can Tho City, which will most likely become one of the main international airports in Vietnam in the near future (Le 2006: 119).

was the outcome of a deliberate policy or merely a function of population distribution and the urban land market.

Vietnam is on the path towards a knowledge-based economy in which Ho Chi Minh City and the Mekong Delta are playing decisive roles. Our data show that the number of knowledge-producing organisations is continuing to grow. Still, although there is a great deal of potential in improving the epistemic landscape of this region, it can nevertheless be assumed that the education and research sectors will develop in line with economic prosperity, since it is more likely that wealthier families – especially in urban areas with suitable facilities – will invest more in the higher education of their children.

## 15.3 Resistance to Knowledge Sharing: Limitations on the Development of the Research and Development Sector

As part of our further investigation, we shall analyse different aspects of clustering, e.g. to what extent proximity or clustering have led to inter-organisational networking and knowledge sharing. Through our interviews and survey data, we establish that the situation can be adequately described as one of hierarchical or bureaucratic sharing, insofar as research results are primarily channelled into either government departments or international donor agencies. Intra-organisational knowledge sharing still seems to be in the embryonic stages of development and horizontal research cooperation and knowledge sharing between knowledge-producing organisations hardly take place. There are numerous uncertainties and obstacles of the Vietnamese social, economic and political setting that Vietnamese scientists have to deal with.[15]

The present Vietnamese science policy and its organisational set-up can only be understood by looking back to the mid 1970s. Vietnam was reunified in 1976 and the centrally planned economy and science system already existing in the North was introduced to the South of the country. In the period between 1958 and 1975 the government of North Vietnam implemented the highly centralised science and research system equivalent to the Soviet model (Annerstedt and Nguyen 1996: 226). Numerous Soviet research institutes were imitated in the North of Vietnam, and after reunification these institutes expanded to the South and established equivalent sister organisations. According to the Soviet model, research activities followed a mission-oriented approach in that research tasks were determined in advance and the budget was allocated according to the national Science and Technology (S&T) plan (Bezanson 2000: 13). The absent linkage between R&D and the business sector,

[15]For instance, the Vietnamese culture based strongly on the adherence of hierarchy and seniority hinders young educated researchers to apply their knowledge and to develop ideas. Moreover, the strict censorship and control exercised by the state further block new impulses necessary for research and development.

and the marginal orientation of research activities towards the needs of the productive system, meant that Vietnamese enterprises did not value R&D results as a primary source of innovation. Moreover, the centrally planned economy of Vietnam did not necessarily require the development of new technology because of the fixed profit rates of companies that implied no motivation for cost reduction through technological change. This non-productive way of approaching science and technology, in which R&D scientists were public employees working for the government, led to serious problems – especially for the funding of research activities in the early 1980s. At this time also, foreign financial support from other socialist countries, especially the Soviet Union, decreased (Annerstedt and Nguyen 1996: 242).

In this time of the fiscal crisis, the government released a new decree which allowed research institutes and universities to operate in the market in terms of signing service contracts directly with state enterprises. In fact, this ruling solely legitimised a practice which had already been carried out informally for many years (Marr and White 1988: 22). In 1986, the 'renovation' policy induced significant change for the market economy, which caused an about-face turn in the field of science and technology. From this point on, Vietnam departed from the former Soviet system. S&T had a market price and new actors such as entrepreneurial scientists appeared. At the same time, research expenditures continued falling (Annerstedt and Nguyen 1996: 244). From the end of the 1980s, the Vietnamese government encouraged actively the process of commercialising the S&T system due to a lack of funding and research equipment. Research organisations shifted from R&D funded directly from the state budget to contract research, technical services and consultancy work.[16] Considering that all contract work was given to government organisations, the financial burden of the state could not be dissolved substantially. In fact, it was a form of indirect state financing and not a true market demanding approach (Bezanson 2000: 12ff). Following on from the issue of a series of laws in the last few years, the commercialisation of S&T has made enormous progress. In particular, Decree No. 115/2005/ND-CP, which was issued in 2005, allows universities and research institutes[17] to operate as firms while conducting S&T research. This regulation has led to an increase of direct commercial contracts with industry and the setting up of private research organisations.

The decentralisation of the state budget is increasingly promoted, given the fact that half of the budget expenditure is allocated to the provincial level. However, problems as a direct result of the planning system remain. For instance, investments by the state are made separately from decisions on regular expenditure. This disparity

[16]In 1995, the overall research budget was composed of 90% of national sources – 57.6% state budget and 32.4% contract work – and 10% foreign sources. In 2000, the budget still provided 85% of the total finance for R&D.

[17]Their services include the transfer of technology, consulting services or experimental manufacturing. These new and emerging knowledge-producing organisations are autonomous in financing research activities, but remain under the rule of the Vietnamese government (Ho Chi Minh City, 6 January 2009, researcher from the National Research Institute).

between capital and expenditure leads to financial shortages and insufficient maintenance or the monetary support of public goods. Furthermore, the Ministry of Planning and Investment, as well as the Ministry of Finance, are in charge of preparing annual budgets and long-term investment plans, even though coordination between the two ministries is very limited (World Bank 2004). In general, Vietnamese governmental and ministerial bodies can be characterised by poor inter-agency collaboration, as well as inconsistencies in planning procedures (Waibel 2010: 14). This phenomenon is intensified by the rapid development and re-organisation of knowledge-producing organisations. Many new organisations emerge, others have closed and existing organisations tend to change their names in the course of restructuring measures due to the change of the head organisations or upgrading towards internationalisation, in order to enlarge the research area and thereby attract more international S&T cooperation partners for joint projects. Moreover, research activities continue to be carried out separately, for instance because of internal conflicts in terms of newly merged departments which are not willing to cooperate, while the exchange of information hardly ever takes place on the organisational level.

A further obstacle for coordination among knowledge-producing organisations lies in the difficulty of their localisation.[18] The organisational landscape of Vietnam develops fast and therefore makes coordination difficult. Furthermore, although new regulations make room to open up new research institutes, at the same time it seems that organisations themselves are less interested in inter-organisational coordination. Organisational staff effectively utilise individual networks for managing their work, while formalised interactions among organisations tend to be rather unsuccessful due to the absence of personal relationships.

Besides the poor coordination that exists among state agencies, the complexity of the administrative apparatus and legal framework is very high. For instance, ministerial restructuring measures challenge the performance of knowledge-producing organisations. In the same way, Vietnamese organisations have to struggle with a more complex legal and administrative system nowadays, while officials and civil servants reach their limits[19] due to the low educational level of the managing officials.

According to interview partners, it is increasingly difficult to work under government policy because, annually, the government changes targets and requirements in order to meet increasingly more ambitious plans. Still, the identification and management of research tasks rarely derive from actual socio-economic needs. The selection criteria and the assessment of research projects are insufficient and the quality of research projects cannot be guaranteed at all. Hence, in comparison to neighbouring countries, Vietnam's capacities in science and technology are limited.

---

[18]This phenomenon became apparent when conducting a telephone campaign during the fieldwork. Some respondents indicated that the organisation has been dissolved or moved to another location (Can Tho City, 20 February 2009, research assistant).

[19]The interpretation of new directives remains mainly with local authorities in the provinces, which inform officials on the district level, who then instruct the people in the communes about new regulations (Can Tho City, 15 December 2008, head of office of the DoNRE).

### 15.3.1 Scarce Research Budget – Bureaucratic Struggle for Funding

Even though the Vietnamese government has pushed knowledge-producing organisations towards the autonomy of R&D funds to relieve the burden from the state, educational and research organisations, have to position themselves in the struggle for government budget allocations. In all conducted interviews with knowledge-producing organisations, the interview partners talked about insufficient government salaries and research funds to carry out serious research and ensure the sustainability of their organisation, including expenditure for administration, training and conferences, equipment and expendables, facilities, publications and staff salaries, to mention but a few.

Therefore, alongside the government budget, which is provided by the head organisation, e.g. ministries or national academies, knowledge-producing organisations have supplementary sources of revenue. Projects on the provincial, ministerial and state level provide large budgets. Contracts with local authorities for scientific services, as well as consultancy works with local organisations such as companies and other institutes, also contribute to the organisation's total budget. Additionally, teaching and international S&T cooperation give staff members the chance to earn an additional salary.

Foreign sources provided by foundations, bilateral cooperation agencies or multilateral organisations are an important part of the R&D expenditure of Vietnamese educational and research organisations. In particular, universities and their research institutes are most dependent on international funding. For instance, a study carried out by the European Commission in 2008 showed that only 15.3% of universities' research activities in Vietnam were funded by the government. About 30% of the funding sources were gained through contracts with enterprises and almost 50% provided by international organisations (Schüller et al. 2008: 35). In contrast, an interview partner from a ministerial research institute estimated the contribution of international funding to be only 5%, whereas funding by the ministries with 50–60% and the provinces (30–40%) were more important. Governmental research organisations rely less on international cooperation than universities and their respective affiliated research institutes. This development goes back to the decision of the government to reallocate research responsibilities among knowledge-producing organisations. Prior to Doi Moi, most basic research was conducted by national research institutes of the government. Since the beginning of the 1990s, funding for basic research has declined continuously. Furthermore, according to the new S&T policy, from 1992 basic research was to be transferred to universities which had previously been mainly responsible for teaching. Simultaneously, universities were required to become more demand-driven and to expand their research activities (Bezanson 2000: 13). Thereby, research institutes have focused their activities mainly on applied research supplemented by a mix of basic research and contract work for technological services (Dang 1998: 162). In 2005, national sources constituted 80% of R&D expenditure, of which 70% supported applied research and the remaining

30% basic research (Vu 2007: 15). Consequently, universities and their research institutes were forced to look for even more alternative income sources because they severely lacked funding.

In general, research organizations still lack adequate autonomy on planning, funding and human resources due to the slow re-organization of the research system. Research results come slowly and, if at all, will be hardly ever put to use. In fact, the national research budget is scattered and unfocused on key research fields and technical facilities, while non-state financial sources are not mobilised (MoST 2004). Nevertheless, the problem of linking R&D organisations with the industry limits Vietnam's innovative capacity tremendously. Hence, most technologies still have to be transferred from foreign countries. Even though Vietnam's S&T development has been officially affirmed by government regulations, implementation at sub-national levels is lagging behind. Evers (2005) states that governments in Southeast Asia, generally, have quickly embraced the policies of market deregulation, but have been less enthusiastic in terms of their science and research systems. Accordingly, the Vietnamese government plays a decisive role in setting the framework for knowledge production. Though control has loosened after Doi Moi, the Communist Party of Vietnam remains the gatekeeper and decides what information on sensitive issues is published or withheld from the public.

### *15.3.2 Persisting Coexistence of Knowledge Systems – Weak Cooperation and Knowledge Sharing*

Knowledge systems are composed of a network of individuals, groups and organisations that utilise knowledge for action. Thereby, financial and human resources, cultural factors as well as the institutional framework give these knowledge systems the capacity to handle and shape their work. Existing Vietnamese knowledge systems comprise state management agencies as well as educational and research organisations, all of which generate, disseminate or apply knowledge in the Vietnamese context. The relations among involved organisations and levels of hierarchy strongly determine the exchange of knowledge, as well as inter-organisational collaboration that affects the performance of the Vietnamese S&T system.

Knowledge systems exist next to each other without synergising, collaborating or coordinating their work. In general, the Vietnamese S&T system is characterised by overlapping regulations and blurred responsibilities of ministries. State management agencies and research organisations result in a lack of institutional incentives for cooperation and knowledge exchange, too.

Central- and local-level governments perform their tasks independently. Hence, national guidelines are implemented differently on the sub-national level and lead to the different performance of provinces and organisations; local governments insist on their autonomy from the national level as well as from other provinces. Historically rooted, numerous policy changes have caused an uncertain working environment and a lack of institutional trust, whereby bureaucracy and networking activities have been increased to establish stable relationships within their administrative boundary.

Furthermore, horizontal cooperation among ministries or other organisations rarely takes place. Similar to local governments, it has been shown that ministries, as well as universities and research institutes, insist on their sphere of responsibility to protect their access to financial resources while ignoring similarities with other organisations' work. There is strong competition for funding, which allows organisations to withhold their information and refuse to collaborate with each other.

The institutional framework is still unable to provide clear regulations and precise responsibilities for organisations, which seriously constrains inter-organisational cooperation and active knowledge sharing among different research and knowledge systems. As, in addition, Vietnamese culture promotes informal procedures and a reform of the science system has not yet been undertaken, research results tend to be of lower quality than could have been expected. These states of confusion leave space for obscure bureaucratic procedures and biased non-objective practices for the identification of research tasks and the allocation of funds, which are chronically insufficient. Researchers continue to work in an isolated research environment, which limits the opportunities for highly qualified research tremendously. Given these barriers, Vietnamese researchers and knowledge-producing organisations hesitate to share data and information and very often neglect organisational performance.

Geographical clustering without knowledge sharing has greatly reduced the effectiveness of knowledge production and knowledge output. It therefore remains to be seen whether the attempts by some Vietnamese researchers will bear fruit when they start to work toward more intensive knowledge sharing between organisations. Knowledge clustering needs to be supplemented by networking and the building of knowledge sharing, epistemic communities to produce new knowledge and economically viable innovations. In our terminology the 'knowledge hubs' of networking and knowledge sharing are yet to be fully developed – the epistemic landscape still has to be completed (Bauer 2011).

## 15.4 Access to Scientific Innovations and Knowledge Sharing at the Local Level

While the preceding discussions focused on knowledge production and sharing within the Vietnamese scientific community, the following elaborates on the limited implementation of water-related research and technologies in rural areas. Wild fish scarcity in the context of water infrastructure developments and aquaculture promotion will be taken as a case study. Landless people without access to aquaculture technology and decreasing entitlements to floodplain habitats turn their traditional fishery knowledge into a strategic resource. Knowledge sharing on the local level thus, as in the case of the aforementioned research community, becomes a highly deliberate process.

The designation of the Mekong Delta as "hydraulic society" (Wittfogel 1957; cf. Evers and Benedikter 2009) implies agricultural intensification via a complex irrigation system and by controlling the waters of the Mekong for economic development (see Chap. 3). Water management schemes for channels and dykes cut the open land of the delta into grid-like plots towards which the floodwaters are guided according to human decisions. Before the development of irrigation channels and border dykes, floodwaters spread evenly everywhere and brought with them wild fish stocks – an abundance remembered nowadays only by village elders.

The massive construction of small- and large-scale water infrastructure, however, has destroyed wild fish habitats and negatively affected the sustainability of wild fish resources. Population growth also exerts additional pressure on wild fish stocks. In the pursuit of agricultural modernisation, and in order to live up to the country's ambitious rice export orientation, the agricultural policy for the Mekong Delta foresees a further shift in land-use patterns from scattering to concentration (Nguyen 2006: 3). The concentration of land and mechanised agriculture aggravate the unemployment rate and stabilise the trend of landless households (Akram-Lodhi 2005). Rural industrialisation is yet unable to absorb the massive labour externalities. Finding farm employment or non-farm jobs is a highly competitive endeavour among poor and landless households. Small-scale farmers who lose out in this process of rural transformation add to the numbers of people depending on natural water products as a source of subsistence. The pressure on wild fish resources is thus growing considerably (see Chap. 11).

In response to this situation, Vietnam's agricultural diversification policy strongly promotes aquaculture as a reasonable measure to compensate for decreasing wild fish resources. The Vietnamese Ministry of Fishery (MoFI) emphasises the intensification of fish production and the promotion of technological innovations of diverse aquaculture systems (DFID 2002: 13f). The following case study represents an example of fishery science implementation on the local level as part of current fishery policy.

### *15.4.1 Knowledge for Livelihood Security*

An international non-governmental organisation (NGO) in cooperation with a Vietnamese aquaculture research institute from Ho Chi Minh City introduced the model of integrated rice-fish farming on a larger scale in the research site (Vinh Thanh district, Can Tho City). During the flood season farmers raise culture fish inside the flooded rice fields, enclosed by a surrounding net. Towards the end of the flood season, after 2–3 months, the water is pumped out of the field to prepare the lands for the next crop and to harvest the fish. Traditionally, some farmers apply this model on a very small scale and on an individual basis. The global scientific model of community-based floodplain fishery, however, is replicated in Bali, Cambodia, China, Indonesia and Vietnam, in certain test areas. In contrast to individual small-scale fish farming, the scientific project aimed to join the production lands of 20–30 farming families to make up between 60 and 100 ha of integrated rice-aquaculture. Furthermore, it endeavoured to offer alternative and under-exploited livelihood opportunities for rice farmers during the flood season. From the perspective of poor and landless families in the area, however, they were literally fenced off from the floodplain, which under customary rule they are entitled to access for subsistence purposes. The social complexity of and growing competition over water livelihood approaches were not taken fully into account by the project. It transpired that the farmers did not want the poor or the landless to join the floodplain project because they didn't want their poor neighbours, who were participating only in-kind, to share the benefits without contributing any land or capital. In addition, due to the pressing need for daily returns on their work, poor people could not be involved in a project that would bear fruit only after 2–3 months of work. Closed-off fishing grounds forced them to migrate to other places for the fishing season, so instead of wild fish capture in the flooded fields in the local area, which they could no longer access, they had to organise boats to travel to places further away. The problem with this scenario was that the very poor could not afford the boats to travel to distant places for fishing. Taking all of these aspects into consideration, it turned out to be extremely difficult to integrate the interests of the landowners and the poor (Ehlert 2011).

The example above shows that conflicts arise due to people having divergent capacities and interests in freshwater resources (Siar and Sajise 2009: 92). As wetlands are drawn on for a multitude of users and functions, conflicts between the different resource users are likely to occur (Adger and Luttrell 2000: 78), and culture fish models during the flood season only benefit the landowners. Since the flooded fields are subject to being embanked, landless people can neither participate in such innovative production models nor access the land for traditional floodplain fishing as a means of subsistence. The preceding example proved that pure technology transfer is not enough to make scientific innovations, such as in aquaculture production, relevant for local contexts. By contrast, the lack of the socio-cultural embeddedness of science can even lead to total project failure.

In general, it was further observed that land agglomeration and the promotion of affluent farmers foster technology and capital-intensive, large-scale production. Livelihood models of some people foreclose the alternative income-generating strategies of others. Embankments that on the one hand support rice intensification, catfish farming or integrated rice-fish culture exclude landless people from the floodplain where they traditionally engage in subsistence fishery. In addition, they are detained from capital-intensive fishing technology because they lack investment capital. The social inequality in accessing high-tech livelihood knowledge is deepened by differential command over land and credit opportunities. The agricultural bank, for example, is set up as a private company and issues loans to individual farmers' households which own enough land. Bank-supported water-based models include integrated rice-fish farming, shrimp farming and pond fish raising. Research institutes such as Can Tho University engage in studies on the further integration of agri- and aquaculture and thus on the most effective use of seasonal flood resources. Well and better-off farmers could implement this up-to-date knowledge on their land, whereas the landless are excluded from economically viable technology-based water livelihood innovations (Ehlert 2011).

Since an ever-growing number of landless people are unable to access certain livelihood innovations, they are all the more dependent on their traditionally honed local knowledge. It will be shown that under constraining conditions of livelihood insecurity, landless people, who traditionally base their household economy on wild fish resources and fishing during the flood season as one component of wider livelihood strategies (DFID 2002), cautiously calculate the risk of whether and with whom to share their knowledge. The growing scarcity of wild fish resources has been perceived as a qualitative change in the local natural environment. The strategic and controlled transfer of local fishing knowledge therewith constitutes an adaptive measure to cope with growing fish scarcity and competition.

### *15.4.2 Knowledge Sharing and the Transfer of Knowledge*

In the following we discuss two factors which impede knowledge from being a collective asset in a certain locality. The first is the tendency of actively controlling the transfer of knowledge in the struggle for natural resources under growing scarcity, while the second is the tacit character of local knowledge functioning as an imminent barrier to knowledge sharing. These two factors undermine the ability of knowledge to move freely, but instead sensibly determine to where it actually travels.

The flood season in the research site is characterised by an interesting dualism of rest and peace for the paddy farmers and dynamic action by the landless people who engage in fish harvesting. Paddy farmers typically use the flood season as a time to rest and only engage in small-scale fishing as a food supplement. For landless fishermen, by contrast, the flood season constitutes the peak of annual livelihood activities – with floodplain fishing as important base of a household's subsistence and small-scale economy. During this time of the year, landless fishermen migrate to find rich fishing grounds in the flooded lands owned by the paddy farmers.

In some ways farmers and fishermen behave in a similar way, as described by the government employees and researchers above. They tend to hide activities in order to keep their fishing skills a secret and to capture the specialised knowledge that forms the basis of everyday life for landless people. In the case of consciously holding back one's skills and knowledge from others, strategies of 'not telling' or of 'keeping it short' when asked are applied as evasive forms of communication and interaction. Such typical evasive patterns were illustrated nicely by a paddy farmer, who himself had acquired certain fishing skills – not through knowledge exchange with the landless fishermen but through a very long-term self-instructive process. Within the paddy farmer's commune this man and his family represent a rather exceptional case, since most rice farmers in this area refrain from fishing. However, although for this farmer fishing knowledge serves his subsistence far less than as some small extra economic income, he nevertheless applies certain strategies to actively close off his knowledge from others:

> They [the fishermen] are very creative and experienced. For each kind, such as prawn, abanas, snake-head fish, they have one catching style and one fishing tool. But they also want to hide their technology. They do not want to show it to all people. [...].

According to the farmer, selling wild fish – the fishermen's knowledge- and skills-embedded business product – at a faraway market, for instance, would be one strategy to prevent the local rice farmers from becoming suspicious. He explained that by proactively hiding their catch, the fishermen would prevent rice farmers from becoming interested in the potential resources of their own lands in the flood season. This pro-action shall be referred to as the strategy of 'not telling'. Although the paddy farmers do not represent an immediate risk, the fishermen's action is strategically foresighted to not encourage any kind of conflict around the free access to flooded lands. This represents a strategic way of upholding the free access to the paddy farmers' lands and floodwaters as common property. Access to farmers' land is the prerequisite for fishing in seasonal floodplains, since the seasonal water is managed by evermore complex irrigation systems, which make access to water increasingly dependent on access to land. Nonetheless, the rice farmer, who is well known for his fishing skills among neighbouring rice farmers, applies similar strategies of 'not telling'. He also wants to keep his skills a secret by, for example, 'never showing off' in front of his neighbours or by going out at twilight to place his bamboo traps. He feels that otherwise people in his neighbourhood would envy and 'play tricks' on him, for instance by destroying the traps or learning the trade themselves.

Another form of direct evasive strategy regarding the transfer of livelihood knowledge turned out to be the strategy of 'keeping it short'. One landless woman who, for many years already, has come with three other families from afar to settle down temporarily on the same rice farmer's residential land, articulates the tenor that when being asked for certain fishing tips one should never disclose 'business secrets' but always only very 'briefly' show or tell the questioner about the skills.

Such conscious evasive strategies are based on a fine-tuned deliberation process in which it is decided with whom to share one's knowledge and when to strategically close the conversation. It is evident that trust is the dominant factor that facilitates knowledge sharing. Knowledge transfer is bounded in trustworthy reciprocal relationships. Kinship constitutes the basis for such relationships in which knowledge is passed on. If trust is not guaranteed by blood, then it can grow on the basis of time-consuming investments in personal relationships, requiring strong commitments to mutual reliability (Menkhoff and Gerke 2002).

### 15.4.3 The Strategic Importance of Knowledge

Besides trust as the overarching condition for knowledge sharing, other rationalities come into play. Economic motivation regarding the transfer of local knowledge has to be considered. If two fishermen share the same livelihood status – poor and landless – they most likely would not share their knowledge about the rich fishing grounds. Instead they will not disclose the best fish resources or the names of the landowners who let them stay for free on their compounds during the flood season. Since the other fishermen have similar fishing skills, one would at least try to keep an economic advantage by hiding this kind of business knowledge. Such economic considerations, however, can be offset by trust. Interviews with rice farmers and fishermen showed trust to be the dominant rationality of knowledge sharing. Trust cross-cuts economic status and is based solely on the belief in the 'morality' of the person with whom one shares knowledge. This in turn insures that this knowledge will not be used in any strategic way against the person.

What has been discussed above as the "local strategies of knowledge sharing" demonstrate that knowledge can be consciously held back or shared when it comes to coping with livelihood insecurities. Two strategies have been elaborated so far, namely the strategies of 'not telling' and of 'keeping it short'. Furthermore, the tacit nature of local knowledge in itself has an exclusive moment which is constituted by long-term practical experience, inborn talent and creativity. Since this kind of knowledge is, besides factual knowledge, the precondition for the professionalisation of skills, it has a latent strategic element in itself inasmuch that it naturally regulates its access. In the case of the first two strategies, it is actors negotiating access by consciously holding back knowledge. It now will further be argued that the technical limit of knowledge sharing can also be turned strategically by the actors. This refers to the third strategy of 'telling', even though this openness has an inherent mechanism of demarcation based on the tacit character of skills. The characteristics of tacitness can back up culturally appropriate ways of communication and interaction by simultaneously upholding the exclusiveness of knowledge – in this respect, tacit knowledge can be strategically turned into a proactive form of communication. If somebody asks you about your recipe for fishing success, the only thing you could do is to 'tell' him the facts and 'show' him the procedure. Nevertheless, the process knowledge you could not verbalise, so the other person would have to observe your actions over the long term and try to successfully internalise this process knowledge. Saying nothing when being asked makes you lose face because of your demonstration of not-knowing. At the same time you would make the other person lose face by not paying enough respect when leaving his or her question neglected and the situation uncomfortable.

'Not telling', 'keeping it short' and 'telling' have been discussed as control mechanisms for knowledge flow in the case of seasonal floodplain fisheries in the Mekong Delta. At the moment it is rather the exception that the paddy farmers in the research commune have an interest in harvesting wild fish themselves, but even the very few who have acquired such professional knowledge try actively to hold it back from other paddy farmers in their commune. This certainly indicates competitive behaviour around fishing knowledge as a strategic resource. Possible future scenarios might further stimulate the demand for floodplain fishery knowledge, as growing scarcity makes a resource precious in terms of market value. The decrease in wild fish levels, for instance, will potentially arouse the economic motivation of better-off people in the rural area to strategically seize those scarce assets. The market price for wild fish products increases proportionately to the amount of culture fish on the market, the latter being valued far less for its taste than wild fish. The stabilising trend of landlessness will result in more and more people aiming at the diversification of their reproductive base – here, fishery knowledge could be a complementary asset in achieving livelihood security. The growing conflict around access rights to decreasing wild fish grounds will evidently upgrade such knowledge. Although landless people cannot control the physical accessibility to fish by fencing off their lands, they can at least close off the flow and regulate the accessibility of their fishery knowledge through strategic action.

## 15.5 Conclusion

Our data clearly show that knowledge is a major factor in the development of the Mekong Delta. Nonetheless, knowledge production, dissemination and utilisation need to be supported and managed. A science and research policy or the use of an information system like that developed within the WISDOM project are steps to achieve this aim. As our analysis shows, "knowledge clusters", i.e. areas of a high density of knowledge producing institutions and high-level manpower have developed in Ho Chi Minh City and Can Tho, but the desired positive effects are marred by a lack of knowledge sharing within and between institutions. This also means that the transfer of knowledge to policy-implementing agencies and the local population is severely hampered. We could also show that local knowledge on water-related activities is important on the local level, but here also barriers to knowledge sharing inhibit the flow of knowledge and experience to those that could use it effectively for economic activities and, ultimately, a reduction of poverty levels. On the positive side, it can be concluded that standards of living have risen during the past 20 years and a research infrastructure has been put in place. The Mekong Delta is therefore equipped to deal with the challenging problems arising from unstable world markets, as well as climate change.

## References

Adger WN, Luttrell C (2000) The value of wetlands: landscape and institutional perspectives. Property rights and the utilisation of wetlands. Ecol Econ 35:75–89

Akram-Lodhi AH (2005) Vietnam's agriculture: processes of rich peasant accumulation and mechanisms of social differentiation. J Agrar Chang 5(1):73–116

Annerstedt J, Nguyen TH (1996) Demolishing the Ivory Tower – science, technology and economic development. In: Norlund I, Gates CL, Vu CD (eds) Vietnam in a changing world. Curzon Press, Richmond, pp 219–250

Antweiler C (1998) Local knowledge and local knowing. An anthropological analysis of contested 'Cultural products' in the context of development. Anthropos 93(4–6):469–494

Bauer T (2011) The challenge of knowledge sharing – practices of the Vietnamese Science Community in Ho Chi Minh City and the Mekong Delta. Unpublished dissertation, University of Bonn

Bezanson K (2000) A science technology and industry strategy for Vietnam. Final UNDP-UNIDO report to the Government of Vietnam. Hanoi, UNDP

Boland RJ, Tenkasi RV (1995) Perspective making and perspective taking in communities of knowing. Organ Sci 6(4):350–372

Chong Li Choy (2002) Business environment and opportunities in Vietnam - Ho Chi Minh City and its surrounding region. Verein Business Books & Tools St. Gallen, St. Gallen

Dang DT (1998) From plan to market – S&T policy reform in Vietnam. In: Meske W, Mosoni-Fried J, Etzkowitz H, Nesvetailov G (eds) Transforming science and technology systems – the endless transition? Ios Press, Amsterdam/Berlin/Oxford/Tokyo/Washington, DC, pp 160–170

Dang Mong Lan (2006) Khoa học Việt Nam: Những nghịch lý và một số cách giải (Science in Vietnam: Absurdities and some Solutions). Khoa Học- Giáo Dục. Đi tìm diện mạo mới (Science – Education. Looking for new aspects). N. t. giả. TP. Hồ Chí Minh, Nhà xuất bản trẻ

DFID [Department for International Development] (2002) Poverty and aquatic resources in Vietnam: an assessment of the role and potential of aquatic resource management in poor people's livelihoods. www.streaminitiative.org/Library/pdf/DFID/VietnamPovertyReport_23.pdf. Accessed 22 Nov 2007

DoSTE [Department of Science, Technology and Environment of Ho Chi Minh City] (1998) Giới thiệu các tổ chức hoạt động khoa học và chuyên gia có học hàm, học vị tại TP. Hồ Chí Minh (Introduction of all scientific organisations and experts with an academic degree in Ho Chi Minh City). TP. Hồ Chí Minh, Trung Tâm Thông Tin Khoa học và Công Nghệ

Ehlert J (2011) Living with flood: local knowledge in the Mekong Delta, Vietnam. Unpublished dissertation, University of Bonn

Evers HD (2000) Globalization, local knowledge, and the growth of ignorance: the epistemic construction of reality. Southeast Asian J 28(1):13–22

Evers HD (2005) Global knowledge: the epistemic culture of development. In: Hassan R (ed) Local and global: social transformation in Southeast Asia. Brill, Leiden/Boston, pp 3–17

Evers HD, Benedikter S (2009) Hydraulic bureaucracy in a modern hydraulic society – strategic group formation in the Mekong Delta. Water Altern 2(3):416–439

Evers HD, Genschick S, Schraven B (2009) Constructing epistemic landscapes: methods of GIS-based mapping. ZEF working paper series 44. Centre for Development Research (ZEF), University of Bonn, Bonn

Evers HD, Gerke S, Menkhoff T (2010) Designing epistemic landscapes of knowledge clusters and knowledge hubs for development. J Knowl Manag 14:678–689

Evers HD, Gerke S, Menkhoff T (2011) Knowledge hubs and knowledge clusters: a knowledge architecture for development. In: Menkhoff T, Evers HD, Yue Wah Chay, Eng Fong Pang (eds) Beyond the knowledge trap: developing Asia's knowledge-based economies. World Scientific, Hackensack/London/Singapore/Beijing, pp 27–46

Gerke S, Ehlert J (2011) Local knowledge as a strategic resource: fisheries in the seasonal floodplains of the Mekong Delta, Vietnam. In: Menkhoff T, Evers H-D, Yue Wah Chay, Eng Fong Pang (eds) Beyond the knowledge trap: developing Asia's knowledge-based economies. World Scientific, Hackensack/London/Singapore/Beijing, pp 383–410

Gerke S, Evers HD (2006) Globalizing local knowledge: social science research on Southeast Asia, 1970–2000. J Soc Issues Southeast Asia 21(1):1–21

GSO [General Statistics Office of Vietnam] (2007) Statistical yearbook of Vietnam 2006. Statistical Publishing House, Hanoi

Le T (2006) Các tỉnh và thành phố Đồng Bằng Sông Cửu Long (Geography of all provinces and cities in the Mekong Delta). Nhà xuất bản giáo dục, Hà Nội

Marr DG, White CP (eds) (1988) Postwar Vietnam: dilemmas in socialist development. Cornell Southeast Asia Program, Ithaca

Menkhoff T, Gerke S (eds) (2002) Chinese entrepreneurship and Asian business networks. Routledge, London/New York

Menkhoff T, Evers HD, Chay YW, Pang EF (eds) (2011) Beyond the knowledge trap: developing Asia's knowledge-based economies. World Scientific, Singapore/ Hackensack

MoST [Ministry of Science and Technology] (2004) Vietnam education and training directory. Education Publishing House, Hanoi

MoST [Ministry of Science and Technology] (2008) Những Điều cần biết về tuyển sinh đại học và cao đẳng năm 2008 (Everything you need to know about enrolment in universities and colleges 2008). Nhà xuất bản giáo dục, Hà Nội

Neubert D, Macamo E (2002) Entwicklungsstrategie zwischen lokalem Wissen und globalern Wissenschaft. Geographische Rundschau 54(10):12–17

Nguyen ND (2006) Farmers, agriculture and rural development in the Mekong Delta of Vietnam. Education Publishing House, Ho Chi Minh City

Pham Duy Hien (2006) 150 triệu USD mà cỗ xe khoa học công nghệ vẫn chạy chậm (150 Mio USD but science and technology still develops slowly). Khoa Học- Giáo Dục. Đi tìm diện mạo mới (Science – Education. Looking for new aspects). N. t. giả. TP. Hồ Chí Minh, Nhà xuất bản trẻ

Schüller M, Gruber F, Trines R, Shim D (2008). International science and technology cooperation policies of South East Asian Countries. Consultation paper prepared for the EU Commission. First Bi-Regional Science and Technology Policy Dialogue, EU ASEAN, Paris. https://www.zsi.at/attach/ConsultationPaper.pdf. Accessed May 2011

Siar SV, Sajise PE (2009) Access rights for sustainable small-scale aquaculture and rural development. In: FAO [Food and agriculture organization of the United Nations] (ed) Measuring the contribution of small-scale aquaculture. An assessment. FAO Fisheries and Aquaculture and Technical Paper 534, Rome

Truong Thi Minh Sam (2007) Chuyển dịch cơ cấu kinh tế khu vực dịch vụ ở thành phố Hồ Chí Minh trong quá trình công nghiệp hóa, hiện đại hóa (Transfer of economic service sector mechanisms in HCMC during the process of industrialization and modernization). Nhà xuất bản khoa học xã hội, Hà Nội.*United Nations High Commissioner for Refugees* (UNHCR) (2000). The State of the world's refugees 2000 fifty years of humanitarian action. www.unhcr.org/4a4c754a9.html. Accessed Aug 2009

Vu Le Thai Hoang (2007) Vietnams Suche nach Einfluss in Südostasien und die Folgen für die Grenzstreitigkeiten mit Laos und Kambodscha. Südostasien Aktuell 2007, (2):6–37

Waibel G (2010) State management in transition: understanding water resources management in Vietnam. ZEF working paper series, 55

Wittfogel KA (1957) Oriental despotism – a comparative study of total power. Yale University Press, New Haven

World Bank (1999) Vietnam development report 2000: attacking poverty. World Bank, Hanoi

World Bank (2004) Vietnam development report 2005 – governance, rethinking the role of government in Vietnam. World Bank, Hanoi

Yellow Pages (2007) Tập Đoàn bưu chính viễn thông Viêt Nam – Yellow Pages 2007 Ho Chi Minh City. Niên giám điện thoại những trang vàng và những trang trắng. TP. Hồ Chí Minh, Nhà xuất bản bưu điện

# Chapter 16
# A Water-Related Web-Based Information System for the Sustainable Development of the Mekong Delta

**Verena Klinger, Thilo Wehrmann, Steffen Gebhardt, and Claudia Kuenzer**

**Abstract** One goal of the WISDOM (Water-related Information System for the Sustainable development of the Mekong Delta) project lies in the development and implementation of an innovative water information system containing all the outcomes and results of the different research disciplines involved in the bi-lateral German-Vietnamese project. The topics stem from natural and social sciences and cover all major issues that come into play in an Integrated Water Resources Management (IWRM) system. In line with the IWRM principles, the envisaged information system communicates all research results to local stakeholders and decision makers. It is a web-based geo information system, additionally incorporating non-geographic data such as statistics, literature and reports, legal documents, institutional mappings and others. One major challenge, therefore, lies in the management of heterogeneous data for easy access, so that non-GIS expert users can easily find information to support them in their decision-making. This chapter gives an overview of the main requirements that constitute the system's architecture and describes the layered architecture with its data management tier, logic tier and presentation tier. A short discussion on the sustainable operation, after the project ends, is also given in the outlook section.

---

V. Klinger (✉) • T. Wehrmann • C. Kuenzer
German Aerospace Center (DLR), Earth Observation Center (EOC), German Remote Sensing Data Center, Oberpfaffenhofen, 82234 Wessling, Germany
e-mail: Verena.Klinger@dlr.de; thilo.wehrmann@dlr.de; claudia.kuenzer@dlr.de

S. Gebhardt
Department of Remote Sensing, University of Würzburg,
Muenchner Str. 20, Oberpfaffenhofen, 82234 Wessling, Germany
e-mail: steffen.gebhardt@cimonline.de

F.G. Renaud and C. Kuenzer (eds.), *The Mekong Delta System: Interdisciplinary Analyses of a River Delta*, Springer Environmental Science and Engineering,
DOI 10.1007/978-94-007-3962-8_16, 

## 16.1 Introduction to Information Systems

> An information system is a set of people, procedures and resources that collects data which it transforms and disseminates. (Boddy et al. 2005, p. 10)

This statement demonstrates that an information system cannot be only reduced to software and hardware, but that there is more involved. The dependencies between data, information generation, people with their individual skills and experiences, and knowledge generation within defined organizational structures are also part of an information system. This broad definition accounts for the need to understand processes of the organizational environment in order to successfully implement an information system.

Information systems today are used in many different contexts, in research as well as for administrative tasks. They are applied to support management tasks and to process, visualize, analyze, and disseminate data and information. Information systems can be more than just expert systems tailored to fulfill a special task used by a specialized user group. In contrast information systems can be platforms to bundle different topics or themes and serve a large variety of users. The input data range from alphanumeric, unstructured, one-dimensional data to multi-dimensional, complex information with varying spatial and temporal properties.

A special category of information systems are geographical information systems (GIS) to which the system described here belongs to.

> A geographic information system (GIS) integrates hardware, software, and data for capturing, managing, analyzing, and displaying all forms of geographically referenced information. GIS allows us to view, understand, question, interpret, and visualize data in many ways that reveal relationships, patterns, and trends in the form of maps, globes, reports, and charts. (ESRI 2011)

All different types of geographical information systems can be categorized based on their functionality, intended target users, or use of commercial versus non-commercial components of its architecture.

One differentiating factor is the use of commercial software components in contrast to the use of open source components in the system's architecture (Rao and Vinay 2009). If the components and layers of architecture are loosely coupled hybrid architectures are possible. For example, a commercial database management system such as Oracle (www.oracle.com) can be used, but the services retrieving data from the data base are based on open source products such as UMN Mapserver (http://mapserver.org) or GeoServer (http://geoserver.org). Determining which component to choose depends largely on the project's requirements.

Another differentiating factor is the use of standards to decouple the components of a specific architecture. Especially in the context of GIS the Open Geospatial Consortium (OGC) (www.ogc.org) plays a major role, because it defines different, internationally accepted standards and interfaces to access geo data. Examples here are visualization interfaces such as the Web Mapping Service (WMS), interfaces to realize download services such as Web Feature Service (WFS) or Web Coverage

Service (WCS), interfaces to search for data such as the Catalogue Service for the Web (CSW), and interfaces for processing services such as Web Processing Service (WPS). The driving idea of using services, which are available via the internet or intranet, is interoperability and the design of open architectures, which leads a GIS towards a spatial data infrastructure allowing an easy and efficient exchange of data (Rajabifard and Williamson 1999; Maguire and Longley 2005). This also implies the use of metadata standards to describe data, as otherwise the process of searching and finding relevant data cannot be guaranteed.

Systems applying the above described interfaces can be divided into those using thin clients or thick clients as user interfaces. A thin client is an application that connects to another computer (server) in order to access services and resources. The server bears the main load in this scenario. For example, in a web-based GIS the browser (Internet Explorer, Firefox) connects to a web server to retrieve data and processing results. In contrast, a thick client communicates with a server, but retrieves raw data to process them locally on its own hardware. This is an efficient method, provided that the local computer has enough hardware resources. Both alternatives can incorporate the same standardized interfaces. There are studies and publications in this field (Juanle et al. 2004; Steiniger and Hunter 2011) which discuss the advantages and disadvantages of the alternatives.

The next section sketches the main idea of the envisaged information system. Then an overview over the development gives an insight in how the information system came into being. Guiding requirements are then outlined. As the information system has to handle a vast amount of data, data standardization is described in separate section, followed by an overview of the architecture, looking at all the layers of the system.

## 16.2 Main Idea of the WISDOM Information System

The acronym WISDOM stands for 'Water related Information System for the Sustainable Development of the Mekong Delta' which is a multi-disciplinary project addressing the principles of Integrated Water Resources Management (IWRM). One goal of WISDOM consists in the development and implementation of an innovative water information system containing all the outcomes and results of the different research disciplines involved in the project. Figure 16.1 gives an overview of all the research fields contributing to the contents of the information system, coming from natural and social sciences.

The users of the WISDOM system can be divided into two categories. The first category comprises researchers, using the system as a platform to gain visibility for their work. They can be considered to be, and will hereafter be called, *information producers*. But they are also data consumers, as they have the option to visualize and download data from other institutions contributing to the information system. This implies the will to share data, as an information producer participating in

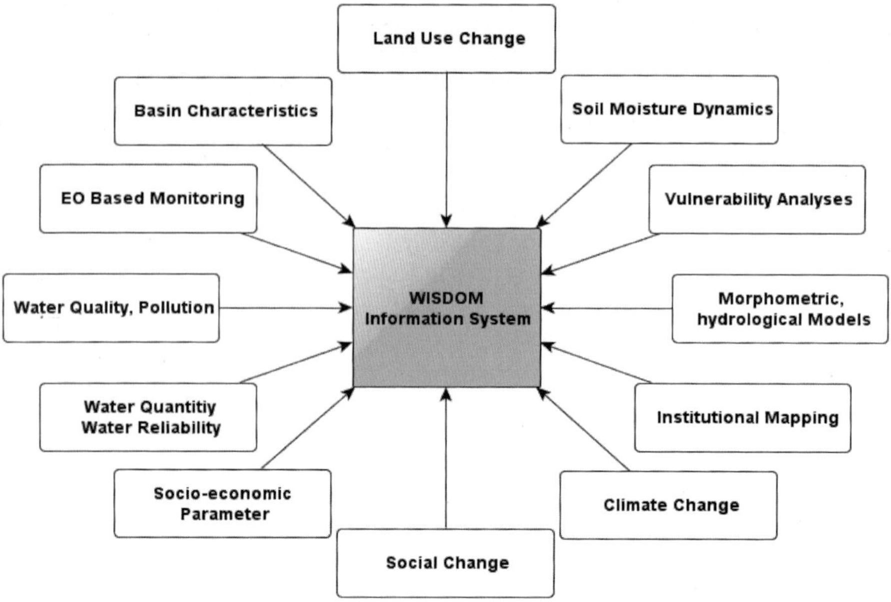


**Fig. 16.1** All the different fields of research deliver data and information products for integration into the WISDOM Information System

WISDOM not only contributes data but also benefits from other researchers' outputs. In addition researchers outside the WISDOM project are of course potential users. On the other hand, decision-makers, who form the second category of users, shall utilize the information system as a tool to better grasp and understand all the different water-related processes in the Mekong Delta. This will support them in their decision-making process through the supply of up-to-date, interdisciplinary information products. This category will be called *end user* hereafter.

## 16.3 Road Map for the Development of the Information System

Predictive software engineering models, which focus on planning a software project in detail from the very beginning to the final system, did not fit the development process envisaged here. In order to integrate feedback from the potential users, we followed the approach of vertical prototyping. In a first step, a thorough user requirements analysis was conducted, which clarified what the users wanted, and what they needed. Based on the outcomes, a first prototype with reduced functionality was developed, which the users could test and give feedback on. This way, the users got a feeling of the potentials the system has and how it might look like, and also identify drawbacks and missing functionality. The prototype was then modified, and expanded in order to fit to the users' needs. Furthermore, the developers could better react to changing circumstances. During prototype development we followed the principles formulated by Kierulf et al. (1990): First, key components were designed and implemented in an early stage to provide key functionality. Other components were successively added later on. Second, an active prototype had to be operative early on in order to ensure constant feedback from the users. Third, the architecture has been kept open for future developments. Fourth, during the development process the development team stayed informed on similar projects in order to derive best practices and evaluate alternative techniques. This way of "prototyping" is suggested especially for projects, in which requirements are not clear from the very beginning and are subject to change over time (Kierulf et al. 1990; Balzert 1996; van Vliet 2008). An agile software development process (Tate 2005; Pham and Pham 2011) was introduced to always have a functional intermediary status of the prototypical information system and to increase flexibility if requirements changed.

In parallel to this rather technical software development process, topics like data standardization, data management, and data integration had to be addressed in a long interaction between all participating project partners. The definition of a project wide data standard, the WISDOM Geodata Exchange Format (WGEF) was defined. Finally, the WISDOM data model is able to efficiently store and manage heterogeneous planning relevant data (Gebhardt et al. 2010).

## 16.4 Main Requirements for the WISDOM Information System

This section describes the most important non-functional requirements for the Information System, which had the biggest impact on architecture and design. The presentation of a full list of requirements goes beyond the scope of this chapter.

### *16.4.1 Usage of Open Source Technology*

The Vietnamese Government strives to strengthen the adoption of open source software and open source developments (VietnamNet Bridge 2005, 2009). Therefore, within the WISDOM project context, the information system had to be based on open source software and libraries, which dictated which third party software could be used. The utilization of Open Source Technology avoids future license fees, grants the opportunity to install the architecture on a multitude of platforms, and allows exploring a variety of solutions for the different architectural components of the system. However, at the same time open source solutions can, but might not always reach the same performance as costly licensed tools.

### 16.4.2 Usability and Simple Access to WISDOM

The users themselves are a heterogeneous group with different levels of experience in using geographical information systems, or using computers in general. Thus, the graphical user interface of the system has to be simple to use, and user trainings facilitate the introduction of the information system.

To make the information system easily accessible, a web-based graphical user interface is required. This also avoids long installation procedures on local PCs and the users can utilize the system right away with a standard browser. As the IT infrastructure at local planning offices is not always up to date and PCs are sometimes not powerful enough to handle big memory-consuming programs, the main processing load is on the server. As a result, the web-based information system is usable with a thin client interface rather than a thick client interface.

### 16.4.3 Management of Heterogeneous Data

All the different research foci, work procedures, and experience of information producing partners in the project leads to heterogeneous data generation with respect to content, types, formats, and documentation. The data management component, therefore, has to be capable of managing data from the different fields given below:

- Geo base data and other data in vector format: country boundaries, major cities in Vietnam and neighboring countries, digitized residential areas in cities such as Can Tho, river networks on a national, provincial and district level, airport locations, hydropower projects, road networks, administrative boundaries with different temporal validity, and others.
- Remote sensing data and products: these can be raw data such as satellite images from different sensors with different resolution. But most of the data sets are value added products, such as land cover and land use classifications based on different sensors, inundated areas are derived on a regular basis, soil moisture data, and value added products depicting surface sealing, water quality, and many more are derived. Products available as time series data have to be stored efficiently for a quick retrieval later on.
- Hydrological and hydraulic modeling results: Hydrological models were developed and calibrated at the corresponding research institutes and the results were integrated into the information system as raster data, e.g. as flood maps from a hydraulic model for the whole Mekong Delta depicting the maximum flood extent under different scenarios.
- In-situ sensor measurements from different station networks such as water level and discharge, turbidity (sediment load), electric conductivity (salinity), but also manually sampled measurements, such as endocrine disruptors and pesticides were integrated. Some of the automatic stations took measurements every 15 min, which results in a great amount of data points that have to be stored.

- Field data: field data were collected to train and validate classification algorithms or to calibrate models. Geographic coordinates of these in-situ sampling sites were collected with GPS. Pictures from all directions (north, east, south, and west) and class information related to these locations were also acquired.
- Organizations in the water sector: a database storing addresses and affiliations of organizations and institutions active in the water sector, especially in the Mekong Delta, was incorporated. All issued legal documents and other relevant publications released by a certain institution are stored, some in English as well as in Vietnamese.
- Statistical data on different topics for different years and varying administrative levels (national, provincial, district) were gathered by project partners and stored so that the data values can be mapped to administrative units for later visualization. Statistical data on aquaculture, and agriculture, forestry and industry, as well as data on population demographics were integrated.
- A literature database was also incorporated into the data model, enabling the user to search for literature related to the Mekong Delta and water resource management in general.

The data management component was designed to manage these different data sets so that they can be found and retrieved in a user friendly way.

## 16.5 Data Standardization

In order to harmonize all the different data types, a common WISDOM Geospatial Exchange Standard (WGEF) was developed, which defines mandatory map projections, specific data formats for vector- and raster data, file naming conventions, and metadata, as well as the actual data exchange format. The WGEF standard is a data container which bundles all important aspects of the data within the WISDOM system, including data, metadata, and visualisation. A workshop on data integration together with documentation helped to introduce the standard definitions to all partners and facilitated data import (Gebhardt et al. 2010).

### *16.5.1 Data Formats*

Geo data informal standards such as *Shapefile* for vector data or *GeoTIFF* for raster data were used in the WISDOM project. ESRI developed shapefile with ArcView version 2 in the early 1990s, but it can now be read and written with a number of programs (http://en.wikipedia.org/wiki/Shapefile). With the public domain metadata standard GeoTIFF, georeferencing information such as map projection, coordinate reference systems, ellipsoids, can be embedded within "tiff" images (http://trac.osgeo.org/geotiff/). The OGC Styled Layer Descriptor (SLD) can be used to portray the output of Web Map Servers, Web Feature Servers and Web Coverage Servers (http://www.opengeospatial.org/standards/sld). SLD is an XML schema for describing the appearance of map layers capable of describing the rendering of vector and raster data. With SLD visual parameters of each map layer component either vector objects or raster data can be specified. The WISDOM system utilizes these style definitions for spatial data for graphical presentation within maps.

For in-situ sensor data, a proprietary exchange format was defined based on comma separated values (csv) for data integration. The same complies for statistical data. No information producer was serving sensor data using OGC's Sensor Observation Service (SOS) at the time of writing although this might change in the future.

Literature can be integrated semi-automatically using the BIBTEX format. Users can download literature and reports that are publicly available. For others, a download link to, e.g. the Science Direct portal is provided.

### *16.5.2 Coordinate Reference Systems and Projections*

Every geographic data set is coded towards a coordinate system and map projection. The standard Coordinate Reference System (CRS) within the information system is UTM, Zone 48; WGS 84. This CRS is widely used in the South-East Asian region. It is suited for mapping in large and medium scale. The Geodetic datum is the World Geodetic System 1984 with the respective WGS 84 ellipsoid. The map projection is the Universal Transverse Mercator (UTM) projection. The UTM zone 48 refers to the reference meridian at 105° East. However, the system is able to manage data in various CRS. A worldwide unique code for every CRS exists, namely the EPSG code, which has to be given for every geographic data set to be integrated into the information system. This is especially important for vector data in Shapefile format for which the EPSG code is defined in the .prj file with the Well Known Text (WKT) format. For data in the VN-2000 Zone 48 North CRS the following WKT was used:

```
PROJCS["VN-2000 / UTM zone 48N",GEOGCS["VN-2000",
DATUM["Vietnam_2000", SPHEROID["WGS 84", 6378137, 298.257223563,
AUTHORITY["EPSG","7030"]], TOWGS84[-192.873, -39.382, -111.202,
0.00928836, -0.01975479, 0.00427372, 1.000000252906278],
AUTHORITY["EPSG","6326"]], PRIMEM["Greenwich", 0,
AUTHORITY["EPSG","8901"]],UNIT["degree",0.01745329251994328,AUTHORITY
["EPSG","9122"]],AUTHORITY["EPSG","4326"]],PROJECTION["Transverse_Mer
cator"],PARAMETER["latitude_of_origin",0],PARAMETER["central_meridian
",105],PARAMETER["scale_factor",0.9996],PARAMETER["false_easting",500
000],PARAMETER["false_northing",0],UNIT["metre",1,AUTHORITY["EPSG","9
001"]],AUTHORITY["EPSG","3405"]]
```

### *16.5.3 Definition of a WISDOM Metadata Profile*

Metadata in GIS provide information about the identification, the extent, the quality, the spatial and temporal scheme, spatial reference, and distribution of digital geographic data in an XML Extensible Markup Language file. There are competing standards, like the "Contend Standard for Digital Geospatial Metadata (CSDGM)" of the Federal Geographic Data Committee (www.fgdc.gov) in the USA, or the standard resulting from a European initiative INSPIRE (http://inspire.jrc.ec.europa.eu), which is based on the guidelines of ISO 19115 and ISO 19119. WISDOM created an own profile compliant to ISO 19115, including all mandatory fields and extracting the most important optional fields, like metadata author information, dataset point of contact, dataset identification and abstract, keywords for theme, region, discipline and temporal validity. Currently, a standalone tool to generate metadata based on this WISDOM profile helps the information producers to create compliant metadata for their datasets. The introduction of metadata generation was accompanied with trainings in Vietnam (Gebhardt et al. 2010).

## 16.6 Architecture of the WISDOM Information System

According to different sources, the number of internet users in Vietnam has rapidly increased during the last years, e.g. from 12.8% users in 2005 to 31.5% in February 2011 (MOIC 2011a; Internet World Stats 2011). The Vietnamese government recently approved a program entitled "Developing High Technology Till 2020". This program aims at supporting high-tech companies centering on microchips, basic software, and technologies that ensure a modern information and communication technology (ICT) and "new-generation" internet (MOIC 2011b). As a consequence the ICT infrastructure is rapidly developing in Vietnam, and there are more and more users familiar with internet applications. Therefore, a web-based information system was envisaged as users can easily access the system with a standard browser and do not have to download and install a software package.

The information system is based on open source software components and libraries, in order to be compliant to the Vietnamese open source strategy. It adopts strategies from a Service Oriented Architecture (SOA), incorporating standardized interfaces. The layered architecture, which is depicted in Fig. 16.2 consists of three layers or tiers: the data management tier comprises all components to efficiently store and manage the heterogeneous project data. The logic tier with its services then accesses the data management tier through defined interfaces. The presentation tier then connects to the services of the logic tier and visualizes all the information retrieved in a user friendly way. The graphical user interface runs in a standard browser.

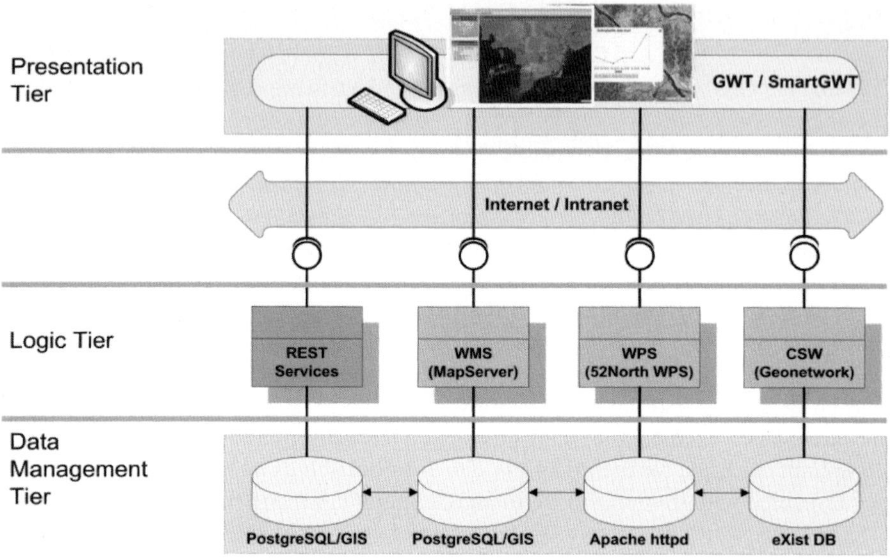


**Fig. 16.2** Simplified architecture of the WISDOM information system. All components are explained in the main text

## *16.6.1 Data Management Tier*

The next subsection deals with the data management tier and data integration. Only the most important aspects are described here.

### 16.6.1.1 Managing Aspects of Data

A data model was developed to describe datasets in detail and to define how data is stored. In PostgreSQL (http://www.postgresql.org), which is a powerful open source relational data base management system that can be extended with PostGIS to store vector data, most of the model is implemented. As raster data cannot be stored inherently in PostgreSQL, GeoTIFFs are stored in a file system, managed by Apache HTTP Server (http://httpd.apache.org). Metadata and SLDs are XML data, therefore the XML database eXist (http://exist.sourceforge.net) was chosen to manage those. The components can be accessed via protocols such as Java database connectivity (jdbc), hypertext transfer protocol (http) and web-based distributed authoring and versioning (webdav).

The main challenge is to efficiently store data in a way to be able to persistently store all necessary aspects of data and to find relevant data easily again. Each data set is described with a set of attributes, giving information on date of creation, provider, method of generation, geographical location and others. To let the user

- **Thematic Reference**
  - **Base Data**
    - **Elevation**
      - Digital Elevation Model
      - Elevation Contours
    - **Maps**
      - Geological Maps
      - Topographic Maps
    - **Satellite Images**
      - Optical Sensor Images
      - Radar Sensor Images
  - **Climate**
    - **Climate Station Network**
    - **Evapotranspiration**
    - **Rainfall**
    - **Temperature**
  - **Fauna**
    - **Aquatic**
    - **Terrestric**
  - **Governance and Administration**
    - **Actors**
    - **Financing**
    - **Legal Framework**
    - **Policy and Planning**
    - **Political Boundaries**
  - **Hydrology**
    - **Constructions**
    - **Flood Scenarios**
      - Water Depth
      - Water Distribution
    - **Inundation Monitoring**
      - Water Depth
      - Water Distribution
    - **Rivers, Canals and Lakes**
    - **Station Network**
    - **Water Quality Monitoring**
      - Chemical
      - Organic
      - Physical
  - **Infrastructure**
    - **Commercial Areas**
    - **Industry**
    - **Public Places**
    - **Residential Areas**
    - **Settlements**
    - **Transportation**
  - **Land Cover and Land Use**
    - **Agriculture**
    - **Animal Production**
    - **Aquaculture**
    - **Ecology**
    - **Forestry**
    - **Land Cover**
    - **Land Use**
    - **Soil**
  - **Natural Disasters**
    - **Earth Quakes**
    - **Floods**
    - **Forest Fires**
    - **General**
  - **Population and Economy**
    - **Economy**
    - **Education**
    - **Livelihoods**

**Fig. 16.3** Thematic reference used to search data sets in WISDOM. Every data set is registered to this tree via its product group

efficiently search and retrieve data sets, a thematic reference was defined to which a data set can be registered, independently of whether it is an article as pdf, a vector dataset, or statistics. This thematic reference scheme was developed in a joint German-Vietnamese workshop. All themes are organized in a tree-like structure, see Fig. 16.3.

Every data set is additionally tagged with a *product group*, e.g. "watermask from optical sensors", "inundation scenario from raster", "water depth from raster", "water chemical substances", "rainfall", "soil moisture", and many more. Every

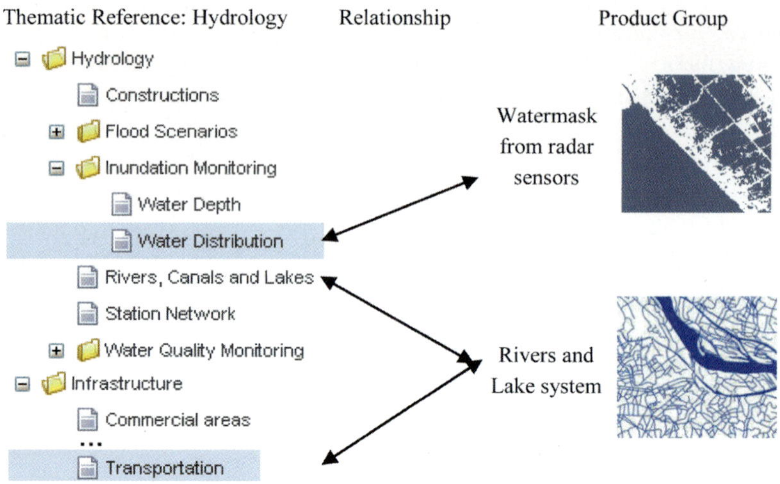


**Fig. 16.4** Relationship between thematic reference and product group

product group is mapped to the thematic reference tree given above. An inundation map with product group "watermask from radar sensor", e.g., showing inundated areas derived from satellite imagery, can be found under Hydrology > Inundation Monitoring > Water Distribution. This mapping can be done unambiguously here in a n:1 relationship, which means that this product group is mapped only once to the given entry in the thematic reference. But more than one product group can be mapped to one entry in the thematic reference, see Fig. 16.4 top. A vector data set showing canal systems of a province, in contrast, can be registered to Hydrology > River, Canals and Lakes and also to Infrastructure > Transportation. To be able to double register one data set, a m:n relationship between thematic reference and product group was set up, see Fig. 16.4 bottom. Statistical data, reports, sensor measurements, etc. are referenced in the same way.

Furthermore, a concept of spatial reference objects was developed to allow a mapping from datasets in the system to spatial administrative units (Gebhardt et al. 2010). A table in the PostgreSQL database holds the actual geometry of all the administrative units for each level. Starting with the boundary of Vietnam, there are regions, subdivided by provinces which are themselves subdivided by districts. This is again a tree-like structure, which was mirrored in the model as a parent child relationship, see Fig. 16.5. As administrative boundaries changed over time, changed geometries of affected units have to be stored more than one time, accompanied by the date from which the geometry is valid. The extent of a dataset then is mapped to this regional reference tree, resulting in the relationship *dataset to administrative unit*, and consequently giving the possibility to search for datasets in the province of Can Tho, to give an example. The dataset's extent can be calculated automatically for spatial data from the bounding box information, but for literature and reports, it has to be defined manually supplying metadata.

| code text | parentcode text | name text | level integer | date_from date | the_geom geometry |
|---|---|---|---|---|---|
| w1c05cn42 | w1c05 | Việt Nam | 3 | 0001-01-01 BC | |
| w1c05cn42r08 | w1c05cn42 | Mekong Delta | 4 | 0001-01-01 BC | |
| w1c05cn42r08p12 | w1c05cn42r08 | Cần Thơ | 5 | 2004-01-01 | 0106000020E6 |
| w1c05cn42r08p13 | w1c05cn42r08 | Hậu Giang | 5 | 2004-01-01 | 0106000020E6 |
| w1c05cn42r02p03 | w1c05cn42r02 | Lai Châu | 5 | 2004-01-01 | |
| w1c05cn42r06p04 | w1c05cn42r06 | Đắc Nông | 5 | 2004-01-01 | |
| w1c05cn42r06p05 | w1c05cn42r06 | Đắc Lắc | 5 | 2004-01-01 | |
| w1c05cn42r02p04 | w1c05cn42r02 | Điện Biên | 5 | 2004-01-01 | |
| w1c05cn42r08p07d04 | w1c05cn42r08p07 | Bình Tân | 6 | 2004-01-01 | 0103000020E6 |

**Fig. 16.5** Administrative units in Vietnam (example) as they are stored in the spatial reference table

#### 16.6.1.2 Data Integration

To upload a new vector or raster dataset, the information producer has to create a WGEF archive, containing all necessary files that are mandatory for ESRI Shapefile or GeoTIFF respectively, containing appropriate projection information. Furthermore, the metadata file as XML has to be provided. An SLD should be provided to define the looks of the dataset when it is visualized later on.

This zipped archive serves as input parameter for the Java tool Data Entry Portal (DEP) that realizes the integration automatically. This tool automatically (i) validates the input data structure, (ii) unzips the archive, (iii) analyses the data in terms of data format, map projection and metadata, (iv) cleans the data and extracts some fields for metadata automatically, renders a preview image based on the given SLD, and (v) disseminates the data to the database management system, see Fig. 16.6 for an overview. There is no further interaction needed. After the integration of a dataset, it can be directly viewed in the information system.

### *16.6.2 Logic Tier: OGC Services and REST*

The logic tier incorporates OGC CSW and WMS standards. With an instance of Geonetwork, an OGC compliant CSW implementation, an interface is provided for the presentation tier to let the user search for data doing a metadata search. A WMS provides a simple interface to retrieve images from a spatial dataset; here the UMN MapServer is used. In a so called mapfile it is configured where MapServer finds the raw data and how it shall be rendered. For vector data an SQL statement pointing to the geometry information stored in PostGIS is given in a mapfile, whereas the path to the tif file is stored for a raster layer. Therefore, the WMS communicates with the data tier directly via SQL and the local file system.

Other services which cannot be implemented by using OGC standards are accessible via REST (Representational State Transfer), which is based on the HTTP

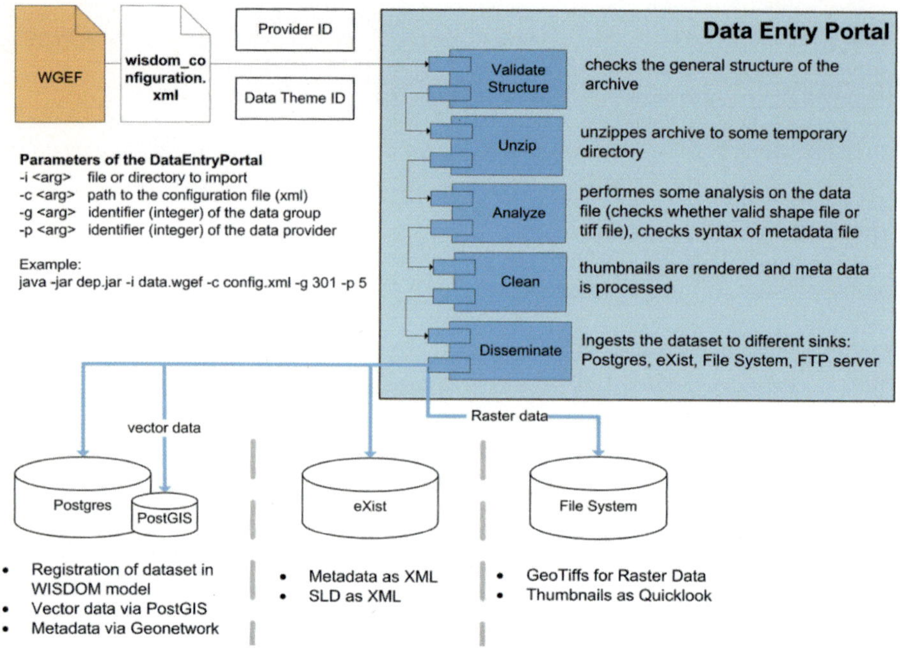


**Fig. 16.6** Steps in the Data Entry Portal and dissemination paths

protocol (Fielding 2000). RESTful services are used to decouple the presentation tier from the logic tier in the WISDOM architecture (Allamaraju and Amundsen 2010; Klinger et al. 2010). The implementation is done with Java 1.6, based on the restlet api (www.restlet.org). There exist services to:

- Search for organizations in the water sector
- Search for literature based on search string, keyword and author information
- Retrieve the thematic and regional reference trees
- Retrieve available years and levels (national, provincial, district) of a statistics
- Search for in-situ measurements for a given theme and period of time, station-wise

Figure 16.7 gives an example how the graphical user interface is connected to the services and what kind of results there are to be expected by the presentation tier. With the user clicking on the button 'Start', a GET request is triggered in the web application on the client side. The corresponding REST service in the logic tier processes this request and sends back the response as JSON (JavaScript Object Notation), which is depicted in Fig. 16.7 by the dotted arrow from right to left. The response is interpreted by the client and displayed in graphical elements.

The REST services themselves communicate with the data tier via the architectural pattern Data Access Objects (DAOs) (Java Sun 2011). A DAO is responsible for hiding the communication with the data source, which in this case is the data

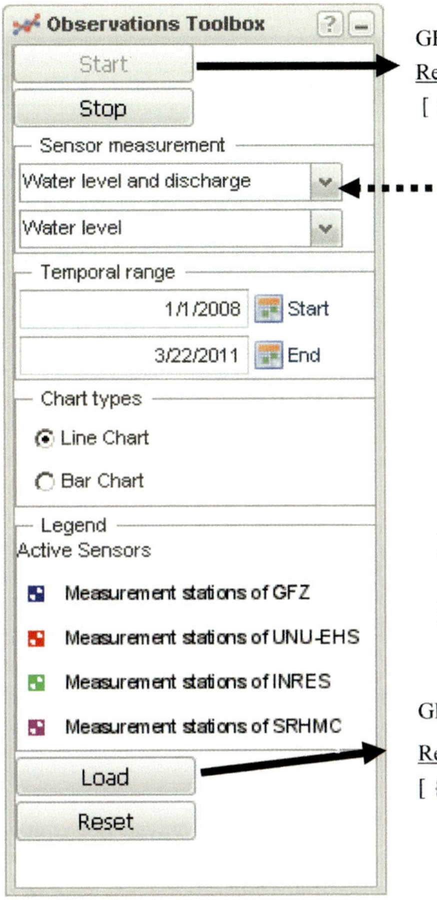


GET /elvis_rest/sensor/themes.json
Response:
[ { "name": "Water level and discharge",
"code": 112,
"moduleList": [ {
"name": "Average Water level",
"unit": "m", "acronym": "ha3"
}, {
"name": "Average Water level (HI)",
"unit": "m", "acronym": "hi3"
}, {
"name": "Water level",
"unit": "m",
"acronym": "wl"
} ]
},{ "name" : "Water chemical substances",
...
},{..}]

GET /elvis_rest/wl/01-01-2008/22-03- 2011/list.json
Response:
[ { "name": "H5",
"id": "GFZ_H5",
"lat": 105.37933082541238,
"lon": 10.760394364634218 },{ ...} ]

**Fig. 16.7** With a click on the start button (*left side*), a GET request is performed to query corresponding REST service (*right side*). The response is transported as JSON. With click on load, all relevant stations for chosen topic and time range are retrieved, here water level, starting 1st of August 2008 until 22nd of March 2011

base PostgreSQL but can be any other source. The object-relational mapping which translates results from the data base into Java objects is done in the DAO. These objects can then be used in the resource classes of the services.

Additionally, the OGC standard WPS is incorporated into the logic tier. A WPS serves as a wrapper for GIS functionality defining input and output parameters for the underlying process. Again, a WPS can be requested via HTTP with both GET and POST methods. The input parameters defined during a WPS call are delegated to the underlying process by the service. The server implementation of 52North (http://52north.org/communities/geoprocessing/wps/) is used to register all WPS services to the logic layer. The already mentioned Data Entry Portal (DEP) (see above) was wrapped as a WPS to make it available from the internet. Therefore, data integration can be triggered via the graphical user interface (see below).

Automatic data processing is also possible in the logic tier in a more complex way. As inundated areas are derived regularly from satellite imagery, the necessary algorithms implemented in Matlab, Python or Java had been wrapped as WPS and chained together (Wehrmann et al. 2010). A prototypical process orchestration was developed to execute the different processes one after another guaranteeing the correct chaining according input and output parameters. The processing chain is triggered by email notification and runs fully automatically. The last process is the DEP WPS which ensures that the newly derived dataset is integrated in to the data model. Metadata generation is done automatically based on a template and information extracted from the input dataset.

Being able to wrap all kinds of algorithms as WPS and chain them together with the orchestration engine gives a powerful tool to offer functionality to the user, even if they have limited local processing power.

### *16.6.3 Presentation Tier: The graphical User Interface*

As this system is a web-based information system, the user can access all functions via a web browser. Google Web Toolkit (GWT) was used to develop a platform independent web application. SmartGWT, which provides an extensive widget library, helped to make the application user-friendly and professional. OpenLayers was integrated for loading and rendering maps.

As indicated above, every dataset is mapped to the thematic reference tree via its product group and to the regional reference tree via its extent. The user can search for data based on a theme, an administrative unit, and time constraints. The list of results not only contains spatial datasets but also literature and reports and statistics. The user can view the results for these three data types by switching between tabs, see Fig. 16.8, the tabs are marked in boxes. Depending on the results the user can interact with the retrieved data: spatial data and statistics can be loaded to a mapping client (see Fig. 16.9 for an example), literature and reports can be either downloaded directly or a link is presented, in-situ sensor measurements can be visualized in a graph and also be downloaded, see Fig. 16.10.

Besides these functionalities the user can overlay layers, visualize legend information for each layer and turn on and off the background layer, which is very useful when small datasets and high resolution data are observed.

Furthermore, organisations in the water sector can be searched, and legal documents and other issued documents can be retrieved, see Fig. 16.11.

To facilitate the upload process, a graphical user interface for data integration was developed that can be used by the information producers to upload new data sets easily (Ahrens 2010). The user does not have to generate a WGEF manually, but is guided through all the different steps during which user input is validated on the fly. A WGEF archive is generated in the background and sent to the DEP WPS for integration.

**Fig. 16.8** Result list of a query by theme, region and time range. The user can switch between result types by tabs (*marked with boxes*)

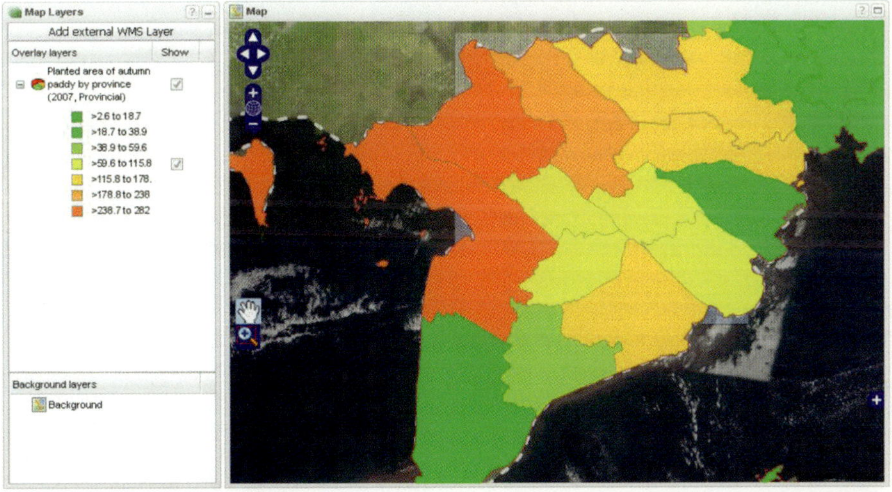


**Fig. 16.9** Statistical data is mapped to the map client. The statistical values are related to administrative units, resulting in the coloring of provinces. The legend information is given on the *left-hand side* showing the seven color classes and their boundaries

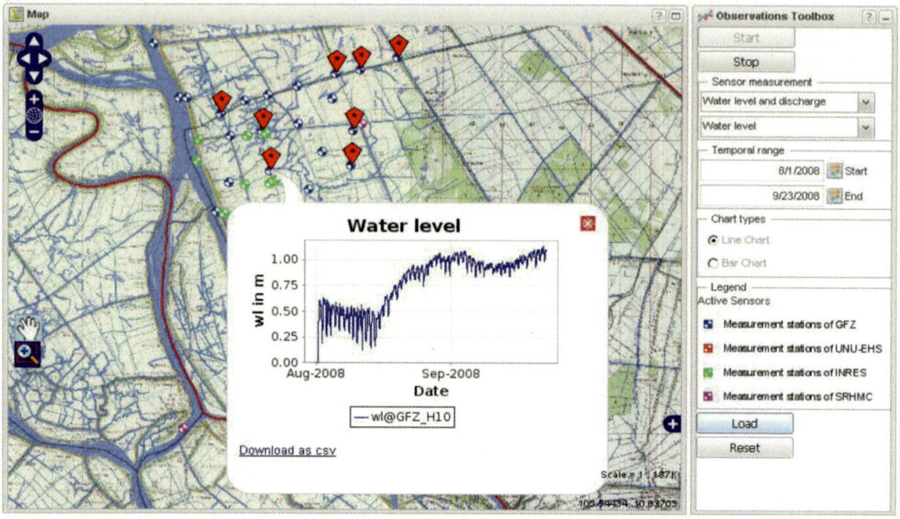


Fig. 16.10 Water level information for one measurement station

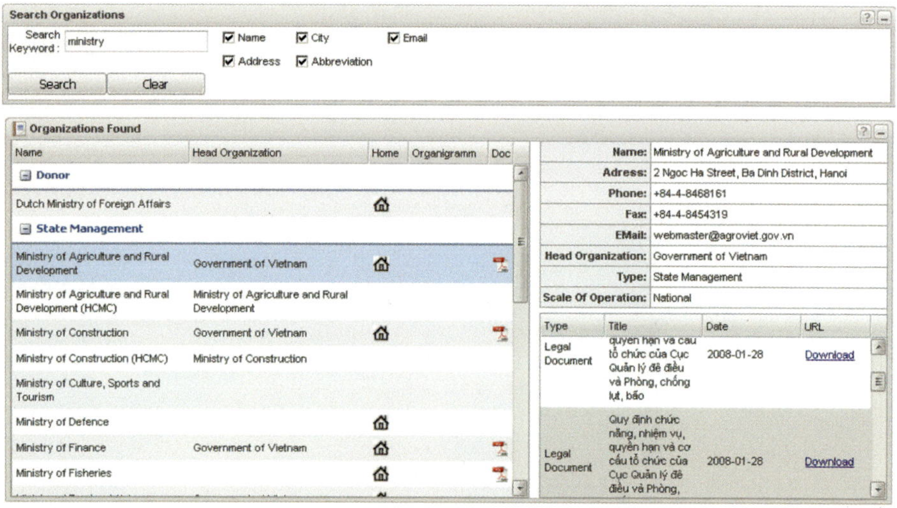


Fig. 16.11 Search for organizations in the water sector with accompanied issued documents

## 16.7 Outlook

The envisaged information system will be able to support end users in giving answers to their questions concerning flooding, water consumption and water demand, water quality and pollution, water quantity, vulnerability, hydraulics,

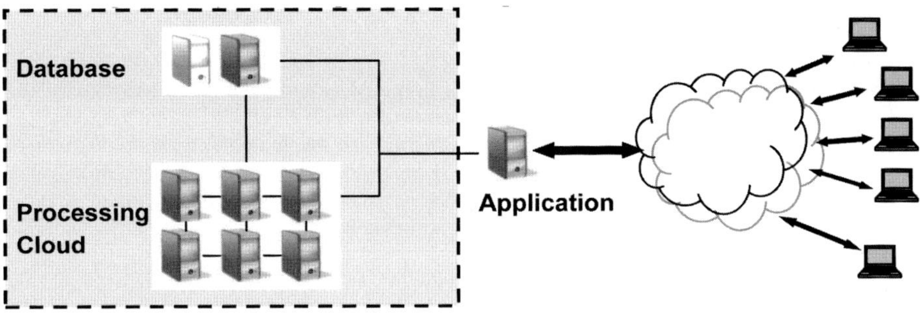


**Fig. 16.12** Information system installation scheme, an example

and so forth. The user will be able to click on predefined questions. For example, how many people are affected by the current flood in Can Tho province? Did water quality change over time in Tam Nong district? How did the land cover change on the coast line, how many mangrove forests were converted to other land uses? The information system is easy to use, so that non-GIS experts can use it right away.

Different scenarios for system installation are possible in order to increase performance. As the system comprises different layers, the system can be distributed among different servers to distribute load. One scenario is shown in Fig. 16.12: A dedicated database server hosts the database, a server cloud realizes processing if needed, an application server hosts the web application to which the browsers connect to.

To maintain the system, a thorough documentation of its architecture and used software components will ensure that experts in Vietnam understand the system. Ongoing trainings and meetings enable experts to understand the system log files, to reconfigure the system components and to perform crucial maintenance tasks.

Ideally, the information system with most of its results will be publicly available for a widespread community after the project ends in 2013 (Künzer 2010). In the already ongoing development of a concept of operations, roles are defined to ensure that the system is fed with new datasets. Without a functioning update procedure the information system will gradually loose its value as the data will be rapidly outdated. Workflows for data integration will be set up, ensuring a sound, up-to-date and high quality data basis. Demand-driven data and product generation is highly important, because the covered topics must be relevant to planning in order to meet the end users' needs (Künzer 2010). On the other hand, the role of system maintainer has to be filled, which can be a service company in Vietnam or any other research institute having the capabilities to run such an information system operationally. To really implement the concept of operations, the financial situation has to be clarified for the operational system after 2013 (Künzer 2010).

# References

Ahrens M (2010) Integration raumbezogener Daten in ein Umweltinformationssystem - Konzeption und Implementierung eines grafischen Assistenten für die Eingabe, Zusammenstellung und Integration von Geodaten in ein WebGIS. Erweiterung der Integrationssoftware für den Import von in-situ Sensormessdaten. Diplomarbeit. FH Karlsruhe

Allamaraju S, Amundsen M (2010) RESTful web service cookbook. O'Reilly Media

Balzert H (1996) Lehrbuch der Software-Technik, Software-Entwicklung. Spektrum Akademischer Verlag, Heidelberg

Boddy D, Boonstra A, Kennedy G (2005) Managing information systems: an organisational perspective. Prentice Hall, Harlow

ESRI (2011) What is GIS? http://www.gis.com/content/what-gis. Accessed Mar 2011

Fielding RT (2000) Architectural styles and the design of network-based software architectures. University of California, Irvine

Gebhardt S, Wehrmann T, Klinger V, Schettler I, Huth J, Künzer C, Dech S (2010) Improving data management and dissemination, in web based information systems by semantic enrichment of descriptive data aspects. Comput Geosci 36(10):1362–1373

Internet World Stats (2011) Vietnam internet usage stats and marketing report. http://www.internetworldstats.com/asia/vn.htm. Accessed 25 Feb 2011

Java Sun (2011) Core J2EE patterns – data access object. http://java.sun.com/blueprints/corej2eepatterns/Patterns/DataAccessObject.html. Accessed 22 Jan 2011

Juanle W, Songcai Y, Chuanjie X, Yuanqiang Z (2004) Study in web-oriented Geo-data Sharing Platform System Structure. CNKI Geo-information Science 2004 6(4):62–65

Kierulf A, Ken C, Nievergelt J (1990) Smart game board and go explorer: a study in software and knowledge engineering. Commun ACM 33(2):152–166

Klinger V, Wehrmann T, Gebhardt S, Künzer C (2010) RESTful WISDOM (Water-related Information System for the sustainable development of the Mekong Delta). In: Proceedings of the 31st Asian remote sensing conference, Hanoi, Vietnam, 1–5 Nov 2010

Künzer C (2010) WISDOM: water related information system for the sustainable development of the Mekong Delta – thoughts on sustainability and long term implementation. In: Proceedings of the 3rd Indo-German conference on research for sustainability: water and waste management, New Delhi, India, 3–4 Feb 2010

Maguire D, Longley P (2005) The emergence of geoportals and their role in spatial data infrastructures. Comput Environ Urban Syst 29(1):3–14

Ministry of Information and Communication (MOIC) (2011a) Report on internet statistics of Vietnam. http://www.thongkeinternet.vn/jsp/trangchu/index.jsp. Accessed 25 Feb 2011

Ministry of Information and Communication (MOIC) (2011b) Vietnam to invest more on hi-tech development. http://english.mic.gov.vn/tintucsukien/Trang/VNtoinvestmoreonhi-techdevelopment.aspx. Accessed 25 Feb 2011

Pham A, Pham PV (2011) Scrum in action: agile software project management and development. Course technology PTR

Rajabifard A, Williamson I (1999) Spatial data infrastructures: concept, SDI hierarchy and future directions. Suitability of internet technologies for access, transmission and updating digital cadastral databases on the web. In: Proceedings of AURISA. 97, Christchurch

Rao S, Vinay S (2009) Choosing the right GIS framework for an informed Enterprise Web GIS Solution CIESIN, Columbia University & NASA

Steiniger S, Hunter AJS (2011) Free and open source GIS software for building a spatial data infrastructure. In: Bocher E, Neteler M (eds) Geospatial free and open source software in the 21st century: proceedings of the first open source geospatial research symposium, 2009, LNGC. Springer, Heidelberg (in press, to be published in 2011)

Tate K (2005) Sustainable software development: an agile perspective. Addison-Wesley Professional, Upper Saddle River

van Vliet H (2008) Software engineering: principles and practice. Wiley, Chichester/Hoboken

VietnamNet Bridge (2005) Intel assists local open-source code. http://english.vietnamnet.vn/tech/2005/04/413581/. Accessed 18 Feb 2011

VietnamNet Bridge (2009) Vietnam to widely use open source software. http://english.vietnamnet.vn/tech/2009/01/822425/. Accessed 18 Feb 2011

Wehrmann T, Klinger V, Gebhardt S, Künzer C (2010) Web services enabled architecture coupling data and processing resources. In: Proceedings of the ISPRS technical commission IV, geospatial and geovisualization: environment, security and society, Orlando/FL, USA, 15–19 Nov 2010

# Chapter 17
# The Water-Development Nexus: Importance of Knowledge, Information and Cooperation in the Mekong Delta

**Fabrice G. Renaud and Claudia Kuenzer**

**Abstract** As most world deltas, the Mekong Delta [or the "delta" hereafter (in this chapter we exclusively concentrate of the Vietnamese portion of the delta which covers roughly four-fifths of the Mekong Delta)] is highly dynamic both from environmental and socio-economic perspectives. The delta plays a crucial role for the Vietnamese economy in terms of agricultural and aquaculture production, is rapidly urbanising, and has seen an important development of its infrastructures, particularly those related to water resources management. Water plays a central role with respect to the development of the region but, as described throughout this book, water resources are also threatened by anthropogenic processes such as pollution, infrastructure development, and over-abstraction, as well as by global environmental change including the impacts of climate change. Knowledge generation, knowledge sharing and stakeholder cooperation with respect to water resources management are still limited in the region and this should be addressed rapidly in order to allow for efficient decision-making in the face of rapid societal, economic, climate and environmental change.

F.G. Renaud (✉)
United Nations University - Institute for Environment and Human Security (UNU-EHS), UN Campus, Hermann-Ehlers-Str. 10, Bonn 53113, Germany
e-mail: renaud@ehs.unu.edu

C. Kuenzer
German Aerospace Center (DLR), Earth Observation Center (EOC), German Remote Sensing Data Center, Oberpfaffenhofen, 82234 Wessling, Germany
e-mail: claudia.kuenzer@dlr.de

F.G. Renaud and C. Kuenzer (eds.), *The Mekong Delta System: Interdisciplinary Analyses of a River Delta,* Springer Environmental Science and Engineering,
DOI 10.1007/978-94-007-3962-8_17, © Springer Science+Business Media Dordrecht 2012

## 17.1 Water at the Centre of the Delta's Development

The development of the Mekong Delta relies to a very large extent on its water resources and infrastructure, whether we talk of the construction of navigable waterways for transport of goods, the excavation of canals for drainage and irrigation of agricultural land, the supply of freshwater to communities, or the protection measures against floods and salinity intrusion. Despite the abundance of water, the development of the delta is often hindered by too much water during the flood season, too little during the low flow season, or too polluted water. The impacts vary depending on the geographical location in the delta but also on the socio-economic circumstances of individual households, as the delta is far from being homogeneous from both physical and societal perspectives. This chapter synthesizes some of the key points made in previous chapters of the book, linking water resources management, knowledge management, and development.

### *17.1.1 Water Infrastructure, Rural Development and Poverty*

In the last few decades, agricultural development in the delta has proceeded rapidly principally through the impact of the Green Revolution, infrastructure development and improved agricultural practices (Nguyen Duy Can et al. 2007; Chap. 3). Three main periods of policy-driven changes with respect to the delta's agricultural systems can be identified (see Nguyen Duy Can et al. (2007: 78–79) and Garschagen et al. – Chap. 4). A first shift was from extensive rice production systems in the pre-*Doi Moi* period to rice expansion and intensification in the 1990s which was partly promoted thanks to infrastructure development (e.g. dikes, irrigation canals and sluice gates). The second shift was triggered in the 2000s through "effective land use" policies which encouraged the farmers to diversify their agricultural systems (e.g. integrated rice-fish production systems, shrimp farming – see also Ngo Thi Phuong Lan 2011). The third shift was induced by taking stock of the lessons learnt from previous agricultural intensification practices, forcing the government to reconsider giving priority to rice production in some areas of the delta and to look further at alternative land uses. These relatively rapid transformations have not been uniform in the delta, and there are large disparities in terms of development at the provincial level.

Infrastructure developments had both negative and positive effects in terms of management of water resources and land use changes (Dang Kieu Nhan et al. 2007, Chap. 6). Economic benefits included possibilities for rice monoculture intensification, agriculture diversification, rural road improvement providing among others, better access to markets, the development of aquaculture production zones, and the reduction of flood damages. Negative impacts encompassed the destruction of natural wetlands, the deforestation of protective mangrove belts along the coastline (Chap. 12), degradation of river ecology and natural fish stocks, soil fertility reduction, soil erosion, and conflicts between upstream-downstream users, as well as within or between agricultural, urban, and industrial activities (see e.g. Biggs et al. 2009). In addition to internal transformation processes, the delta, as the most downstream region of the Mekong River Basin, has to face the consequences of all activities taking place at the scale of the basin without having the political power to substantially influence decision-making in the other riparian sovereign nations. The region is therefore facing many uncertainties linked to internal transformations, development activities within Vietnam and in upstream riparian areas (Chap. 5), and global economic and environmental changes, including climate change (see also Biggs et al. 2009).

One of the main objectives of infrastructure development in the delta was (and remains to be) the social and economic development of the region and the country as a whole and, linked to this, the reduction of poverty. The Mekong Delta has seen rapid development which led to an impressive reduction in overall poverty (Chap. 4), but a part of its rural population remains poor. Recognizing the problem, the Government of Vietnam has recently set ambitious targets in terms of national food security for the period 2020 (GoV 2009) which also addresses rural poverty. These targets include increases or stabilisation in production of rice, of other crops, of fisheries, and of aquaculture; they address the nutritional needs of the population and children in particular and, perhaps more critically, aim to improve the profitability of agriculture, particularly for rice farmers. Although rice production plays a central role in the delta, many rice farmers remain in precarious situations (Biggs et al. 2009). Coclanis and Stewart (2011) illustrated this fact by discussing "precarious paddies", highlighting the uncertainty, instability, insecurity, and risk burden characterising rice farming in the delta. They reviewed various historical, environmental, and developmental factors affecting rice farming in the region and showed that political reforms and scientific innovations (e.g. more productive rice varieties) have not drastically reduced the uncertainties under which rice farmers operate.

Garschagen et al. (Chap. 4) showed that a consequence of poverty in the region is a net negative migration flow of people, principally due to out-migration towards more economically dynamic regions. As noted by Huynh Truong Huy and Le Nguyen Doan Khoi (2011), migration and development complement each other as development induces migration and migration contributes to development. Contemporary migration is from rural to urban areas with nearby Ho Chi Minh City being one of the main receiving cities with ca. 50% of migrants moving there, but also Can Tho City. According to Huynh Truong Huy and Le Nguyen Doan Khoi (2011), this migration is mainly driven by poverty, natural disasters, failures in production and the markets, socio-economic background, and ethnicity in the commune of departure.

Through the enactment of various policies, the Government of Vietnam has favored agricultural diversification. In various provinces of the delta, aquaculture, including shrimp farming, is replacing the traditional rice crops as farmers see higher income opportunities even if production risks are also higher. For some farmers, this represents one strategy to step out of chronic poverty although here again, inequality can prevail as the poorest farmers have difficulties in meeting the high investment costs required to shift from a rice to a purely aquaculture production system. Ngo Thi Phuong Lan (2011) and various ongoing research in the coastal provinces of the delta indicate that the shift from a freshwater rice production system to saltwater shrimp cultivation systems contributes to long-term changes in the social (e.g. power relations, shifts in main activities), economic (possibilities for higher returns on investment, but also of financial failure as has been experienced by many farmers already), and of course ecological (shift from a freshwater to a saline water environment) systems. These changes are directly linked to and impacting water resources in the delta as the ultimate choice of farming system has consequences both in terms of water use and water pollution. Rice and aquaculture systems require relatively large amounts of water albeit of different qualities (freshwater vs. saline water) and will also have different pollution impacts. These impacts are not just in terms of further agricultural water use but also on freshwater availability for human consumption, and other domestic uses. Policies to encourage shifts in agricultural systems therefore need to consider the related environmental, social and economic consequences thoroughly as failure (e.g. of intensive shrimp farming) will have long term repercussions on farmers' direct livelihoods and on the ecosystems and their services. Some researchers now suggest that in terms of aquaculture, rice-shrimp and extensive shrimp systems are more sustainable in terms of soil and water quality impacts when compared to intensive shrimp systems (Chap. 14) and therefore offer fewer risks to farmers' livelihoods. Agricultural production systems in the delta will have to adapt to the effects of climate change and in particular with increased salinisation of coastal areas induced by a combination of lower river flows during the dry season and of sea-level rise. Various development pathways are possible including (but certainly not limited to) considering shrimp farming systems as an adequate adaptation strategy and/or the building of additional infrastructures such as sluice gates at river mouths to preserve rice production in regions not yet affected by salinity (Marchand et al. 2011; see also Chap. 2 where other adaptation strategies are discussed). Each development pathway will lead to different economic and social development outcomes and carries different levels of risk.

### *17.1.2 Pressure on Water Resources*

With intensification and diversification of land use, new pressures have started to be exerted on water resources in the delta. The Ministry for Agriculture and Rural Development (MARD) is officially responsible for irrigation infrastructure development and for hydro-climatic disaster prevention (Decree 01/2008/ND-CP, GoV 2008), and for climate change adaptation in agriculture (Decision 2730/QD-BNN-KHCN, MARD 2008), among others. Yet, it is the Ministry of Natural Resources and Environment (MoNRE) which is responsible for organising and directing the National Water Resources Strategy towards the year 2020 (GoV 2006) and is the focal point for coordinating climate change adaptation activities. In this context, MoNRE noted that water resources with respect to sustainable development and health in Vietnam were under-appreciated and not well protected. MoNRE recognised that adding infrastructures to reduce, for example, drought occurrences (which negatively impact agricultural production and livelihoods during the low-flow season) is only part of the solution, and that more integrated approaches are required to solve the dry-season water deficits, or general variations in flood pulses. The Ministry also recognised that giving access to sufficient clean water is of absolute necessity to achieve sustainable development and that water resources management activities are dispersed, poorly coordinated, and overlapping responsibilities exist between various stakeholders. There is a lack of cross-sectoral cooperation and a lack of applied sub-basin management (MoNRE 2006; GoV 2006; Chap. 6).

The rapid development of the delta has brought about many environmental changes and disruptions (Fabres 2011), and ecosystem services are negatively affected in many ways (Vo Quoc et al. 2012). Historically, from an environmental perspective, the main challenges for the delta are seasonal floods[1] during the rainy season, as well as drought and salinity intrusion during the dry, low-flow season. Frequency and extent of floods and salinisation are reviewed by Le Anh Tuan et al. (2007). All these hazards are now predicted to be aggravated by the consequences of climate change (Chap. 2) and infrastructure development upstream of the delta and within the delta can either reduce or worsen flood impacts (Chap. 5). The region is also experiencing rapid urbanization with insufficient urban planning (Garschagen et al. 2011) as well as industrialization. All these developments have increased the pressure on water and land resources in many ways:

- Surface and groundwater pollution (Chaps. 7, 13, 14): Pollution of water resources has many origins, some driven by the (i) intensification of agriculture and aquaculture production – e.g. pollution by pesticides, antibiotics, nutrients, acidity (the latter originating from the management of acid-sulphate soils), (ii) urban development – e.g. high concentrations of untreated urban wastewater released in river and canal systems or surface runoff from sealed and polluted surfaces, and (iii) industrial development – e.g. industrial chemicals, some of which releasing compounds which can affect the endocrine system. As a portion of the delta population still relies on untreated surface water and on groundwater, risks linked to polluted water on population health could be significant. Furthermore, the same pollution poses an increasing threat to aquatic ecosystems (Chap. 11).
- Rapid land use changes with the removal of most of the native vegetation, particularly in coastal areas but also, through the development of hydraulic infrastructures (Chaps. 2, 3, 12). This threatens the biodiversity of the delta's ecosystems (Chap. 11; Fabres 2011), whereby large regions of the delta are being converted from freshwater systems to saltwater systems because of shifts from rice production to shrimp production. For example, Binh et al. (2005) have illustrated the rapid changes in land use in the Cai Nuoc District of Ca Mau Peninsula in the delta: whereas rice-based farming systems covered more than 50,000 ha in

[1]Floods are also beneficial in many ways for the delta. We refer here only to large flood events which affect people and their livelihoods negatively.

the region in 1968, they had completely disappeared by 2003; in parallel, shrimp monoculture, which was non-existent in 1968, represented ca. 73% of the land cover in 2003. In fact, in 1998, shrimp cultivation systems only represented ca. 15% of the land cover and the rapid increase in shrimp monoculture was a direct result of agricultural diversification policies issued at the national level.[2] The region also suffered from extensive loss of forest cover and diversity with most conversion taking place because of the extension of the rice cultivation area in the 1970s–1990s and the development of shrimp cultivation (Binh et al. 2005; Chap. 12).
- Water over-abstraction, in particular in the case of groundwater. Groundwater is seen by farmers, industries and urban dwellers as a reasonably accessible resource but its management is not optimized (Chap. 7; Reis 2012). Excessive groundwater withdrawal can also lead to slow but steady land subsidence, which would aggravate the effects of sea level rise in coastal provinces (Chap. 2)
- Agricultural land degradation resulting in declining soil fertility induced by intensification of the cropping systems, in particular rice production (Nguyen Duy Can et al. 2007).

Degraded water quality can impact human health directly (consumption of untreated or partially treated surface water) or indirectly (consumption of aquatic products). Large numbers of the delta population relies on surface water for their domestic uses and treatments applied at the household level do not allow eliminating all sources of pollution (for pesticides, see Pham Van Toan 2012). Rural and peri-urban populations are the most exposed to this lack of connection to clean piped water systems, and although many communities harvest rainwater during the rainy season or access groundwater resources, others still rely on surface water for their daily freshwater supplies.

### *17.1.3 Differentiated Impacts from Water-Related Hazards*

In addition to water access and water pollution problems, the delta will have to deal with increased frequency and magnitude of hydro-climatic hazards such as floods (in terms of variability, see Chap. 9) typhoons, salinity intrusion or droughts (Chap. 2). The region has already suffered greatly from some of these hazards (e.g. the flood of the year 2000 or typhoon Linda in 1997), and at the time of writing, the 2011 flood has already caused human losses and important damages to agriculture and infrastructures in the region. The research of Birkmann et al. (Chap. 10) dealing with the assessment of vulnerability of households exposed to various hydro-climatic hazards in the delta clearly showed that vulnerability in the delta is socially differentiated and is not only limited to a problem of exposure. Some of the vulnerabilities are shaped by factors that are hazard independent (e.g. land tenure, access to loans, presence of

---

[2]Principally resolution 09/NQ-CP passed in 2000.

social networks), while others are more hazard specific (e.g. type of housing). The mechanisms available to cope with the hazards vary greatly depending on the actual impact and wellbeing of the affected households. One way of coping, frequently undertaken by the poorer households, is temporary or permanent migration of family members (see Dun 2011 in the context of floods; Chap. 10). Another alternative is resettlement (Birkmann et al. in press; Dun 2011). With respect to resettlement, in an attempt to reduce the impacts of the most severe floods, the Government of Vietnam introduced the "living with floods" programme, which consists of the relocation of exposed households into new settlements which are protected from flood impacts. This is in a way a forced migration situation which has both benefits and disadvantages for the concerned population. Vo Thanh Danh and Mushtaq (2011) reviewed some of these and, for a case study in An Giang Province, identified changes in land use/ownership, a decrease in animal husbandry activities, changes in employment (e.g. less engagement in agricultural activities, increase in unemployment) with changes in income sources, and improvement in social conditions (such as greater access to public services including education, particularly during the flooded periods of the year). Furthermore, forced relocation often goes along with a financial burden for the people affected in case they want to maintain their standard of living with respect to their previous living place. Changes are felt differently depending on each household's circumstances prior to relocation, implying that relocation is not necessarily the most appropriate solution for everyone.

The above concerns and additional discussions by Fabres (2011) and in several chapters of this book indicate that the Government of Vietnam is a main actor when it comes to addressing water-related threats and considering the links between water and development in the country in general, and in the delta in particular. Yet, public investments to reduce these threats remain relatively low. Many laws and decrees are in place (see e.g. Chaps. 2 and 6), but progress towards good environmental and integrated water management is still a challenge, particularly in terms of clarifying responsibilities and cooperation at various governance levels and in terms of enforcement. Furthermore, water resources planning in the Mekong Delta is still "top-down". Villagers and farmers play a very minor role in water planning (Le Anh Tuan et al. 2007). This can lead to poor planning or mismanagement of water-related infrastructures, both in terms of water supply for irrigation and domestic use, but also in terms of protection against environmental hazards. The top down approach does not allow capturing and addressing adequately the socially differentiated needs of the population at the local level, as indicated in Chap. 10 in the context of hydro-climatic hazards.

Future management of water resources will determine whether or not it will continue to be an agent of development for the region as a whole or whether its continued unsustainable use and pollution will increase the pressure on agricultural and industrial development and on the livelihoods of the poorest segments of the delta population. There are many opportunities for a sound management of water resources in the region, and we focus the rest of the discussion on two particular aspects: knowledge generation and sharing and increased cooperation between the many stakeholders influencing/involved in water resources management in the delta.

## 17.2 Knowledge and the Need for Increased Cooperation

### *17.2.1 Knowledge Generation and Sharing*

Given the context of agricultural intensification, burgeoning industrialisation, and environmental and climate change, access to information will be crucial for planning purposes in the Mekong Delta. Le Anh Tuan and Chinvanno (2011) noted that, among other factors, the development of policies for information sharing is very important. However, knowledge generation and sharing is a necessary but certainly not sufficient condition for good policy-making. Many factors are at play in the policy-making process that do not necessarily follow rational processes regardless of the amount of knowledge policy-makers have at hand. We nevertheless argue here that policy-making and the development of IWRM in the delta and upstream can certainly be better informed through the generation of additional, pertinent knowledge and the sharing of this knowledge between various stakeholders.

The amount of research and studies that has recently been conducted in the delta and those that are ongoing is considerable. Of critical importance is the knowledge generation by Vietnamese universities and organizations and, according to the work of Gerke et al. (Chap. 15), local-level knowledge generation is on the increase. This trend needs to be further enhanced and additional funds made available so that Vietnamese scientists are able to conduct independent research that could be published in international peer-reviewed publications and that would complement donor-driven collaboration. Knowledge clusters, which are defined as areas of a high density of knowledge producing institutions and high-level manpower, have developed in Ho Chi Minh City and Can Tho (Chap. 15; Bauer 2011). While Can Tho University and neighbouring institutes such as the Mekong Delta Rice Research Institute are clusters of knowledge production in the delta, other delta provinces host smaller universities whose student numbers are on the rise. However there continues to be a lack of knowledge-sharing within and between institutions, limiting the transfer of knowledge to policy-implementing agencies. Gerke et al. (Chap. 15) showed that this is problematic at various spatial scales, from the very local (e.g. farmers not sharing knowledge with other farmers to keep a productivity advantage) to the national (e.g. central- and local-level governments performing their tasks with little coordination). One of the implications of the latter is that national policies are implemented differently at the sub-national level, and in particular, each province insists on its autonomy which can have negative implications with respect to the implementation of Integrated Water Resources Management - IWRM (Chap. 6). Knowledge-sharing between international organisations working either in research or development activities in the delta is also relatively poor: there usually are very little exchanges, and actions are often driven by agencies interests and funding opportunities rather than strategically, in an attempt to build synergies between projects. Finally, data and knowledge exchange and transboundary cooperation between the riparian countries could also be significantly improved (Le

Anh Tuan et al. 2007; Le Anh Tuan and Chinvanno 2011) if a true IWRM strategy at the basin-scale is to ever materialise (Chap. 5).

As noted by Fabres (2011), taking advantage of and contributing to international information systems is critical. Several agencies and initiatives are trying to actively promote the sharing of information. These are the Mekong River Commission (MRC) which is supplying general knowledge of the Mekong Basin as well as water level and geodata via its website; programmes like M-Power or Mekong Basin related knowledge expansion initiatives of the World Wildlife Fund, the International Union for the Conservation of Nature, and other Non Governmental Organisations. However, the difficulty remains that much of this knowledge will not arrive at the stakeholder level, or will even be observed with scepticism, as it is assumed that the mentioned – often non political – organizations actually have their own political agenda. For the MRC the future pathway is currently unclear, as the leadership has been handed back to the riparian countries, and staff policies for external experts, future fields of focus, funding and strategic redirection remain unclear. However, openly available scientific knowledge of high – reviewed – quality is urgently needed, and the biggest challenge will be to produce, present and communicate this knowledge in a manner that it reaches stakeholders at all levels so that crucial information for the basin or the delta can be incorporated into decision making.

An example of knowledge sharing is the water-related information system developed through the WISDOM project[3] for the region of the Vietnamese Mekong Delta. A major focus of the project is to establish a comprehensive, web-based information portal on the water resources in the delta. This bilingual (English and Vietnamese) water related information system for the sustainable development of the Mekong Delta (Chap. 16) does not only contain a broad range of geo-data, ranging from geophysical information on soil moisture, inundation, water quality, or land use to socio-economic data on demographic patterns, or economic production, but also contains an extensive literature database, a comprehensive database with relevant legal documents, as well as a database containing all stakeholders and decision-making organizations involved in water resources management (Kuenzer 2010). The system is designed as a web-based platform, which can be accessed by all project partners and decision makers involved in the project, as well as by external scientists, stakeholders and interested experts requesting a login. The current phase of the project is focussing on the improved adaptation of the system to the user's needs in the delta, which are especially the provincial offices of MoNRE and MARD. The biggest challenge with information systems in general is to ensure local ownership in a way that they are used, maintained, and that relevant data on natural resources, as well as socio-economic information are being updated in the future (Kuenzer 2010). Therefore, the project, which is now running in its second phase until the end of 2013, is undertaking large efforts in terms of capacity development

[3] Water-related information system for the sustainable development of the Mekong Delta in Vietnam (see Chap. 1 and http://www.wisdom.caf.dlr.de)

in the Mekong Delta provinces through 'trainings of trainers' to ensure the long term sustainability of the system. Furthermore, financial resources need to be mobilized to fund personnel, hardware and software to operate the information system after the end of the WISDOM project.

### 17.2.2 Cooperation Between All Stakeholders

At the international level, there is a great need, among other factors, to share data and information to improve risk preparedness with respect to floods and droughts and there needs to be further consultation when it comes to the development of the Mekong River (see e.g. Grumbine and Xu 2011). This would allow joint co-development of the basin for the benefit of most. Examples of tensions arising from poor cooperation are the development of dams in upstream countries which have been blamed (perhaps wrongly, see Chap. 5) for droughts in the lower reaches of the Mekong River or the recent controversies linked to the development of the Xayaburi Hydroelectric Power Project in Laos, where consultations seemed to go in the right direction until they fell apart in 2011 with Cambodia deciding to go ahead with construction plans (Stone 2011). However, should a compromise be found on the development of this dam, Grumbine and Xu (2011) suggest that this would then become a prime example of successful integrated, transboundary river basin management. It is clear that each of the Mekong Basin riparian countries have their own development needs and objectives and that the Mekong River can be the backbone for such development. However, as in all transboundary basins, individual countries will only truly reap benefits from the development of the water resources if an international strategy is put in place, not sector by sector as still being emphasised by many, but in an integrated way from the beginning. This applies to the Mekong Delta which is often seen as the region that has to suffer the consequences of development processes in upstream countries. However, Vietnam also develops infrastructures on the tributaries of the river and is interested in purchasing electricity generated from the Mekong River in upstream countries (Methonen 2008). Vietnam is therefore both driver and recipient of changes taking place in the Mekong Basin and needs to implement a coherent approach to its own development projects, in collaboration with the other riparian countries.

As mentioned above, one available platform for this is of course the MRC which could be strengthened in its ability to deal with and negotiate complex matters such as the development of hydropower on the Mekong River and its tributaries (Grumbine and Xu 2011). This would constitute a shift with respect to the MRC's current limited ability to influence national-level processes and gaining access to non-member, upstream countries – this might change with shifts in the organisation's working structure (Ha 2011). Another exchange platform between the riparian Mekong countries – as driven by economic interests – is the Asian Development Bank funded Greater Mekong Subregion Initiative (GMS). This initiative includes the four member countries of the MRC, as well as China and Myanmar, and aims at regional and sub-regional economic cooperation mainly via investment into the development of infrastructure (Nikula 2008). These activities are backed up by the Association of Southeast Asian Nations (ASEAN), which in 2002 agreed jointly with China to create the world's biggest free trade zone. Here investment and development of the Mekong River Basin are key priorities for cooperation in the region. The very strong engagement of China with the GMS and ASEAN is spurred by important financial incentives.

As reviewed in the previous section and in Chap. 6, improved cooperation is also required at the national level. Most national-level policies and directives identify multiple actors for their implementation, particularly at the ministerial level. However, horizontal cooperation among ministries or other organisations remains weak (e.g. Chap. 15). This is particularly the case for all the directives related directly or indirectly to water resources management. This lack of cooperation at the Ministerial level is not specific to Vietnam and is observed worldwide. But it is at least openly acknowledged in Vietnam whereby MoNRE recognises that when it comes to the current national water resources strategy, the management, protection, and development of water resources must reflect the integrated nature of river basins and not be separated by administrative boundaries (MoNRE 2006). This is unfortunately easier said than done although mechanisms have already been put in place in Vietnam, such as the River Basin Organisations, which have representatives from provinces and national-level ministries but which, by and large do not fulfil their role adequately as discussed by Waibel et al. (Chap. 6). Collaboration at the national level but also with sub-national governance levels is essential to ensure that decision-making considers not only the national targets, but also the needs of local populations and the trickle-down-effects of policy decisions. Lack of joint ownership of project development can be detrimental not only locally but also have off-site consequences. For example, Birkmann et al. (Chap. 10) show that dyke construction in the province of Dong Thap to protect populations against flooding can actually generate serious flood problems further downstream. This can be avoided with the implementation of IWRM principles.

Many actors are also directly or indirectly involved in water resources management and protection at the sub-national level. As previously discussed though, there is competition at these governance levels, in particular at the level of provinces. At the sub-provincial level, the People's Committees also exert great power on decision making and implementation of national-level policies can vary greatly from one locality to another. As one of the last group of stakeholders in this hierarchal governance chain, farmers actually have little to no voice when it comes to conflicts pertaining to national initiatives (Biggs et al. 2009). More participatory decision-making is therefore required within and at the interface between governance levels and stakeholder groups so that the needs of all can be considered when it comes to water resources management. As indicated by numerous national newspaper articles in 2011 alone, despite policies and legislation, people and industries excessively withdraw or pollute water resources in the delta and elsewhere with few if any legal consequences. The impacts are felt by different stakeholders in terms of freshwater provisioning for domestic and production uses. It should therefore be a priority for the Government to ensure that legislation is properly enforced as, if the status quo remains, defiance will prevail between various stakeholders, making it impossible to apply IWRM principles. The new water law which is currently being drafted may be a good step in the right direction.

## 17.3 Conclusions

In order to benefit further from its extensive water resources, generation of (local) knowledge related to water resources and their management in the Mekong Delta needs to be further enhanced and this knowledge needs to be shared more widely than it currently is. In addition, a nested institutional approach is required to understand all social and ecological implications of water resources management in the region: one covering the basin scale whereby decision-making on water and land resources management needs to be more integrative and participatory; one at the national and regional (or better, local basin) levels where directives and decisions need to be more integrated with each other and where responsibilities are streamlined to ensure greater effectiveness in addressing all developmental-related issues of water resources (e.g. at the basin level by providing a stronger mandate and resources to RBOs); and one at the provincial and local levels where cooperation instead of competition should be encouraged.

Water resources in the delta are complex both in their bio-physical characteristics and in their social representation and many research gaps need to be addressed in order to implement a true IWRM strategy in the longer run. These cover the resource base *per se*, such as understanding better extreme water and sediment flows (see Chap. 8) as affected by climate change and infrastructure development; developing decision support systems to find solutions (technical, environmental, social and economic) for more effective water resources management and disaster risk reduction; characterizing the multiple pollution sources generated within the delta but also transferred from riparian countries and identifying and enforce adequate technical and political solutions to reduce and remediate these pollution; improved understanding and valuation of the ecosystem services provided by the extensive freshwater system which dominates the region. Research gaps also need to be addressed, in parallel and/or conjointly with the above, on the social and institutional side of the equation. These include, for example, the development of quality criteria for climate change adaptation; scenario-planning for the delta which incorporates not only projections from the perspective of the physical environment but which also address future social and economic changes; and integrated vulnerability and hazard mapping.

Enhanced cooperation and generation and sharing of knowledge are crucial to ensure equitable use of water resources in the region, protection of individuals and ecosystems against environmental (including climate change related) and anthropogenic pollutions and hazards, and ensuring the sustainable development of the region and its increased resilience to external shocks. As the Mekong Delta is highly vulnerable to the consequences of climate and environmental change it is urgent to put in place a mechanism were politicians, decision-makers, scientists, and the many development actors can meet and discuss on a regular basis the broad issue of the water-knowledge-development nexus of the region.

## References

Bauer T (2011) The challenge of knowledge sharing. Practices of the Vietnamese Science Community in Ho Chi Minh City and the Mekong Delta. LiT Verlag, Berlin

Biggs D, Miller F, Hoanh CT, Molle F (2009) The delta machine: water management in the Vietnamese Mekong delta in historical and contemporary perspectives. In: Molle F, Foran T, Käkönen M (eds) Contested waterscapes in the Mekong Region: hydropower, livelihoods and governance. Earthscan, London, pp 203–225

Binh TNKD, Vromant N, Hung NT, Hens L, Boon EK (2005) Land cover changes between 1968 and 2033 in Cai Nuoc, Ca Mau Peninsula, Vietnam. Environ Dev Sustain 7:519–536

Birkmann J, Garschagen M, Fernando N, Hettige S, Tuan VV, Oliver-Smith A (2012) Relocation: reduction in exposure – increase in susceptibility? Measuring dynamics of vulnerability before, during and after relocation processes in the context of disasters and natural hazards. In: Birkmann J (ed) Measuring vulnerability to natural hazards. Towards disaster resilient societies, 2nd edn. UNU Press

Coclanis PA, Stewart MA (2011) Precarious paddies: the uncertain, unstable, and insecure lives of rice farmers in the Mekong Delta. In: Stewart MA, Coclanis PA (eds) Environmental change and agricultural sustainability in the Mekong Delta, vol 45, Advances in global change research. Springer, Dordrecht, pp 103–114

Dang Kieu Nhan, Nguyen Van Be, Nguyen Hieu Trung (2007) Water use and competition in the Mekong Delta. In: Tran Thanh Be, Bach Tan Sinh, Miller F (eds) Challenges to sustainable development in the Mekong Delta: regional and national policy issues and research needs. The Sustainable Mekong Research Network, Bangkok, pp 143–188

Dun O (2011) Migration and displacement triggered by floods in the Mekong Delta. Int Migr 49(S1):e200–e223

Fabres B (2011) Think global, act global in the Mekong delta? Environmental change, civil society, and NGOs. In: Stewart MA, Coclanis PA (eds) Environmental change and agricultural sustainability in the Mekong Delta, vol 45, Advances in global change research. Springer, Dordrecht, pp 7–34

Garschagen M, Renaud FG, Birkmann J (2011) Dynamic resilience of peri-urban agriculturalists in the Mekong Delta under pressures of socio-economic transformation and climate change. In: Stewart MA, Coclanis PA (eds) Environmental change and agricultural sustainability in the Mekong Delta, vol 45, Advances in global change research. Springer, Dordrecht, pp 141–163

GoV (2006) Decision 81/2006/QD-TTg: approving the national strategy on water resources to 2020. Socialist Republic of Viet Nam, Hanoi, 13 p

GoV (2008) Decree 01/2008/ND-CP defining the functions, tasks, powers and organisational structures of the Ministry of Agriculture and Rural Development. Socialist Republic of Viet Nam, Hanoi

GoV (2009) Resolution 63/NQ-CP on National Food Security. Socialist Republic of Viet Nam, Hanoi. Unofficial English translation accessed on 02/09/2011 at http://asemconnectvietnam.gov.vn/lawdetail.aspx?lawid=1735

Grumbine RE, Xu J (2011) Mekong hydropower development. Science 332:178–179

Ha M-L (2011) The role of regional institutions in sustainable development: a review of the Mekong river commission's first 15 years. Cons J Sustain Dev 5:125–140

Huynh Truong Huy, Le Nguyen Doan Khoi (2011) Analysis of labour migration flows in the Mekong Delta of Vietnam. In: Stewart MA, Coclanis PA (eds) Environmental change and agricultural sustainability in the Mekong Delta, vol 45, Advances in global change research. Springer, Dordrecht, pp 115–140

Kuenzer C (2010) WISDOM: Water related Information System for the Sustainable Development of the Mekong Delta – thoughts on sustainability and long term implementation. In: Proceedings of the 3rd Indo-German conference on research for sustainability: water and waste management, New Delhi, India, 3–4 Feb 2010

Le Anh Tuan, Chinvanno S (2011) Climate change in the Mekong River Delta and key concerns on future climate threats. In: Stewart MA, Coclanis PA (eds) Environmental change and agricultural sustainability in the Mekong Delta, vol 45, Advances in global change research. Springer, Dordrecht, pp 207–217

Le Anh Tuan, Chu Thai Hoanh, Miller F, Bach Tan Sinh (2007) Flood and salinity management in the Mekong Delta, Vietnam. In: Tran Thanh Be, Bach Tan Sinh, Miller F (eds) Challenges to sustainable development in the Mekong Delta: regional and national policy issues and research needs. The Sustainable Mekong Research Network, Bangkok, pp 15–68

Marchand M, Dam R, Bucx T (2011) Towards a Mekong Delta plan. Synthesis of water sector assessment. Prepared by Deltares in commission of the Delta Coordinator of Directorate General Water in the framework of the Mekong Delta programme. http://www.partnersvoorwater.nl/wp-content/uploads/2011/09/Annex5_mekongdelta-assesments_synthesis.pdf. Accessed 14 Nov 2011

MARD (2008) Action plan framework on climate change adaptation of the agriculture and rural development sector in period 2008–2020 (Decision No. 2730/QD-BNN-KHCN), 30 p

Methonen K (2008) Do the downstream countries oppose the upstream dams? In: Kummu M, Keskinen M, Vairs O (eds) Modern myths of the Mekong. A critical review of water and development concepts, principles and policies. Helsinki University of Technology: Water & Development Publications, Espoo, 187 pp

MoNRE (2006) National water resources strategy towards the year 2020. Ministry of Natural Resources and Environment, Hanoi, 26 p

Ngo Thi Phuong Lan (2011) From rice to shrimp: ecological change and human adaptation in the Mekong Delta of Vietnam. In: Stewart MA, Coclanis PA (eds) Environmental change and agricultural sustainability in the Mekong Delta, vol 45, Advances in global change research. Springer, Dordrecht, pp 271–285

Nguyen Duy Can, Le Thanh Duong, Nguyen Van Sanh, Miller F (2007) Livelihoods and resource use strategies of farmers in the Mekong Delta. In: Tran Thanh Be, Bach Tan Sinh, Miller F (eds) Challenges to sustainable development in the Mekong delta: regional and national policy issues and research needs. The Sustainable Mekong Research Network, Bangkok, pp 69–98

Nikula J (2008) Is harm and destructions all that floods bring? In: Kummu M, Keskinen M, Vairs O (eds) Modern myths of the Mekong. A critical review of water and development concepts, principles and policies. Helsinki University of Technology: Water & Development Publications, Espoo, 187 pp

Pham Van Toan (2011) Pesticide use and management in the Mekong Delta and their residues in surface and drinking water. Unpublished PhD dissertation, Faculty of Agriculture, University of Bonn

Reis N (2012) Tracing and making the state. Policy practices and domestic water supply in the Mekong Delta. ZEF Development Series. Lit Verlag, Berlin

Stone R (2011) Mayhem on the Mekong. Science 333:814–818

Vo Quoc T, Kuenzer C, Vo Quang M, Oppelt N (2012). The concept of ecosystem services – a review. J Ecol Indic (accepted for publication)

Vo Thanh Danh, Mushtaq S (2011) Living with floods: an evaluation of the resettlement program of the Mekong Delta of Vietnam. In: Stewart MA, Coclanis PA (eds) Environmental change and agricultural sustainability in the Mekong Delta, vol 45, Advances in global change research. Springer, Dordrecht, pp 181–204

# Index



F.G. Renaud and C. Kuenzer (eds.), *The Mekong Delta System: Interdisciplinary Analyses of a River Delta,* Springer Environmental Science and Engineering, DOI 10.1007/978-94-007-3962-8, © Springer Science+Business Media Dordrecht 2012

Printed by Printforce, the Netherlands